Ukrainian School of Mining Engineering XII International Scientific and Practical Conference 2018

E3S Web of Conferences Volume 60 (2018)

Berdiansk, Ukraine
4 – 8 September 2018

Editors:

V. Bondarenko
I. Kovalevska
R. Lysenko
O. Malova
F. Cawood
M. Hardygora

ISBN: 978-1-5108-7317-9

Printed from e-media with permission by:

Curran Associates, Inc.
57 Morehouse Lane
Red Hook, NY 12571

Some format issues inherent in the e-media version may also appear in this print version.

Printed by Curran Associates, Inc. (2018)

For additional information, please contact EDP Sciences – Web of Conferences
at the address below.

EDP Sciences – Web of Conferences
17, Avenue du Hoggar
Parc d'Activité de Courtabœuf
BP 112
F-91944 Les Ulis Cedex A
France

Phone: +33 (0) 1 69 18 75 75
Fax: +33 (0) 1 69 28 84 91

contact-edps@webofconferences.org

Additional copies of this publication are available from:

Curran Associates, Inc.
57 Morehouse Lane
Red Hook, NY 12571 USA
Phone: 845-758-0400
Fax: 845-758-2633
Email: curran@proceedings.com
Web: www.proceedings.com

TABLE OF CONTENTS

2018

UKRAINIAN SCHOOL OF MINING ENGINEERING

XII INTERNATIONAL SCIENTIFIC AND PRACTICAL CONFERENCE

BERDIANSK, UKRAINE
September 4 – 8
2018

OUR PARTNERS

HIGHER EDUCATIONAL INSTITUTIONS

NATIONAL ACADEMY OF SCIENCES OF UKRAINE

PRODUCTION ENTERPRISES

ORGANIZATIONAL COMMITTEE

Dear participants of the conference!

It is a great honor for us to host guests at the XII International Scientific and Practical Conference "Ukrainian School of Mining Engineering – 2018" in the pearl of the Azov Sea – the sunny Berdiansk.

Undoubtedly, the present meeting is an outstanding event in the life of our School and will encourage innovative development, uniting efforts in discussing topical issues of efficiency and safety of mining operations, solving complex problems of mining and processing industries of the world, an effective impetus for the implementation of new, interesting ideas for optimization of our collaboration.

We are convinced that the professional and representative level, knowledge, experience and high human qualities of the participants of our meeting will enable us to make suggestions and recommendations that will increase the efficiency of further development of the mining industry.

This meeting is a unique opportunity not only to get together, share experiences, new developments, achievements, but also discuss the pressing issues that concern leading mining specialists, progressive research and development workers, laboratories, commercial organizations and universities with advanced mining profiles. After all, it is quite clear that the reform of the industry in Ukraine makes science and enterprises to further develop and modernize.

Therefore, we would like to express our special gratitude to the conference participants for the opportunity to communicate with colleagues and to enrich themselves with new knowledge.

Regards,

Organizational Committee of the School of Mining Engineering

edp sciences Web of Conferences

Statement of Peer review

In submitting conference proceedings to *Web of Conferences*, the editors of the proceedings certify to the Publisher that

1. They adhere to its **Policy on Publishing Integrity** in order to safeguard good scientific practice in publishing.
2. All articles have been subjected to peer review administered by the proceedings editors.
3. Reviews have been conducted by expert referees, who have been requested to provide unbiased and constructive comments aimed, whenever possible, at improving the work.
4. Proceedings editors have taken all reasonable steps to ensure the quality of the materials they publish and their decision to accept or reject a paper for publication has been based only on the merits of the work and the relevance to the journal.

Title, date and place of the conference

Ukrainian School of Mining Engineering;
from 04 to 08 of September, 2018;
Berdiansk, Ukraine

Proceedings editor(s):

Prof. Volodymyr Bondarenko, National Mining University, Ukraine
Prof. Iryna Kovalevska, National Mining University, Ukraine
Res. Ass. Roman Lysenko, National Mining University, Ukraine
Res. Ass. Olha Malova, National Mining University, Ukraine
Prof. Frederick Cawood, Wits Mining Institute, S. Africa
Prof. Monika Hardygora, Wroclaw University of Science and Technology, Poland

Date and editor's signature

19.09.18 21.09.18

20.09.18 26.09.18

20.09.18 27.09.18

Maintenance of reusable mine workings during the augering mining of coal seams

Volodymyr Bondarenko[1], *Hennadii* Symanovych[1], *Iryna* Kovalevska[1*], and *Viktor* Chervatiuk[2]

[1]National Mining University, Department of Underground Mining, 19 Yavornytskoho Ave., 49005 Dnipro, Ukraine
[2]LLC "DTEK Energy", Department on Coal Production, 57 Lva Tolstoho St, 01032 Kyiv, Ukraine

Abstract. The augering mining of thin and very thin coal seams has been proposed for using the efficiency of this task, related to the stability of reusable mine workings. The interrelation has been established and optimized between the support loading parameters in the extraction working during the augering mining of thin and very thin coal seams with the stress-strain characteristics of the protective monolithic strip of variable rigidity. The strength properties of sand-cement stone for the erection of rigid and yieldable parts of the protective strip are studied. The regression equations are obtained, which allow to determine the rational force, deformation and geometric parameters of the "support – protective strip" system elements during the extraction workings maintenance.

1 Introduction

One of the main directions of production processes intensification of coal mining in thin and very thin seams, which were developed in the international practice, is the augering technology [1], which provides coal extraction without the breakage face support setting and the presence of people in it [2].

Many of these seams have already been entered and partially or fully prepared for extraction, so that the capital costs for these processes can be reduced significantly. Furthermore, the augering technology can also be applied to the development of established previously coal pillars of various purpose. The development of this direction will increase the recovery ratio of coal [3] also due to the partial extraction of non-commercial reserves and protective pillars. However, the efficiency of the augering mining is largely related to the stability of reusable mine workings [4 – 7], some of which should be maintained for a long period in the zone of coal-face work influence [8 – 11]. This problem is solved in a complex on the basis of linking the issues of fastening and protection of extraction drift as a single and mutually influencing system, which is proposed to be called as "support – protective strip". The maintenance of the extraction workings is achieved by low-cost methods based on bearing pressure control through the protective monolithic strips of variable rigidity (yield) [12].

Directly to the mine working the yieldable part of the strip with width of L_y adjoins,

* Corresponding author: kovalevska_i@yahoo.com

which is followed by a rigid part with width of L_r. The rigid part takes up the bulk of the bearing pressure $\Phi(\gamma H)$, but the yieldable part pushes back the maximum of the bearing pressure from the contour of the mine working deep into the mined-out space. An unloading zone is formed around the mine working, the tasks of which include: on the one hand, to reduce the rock pressure to a minimum; on the other hand – to prevent formation of a significant volume of unstable rocks which during collapse will form high vertical and lateral loading on the support. These tasks are realized by choosing the necessary width of the yieldable part of the strip and the required value of its yield.

Traditionally, the applied technology of a protective strip erection by the dry shotcreting method [13] has been improved [14] in terms of improving the quality by addition of water to a mix and air dedusting, reducing the rebound of the solid phase of the shotcrete flow by separating it on exit from the nozzle onto the solid and air phases, reducing the wear of the pipeline end section. For these purposes, more advanced designs of auxiliary equipment have been manufactured and used.

2 Main part

2.1 Loading of the "support – protective strip" system elements

The peculiarity of the coal-overlaying strata movement during the augering coal mining is that the rocks of the main roof smoothly descend to the interwell pillars and rock cushion in the exhausted wells with the discontinuity of the seam only mainly in the tensile stress area of the rock beam, thus forming a thrust system of large rock blocks, which takes up the weight of the overlying strata of rocks to the daylight area surface. Because of these reasons, the height h of the disturbed rocks is quite limited and, as a rule, not less than one order of magnitude smaller than the depth H of the coal seam occurrence [15, 16]. Accordingly, the weight of the hanging console rocks will be at least one order of magnitude less than the hydrostatic pressure γH. In addition, this small loading on the protective element repeatedly decreases due to the support of disturbed rocks, which is caused by the effect of their loosening and increase in volume.

The yieldable part of the strip is intended to provide a zone for the rock massif unloading in the vicinity of the mine working with width of $2(L_y + r_w)$, where r_w is the radius of mine working arch. The height h of unloading zone is determined by the height of the stable rock layer occurrence, which is determined by the thickness m_1, strength properties, span value l_1 of the natural equilibrium of rocks and by forces on the contact q_1 with the rocks of unloading zone. In such a case, the length l_1 of a span is less than the value $2(L_y + r_w)$ due to the formation of dome of natural equilibrium over the mine working in the off-loaded rocks. On the contact with mine working support the support reaction q acts on the rocks of unloading zone, as a resultant value of vertical $q_{ver} = \gamma h + q_1$ and lateral $q_{lat} = (\gamma h + q_1) tg^2 \dfrac{90 - \varphi_r}{2}$ components. The lateral reaction of support acts also on the adjacent yieldable part of the protective strip.

The decrease in values of loading q_{ver} and q_{lat}, is carried out in two ways: by reducing the height h of the unloading zone through decreasing the width of the yieldable part of the strip or by the loading q_1 elimination, providing that the support and yieldable part of the strip are displaced in the vertical direction to the required value, when the rock pressure is applied to them. Therewith, the load-bearing capacity of support is significantly dependent on its vertical q_{ver} and lateral q_{lat} components ratio. Therefore, the two force criteria were formulated for support: the total loading along the mine working contour must tend to the minimum possible value and the ratio $\beta = q_{ver} / q_{lat}$ of vertical and lateral loadings should be such, wherein the support reaches its maximum load-bearing capacity Q. In addition, to avoid the negative impact of backing forces q_1 on the value of vertical yield of support U_{ver} and the yieldable part of the strip δ, the following condition is imposed: the yield of the strip δ at the value of the loading on it γh compensates for the increase in the volume of loosened rocks in height h.

As for the mine working support, then the predicted vertical U_{ver1} and lateral U_{lat1} displacements of rocks are also considered, which occurred before the beginning of coal-face works, that means $U_{ver} \geq U_{ver1} + \delta$, $U_{lat} \geq U_{lat1} + \delta tg\frac{90-\varphi_r}{2}$ [17].

The set criteria complex has determined the requirements to the stress-strain characteristics of support and yieldable part of the protective strip. The width L_r of the rigid part of the strip is chosen from the condition of its non-destruction $Q_r \leq Q_{ad}$, where Q_r and Q_{ad} – acting and admitted loadings onto the rigid strip $Q_r = \gamma H\left(r_w + L_y + L_r\right)$. The load-bearing capacity of the rigid part of protective strip is determined according to the ultimate rock equilibrium theory, provided that all the material of the strip changes to the limiting state, described by the rectilinear envelope of Mohr circles:

$$Q_{ad} = 2C_r L_r\left[2 + F\left(\frac{L_r}{m}, \varphi_r\right)\right],$$

where C_r and φ_r are adhesion and internal friction angle of material of the rigid part of a strip; $F\left(\frac{L_r}{m}, \varphi_r\right)$ is tabulated function, which considers combined stress state in the rigid strip, is determined by [18].

From the indicated conditions, the minimum permitted width of the rigid part of a protective strip is determined:

$$L_r = \frac{\gamma H\left(r_w + L_y\right)}{2C_r\left[2 + F\left(\frac{L_r}{m}, \varphi_r\right)\right] - \gamma H}.$$

The numerical analysis of this expression led to the conclusion that for the depths of the mine workings embedding to 800 m and the mechanical properties of the hardened sand-cement mixture of the traditional formulation, the relative width L_r / m of the protective strip rigid part does not exceed, as a rule, one and half thicknesses of the coal seam m. For reliable protection of extraction drifts, primarily of interest is the dependence of the ratio of vertical and lateral loadings onto mine working support on the parameters of the protective strip yieldable part. The main of them are the ratio of the width L_y and height m of the strip, as well as internal friction angle φ_y of its material. The essence of this connection is that the finite width of the yieldable part of the strip affects significantly the value of the lateral loading and its ratio to the vertical one. The curves on Fig. 1 show that the function

$\beta(L_y / m)$ has minimum with variable internal friction angles of material of the protective strip yieldable part. At a narrow width ($L_y / m < 0.15 - 0.35$) of the yieldable strip the parameter β increases rapidly ($\beta > 5$). The same, but more smoothly, occurs when the strip width is increased ($L_y / m > 1 - 1.5$). It is quite natural a strong influence of the yieldable strip material on the parameter of the internal friction angle φ_y, as a parameter that determines the restraining tangential stresses along the slip line of the cleavage prism. From Fig. 1 it follows that the parameters L_y / m and φ_y of the protective strip yieldable part influence significantly on the ratio β of vertical and lateral loadings onto the support, on which its load-bearing capacity depends greatly. The research has established a rational range of the parameter $\beta - 2 - 3$ variation in terms of the maximum load-bearing capacity of the support. The required value of the parameter β is achieved by changing the width L_y / m of the protective strip yieldable part, with the internal friction angle of $\varphi_y \leq 30°$. At $\varphi_y > 30°$, regardless of the width L_y / m, the parameter β extends beyond the range of rational values. With decrease in φ_y, the rational width L_y / m decreases and amounts to $L_y / m = 0.5 - 0.9$ at $\varphi_y = 10 - 30°$.

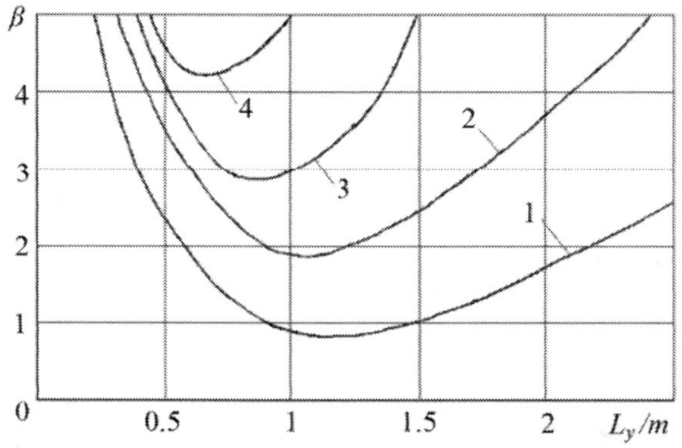

Fig. 1. Dependence of the parameter β on the relative width L_y / m of the protective strip yieldable part: 1 – $\varphi_y = 10°$; 2 – $\varphi_y = 20°$; 3 – $\varphi_y = 30°$; 4 – $\varphi_y = 40°$.

An analysis of the research results of the yield mode of the "support – protective elements" system in interaction with the enclosing rocks has determined the minimum required lateral and vertical yield of the drift support:

$$U_{lat} = w_{y_{lat}} = hL_y\left(\frac{\Delta}{m}\cdot\frac{\sqrt{h^2 + r_w^{\ 2}} - r_w}{\sqrt{4h^2 + l_1^2}} - 4.6\cdot10^{-4}\gamma\right); U_w = (h - r_w)\frac{\Delta}{2},$$

where Δ is coefficient of volumetric loosening of rocks.

2.2 Determination of the required mechanical properties of the protective strip

To perform a full calculation of rational parameters of the "support – protective elements" system, the laboratory studies have been conducted of the mechanical properties of the recommended combinations of rigid and yieldable parts of the protective strip in accordance with the requirements imposed on them: the dependences of adhesion C, internal friction angle of the strip material on different hardening mixture composition (Wat / Cem, $Sand / Cem$) and the age t of sand-cement stone, are studied. And based on the processing of test results by methods of correlation-dispersion analysis, the expressions were obtained for the strength properties determination of the rigid part of a strip:

$$C_r = 10.9\,exp\left(-0.28\frac{Sand}{Cem}\right)\left(\frac{Cem}{Wat} - 0.5\right) lg\left(t+1.5\right);$$

$$\varphi_r = \left(13.7 + 17.25\frac{Wat}{Cem}\right) exp\left(-0.54t\right) - \frac{Sand}{Cem} + 34.$$

A significant dependence of the sand-cement stone compressive strength σ_{compr} on Add / Cem porous-filling additive content (aluminium powder) has been revealed. The general pattern is that with an increase in the content of the additive Add / Cem, the strength of the samples decreases, since the volume of pores and voids in the structure of the sand-cement stone increases. The highest intensity of the strength σ_{compr} reduction occurs in the range $Add / Cem = 0 - 0.5\%$. Then, with an increase in content of the additive (up to $Add / Cem = 2\%$), the compressive strength intensity of the samples decreases asymptotically approaching to a certain value in the range up to 0.1 σ_{compr} of non-porous-filled sand-cement stone.

When processing the obtained results and using the methods of correlation-dispersion analysis, the regression equation was determined in the form:

$$\sigma_{compr} = 27.2\,exp\left(-15.8\sqrt{\frac{Add}{Cem}}\right) exp\left(-0.23\frac{Sand}{Cem}\right)\left(\frac{Cem}{Wat} - 0.5\right) lg\left(t+1.5\right).$$

On the basis of this equation, with account of the condition $\sigma_{compr} \le \gamma H$ for achieving a yield mode of the strip operation, the minimum required content of the porous-filling additive has been determined:

$$\frac{Add}{Cem} \ge 4\cdot 10^{-3}\, ln^2\left[\frac{27.2}{\gamma H} exp\left(-0.23\frac{Sand}{Cem}\right)\left(\frac{Cem}{Wat} - 0.5\right) lg\left(t+1.5\right)\right].$$

In the tests of a porous sand-cement stone the internal friction angle φ_{Add} has also been determined which depends on the porous-filling additive content. The general tendency is that with an increase in the content of the additive the internal friction angle of the porous sand-cement stone decreases. This fact is explained by the occurrence of pores and voids, which at the initial moment of the tested sample material shear are filled with the destroyed skeleton of the sand-cement stone and the friction forces along the slip line decrease.

By numerical analysis of the experimental data and with the use of the methods of correlation-dispersion analysis, the regression equation has been obtained:

$$\varphi_{Add} = \left[\left(13.7 + 17.25\frac{Wat}{Cem}\right) exp\left(-0.54t\right) - \frac{Sand}{Cem} + 34\right]\left[0.53 + 0.47\,exp\left(-317\frac{Add}{Cem}\right)\right].$$

The studies have confirmed the necessity to consider the change in the internal friction angle φ_r in time t when calculating the parameters of the rigid part of the protective strip. And based on the results obtained, a regression equation was formed for the dependence on time of the load-bearing capacity growth of the protective strip rigid part:

$$Q_{adm}(t) = 21.8\,exp\left(-0.28\frac{Sand}{Cem}\right)\left(\frac{Cem}{Wat} - 0.5\right)lg(t+1.5)L_r \times$$

$$\times\left\{2+\left(2\frac{L_r}{m} - 1.87\right)exp\left\{0.074\left[\left(13.7+17.25\frac{Wat}{Cem}\right)exp(-0.54t) - \frac{Sand}{Cem} + 34\right]\right\}\right\}.$$

In this equation, as an argument, it is possible to use the distance X from the breakage face according to the known daily velocity V of its advance.

Thus, the calculation of rational parameters for extraction drifts fastening and protection scheme has been developed, based on consideration of its separate load-bearing elements as a single and mutually influencing system, in which the geometric and mechanical parameters of the protective strip of variable rigidity are the tools for formation of a favorable loading on the support.

3 Conclusions

It is established that the arched yieldable support with the circular arch has the maximum load-bearing capacity at the ratio of vertical and lateral loads in the range of $\beta = 2 - 3$, which is practically constant for most of typical cross-sections of mine workings.

The study of the yieldable part deformation of the protective strip with account of the slip prisms gradual cleavage and accompanying redistribution of the loading from the weight of the softening zone rocks, has indicated that the achievement of reasonable ratio of vertical and lateral loadings onto support is provided with the relative width of the yieldable part of a strip $L_y / m = 0.5 - 0.9$ with an internal friction angle $\varphi_y \leq 30^{\circ}$ of its material.

The control of the internal friction angle φ_y of the protective strip yieldable part is provided by addition of the required amount of porous-filling additive into hardening sand-cement mixture for adhesion of armour stone. The obtained dependence of the hardening mixture formulation and the internal friction angle φ_y of the hardened stone indicates its changes in the range of $\varphi_y = 10 - 30$ degrees with the addition of a porous-filling additive in an amount $1 - 4\%$ of the volume of grade M400 Portland cement at a sand-cement ratio $Sand / Cem = 3/1$ and water-cement ratio $Wat / Cem = 0.4$.

The calculation of the protective strip rigid (supporting) part parameters for strength shows that for typical mine working cross-sections, with the calculated width of the yieldable part of the strip at depths of $H = 300$ to 600 m, the reasonable relative width of the rigid part ranges from $L_r / m = 1 - 1.5$ with the use of a sand-cement hardening mixture $Sand / Cem = 3/1$ and $Wat / Cem = 0.4$ formulation.

The authors express their gratitude to senior management of "Pavlohradvuhillia DTEK" PLC and Mine Administration for the assistance in research development.

References

1. *Izyskanie sposobov razrabotki ugol'nykh plastov bez postoyannogo nakhozhdeniya rabochikh v zaboyakh*. (1960). Pod redaktsiey A.P. Sudoplatova. Moskva: Institut gornogo dela im. A.A. Skochinskogo.
2. Kovalevska, I. (2006). Minimizing coal losses when extracting thin coal seams with the use of auger mining technologies. In *International Mining Forum* (pp. 27-34). Cracow: Printed and

bound PATRIA. https://doi.org/10.1201/noe0415401173.ch4

3. Korzh, P.P. (1997). Puti povysheniya polnoty izvlecheniya zapasov pri buroshnekovoy vyemke uglya. In *Ekologicheskie problemy razvedki, razrabotki i obogashcheniya poleznykh iskopayemykh Ukrainy* (p. 109). Dnipropetrovsk: Systemni tekhnolohii.
4. Majcherczyk, T., & Niedbalski, Z. (2010). Numerical modeling used for designing of coal mine roadway support. *New Techniques and Technologies in Mining*, 77-82. https://doi.org/10.1201/b11329-14
5. Khomenko O.Ye., Sudakov, A.K., Malanchuk, Z.R., & Malanchuk, Ye.Z. (2017). Principles of rock pressure energy usage during underground mining of deposits. *Naukovyi Visnyk Natsionalnoho Hirnychoho Universytetu*, *2*(158), 34-43.
6. Kovalevska, I.A., Malykhin, O.V., Gusiev, O.S., & Movchan, V.S. (2015). Research and calculation of side anchors that are set at a height of roof brushing in excavation. *Mining of Mineral Deposits*, *9*(3), 313-317. https://doi.org/10.15407/mining09.03.313
7. Sdvizhkova, Ye.A., Babets, D.V., & Smirnov, A.V. (2014). Support loading of assembly chamber in terms of western donbas plough longwall. *Naukovyi Visnyk Natsionalnoho Hirnychoho Universytetu*, (5), 26-32
8. Kovalevs'ka, I., V. Vivcharenko, V., & Snigur, V. (2013). Specifics of percarbonic rock mass displacement in longwalls end areas and extraction workings. *Annual Scientific-Technical Collection – Mining of Mineral Deposits*, 29-33. https://doi.org/10.1201/b16354-7
9. Prykhodko, V.V., & Ulanova, N.P. (2018). Modeling of stress-strain state of fractured rock mass nearby of conjugated workings. *Naukovyi Visnyk Natsionalnoho Hirnychoho Universytetu*, (1), 5-11.
10. Symanovych, G.A, Chervatiuk, V.G., Snigur, V.G., & Malykhin, O.V. (2015). Displacement mechanism of above-the-coal strata and loading on support along extraction mine working behind the longwall. *Mining of Mineral Deposits*, *9*(3), 299-306. https://doi.org/10.15407/mining09.03.299
11. Kovalevska, I., Barabash, M., & Gusiev, O. (2016). Research into stress-strain state of reinforced marginal massif of extraction mine working by combined anchoring system. *Mining of Mineral Deposits*, *10*(1), 31-36. https://doi.org/10.15407/mining10.01.031
12. Dunayev, A.G. (2001). Obosnovanie konstruktsii i sposoba vozvedeniya okhrannykh tselikov pri buroshnekovoy vyemke ugol'nogo plasta. *Naukovyi Visnyk Natsionalnoi Hirnychoi Akademii Ukrainy*, (3), 133-135.
13. Kovalevskaya, I.A., Simanovich, G.A., & Koval', A.I. (2007). Tekhnologiya vozvedeniya u vyyemochnogo shtreka okhrannoy litoy polosy s razgruzochnoy polost'yu. *Naukovyi Visnyk Natsionalnoho Hirnychoho Universytetu*, (1), 3-4.
14. Bondarenko, V.I., Simanovich, G.A., Kovalevskaya, I.A., & Dunayev, A.G. (2001). Sovershenstvovanie konstruktivno-tekhnologicheskoy skhemy vozvedeniya iskusstvennykh tselikov. *Naukovyi Visnyk Natsionalnoi Hirnychoi Akademii Ukrainy*, (4), 3-5.
15. Bondarenko, V.I., Kovalevskaya, I.A., & Dunayev, A.G. (2001). Povyshenie ustoychivosti uchastkovykh vyrabotok pri buroshnekovoy otrabotke ugol'nykh plastov. In *Geotekhnologii na rubezhe XXI veka* (pp. 70-74). Donetsk: Donetskyi derzhavnyi tekhnichnyi universytet.
16. Bondarenko, V.I., Kovalevskaya, I.A., Simanovich, G.A., & Koval', A.I. (2007). Geomekhanika vzaimodeystviya okhrannoy konstruktsii vyyemochnoy vyrabotki s razuprochnennym porodnym massivom. *Naukovyi Visnyk Natsionalnoho Hirnychoho Universytetu*, (5), 3-8.
17. SOU 10.1.00185790.011:2007. (2007). *Pidhotovchi vyrobky na polohykh plastakh. Vybir kriplennia, sposobiv okhorony*. Kyiv: Ministerstvo vuhilnoi promyslovosti Ukrainy.
18. Ruppeneyt, K.V. (1954). *Nekotorye voprosy mekhaniki gornykh porod*. Moskva: Ugletekhizdat.

Interaction of deformation-strength characteristics of the support load-bearing elements in the preparatory workings

Iryna Kovalevska[1*], *Mykhailo* Barabash[2], *Oleksandr* Husiev[3], and *Vasyl* Snihur[4]

[1]National Mining University, Department of Underground Mining, 19 Yavornytskoho Ave., 49005 Dnipro, Ukraine
[2]LLC "DTEK Energy", Department on Coal Production, 57 Lva Tolstoho St., 01032 Kyiv, Ukraine
[3]MM "Dniprovske", PJSC "DTEK Pavlohradvuhillia", 76 Soborna Ave., 51400 Pavlohrad, Ukraine
[4]MM "Pershotravenske", PJSC "DTEK Pavlohradvuhillia", 76 Soborna Ave., 51400 Pavlohrad, Ukraine

Abstract. The objective of this work is to study the relationship of the deformation-strength characteristics of the fastening system with its displacement value. The tendencies of influence are divided into two groups. The multivariate computational experiments have been performed in the conditions of coal-face works. It is shown the influence of geomechanical conditions of the mine working maintenance on the fastening system operating mode change when choosing its rational deformation-strength characteristics. On the basis of the performed studies, it was concluded that it is necessary to adjust the fastening elements of each specific support setting scheme. This is necessary to provide recommendations for the maintenance schemes optimization of reusable extraction workings.

1 Introduction

The deformation-strength characteristics of the fastening system in the preparatory working is represented in the form of a function $q(u)$, which indicate the relationship between the bearing reaction q of support and its displacement value, and under the influence of the rock pressure manifestations during coal-face works. Therefore, the research was carried out from the point of view of the evaluation of the relationship between the deformation-strength characteristics of the preparatory working fastening system and the degree of its stability during the breakage face driving in the area from the bearing pressure zone in front of the longwall to the zone of the displacement processes stabilization in the coal-overlaying strata behind the longwall. At the same time, the relations of different forms of rock pressure manifestations were taken into account. For example, weakening of extensive rock volumes in the sides of mine working leads to a more intense lowering of the roof rock layers, and the same weakening of the side rocks and the formation of increased lateral rock pressure deforms the frame prop stays that leads to sudden drop of its load-bearing

* Corresponding author: kovalevska_i@yahoo.com

capacity. This, in turn, increases the loading on the reinforcement support prop stays with the probability of subsequent loss of their stability [1]. Thus, the revealed diversity of the rock pressure manifestation features requires consideration of the relationship between separate geomechanical processes in the following fields of study:

– interaction of rocks weakening in the sides of mine working and the processes of stratification and lowering of the roof rock layers;

– coupling of the bolting strengthening in the sides of mine working with that in its roof and setting the rational resistance modes of the formed armour-clad and rock structure as one of the fastening system elements;

– providing the operating modes synchronization of the fastening system constituent elements with each other and the search for its rational deformation-strength characteristics, depending on the geomechanical conditions for the extraction working maintenance in the zone of coal-face works influence.

The variety of mining and geological conditions for the maintenance of the preparatory workings and their fastening systems structures intensifies the need for a complex of multivariate computational experiments to reveal the regularities which are of interest to us [2, 3]. A wide range of trends in the influence of deformation-strength characteristics $q(u)$ of the fastening system elements on their state may be divided into two groups:

– patterns of the general function $q(u)$ relation for the fastening system as a whole and for its individual elements $q_i(u_i)$ with the main geomechanical factors of the mine working maintenance (mining depth, stress-strain properties of rocks in the adjacent coal-bearing strata, its structure and the structure of roof rocks in the mined-out area);

– patterns of mutual influence on each other of functions $q_i(u_i)$, making up the fastening systems of elements (i – element number).

Such a division, in our opinion, will enable a more systematic approach to assessing the effect of the fastening system operating mode on the extraction working stability and the prospects for its reuse.

2 The research problem setting

By now, a great deal of results of computational experiments have been accumulated which represent the trends of the geomechanical factors influence on the mine working support state [4, 5]. But our task involves consideration of patterns from a slightly different point of view, namely: the manner in which the geomechanical conditions of the mine working maintenance influence on the fastening system operating mode change in terms of determination of its rational deformation-strength characteristics [6]. Here it should be kept in mind the following. In normative documents [7, 8], the prediction of displacements does not consider the support reaction to restraint of displacements, and the loading value P upon the support is determined from the displacement coordinate $u = u_{max}$ with a certain margin in the direction of increasing. Any computational experiments using the finite element method (FEM) consider automatically the fastening system reaction to the mine working rock contour displacements. But, what is more, each calculation is carried out for a specific bearing reaction q at a specific yielding value u of each element of the fastening system. In such a way, we have not the entire picture of the loading P relationship with the deformation-strength characteristics $q(u)$. Hence, the problem of support operating mode optimization (even for mine working outside the zone of coal-face works influence) still requires its final solution. The complexity of solving the problem for the fastening system operating mode optimization in the zone of coal-face works influence is caused by two factors:

– firstly, the large volumes of coal-bearing massif with extensive areas of complete destruction, weakening and change of the original lithotypes structure are many times involved in the process of enclosing rocks displacement;

– secondly, the fastening system usually includes more elements and much more diverse; a safe-guard system with its structural features is added to it.

In view of the above mentioned factors, the complexity of the study and description of the geomechanical processes of the coal-overlaying strata displacement, the method of multivariate computational experiments performance is truly effective (in terms of adequacy). On the one hand, this method only gives a particular result in the calculation of the specific geomechanical model. But, on the other hand, having carried out a number of computational experiments with variation of any parameter, we can get a certain tendency of this parameter influence, and it can be expressed in the form of a certain rational interval or directly as a regression equation of the studied parameter relationship with other parameters in the geomechanical system.

In this respect, the results of both separate computational experiments and their purposeful series are of interest, which are stated, for example, in the works [9, 10 – 13]. Thus, in studies [9], the development of rock pressure manifestations (in the vicinity of preparatory workings) was studied in the process of breakage face advance: the tendencies were traced of the parameters change in the unloading zone and the bearing pressure zone, the growth of the loading on the elements of fastening system and safe-guard system of the drift. The logical result of research was the creation of calculation methods and methods for parameters choice of support setting means and extraction drift safe-guard with the purpose of their reuse.

Taken together, the two analysed works [9, 10] created the basis for the technical documentation development for reused extraction drifts maintenance in the Western Donbas mines. But, there is a common drawback: there are no criteria to optimize the deformation-strength characteristic for both the maintenance scheme individual elements, and the operating modes of the fastening and safe-guard system as a whole. That is, the issues of the function $q(u)$ relationship with geomechanical factors and the choice of the most effective operating mode of the maintenance scheme elements in specific mining and geological conditions, are not considered.

This setting is more complex and requires multiple expansion of the computational experiments variants. Therefore, attention is focused on separate fragments of the study of the relationship between the fastening system operating mode (and its elements) and the geomechanical parameters of the maintenance conditions for the extraction workings after the breakage face driving.

3 Mining depth influence on the fastening system state in the preparatory working

At the first stage, the fragments of influence of the mine working location depth H on the change of the load-bearing elements state for its maintenance under the angle of estimating the degree of rationality of these elements deformation-strength characteristics. Two variants of computational experiments with unchangeable structure and properties of the coal-bearing strata, the mine working dimensions and the design parameters of its maintenance scheme, have been carried out. The only difference is in mine working location depth: first calculation – $H = 300$ m, second calculation – $H = 450$ m; that is, the compared variants differ from each other by an increase in the mine working location depth by 50%.

The analysis of the state of mine working maintenance scheme elements has been started with consideration of the vertical stresses curves σ_{coal} (Fig. 1).

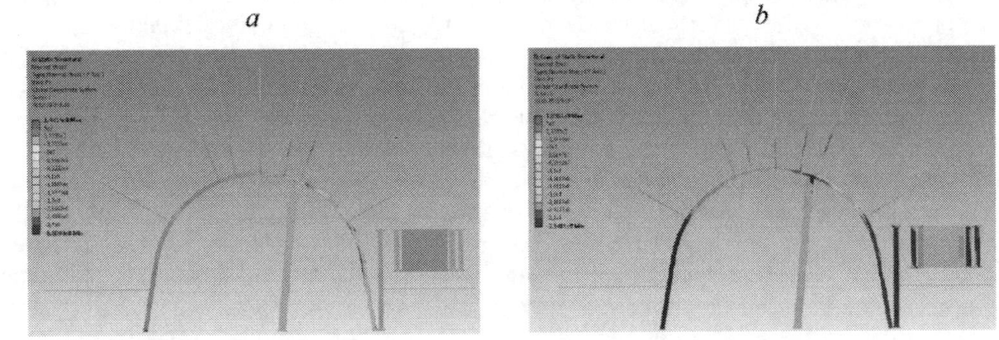

Fig. 1. Example of vertical stresses σ_{coal} distribution in fastening and safe-guard system of the preparatory working in dependence of the depth H of its location: a – $H = 300$ m; b – $H = 450$ m.

The component σ_{coal} distribution in the frame support (regardless of depth H) represents the specifics of load action in conditions of mine workings maintenance in the Western Donbas mines to the full extent: mainly unloaded state of the frame cap at high σ_{coal} in its prop stays. In the frame cap at $H = 300$ m (see. Fig. 1a) low σ_{coal} from 20MPa of tensile to 30MPa of compression are observed. The exception is the part of the cap length in the area of contact with the central wooden prop stay of the reinforcement support; here the compressive σ_{coal} increases up to $80 - 100$ MPa, and tensile σ_{coal} increases up to $50 - 70$ MPa. With an increase in the mine working location depth ($H = 450$ m) for the most part of its length, the frame cap is still unloaded with insignificant changes of σ_{coal}. But the part of the cap length, where there is an impact of the central prop stay of the reinforcement support, have expanded significantly; there the fluctuations in the absolute value σ_{coal} increased, especially in the compression area – to $250 - 300$ MPa, and this causes the formation of creepage zones in the cap. In addition, the effect of alternating-signs of the bending moments is clearly manifested (4 maximums along the length to 1.1 m) with the greatest value along the length of a contact with the wooden prop stay. And this value causes plastic bending of a cap (see Fig. 1b).

The given results of the H parameter influence are consistent with the existing ideas about the geomechanical processes development in the vicinity of mine workings. But we focus on the above mentioned trends in terms of the rationality of the deformation-strength characteristics $q(u)$ of the frame cap. The mainly unloaded state of the cap can be explained by the action of a number of factors, two of which are predominant.

The first is strengthening of adjacent roof layers with bolting systems, as a result of which the formed load-bearing armour-clad and rock plate rests on the side rocks in the virgin massif, from one side, and on the safe-guard structure from the side of the worked out area. With the relative integrity of the armour-clad and rock plate, the deflection of the central part of its span is low and in joint deformation mode with a cap there are no significant vertical forces on the contact. It happens, specifically, because of the damping effect of the easily deformed weakened rocks in the fastened space and simply due to the gaps which appear between the frame cap and the rock contour. Consequently, the frame cap is always underloaded without tight contact with the armour-clad and rock plate.

The second is a cap, which due to the yielding joints actuation "leaves" from the rock pressure in the roof until the yield margin is used up, and its components are: construction yield of the frame support (for example, TSYS series), pressing-in of the frame prop stays into the soft bottom rocks, gaps between the cap and rock contour of the mine working, and others. The total value of vertical yield can reach 0.5 – 0.8 m (sometimes even more) and it is quite enough to compensate for any kind of armour-clad and rock plate deformations if it is in a stable state. The only alternative for the full loading of the frame cap is the formation of ultimate rock equilibrium from the destroyed incoherent rocks in the vast arch roof. These rocks with their weight, in case of collapse behaviour, form a vertical pressure which does not depend on the frame yield.

The scheme of this process is shown in qualitative terms in Fig. 2 and is as follows. In accordance with the classical ideas, the loading on the support is formed as a result of the interaction between the adjacent rock massif, which is weakened and increased in volume (lines $1f$ and $1d$), and the formation of an unstable volume of rocks within the dome of natural equilibrium (lines $2f$ and $2d$). The index "f" was assigned to a more "favourable" variant of mine working maintenance at the depth of $H = 300$ m; index "d" refers to more "difficult" conditions and corresponds to $H = 450$ m. The points A_f and A_d characterize the minimum possible loads P_f and P_d, respectively, in favourable and difficult mining and geological conditions.

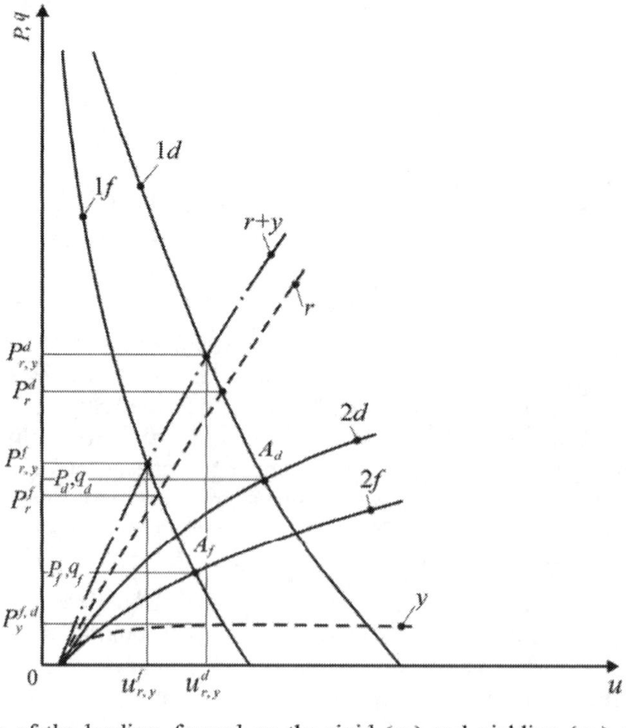

Fig. 2. The scheme of the loading, formed on the rigid (r) and yielding (y) components of the general fastening system in the mine working: $1f$ and $1d$ – deformation-strength characteristics of the weakening massif near the mine working contour in favourable (f) and difficult (d) mining and geological conditions; $2f$ and $2d$ – deformation-strength characteristics of natural equilibrium dome rocks; – – – deformation-strength characteristics of the rigid (r) and yielding (y) components of the fastening system; – · – general deformation-strength characteristics of the fastening system.

Dotted lines show the deformation-strength characteristics $q(u)$ for the rigid (index "r") and yielding (index "y") components of a single fastening system of the mine working, but chain-dotted line shows the general deformation-strength characteristics of the fastening system. As is obvious, the functions $q(u)$ for the rigid and yielding components are far from the optimal and do not pass through points A_f and A_d. It is expected that the rigid component takes over the bulk of the loading under any conditions, and their deterioration results in the loading growth ($P_r^d > P_r^f$) practically on a rigid component, whilst the yielding component is in approximately the same ($P_y^f \approx P_y^d$) conditions of load action.

The significant curve σ_{coal} changes in the frame cap happen in the area of setting the reinforcement support central prop stays: they abruptly increase the rigidity of a cap in the vertical direction and the component σ_{coal} is immediately amplified with an increase in H (see Fig. 1).

The very central prop stay is a rather rigid element, even with account of the crushing of wooden backings under the cap of the frame and under the footing of the prop stay. And if so, an increase of H by 50% (according to [5], a similar increase in the lowering of the roof rocks is predicted) should cause a sudden increase in stresses σ_{coal} in accordance with the developed scheme in Fig. 2.

When analyzing the curve σ_{coal} (see Fig. 1) in the central prop stays of the reinforcement support, it should be noted that at $H = 300$m $\sigma_{coal} \leq 30$ MPa; at $H = 450$ m $\sigma_{coal} \leq 83$ MPa. That is, there is an increase in stresses σ_{coal} by $2.7 - 2.8$ times with an increase in H by 1.5 times. Therefore, there is a more intensive growth of σ_{coal} in the rigid element (in this case in a wooden prop stay), than an increase in the depth of mine working location. Although, for the sake of objectivity, it is necessary to consider that $\sigma_{coal} = 50 - 60$ MPa are already destructive for such wood, as redwood: most likely there is "fringing" of the wooden prop stay butt ends and its breakdown caused by bending.

A further fastening element in assessing of σ_{coal} at different H is the frame support prop stays. The common feature (regardless of H) is as follows:

– gradual growth of σ_{coal} happens in the frame prop stays, beginning with the yielding joint, and already at the end of the curvilinear part of the stress it reaches a maximum;

– in the prop stay from the side of virgin massif the maximum values of compressive σ_{coal} are distributed uniformly both along the length and in the cross section of SCP, and this indicates that there is no significant bending moment;

– in the prop stay from the side of mined-out area, the effect of a number of bending moments is clearly manifested (by its height): maximal one is in the curvilinear part of the prop stay and minimal one – in its footing area.

In quantitative terms: at $H = 300$ m maximal value of compressive σ_{coal} amounts to $190 - 210$ MPa, at $H = 450$ m σ_{coal} increase to $270 - 300$ MPa. That is, an increase from 40 to 45% is observed, which corresponds approximately to 50% increase in the depth of mine working location. These results will be analysed from the point of view of the frame prop stays operating mode. On the one hand, a loading is delivered on the prop stays, which is "collected" by the frame cap, and in this regard we can speak of a sufficient its independence from *H*, that is, of the yielding operating mode. But, on the other hand, the TSYS support construction is characterized by a certain inclination of the prop stays to the central axis of the drift; the vertical load acts on the width of this slope, and the prop stay, as such, does not already have a yielding node, because it is located higher – in the junction with the cap. Consequently, the frame prop stay operates in a rigid mode, but only in relation to the specified part of the loading on it. In general, the operating mode of the frame prop stay can be called "intermediate" by nature, as it is anywhere from rigid to yielding mode.

As a result, three operating modes of the fastening elements have been identified only by one SCP frame with the reinforcement support prop stay: rigid, yielding and intermediate. Obviously, the fastening elements, which operate in three different modes, are difficult to optimize from the point of view of achieving the conditions for forming a minimum loading on the drift fastening system.

In addition to the studied elements, there are two more elements in this fastening system – resin-grouted roof bolts and lateral prop stays of the reinforcement support, set on the border of the berm. These prop stays have a rigid working characteristic and the ratio σ_{coal} in favourable and difficult mining and geological conditions, which is similar to that for rigid structures.

A certain specificity of tensile σ_{coal} growth (in the range of $H = 300 - 450$ m) is observed in resin-grouted roof bolts, set in the mine working roof. The very tension area in the roof bolt gives evidence of its resistance to roof rocks stratification in the process of armour-clad and rock plate formation. When comparing the tension areas σ_{coal} length, their growth in all the roof bolts of the roof with an increase in the depth of the mine working location, is clearly manifested.

In conclusion of the tendencies analysis of the connection between the fastening system elements operating modes and mine working location depth, the stress intensity curves σ have been studied, since one curve σ_{coal} does not fully characterize the stress condition.

First of all, we can note that there are preserved all above mentioned tendencies of fastening system elements stress growth with an increase in H.

In the frame cap at $H = 450$ m an area of plastic deformations appears in the place of contact with the reinforcement support prop stay; here the cap bending is possible, though at $H = 300$ m this is not observed. The tension in the cap increases more significantly, in spite of its yielding mode of operation. Here, the horizontal stresses σ_x facilitate, since the cap with resistance to horizontal forces has a sufficiently rigid deformation-strength characteristic.

In the central reinforcement support prop stay there is an approximately threefold increase of σ with the growth of H by 50% due to the reason of the prop stay rigid mode of operation. A similar tendency is observed for a rigid lateral prop stay of the reinforcement support.

In the prop stays of the frame support, the increase in stress intensity approximately follows the patterns described for vertical stresses σ_{coal}.

The resin-grouted roof bolts, set in the roof, increase their resistance to stratification in such a way, that in the armature of one of them a plastic state occurs.

Based on the conducted studies results, it is necessary to draw such a basic conclusion: the deformation-strength characteristics $q(u)$ of a fastening element determines significantly the intensity of its loading growth with an increase in depth H of extraction working location; with that, some elements are overloaded, while loading of others is far from the ultimate loading. This different degree of "reaction" to the change in mining and geological conditions for mine working maintenance (shown in qualitative terms in the scheme Fig. 2) indicates the need to adjust the operating modes of the various elements, making up a particular fastening scheme.

4 Conclusions

Increased contact stresses in the areas of different fastening elements junction reduce their load-bearing capacity, since they cause the occurrence of the limiting and super-limiting state of the elements materials, while, in the rest areas of these elements there is a pre-limiting deformation. Thus, the bulk of the fastening elements materials remains underloaded and their load-bearing capacity is reduced. In some ways, we can speak of unequal strength of each fastening element by itself.

The increase in the level of strength uniformity of each of the conjugate fastening elements is intimately associated with the degree of their deformation-strength characteristics compatibility. And in this perspective this task is of primary importance. It is necessary to find a compromise between the two tendencies of impact of deformation-strength characteristics of separate fastening elements on their condition: on the one hand, – reduction of stress concentrations in dangerous areas of the structure; on the other hand, – attempt to minimize the loading on the fastening system as a whole and its separate elements.

Summing up the research results of coal-bearing massif interaction with the fastening system of preparatory workings in the zone of coal-face works influence, it should be noted that qualitative schemes for their operating modes optimization should be transformed into a set of quantitative patterns of interaction between geomechanical factors and the deformation-strength characteristics of fastening elements. And this set is the base to provide recommendations for the maintenance schemes optimization of reusable extraction workings.

This work was supported by the Ministry of Education and Science of Ukraine under the project "Theoretical and practical foundations for the management of unstable geomechanical systems "massif – support" of underground mine workings" (State registration No. 0117U001131). The authors appreciate the help of the head of Underground Mining Department, National Mining University, Professor Volodymyr Bondarenko in consultations and assistance in conducting research.

References

1. Bondarenko, V., Kovalevs'ka, I., Svystun, R., & Cherednichenko, Yu. (2013). Optimal parameters of wall bolts computation in the united bearing system of extraction workings frame-bolt support. *Mining of Mineral Deposits*, 5-9. https://doi.org/10.1201/b16354-3
2. Kovalevs'ka, I., Symanovych, G., & Fomychov, V. (2013). Research of stress-strain state of cracked coal-containing massif near-the-working area using finite elements technique. *Annual Scientific-Technical Collection – Mining of Mineral Deposits*, 159-163. https://doi.org/10.1201/b16354-28
3. Prusek, S. (2018). Perspektywy stosowania samodzielnej obudowy kotwowej w kopalniach JSW S.A. In *XXVII Szkola eksploatacji podziemnej*. Krakow.
4. Kovalevska, I., Barabash, M., & Snihur, V. (2018). Development of a research methodology and analysis of the stress state of a parting under the joint and downward mining of coal seams. *Mining of Mineral Deposits*, *12*(1), 76-84. https://doi.org/10.15407/mining12.01.076
5. Bondarenko, V., Hardygora, M., Symanovych, H., Sotskov, V., & Snihur, V. (2016). Numerical methods of geomechanics tasks solution during coal deposits' development. *Mining of Mineral Deposits*, *10*(3), 1-12. https://doi.org/10.15407/mining10.03.001
6. Bondarenko, V., Kovalevska, I., Symanovych, H., Barabash, M., & Vivcharenko, O. (2018). *Geomechanics of mine workings support systems*. Dnipro: LizunovPress.
7. SOU 10.1.001.85790.011:2007. (2008). *Pidhotovchi vyrobky na polohykh plastakh. Vybir kriplennia, sposobiv i zasobiv okhorony*. Standart Minvuhlepromu Ukrainy. Donetsk: Donetskyi vuhilnyi instytut.

8. KD 12.01.01.201-98. (1998). *Raspolozhenie, okhrana i podderzhanie gornykh vyrabotok pri otrabotke ugol'nykh plastov na shakhtakh.* Normatyvnyi dokument Ministerstva vuhilnoi promyslovosti Ukrainy. Kyiv: Ukrainskyi naukovo-doslidnyi marksheiderskyi instytut.
9. Bondarenko, V.I., Kovalevskaya, I.A., Simanovich, G.A., Barabash, M.V., & Gusiev, A.S. (2015). *Vzaimodeystvie gruzonesushchikh elementov krepezhnoy sistemy vyemochnykh vyrabotok "massiv – rama – anker"*. Dnipropetrovsk: Litohraf.
10. Bondarenko, V.I., Kovalevskaya, I.A., Simanovich, G.A., Martovitskiy, A.V., & Kopylov, A.F. (2010). *Metody rascheta peremeshcheniy i uprochneniya prikonturnykh porod gornykh vyrabotok shakht Zapadnogo Donbassa*. Dnipropetrovsk: Driant.
11. Sdvizhkova, Ye.A., Babets, D.V., & Smirnov, A.V. (2014). Support loading of assembly chamber in terms of western donbas plough longwall. *Naukovyi Visnyk Natsionalnoho Hirnychoho Universytetu*, (5), 26-32.
12. Symanovych, G.A, Chervatiuk, V.G., Snigur, V.G., & Malykhin, O.V. (2015). Displacement mechanism of above-the-coal strata and loading on support along extraction mine working behind the longwall. *Mining of Mineral Deposits*, *9*(3), 299-306. https://doi.org/10.15407/mining09.03.299
13. Kovalevska, I.A., Malykhin, O.V., Gusiev, O.S., & Movchan, V.S. (2015). Research and calculation of side anchors that are set at a height of roof brushing in excavation. *Mining of Mineral Deposits*, *9*(3), 313-317. https://doi.org/10.15407/mining09.03.313

Fire resistance evaluation of reinforced concrete floors with fire-retardant coating by calculation and experimental method

Andrii Kovalov[1*], *Yurii* Otrosh[2], *Olha* Ostroverkh[2], *Oleksandr* Hrushovinchuk[3], and *Oleksandr* Savchenko[7]

[1]Cherkasy Institute of Fire Safety named after Chornobyl Heroes of National University of Civil Protection of Ukraine, 8 Onopriienko Ave., 18034 Cherkasy, Ukraine
[2]National University of Civil Defence of Ukraine, Faculty of Civil Protection, 94 Chernyshevska Ave., 61023 Kharkiv, Ukraine
[3]State Certification Center of the Ukrainian SES, 54 Yevhen Sverstiuk Ave., 02002 Kyiv, Ukraine

Abstract. With the help of a previously developed technique based on the solution of inverse and direct problems of heat conductivity, the fire resistance of a hollow-core reinforced concrete floor with "Neosprei" fire-retardant plaster coating was estimated. The thermal and physical characteristics, as well as characteristics of the fire-proof ability of "Neosprei" fire-retardant plaster coating, have been determined. The conclusion has been made on the effectiveness of this coating and on the boundaries of the fire-retardant coatings use to ensure the normed values of the fire resistance degree of hollow-core reinforced concrete floors, including in the mining industry.

1 Introduction

More than 65.000 fires occur in Ukraine annually, with losses amounting to about 2 billion hrn. Due to the fire hazards impact, from 2.500 to 3.500 people die, and more than 20.000 buildings and structures are destroyed.

Currently, from 30 to 50% of mine workings in Ukraine are accepted as dangerous for the movement of people. Unsatisfactory fastening condition is one of the main causes of high rate of miners accidents [1, 2].

The reliability and safety of mine workings are largely solved by the efficiency of interframe fences (laggings). Despite the large number of their types, only wooden and flat reinforced concrete structures have found wide application in domestic industry. The load-bearing structures of interframe fences such as laggings are usually used in the preparatory mine workings. These laggings may be rigid (reinforced concrete structures), semi-elastic (steel, wooden, polymeric laggings), and elastic (cyclone fencing, rolled glass fibre cloth, etc.).

More rarely, and usually in the main and capital mine workings, a flat reinforced concrete lagging is used, which is a flat plate with rectangular cross-section, made of B15

* Corresponding author: kovalev27051980@gmail.com

and B25 grade of concrete, reinforced with a welded mesh of wire with St3 3 – 8 mm in diameter. The average volume of its application does not exceed 20%, and in the mines of the Central region of Donbas – 10 – 12%. The flat reinforced concrete lagging has several disadvantages: large weight (up to 24 kg), non-fixed positions of reinforcement cage, the lack of reinforcement which takes up the forces during transportation and setting up.

However, the insufficient load-carrying capacity (breakage of the lagging occurs at a load of 10 – 15 kPa) and a short service life (less than 0.5 – 3 years as a result of wood rotting in the mine environment) are the causes for mass lagging failures in the form of a brittle fracture with superlimiting deformations. This leads to the rockfall between the support frames in the mine workings, underground workers injuries, accidents at the mine transport, and other negative consequences.

A great deal of those used in modern building constructions, including in the mining industry, are hollow-core reinforced concrete floors of different types. Since the hollow-core reinforced concrete floors have restricted fire resistance, it is actual to ensure their fire protection and increase the fire resistance degree with the use of various substances and materials. One way to improve fire resistance of such structures is the use of fire-retardant plaster compositions of large thickness.

2 The research problem setting

The results of research on the fire resistance of both building structures with fire-retardant coatings and without them, as well as the calculation of nonstationary heat conductivity are represented in the works [3, 4] of such scientists as Roitman M.Ya., Fomin S.L., Romanenko P.N., Demchyna B.G., Krukovskyi P.G., Shmukler V.S., Givliud M.M., Novak S.V., Dovbysh A.V., Kovaliov A.I., Bartelemi B., Segerlind L.J., Huebner K.N., Zeinkiewicz O.C., and others. Today, however, without consideration of these researchers are the questions on the identification of the mutual interaction between the parameters of fire-retardant plaster coating and the fire resistance degree of hollow-core reinforced concrete floor, as well as the conditions for the fire protection efficiency increase of reinforced concrete building structures with the use of fire-retardant plaster coatings. Theoretically, this question was studied in [5], but the presented results were in need of experimental verification.

Therefore, the use of a scientific approach to determine the fire-proof ability of "Neosprei" fire-retardant plaster coating, scientific substantiation of the thermal and physical characteristics, and characteristics of the fire-proof ability of this coating under the influence of high temperatures in case of fire is an actual scientific and technical task. This conditioned this work, the decision of which will positively affect the fire safety requirements assurance for the construction, reconstruction, replanning of facilities for various purposes.

The objective of this work is to identify the mutual interaction between the parameters of "Neosprei" fire-retardant plaster coating and the fire-proof ability of the hollow-core reinforced concrete floor, as a scientific basis for their use in construction with account of fire safety requirements.

3 Materials and methods used to study the fire resistance of hollow-core reinforced concrete floors

Today in Ukraine, to determine the fire resistance degree of any reinforced concrete structure (columns, beams, floors, coatings, partition walls) with a fire-retardant coating (cladding) of a certain thickness, it is necessary to conduct each time a fire-resistance test. For example, for floors the tests are carried out according to DSTU B V.1.1–20:2007 *"Fire protection. Floors and coatings. Method of test for fire resistance"*, for columns – DSTU B V.1.1–14:2007 *"Fire protection. Columns. Method of test for fire resistance"*. This is done in order to ensure that the chosen thickness value of the fire-retardant coating (cladding) will provide the normed value of the fire resistance degree. Such an approach requires that the structure should be tested for fire resistance if it is necessary to determine the fire resistance with a different value of the fire-retardant coating thickness. It requires significant material and financial costs, because each structure with different values of the fire-retardant coating thickness and the concrete protective layer must be tested to determine the fire resistance degree.

Therefore, in this study, a computational and experimental method will be applied which uses the solution of direct and inverse heat conductivity problems [6 – 9]. Appropriate methods allow to determine the thermal and physical characteristics (TPC) of fire-retardant materials containing and not containing water, as well as the thermal and physical properties of materials which can sprout. In these methods the models of nonlinear nonstationary heat conductivity are used in one- and two-dimensional setting. At the same time, the methods have been developed not only for heat conductivity determination, but also heat capacity as a function of temperature. These methods have the flexibility and versatility of computational methods and, at the same time, the laboratory experimental methods performance raises the reliability of the calculation.

The computational and experimental method (CEM) is defined as a set of experimental and calculation procedures, which make it possible to determine the necessary characteristics of the studied object, in particular, the dependence of the coating thickness on the thickness of the concrete protective layer. As it was noted above, the CEM is sufficiently effective and accurate method in comparison with an experiment or calculation methods.

4 The fire resistance test results of reinforced concrete floors with fire-retardant plaster composition

The determination of the fire resistance of reinforced concrete floors was carried out according to [10, 11] until one of the normed for floors boundary conditions in fire resistance states occurred on the grounds of loss of integrity, load-carrying capacity or thermal insulating capacity. The method consisted in the time interval determination from the start of the test under the standard temperature condition, according to [10], of the floors samples (hereinafter – samples) with the fire impact on the sample from below.

Two samples of the PK 48-12-8 hollow-core reinforced concrete floor were put to test. The samples of floors from below and on the sides were covered with a layer of "Neosprei" composition, fire-retardant coating (plaster), with the average total thickness of 25.9 mm. It was performed with the use of a plastering machine "M-tec DUOMIX".

Before to apply the substance, the lower surface of the floors was cleaned with a metal brush and treated with a "Fenix Contact" primer, diluted 1:6 with water. The prime compound flow was 200 – 300 g/m^2.

The PK 48-12-8 hollow-core reinforced concrete floor with sizes 4780×1190 mm and width 220 mm. The floor has a carrying steel frame, which consists of five lower longitudinal bearing bars of 12 mm (the bearing bar grade is not specified) and BP1 reinforcement wire of 4 mm. Concrete B15. The average value of thickness of the concrete protective layer till the lower bearing bar is 20 mm. The fire resistance degree of the floor, according to the manufacturer, is REI 45.

The samples were set on a horizontal furnace (according to order, a statically indeterminate support scheme) with support at the edges through compensating supports made of basalt plates with a thickness of 40 mm.

 https://doi.org/10.1051/e3sconf/20186000003

The application of loading was performed with calibrated loads in the form of concrete blocks installed on the samples through compensating supports. The actual loading on the samples is set, based on the creation in the plates of stresses, which corresponds to stresses from the specific distribution load of 570 kg/m^2 (Fig. 1).

a *b*

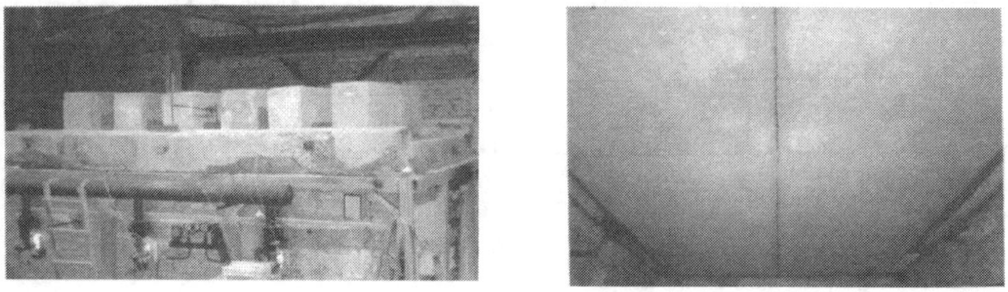

Fig. 1. The view of floors before testing (a) and samples from the furnace side before testing (b).

The deflection limit value of the samples is 220 mm (run $b = 4400$ mm, effective thickness of floors is 220 mm), and the boundary value of the deformation increase rate – 9.8 mm/min.

To measure the average and maximum temperature from the unheated surface of each sample, 5 thermocouples (T1 – T5) were set: one thermocouple (T1) was set in the center of the sample and four in the geometric centers of quarters of the sample (Fig. 2).

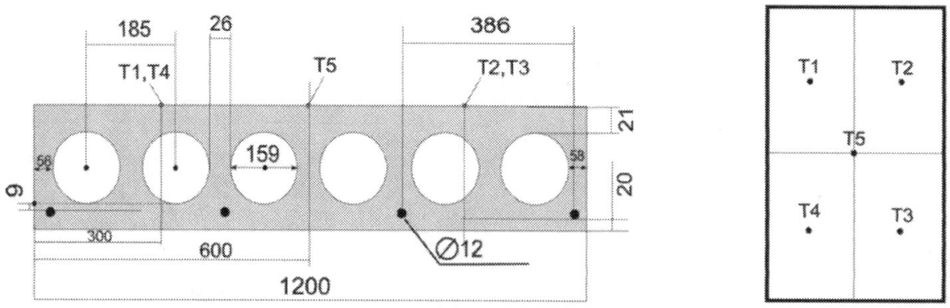

Fig. 2. The scheme of thermocouples placing from the unheated surface of reinforced concrete floor.

Test conditions: air temperature 12°C, relative air humidity 62%. A horizontal test furnace was used for the tests. The results of temperature measurements in the fired furnace are shown in Fig. 3. The temperatures resulted from the tests for fire resistance of reinforced concrete floors with a fire-retardant coating were subsequently used in determination of the thermal and physical characteristics of concrete, "Neosprei" fire-retardant plaster coating. They also were used to determine the dependence of the coating thickness on the thickness of the concrete protective layer for different degrees of fire resistance of the hollow-core reinforced concrete floors (Fig. 4).

The temperature and pressure in the furnace were complied with requirements regulated in accordance with the standard. The excess pressure in the furnace at the 5th minute was 8 Pa, and starting from the 10th minute – 12 Pa. According to order, tests lasted 242 min. The deflection value and the deformation increase rate of the samples after 242 min of tests were 42 mm and 0.4 mm/min (sample No. 1), respectively, and 46 mm and 0.4 mm/min (sample №2), respectively. During the tests, there were no losses of integrity, thermal insulating capacity, and load-carrying capacity of both samples. The test accuracy during testing was 0.42 min. As a result of fire resistance tests, it is established that the fire resistance degree of the above mentioned floor, covered from below with "Neosprei" fire-retardant substance with a thickness of 25.9 mm, is not less than 242 min (fire-resistance rating – REI 240).

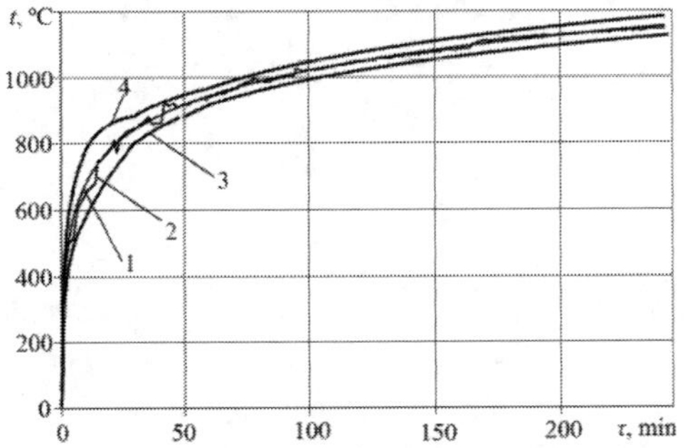

Fig. 3. Dependence of the temperature in the furnace on the time of fire impact on the heated surface of a hollow-core reinforced concrete floor with a thickness of 220 mm, treated with a plaster composition: curve 1 – standard temperature condition curve; curve 2 – real curve of temperature variation in the furnace; 3 – permissible minimum temperature values in the furnace in the process of testing; 4 – permissible maximum temperature values in the furnace in the process of testing.

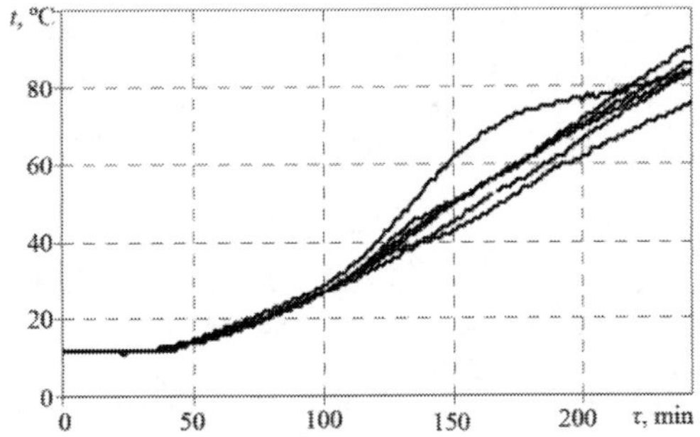

Fig. 4. Dependence of the temperature on the time of fire impact on the unheated surface of a hollow-core reinforced concrete floor with "Neosprei" plaster coating in different places of temperature measurement.

5 The research results of fire resistance of floors and fire-proof ability of coatings

To simulate the thermal state of the floor (see Fig. 1), a physical and computer models were developed. The physical model includes an equation system and boundary conditions, which consider the heat conductivity in the coating and concrete, as well as the complex convective-radiative heat exchange in the air cavities (see Fig. 5).

The mathematical model of the heat conductivity process in such a system in the Cartesian coordinate system, which describes the discussed above physical model, has been repeatedly studied in the literature [12]. It is a one-dimensional heat conductivity equation with a combination of radiant heat exchange and boundary conditions of the third kind on the heated surface, boundary conditions of the third kind on the unheated surface which consider the environment temperature. The mathematical model of such a model of the thermal state of the floor consisted of equations that describe the heat conductivity in a six-layered floor. The second layer in it takes into account natural convection and radiant heat exchange in the cavities by means of effective heat conductivity determined by the solution of inverse heat conductivity problems (IHCP).

The thermal and physical characteristics (TPC) of 1, 3 and 5 concrete layers were assigned from [13], and the thermal and physical characteristics of 2 layer with cavities were found out by solving IHCP (see Fig. 6). The specified thicknesses for 1, 3 and 5 layers for concrete were calculated, and the 2 concrete layer with cavities was chosen as equal to the diameter of the cavities with the available concrete volume.

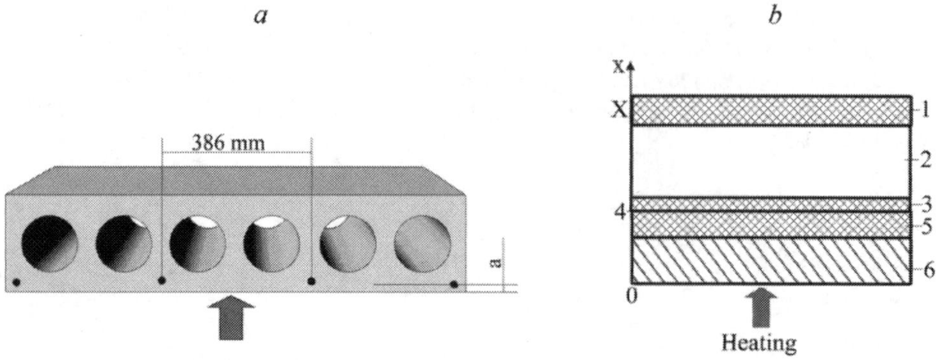

Fig. 5. General view of reinforced concrete floor without fire-retardant coating (a) and the scheme of reinforced concrete floor in a one-dimensional setting (b): 1 – a continuous concrete layer of the floor between the unheated surface and a layer with cavities, 2 – a layer with cavities, 3 – a continuous concrete layer between the cavities and reinforcement, 4 – reinforcement layer, 5 – a continuous concrete layer from the reinforcement till the heated surface, 6 – plaster coating.

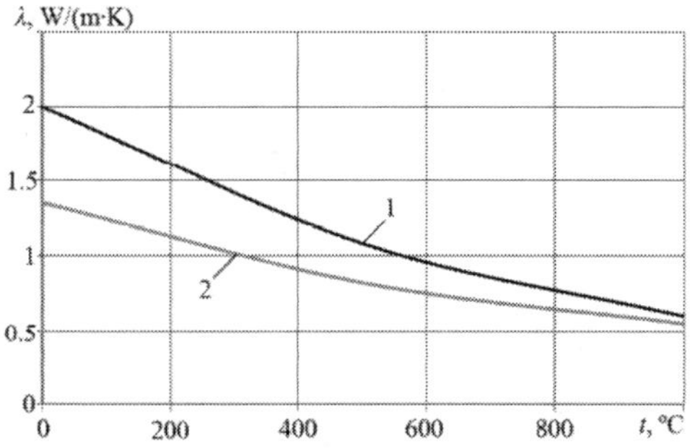

Fig. 6. Coefficient of heat conductivity for concrete: 1 – upper bound; 2 – lower bound.

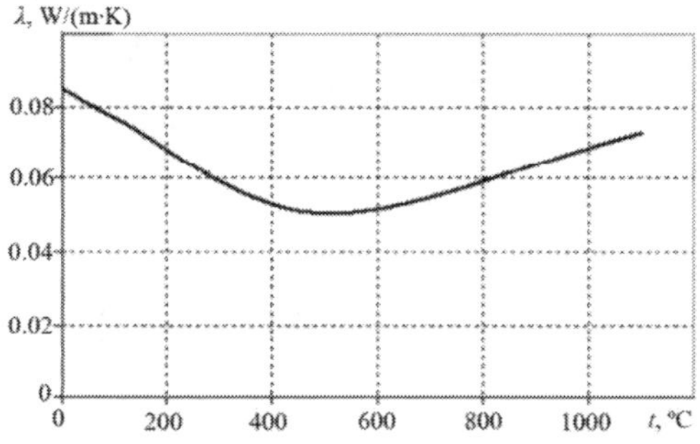

Fig. 7. Dependence of effective heat conductivity coefficient of "Neosprei" plaster coating on the temperature, determined by means of IHCP according to the fire resistance test data.

Specific heat capacity $c_p(\theta)$ of dry concrete ($u = 0\%$) is determined as [8]:

$$c_p(\theta) = 900 \text{ (J/kg·K)} \qquad 20\ ^\circ\text{C} \le \theta \le 100\ ^\circ\text{C};$$

$$c_p(\theta) = 900 + (\theta - 100) \text{ (J/kg·K)} \qquad 100\ ^\circ\text{C} < \theta \le 200\ ^\circ\text{C};$$

$$c_p(\theta) = 1000 + (\theta - 200)/2 \text{ (J/kg·K)} \qquad 200\ ^\circ\text{C} < \theta \le 400\ ^\circ\text{C};$$

$$c_p(\theta) = 1100 \text{ (J/kg·K)} \qquad 400\ ^\circ\text{C} < \theta \le 1200\ ^\circ\text{C}.$$

To determine the TPC of concrete, the computer program FRIEND was used. With its help, it is possible to determine the TPC of fire-retardant coatings of metal, reinforced concrete structures, and other parameters of heat exchange processes from the data of unsteady temperature measurements inside or on the surface of samples.

When determining the thermal and physical characteristics of concrete, as a rule, the method of the inverse heat conductivity problems solution is used. According to this method the determination of the heat conductivity and heat capacity is carried out by comparing the experimental temperatures with the calculated ones.

The criterion for the proximity of these characteristics, most often, is the value:

$$F(P)=\sqrt{\frac{\left[\sum_{j=1}^{m}\left(T_{M,j}-T_{\dot{Y},j}\right)^{2}\right]}{m}}\approx\delta,\qquad(1)$$

where m is the number of experimental measurements used to solve the inverse problem; δ is the mean square error of measurement.

The heat conductivity coefficient of the layer with cavities was calculated as a constant value. The heat conductivity coefficient of 2 layer with cavities was calculated as constant and equal to 5 W/m·K. At the same time, the coefficient of heat transfer from an unheated surface was set according to this law:

T, sec	α_{e2}, W/(m^2·K)
0	0
1500	5
3600	65

The companies-producers of the fire-proof products often indicate the TPC of the developed plaster coatings as a value which does not change with time, that is, it is constant (the heat conductivity in the dry state is 0.11 W/m·K at 20 °C). However, in fact, the TPC of plaster compositions, especially those compositions containing water, vary with temperature. Therefore, within the scope of this work, a determination was made of the heat conductivity coefficient of "Neosprei" plaster composition (see Fig. 7) which depends on the temperature. "Neosprei" fire-retardant plaster coating (hereinafter coating) is a non-flammable, thermal insulating system based on exfoliated vermiculite, inorganic binders, fillers and target additives. For this purpose, the temperatures were used from the unheated surface of hollow-core reinforced concrete floor with fire protection.

As it can be seen from Fig. 7, in the temperature range from 0 to 500 °C), the value of the heat conductivity coefficient of "Neosprei" plaster coating falls and passes through the minimum value of 0.05 W/m·K (at a temperature of 500 °C). This is due to a decrease in density and an increase in the porosity of the coating due to removal of natural and chemically bonding moisture. The increase in the heat conductivity coefficient within the temperature range from 500 to 1100 °C may be explained by the occurrence of the radiation component in the pores of the coating in combination with its high-temperature shrinkage.

When the TPC of a coating has been determined, the specific heat capacity of the plaster coating was defined as constant and equal to $C_v = 1{\cdot}10^6$J/m^3·K.

To assess and predict the fire resistance of hollow-core reinforced concrete floors with fire-retardant coatings, it is necessary to know the thickness of the fire-retardant plaster coating which provides the normed fire resistance degree of this floor. For this purpose, the characteristics found above are used: TPC of concrete, "Neosprei" plaster coating.

In [14], the fire-proof ability of the reinforced concrete floor coating is defined as the dependence of the fire-retardant coating thickness (see Fig. 8) on the thickness of concrete protective layer (the distance between the heated surface and the adjacent reinforcement). At the same time, the normed fire resistance degree of reinforced concrete floor is provided for boundary conditions of fire resistance on the basis of the thermal insulating capacity loss or achieving a critical temperature by the reinforcement with a load level given in the test [15].

With the obtained characteristics, it is possible to determine the thickness of "Neosprei" fire-retardant plaster coating, depending on the concrete protective layer thickness for the required fire resistance of the structure. In such a case, there happens the boundary condition in the fire resistance of construction on the basis of the achieving a critical temperature by the reinforcement at a given load level or thermal insulating capacity of the floor, treated with a fire-proof plaster composition.

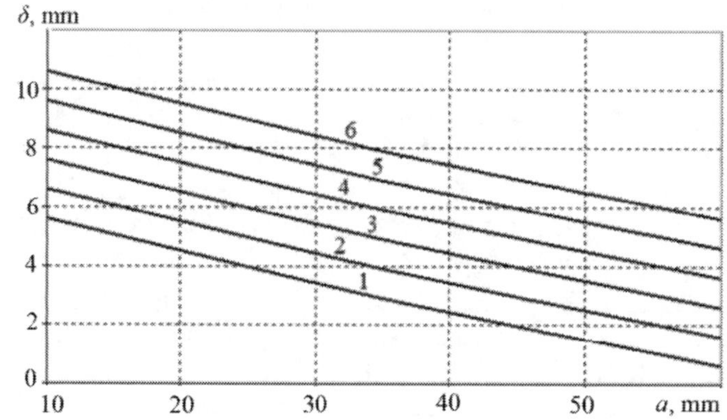

Fig. 8. Dependence of "Neosprey" plaster coating thickness (δ) on the thickness of concrete protective layer (a) of the hollow-core reinforced concrete floor according to the criterion of achieving a critical temperature (500 °C) by the reinforcement for the fire resistance degree: 1 – 60 min; 2 – 90 min; 3 – 120 min; 4 – 150 min; 5 – 180 min; 6 – 240 min.

It should be noted that the accuracy of the fire-proof ability determination is estimated at 16% and can be higher if additional temperature measurements are used not only on the unheated surface, but also on the reinforcement under the fire-retardant coating and inside the floor.

The advantage of the presented work is the use of computational and experimental method for assessing the fire resistance of structures and fire-proof ability of coatings. This method, if compared with the experimental or calculated, is more economical and allows to evaluate the fire-proof ability of coatings and the fire resistance of structures based on the results of one or several experiments (fire tests), by the use of mathematical models. The obtained results will allow a more accurate approach to the assessment of the fire resistance of structures with fire protection in the design of buildings and structures. The studies will be useful for designers, manufacturers of fire-retardant substances, since they allow to calculate such thicknesses of coatings which could provide a normed fire resistance degree of the structure with a given thickness of the concrete protective layer.

6 Conclusions

There was performed an analysis of tests for fire resistance of hollow-core reinforced concrete floors, covered with "Neosprei" plaster composition, manufactured by A PLUS B UKRAINE "PTK", LLC. As a result, it has been established that the fire resistance degree of the PK 48-12-8 hollow-core reinforced concrete floor, covered from below with fire-retardant substance with the thickness of 25.9 mm, consisting of "Neosprei" fire-retardant coating (plaster), is not less than 242 min (fire-resistance rating – REI 240). The test results show the possibility of such structures using in the mining industry with the necessary substantiation of their parameters.

The parameters are substantiated and the effectiveness of "Neosprei" fire-retardant plaster coating is proved. The dependence has been established of the heat conductivity coefficient on the temperature in the conditions of heating in the test furnace of the hollow-core reinforced concrete floor with this coating under the standard temperature condition of a fire.

The mutual interaction has been revealed between the thickness of "Neosprei" fire-retardant plaster coating and the fire resistance of the hollow-core reinforced concrete floors. The necessary minimum thicknesses of such a coating have been calculated to provide the fire resistance degrees of 60, 90, 120, 150, 180 and 240 minutes, regulated by the fire safety requirements.

The authors convey sincere thanks for the help and advice during working on this article to Viktor Rivchachenko, Vice Director of PROMAT UKRAINE, TOV.

References

1. Vivcharenko, O. (2012). Development of coal industry of Ukraine in the context of contemporary challenges. *Geomechanical Processes During Underground Mining: School of Underground Mining 2012*, 1-5. https://doi.org/10.1201/b13157-2
2. Pivnyak, G.G., Pilov, P.I., Bondarenko, V.I., Surgai, N.S., Tulub, S.B. (2005). Development of coal industry: The part of the power strategy in the Ukraine. *Gornyi Zhurnal*, (5), 14-17.
3. Roytman, V.N. (2001). *Inzhenerne resheniya po otsenke ognestoykosti proyektiruyemykh i rekonstruiruyemykh zdaniy*. Moskva: Assotsiatsiya "Pozharnaya bezopasnost' i nauka".
4. Fomin, S.L., & Astakhov, A.A. (2010). Ognestoykost' fragmenta bezrigel'nogo karkasnogo monolitnogo zdaniya. *Naukovyi visnyk budivnytstva*, (61), 122-130.
5. Vasil'chenko, A.V. (2010). Otsenka predela ognestoykosti zhelezobetonnykh plit perekrytiya s ognezashchitnym pokrytiem. *Problemy pozharnoy bezopasnosti*, (27), 45-48.
6. Rubini. P. (2000). *SOFIE – Simulation of Fires in Enclosures*. V 3.0 Users guide. School of Mechanical Engineering, Granfield University (UK).
7. Bek Dzh., V., Blakuell, B., & Sent-Kler, Ch. (1989). Nekorrektnye obratnye zadachi teploprovodnosti. Moskva: Mir.
8. Lie, T.T. (1974). Characteristic temperature curves for various fire severities. *Fire Tachnol*, *10*(4), 315-326.
9. Krukovsky, P. (2005). Fire Safety Analysis for new Safe confinement's Building Constructions of Chernobyl NPP. In *V Międzunarodova Konferencja "Bezpieczeństwo pożarowe budowli"* (pp. 223-227). Warszawa: Instytut Techniki Budowlanej.
10. DSTU B V.1.1-4-98. (1999). *Zakhyst vid pozhezhi. Budivelni konstruktsii. Metody vyprobuvannia na vohnestiykist. Zahalni vymohy*. Kyiv: Ukrarkhbudinform.
11. DSTU B V.1.1-20:2007. (2007). *Zakhyst vid pozhezhi. Perekryttia ta pokryttia. Metod vyprobuvannia na vohnestiikist*. Kyiv: Ministerstvo rehionalnoho rozvytku ta budivnytstva Ukrainy.
12. Kovalov, A.I., Kachkar, Ye.V, & Zobenko, N.V. (2014). Eksperymentalne doslidzhennia vohnezakhysnoi zdatnosti pokryttia "Amotherm Steel Wb" pry temperaturnomu rezhymovi vuhlevodnevoi pozhezhi. *Pozhezhna bezpeka: teoriia i praktyka*, (17), 53-60.
13. EN 1992-1-2: 2004. (2004). *Eurocode 2: Design of concrete structures. Part 1-2: General rules. Structural fire design*. Brussels: The European Union Per Regulation.
14. Kovalev, A.I. (2012). *Usovershenstvovanie metoda otsenki ognezashchitnoy sposobnosti pokrytiy zhelezobetonnykh perekrytiy*. Ph.D. Ukrainskiy NII grazhdanskoy zashchity MChS Ukrainy.
15. Sundstrom, B. (2005). Test Methods and Their use for Fire Safety Engineering. In *V Międzunarodova Konferencja "Bezpieczeństwo pożarowe budowli"* (pp. 141-150). Warszawa: Instytut Techniki Budowlanej.

E3S Web of Conferences **60**, 00004 (2018) https://doi.org/10.1051/e3sconf/20186000004
Ukrainian School of Mining Engineering

A concept to use energy of air flows of technogenic area of mining enterprises

Edgar Cabana[1], *Volodymyr* Falshtynskyi[2], *Pavlo* Saik[2], *Vasyl* Lozynskyi[2*], and *Roman* Dychkovskyi[2]

[1]San Agustin National University Arequipa, Institute of the Center of Renewable Energy and Energy Efficiency, 117 Santa Catalina St., 04000 Arequipa, Peru
[2]National Mining University, Department of Underground Mining, 19 Yavornytskoho, 49005 Dnipro, Ukraine

Abstract. Topical issues concerning the possibilities of effective energy generation on the basis of Ukrainian mining enterprises have been highlighted. Attention is drawn to one of the most challenging tendencies of nontraditional energy sources development in the process of wind-driven powerplant use. The plants consume energy of technogenic air flows. Operation schemes and engineering solutions as for the possibilities to involve nontraditional sources of energy resulting from operation of technogenic objects and industrial structures of enterprises (i.e. stopes, mine dumps, industrial facilities etc.) in power balance of a mining enterprise have been demonstrated. Energy efficiency of the wind-driven powerplants in the context of different operation schemes aimed at utilization of energy of technogenic air flows has been analyzed. Output of a wind-driven system as a component of technological segment of a mining enterprise making it possible to meet 20 to 44% of its power demand has been determined. The objective of the paper is to substantiate process solutions as well as technical and engineering ones as for the obtaining nontraditional energy sources basing upon introduction of wind-driven power plants.

1 Introduction

Energy independence is one of the most important criteria while estimating a level of national safety of a country. It relies upon principles of the reduced volatility, increased energy generation at the expense of proper sources, and increased efficiency of alternative power sources [1, 2]. The abovementioned stipulates sustainable and efficient progress of any industry of modern society [3, 4]. As the authors of [5, 6] mentions, technology is the driving force of industries. It is the powerful factor of achieving and maintaining competitive priority as well as insuring sustainable economic growth [7] as well as ecological problems solutions of post-industrial mining areas [8, 9].

In Ukraine, electrical energy generation by thermal power stations (TPSs), nuclear power stations (NPSs), and hydro power stations (HPSs) is considered as traditional

* Corresponding author: lvg.nmu@gmail.com

process of energy production. Electrical energy generation by TPSs is high on the world list. Nuclear power industry is a separate branch of power industry.

Renewable power generation is a tool of technological leadership of any nation. According to the conclusions by 21st Conference of UNO on the problems of climate, which took place in Paris in December 2015, renewable power industry was determined as the key tool to reduce emissions of greenhouse gases to minimize effects of climatic changes [10, 11].

Currently, renewable energy sources (RESs) are experiencing intensive development stages [12, 13]. Solar energy and wind energy are the most active. A tendency to increase share of RES in the context of total energy balance is worldwide one [14]. The fact can be explained by the growth of investment in renewable power industry (in 2004 – 2016, annual average investment growth was 12.5%).

Ukraine demonstrates confident RESs growth as well. According to the data by the State Statistics Service, in 2004 – 2016, a share of energy generated by RESs increased from 1.7% (2384 thousand tons of o.e.) to 3.9% (3616 thousand tons of o.e.). Ukraine has high wind-energy resource which is sufficient to provide the development of large objects. It should be noted the electric energy, generated by wind power stations (WPSs) is the cheapest among other types of electric energy, generated at the expense of RESs [15]. Thus, according to information by Enerhorynok SE, a share of wind energy is only 2.54% (in terms of 0.66% share of the total electric energy sold to Ukrainian electric power pool in 2016) to be less to compare with solar energy (SE) and the electric energy, generated by small HPSs, which respective shares are 35 and 13% in the context of the total electric energy supply (Table 1).

Table 1. Shares in volumes of electric energy and its cost depending upon primary energy carriers.

Primary energy carriers	Share in the electric energy, %	Share in the cost of electric energy, %
NPSs	53.74	29.32
TPSs	31.78	45.96
HPSs (with exception of small ones)	6.21	5.23
TPPs	7.01	12.51
SHPSs	0.13	0.56
WPSs	0.66	2.54
SE	0.35	3.38
Biomass energy	0.06	0.24
Others	0.06	0.26
Total	100.00	100.00

Introduction of wind-driven power plants, utilizing the energy of natural air flow, provokes a number of problems connected with their operation under storm windflaws, under icing of windwheel and weather equipment, under sharp changes in physical landscape and climate.

Thus, the authors of the paper believe that utilization of energy of technogenic wind flows (TWF) within the territories of mining enterprises is one of the promising tendencies to develop nontraditional energy sources on the basis of wind-driven power plants. Progress of the process solutions as well as technical and engineering solutions concerning energy generation in the while wind-driven powerplants operating makes it possible to prolong life of mining enterprises. That is achieved by means of cardinal reorientation of coal mining; thus, changes in such principle mission of a mining enterprise as “coal mining → energy generation” take place.

2 Mathematical representation and description of using energy of air flows

Introduction of nontraditional energy sources into life cycle of mining enterprises is stipulated by the availability of a number of man-made objects and industrial structures at their territories. Such objects include: mine dumps, headframes, buildings of industrial sites of mining enterprises, shafts, and systems of permanent mine openings. Mining enterprises, terminating their operating life and those, being in a state of stagnation, are preferred [16, 17].

It is common knowledge that circulation of technogenic air flows within underground mine workings of abandoned mines depends upon the depression formed by natural draft at temperature-pressure difference in underground area and at the surface [18 – 20]. Underground air flow is heated owing to contact air-rocks heat exchange [21, 22].

Natural draft depression is the energy, generated by air unit volume of a source causing the natural draft; it is calculated with the help of formula [23]:

$$h_n = \frac{\rho_a \Delta H \left(T_{atm} - T_a\right)}{273 + T_a}, \qquad (1)$$

where ρ_a is specific weight of a mine air, N/m^3; ΔH is difference between marks of a shaft collar, m; T_{atm} is average temperature of atmospheric air entering a mine, in a shaft collar, °C; and T_a is average temperature of mine air, °C.

Mine air flow involves much constant kinetic energy; moreover, it is powerful source to rotate wheels of underground wind-driven power plants with constant velocity and direction. The obtained kinetic energy of technogenic air flows within mine workings is used by means of mounting of a chain of wind-driven powerplants (Fig. 1) [24].

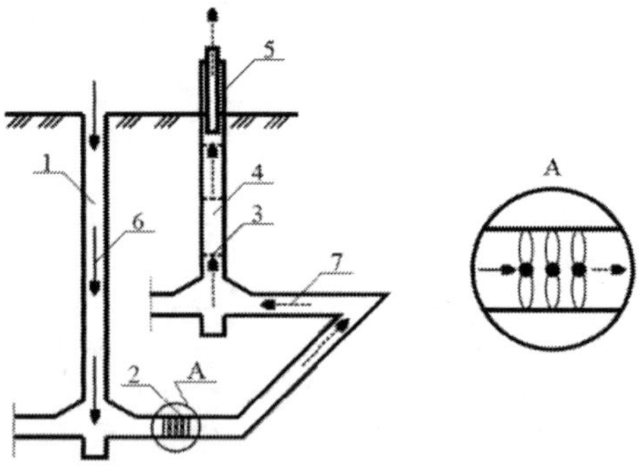

Fig. 1. Procedures to use energy of technogenic air flows within underground mine workings by means of wind-driven powerplants: 1 – supplying ventilation shaft; 2 – block of wind-driven powerplants of a development mine working; 3 – wind-driven powerplants of a shaft; 4 – ventilation shaft; 5 – overshaft tubular helium air heater; 6 – fresh air; and 7 – dusty air.

If wind wheels are located within mine workings, they rotate smoothly without any pulsation and frequency vibration of electric current. Their location in ventilation walls, air doors etc. is the simplest design solution. The authors of paper [25] mention: to evaluate air reserves countrywide, useful wind energy with minimum velocity is used; that is $v = 4.0$ m/s. In this context, the wind energy can be evaluated applying the simplified formula:

$$E = g \cdot v^3 \cdot t \cdot 1.34 \cdot 10^{-7}, \qquad (2)$$

where E is energy efficiency of air flows, kWh/year; g is density of the air flow, kg/m^2; v is velocity of the air flow, m/s; t is time.

Formula (1) can also be used to determine energy efficiency of air flows in coal mines since a chain of wind-driven power plants is mounted in ventilation wells, shafts, drifts, main entries and airways, main gravity inclines and district gravity inclines, and inclines where maximum air velocity may achieve 15 m/s [26].

Hence, generating capacity of wind-driven power plants of a mining enterprise can be calculated on the formula (2) involving useful area of blade faces of wind-driven power plants:

$$E_s = \eta \cdot g \cdot v^3 \cdot t \cdot A \cdot 1.34 \cdot 10^{-7}, \text{ kWh/year}, \quad (3)$$

where E_s is generating capacity of WDPP, kWh/year; η is the efficiency of a wind turbine and electric generator (0.25 for small turbines, and 0.6 for large ones which wattage is 2 MW and more); A is useful area of blades of wind-driven power plants, m^2; E is generating capacity of air flow, kWh/m^2.

Rotation velocities of electric generator and wind wheel are determined through the ratio of diameters of transmission wheels:

$$\frac{v_1}{v_2} = \frac{d_1}{d_2}, \quad (4)$$

where v_1 and d_1 are velocity and diameter of a connecting shaft, m/sec, m; v_2 and d_2 are velocity and diameter of electric generator shaft, m/sec, m.

Analytical research has helped determine the following: if length of a main haulage road is 3.0 to 3.7 km then wind-driven power plants are mounted at the distance of 20 m to extinguish vortexes of air flow behind the wind wheel. Hence, when the velocity of technogenic air flow is 1.65 – 3.16 m/s, then efficiency of the wind-power system may achieve 24.7 – 36.2 kW/hour. Monthly generation of electric energy will be 13.5 – 19.3 MW/hour.

3 Results and discussion

Heat exchangers (electroheating facilities), mounted in nearshaft mine workings, is one of the most helpful technical and engineering solutions [23]. Heated in such a way air flow with 1.19 to 1.04 kg/m^3 density and 2.56 – 4.79 m/s velocity passes through a heating member and moves up quickly along a shaft to the surface rotating 3.5 – 5.2 m diameter wind wheel of wind-driven power plants mounted within the shaft (Fig. 2).

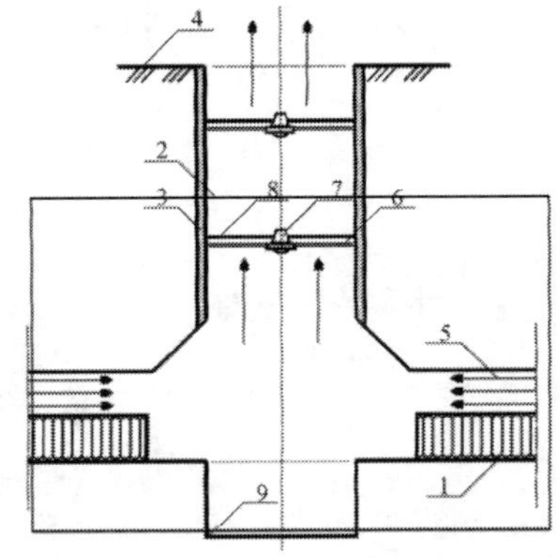

Fig. 2. Mounting scheme of wind-power devices in a mine shaft: 1 – heat exchangers; 2 – mine shaft; 3 – shaft timbering; 4 – earth's surface; 5 – technogenic air flow; 6 – wind wheels of wind generators; 7 – electric generators; 8 – supports of wind generator; 9 – TAF from mine workings; 10 – drainage shaft.

Fig. 3 demonstrates graphs of changes in generating capacity of wind-power device using energy of air flows in a mine shaft under the conditions of natural draft depending upon air flow velocity and a diameter of a wind wheel.

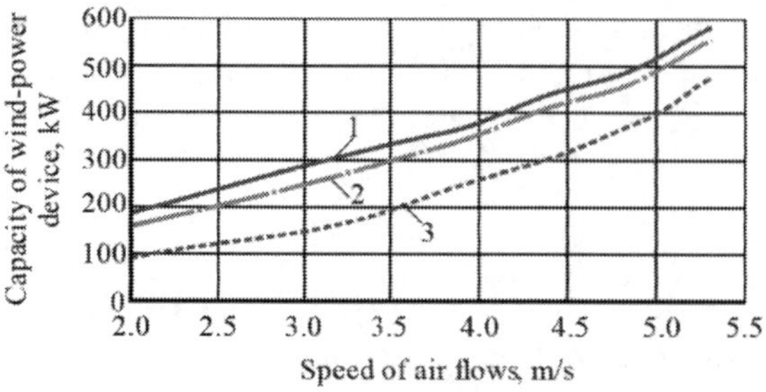

Fig. 3. Graphs of changes in generating capacity of wind-power device using energy of air flows in a mine shaft under the conditions of natural draft depending upon air flow velocity and a diameter of a wind wheel: 1 – 5.2 m; 2 – 5.0 m; 3 – 3.5 m.

Owing to constant directivity and smoothness of changes in the velocity of air flows, efficiency of mechanical energy turning into electrical one at wind-power devices in underground mine workings is 82 – 90%. Use of technogenic air flows, injected by main mine fans, is provided with the help of operation schedule with wind-power units in the shaft of outgoing air flow. In the design of the wind-driven power plants, a wind wheel is three-bladed; depending upon the device remoteness, the wind wheel is mounted in a diffuser to improve density as well as directivity of the kinetic air flow.

Analysis of the data, resulting from full-scale studies and analytical studies concerning air flow injected by main mine fan VOKD-30, has made it possible to determine aerodynamic characteristic of air flow line used to construct graphs demonstrating changes in rational operation parameters of wind-driven power plants mounted on a main mine fan [27].

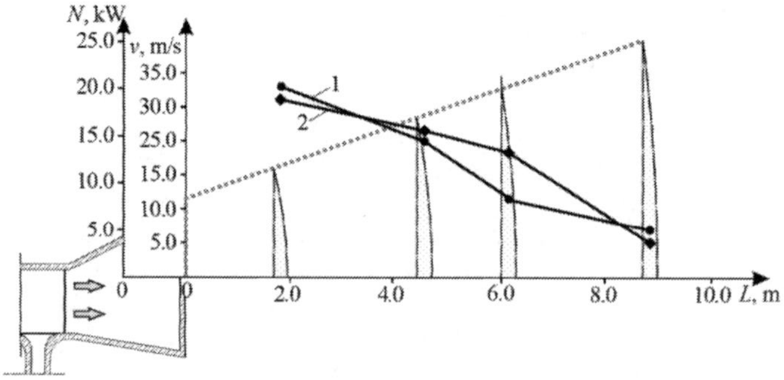

Fig. 4. Graphs of changes in operation parameters of a wind-driven power plant mounted on a main mine fan: 1 – a graph of changes in air flow velocity depending upon windwheel assembling on the outgoing flow of a main mine fan; 2 – a graph of changes in the wind-driven power plant capacity depending upon remoteness of windwheel assembling on the outgoing flow of a main mine fan.

Diffuser provides the improved density as well as directivity of the technogenic air flows on the windwheel within the design of wind-driven power plants. That increases electricity generating indices by 4.6 – 9.5%. Assembling location of the windwheel depends upon its effect on the ventilation system since incorrect selection may result in the significant increase in energy consumption by a main mine fan which in turn will make end product more expensive.

The use of the wind-driven power plants at the territories of mine dumps of mining enterprises is one of the examples of their efficient operation. First of all, rationality of the engineering solution concerning the use of mine dump energy depends upon the mine dump potential as well as availability of the marketable product consumers [28].

Fig. 5 demonstrates operation schedule to utilize air flow being formed at the expense of changes in the density of air flows in heat-generating supply line, mounted at the bottom of a mine dump [29].

Thermal flows, resulting from oxidation of such organic matters as sulphur, coal, carbonaceous rocks, timber etc., form air flows within man-made structures of industrial area of a mining enterprise [30 – 32]. Peculiarity of the technique is as follows: air flow is directed from outside of a mine dump through a pipe, mounted at the bottom of the mine dump. Diffuser is mounted on the outgoing air flow; windwheel of the wind-driven power plant is inside the diffuser.

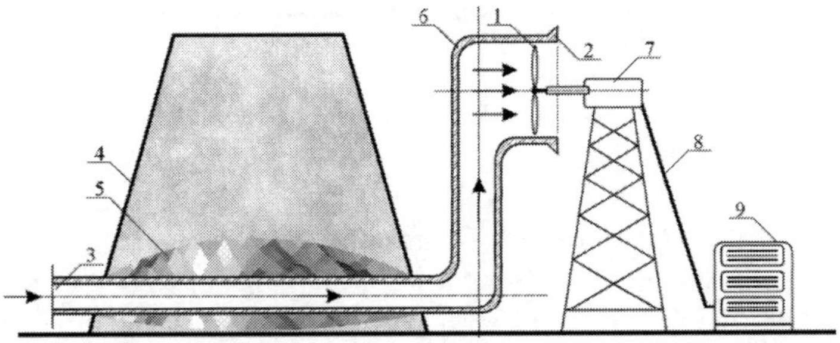

Fig. 5. Operation schedule to utilize energy of technogenic air flow while using heat generated by mine dumps: 1 – windwheel of a wind-driven power plant; 2 – diffuser; 3 – heat-generating pipe; 4 – mine dump; 5 – zone of active oxidation of the mine dump organic mixtures; 6 – thermal protection of the diffuser; 7 – electric generator of the wind-driven power plant; 8 – electric cable; and 10 – control box.

Active motion of air flow through heat-generating pipe is stipulated by pressure difference between ingoing and outgoing flows; it is determined by the formula:

$$\Delta P = H\left(\gamma_{in} - \gamma_{out}\right),\ \text{kg/m}^2, \qquad (4)$$

where H is distance between axes of heat-generating pipe 3 and diffuser 2, m; $\gamma_{in} = f(t)$ is density of technogenic air flow depending upon temperature at the heat-generating pipe entrance 3, kg/m^3; $\gamma_{out} = f(t)$ is density of technogenic air flow depending upon heating temperature on the diffuser exit 2, kg/m^3.

In this context, $\Delta V_{(t)}$ velocity of the heated outgoing from the diffuser 2 airflow is determined using the formula:

$$v = \sqrt{\frac{2\Delta Pg}{\gamma_{in}}} = \sqrt{\frac{2Pg\left(\gamma_{in} - \gamma_{out}\right)}{\gamma_{out}}},\ \text{m/s}, \qquad (5)$$

Capacity of the wind-driven power plant is:

$$N_{B.y} = 0.000481 \cdot d^2 \cdot v^3 \cdot \eta\,, \qquad (5)$$

where d is a diameter of a windwheel, m; υ is velocity of air flow, m/s; η is the wind-driven power plant efficiency.

Capacity of the wind-driven power plant depends upon the velocity of air flow, air density, area of the windwheel, and coefficient of the air flow use by the windwheel involving the diffuser effect. Rated capacity of the heat-generating pipe is 2.8 – 3.6 m^3/s if its diameter is 300 – 350 mm, and air flow velocity is 4.6 – 11.5 m/s. Graph in Fig. 6 demonstrates a characteristic of changes in capacity of wind-driven power plant depending upon windwheel diameter.

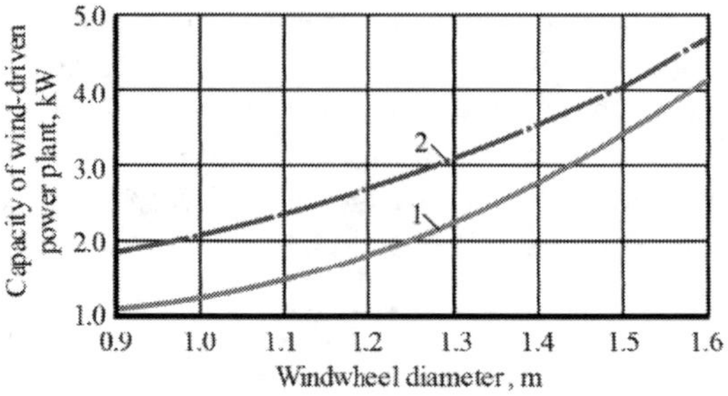

Fig. 6. Graph of changes in capacity of wind-driven power plant depending upon windwheel diameter and air flow velocity: 1 – 4.2 m; 2 – 9.5 m/s.

Relying upon the available minimum dimensions of mine dumps and heat-generating capacity of rocks, it is possible to mount up to 8 wind-driven power plants. The plants can become a part of the unified system of mining enterprise. Total capacity of wind-driven plants will be 11.7 – 45.2 kW; monthly electric energy generation will be 6.3 – 24.3 MW/hour.

Mine dumps, shaped in the form of a cone within a mine land allotment, take considerable area; their height is 42 – 124.5 m with average index of 58.14 m where average velocities of natural wind flows are 6.5 – 10.4 m/s, and maximum velocities are 15.6 – 22.7 m/s. Surfaces of mine dumps are used as areas, where wind-driven power plants are mounted and united into power system taking into consideration energy potential of wind and illiquidity of the dump to be mined as well as under the conditions of complete attenuation [33].

Wind-driven power plant capacity is determined taking into consideration changes in the two basic parameters: windwheel diameter $D_{w.h}$ and wind velocity V_v, m/s (Fig.7).

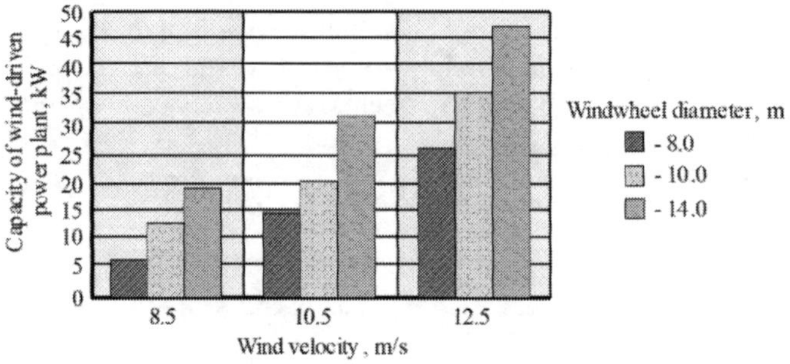

Fig. 7. Changes in the capacity of basic wind-driven device of WDPP depending upon changes in windwheel parameters and wind velocity.

Analysis of data in Fig. 7 makes it possible to select ideal parameters of a windwheel relying upon average determinants of wind velocity. In the context of similar wind velocities and increase in a windwheel dimensions with 2 m step, capacity of the wind-driven power plant experiences its 12% raise. If average dump height is 54 m, then total

https://doi.org/10.1051/e3sconf/20186000004

capacity of wind-driven system of the mine dump will be 15.4 – 71.8 kW depending upon diameter of a windwheel, wind velocity and the number of wind-driven power plants on the mine dump. Theoretical efficiency is 35.2 – 40.1%.

The authors used average data of the basic energy-generating objects (i.e. mine workings, shafts, mine dumps) [34 – 36]. Capacity of a wind-driven system as a component of technological segment of a mining enterprise makes it possible to meet 20 – 40% of its household and industrial requirements as for electric energy. Fig. 8 shows a graph of monthly generation of electric energy by certain segments of wind-driven system.

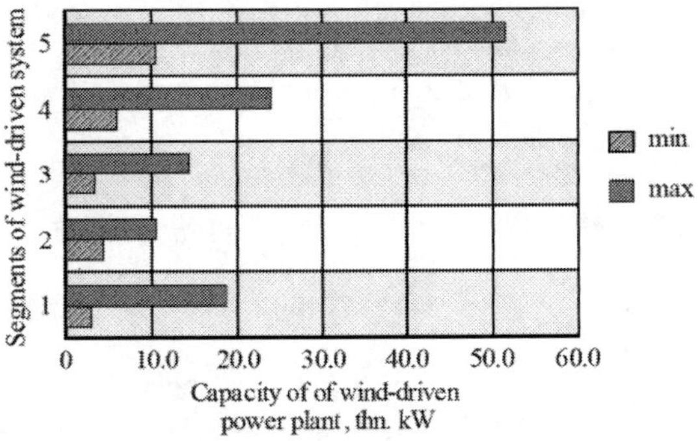

Fig. 8. Monthly electrical energy generation by certain segments of wind-driven system: 1 – wind-driven generators of technogenic air flows of mine workings; 2 – wind-driven generators of air flows of a mine shaft; 3 – wind-driven generators mounted on a main mine fan; 4 – wind-driven generators of thermal technogenic air flows of a mine dump; 5 – wind-driven devices of mine dumps, thousand.

Wind-driven power plants, mounted on mine dumps, are the most active sources to generate electric energy. Such a situation can be explained by significant velocities of air flows to compare with flows, formed underground.

4 Conclusions

The use of wind-driven power plants, basing upon mining enterprises, makes it possible to generate cheap electricity involving reduced prime cost of end product, improved environmental and social indices of the enterprise infrastructure. Formation of a system of wind-driven plants at the territory of basic objects of mining enterprise (i.e. buildings of industrial mine site, mine workings, mine dumps etc.) helps prolong life of the enterprise significantly.

It has been determined that wind-driven power plants, mounted on mine dumps, are the most productive sources to generate electric energy. Efficiency of wind-driven power plants, mounted in mine workings, is 1.2 – 2.1 times higher to compare with those, mounted in mine workings, shafts, and fan diffuser. Efficiency of wind-driven power plants, mounted at a dump surface, is 2.2 – 5 times higher to compare with other facilities. That makes it possible to substantiate the most rational engineering solutions concerning the erection of wind-driven power plants.

This work was conducted within the projects "Scientific basics to form unified system to preserve and generate power of the objects of fuel and energy complex of Ukraine" (State registration No. 0117U001127) and "Scientific substantiation and development of energy saving and low waste technologies of hydrocarbon and mineral raw materials extraction" (State registration No. 0116U008041).

References

1. Fatkhutdinov, V.H., & Bazaluk, O. (2018). The Importance of the Brain Neuro-Programming Technologies in National and Regional Security Strategies. *Philosophy and Cosmology*, (20), 74-82. https://doi.org/10.29202/phil-cosm/20/6
2. Bondarenko, V., Maksymova, E., & Koval, O. (2013). Genetic classification of gas hydrates deposits types by geologic-structural criteria. *Annual scientific and practical collection – Mining of Mineral Deposits*, 115-119. https://doi.org/10.1201/b16354-21
3. Sotskov, V.O., Demchenko, Yu.I., Salli, S.V., & Dereviahina, N.I. (2017). Optimization of parameters of overwoked mining gallery support while carrying out long-wall face workings. *Naukovyi Visnyk Natsionalnoho Hirnychoho Universytetu*, (6), 34-40.
4. Petlovanyi, M.V., Lozynskyi, V.H., Saik, P.B., & Sai, K.S. (2018). Modern experience of low-coal seams underground mining in Ukraine. *International Journal of Mining Science and Technology*. Article in press https://doi.org/10.1016/j.ijmst.2018.05.014
5. Sapytska, I. (2016). Restructuring of mines: technological and economical aspects. *Shid*, *3*(143), 38-42.
6. Rozin, V.M. (2017). Technology as a Time Challenge: Study, Concept and Types of Technology. *Philosophy and Cosmology*, (19), 133-142.
7. Smol, M., Kulczycka, J., & Avdiushchenko, A. (2017). Circular economy indicators in relation to eco-innovation in European regions. *Clean Technologies and Environmental Policy*, 19(3), 669-678. https://doi.org/10.1007/s10098-016-1323-8
8. Pivnyak, G.G., & Shashenko, O.M. (2015). Innovations and safety for coal mines in Ukraine. *Naukovyi Visnyk Natsionalnoho Hirnychoho Universytetu*, (6), 118-121.
9. Gorova, A., Pavlychenko, A., Kulyna, S., & Shkremetko, O. (2012). Ecological problems of post-industrial mining areas. *Geomechanical Processes During Underground Mining*, 35-40. https://doi.org/10.1201/b13157-7
10. Povkhanych, A.Y. (2017). Wind power as a key element of energy strategy. *Naukovyi Visnyk Uzhhorotskoho Natsionalnoho Universytetu*, *13*(2), 81-86.
11. Golinko, V.I., Yavorskiy, A.V., Lebedev, Ya.Ya., & Yavorskaya, Ye.A. (2014). Estimation of frictional sparking effect on firedamp inflammation during fragmentation of gas-saturated rock massif. *Naukovyi Visnyk Natsionalnoho Hirnychoho Universytetu*, (6), 31-37.
12. Caceres, E., & Alca, J.J. (2016). Potential For Energy Recovery From A Wastewater Treatment Plant. *IEEE Latin America Transactions*, *14*(7), 3316-3321. https://doi.org/10.1109/tla.2016.7587636
13. Lozynskyi, V., Saik, P., Petlovanyi, M., Sai, K., & Malanchyk, Ye. (2018). Analytical Research of the Stress-Deformed State in the Rock Massif Around Faulting. *International Journal of Engineering Research in Africa*, (35), 77-88. https://doi.org/10.4028/www.scientific.net/JERA.35.77
14. *Rol i mistse ukrainskoi enerhetyky u svitovykh enerhetychnykh protsesakh.* (2018). Retrived from http://razumkov.org.ua/uploads/article/2018_ENERGY_PRINT.pdf
15. *Vitroenerhetychnyi sektor Ukrainy.* (2016). Retrived from http://uwea.com.ua/uploads/docs/uwea_2016_report_ua_web_open.pdf
16. Falshtynskyi, V., Saik, P., Lozynskyi, V., Dychkovskyi, R., & Petlovanyi, M. (2018). Innovative aspects of underground coal gasification technology in mine conditions. *Min. Miner. Depos.*, *12*(2), 68-75. https://doi.org/10.15407/mining12.02.068
17. Saik, P., Petlevanyi, M., Lozynskyi, V., Sai, K., & Merzlikin, A. (2018). Innovative approach to the integrated use of energy resources of underground coal gasification. *Solid State Phenomena*, (277), 221-231. https://doi.org/10.4028/www.scientific.net/SSP.277.221
18. Lozynskyi, V., Dychkovskyi, R., Saik, P., & Falshtynskyi, V. (2018). Coal Seam Gasification in Faulting Zones (Heat and Mass Balance Study). *Solid State Phenomena*, (277), 66-79. https://doi.org/10.4028/www.scientific.net/SSP.277.66

19. Menshov, O., Kuderavets, R., Popov, S., Homenko, R., & Sukhorada, A. (2016). Thermomagnetic analyzes of soils from the hydrocarbon fields. *Visnyk of Taras Shevchenko National University of Kyiv-Geology*, *2*(73), 33-37. https://doi.org/10.17721/1728-2713.73.05
20. Sotskov, V., & Saleev, I. (2013). Investigation of the rock massif stress strain state in conditions of the drainage drift overworking. *Annual scientific and practical collection – Mining of Mineral Deposits*, 197-201. https://doi.org/10.1201/b16354-36
21. Tabachenko, M. (2016). Substantiating parameters of stratification cavities formation in the roof rocks during underground coal gasification. *Mining of Mineral Deposits*, *10*(1), 16-24. https://doi.org/10.15407/mining10.01.016
22. Kozhevnikov, A.A., Sudakov, A.K., Dreus, A.Yu., & Lysenko, Ye.Ye. (2014). Study of heat transfer in cryogenic gravel filter during its transportation along a drillhole. *Naukovyi Visnyk Natsionalnoho Hirnychoho Universytetu*, (6), 49-54
23. Pivniak, G.G., et. al. (2013). *Tradytsiini ta netradytsiini systemy enerhozabezpechennia urbanizovanykh i promyslovykh terytorii Ukrainy*. Dnipropetrovsk: National Mining University.
24. Kolokolov, O.V., Lapko, V.V., & Falshtynskyi, V.S. (1999). Poluchenie elektroenergii pri utilizatsii vozdushnykh potokov shakht i rudnikov. *Naukovyi Visnyk Natsionalnoho Hirnychoho Universytetu*, (6), 21-23.
25. Tytko, R., & Kalinichenko, V. (2010). *Vidnovliuvalni dzherela enerhii (dosvid Polshchi dlia Ukrainy)*. Varshava-Krakiv-Poltava: OWG.
26. NPAOP 10.0-1.01-10. (2010). *Pravyla bezpeky u vuhilnykh shakhtakh*. Kyiv: Minvuhleprom Ukrainy.
27. Kolokolov, O.V., Tabachenko, M.M., Liashenko, V.I., & Falshtynskyi, V.S. (1997). *Sposib utylizatsii enerhii tekhnohenoho povitrianoho potoku*. Patent No. 15054A, Ukraine.
28. Saik, P. (2017). Study of Methods and Development of Technological Scheme for Heat Removal from Rock Waste Dump. *Advanced Engineering Forum*, (25), 128-135.
29. Tabachenko, M.M., Samusia, V.I., Dychkovskyi, R.O., & Falshtynskyi, V.S. (2012). *Novitni pryntsypy teplonasosnykh ta koheneratsiinykh tekhnolohii vykorystannia vykydnoho tepla*. Dnipropetrovsk: Natsionalnyi hirnychyi universytet.
30. Bondarenko, V., Tabachenko, M., & Wachowicz, J. (2010). Possibility of production complex of sufficient gasses in Ukraine. *New Techniques and Technologies in Mining*, 113-119. https://doi.org/10.1201/b11329-19
31. Sadovenko, I.O., Zahrytsenko, A.M., Podvihina, O.O., & Dereviahina, N.I. (2017). Water balance control within rock mass using the capacity of water-bearing formations. *Naukovyi Visnyk Natsionalnoho Hirnychoho Universytetu*, (4), 19-26.
32. Sukhov, V., Suyarko, V., Niemets, K., & Matveyev, A. (2018). Hydrogeodynamic processes in carbonate rocks. Part II. Karst and its influence on geological environment. *Visnyk of V.N. Karazin Kharkiv national university-series geology geography ecology*. https://doi.org/10.26565/2410-7360-2018-48-15
33. Tabachenko, M., Saik, P., Lozynskyi, V., Falshtynskyi, V., & Dychkovskyi, R. (2016). Features of setting up a complex, combined and zero-waste gasifier plant. *Mining of Mineral Deposits*, *10*(3), 37-45. https://doi.org/10.15407/mining10.03.037
34. Kononenko, M., & Khomenko, O. (2010). Technology of support of workings near to extraction chambers. *New Techniques and Technologies in Mining*, 193-197. https://doi.org/10.1201/b11329-32
35. Khomenko, O., Kononenko, M., & Myronova, I. (2013). Blasting works technology to decrease an emission of harmful matters into the mine atmosphere. *Annual scientific and practical collection – Mining of Mineral Deposits*, 231-235. https://doi.org/10.1201/b16354-43
36. Cherniaiev, O.V. (2017). Systematization of the hard rock non-metallic mineral deposits for improvement of their mining technologies. *Naukovyi Visnyk Natsionalnoho Hirnychoho Universytetu*, (5), 11-17.

Modeling the process of hydromechanical amber extraction

Yevhenii Malanchuk[1*], *Viktor* Moshynskyi[2], *Valerii* Korniienko[3] and *Zinovii* Malanchuk[3]

[1]National University of Water Management and Nature Resources Use, Department of Automation, Electrical Engineering and Computer-Integrated Technologies, 11 Soborna St., 33028 Rivne, Ukraine
[2]National University of Water Management and Nature Resources Use, Department of Land Management, Cadaster, Land Monitoring and Geoiformatics, 11 Soborna St., 33028 Rivne, Ukraine
[3]National University of Water Management and Nature Resources Use, Department of Development of Deposits and Mining, 11 Soborna St., 33028 Rivne, Ukraine

Abstract. The paper represents a process of hydromechanical amber extraction modeling to obtain input data and substantiate operation parameters of mining equipment to develop the improved hydromechanical technique of amber extraction. Intensification of amber mining process is possible when sandy deposit is saturated by water and air. Moreover, mechanical impact by means of vibration is added. Amber displacement within sandy deposit is considered. The deposit is characterized by environmental resistance when influence factors act on the process of amber surfacing. Amber concentration distribution over a deposit surface involving determination of floating periods of different amber fractions in terms of different operation modes as well as computer experiment concerning the amber grades and its distribution over amber-bearing deposit involved the use of computer environment Matlab. Adequate mathematical model to solve one-dimensional boundary problems for systems of parabolic and elliptic differential equations within partial first-order derivatives on one spatial variable and time has been developed. The model describes accurately the behaviour of different amber fractions within amber-bearing deposit in terms of vibration effect as well as water and air supply.

1 Introduction

Ukraine is very rich in mineral resources. Three major classes of mineral resources considered to be important in our country: fuel mineral resource as well as metallic and nonmetallic. Fuels include deposits of black and brown coal, natural gas and peat. At the same time, it is one of the richest places in the world for reserves of manganese ore which is used in the manufacture of high-quality steel [1 – 5]. Despite conventional mineral resources mining, an amber mining is widespread in Ukraine [6, 7].

In the world, amber is deposited at the Baltic Sea in Baltic countries, Poland, Germany, Denmark, Sweden, and Belarus. Until recently, Prymorske deposit in Kaliningrad Region

* Corresponding author: e.z.malanchuk@nuwm.edu.ua

(Russia) has been the largest one [8 – 10]. According to the proven amber reserves, Ukraine ranks third in the world after Poland (700000 tons) and Russia (160000 tons) [11 – 14].

Significant amber deposits have been prospected in the north of Volyn, Rivne, Zhytomyr, and Kyiv Regions (Ukraine). Depth of commercial Ukrainian amber deposits are 2.5 to 10.0 m; their total thickness is 0.5 – 5.0 m. Amber content within the amber-bearing deposits vary from several grams to hundreds grams per cubic meter of rock; total estimated amber Ukrainian reserves are more than 100000 tons. In the context of the deposits, average amber content is more than 50 g/m^3 [12, 15].

Two amber deposits have been prospected in Rivne Region: Klesiv deposit in Sarny District and Vilne deposit in Dybrovytsia District; they are mined currently by the state enterprise "*Burshtyn Ukrainy*". In the context of Rivne region, total area of promising amber-bearing levels is 3810 square kilometers to be 18% of its territory [11, 12].

At the territories, amber is extracted mechanically and hydraulically. Residuals of the mineral with – 5.0 mm size is the extraction techniques disadvantage. Complete amber mining from amber-bearing deposits needs the improvement of its extraction technique with the use of new influence factors (i.e. water, air, and vibration) as well as primary gemological characteristics (refractive index, specific gravity and fluorescence [11, 16].

Thus, the improvement of amber extraction method should involve extra research and mathematical modeling. Mathematical modeling was performed to develop adequate formula concerning flotation of different amber fractions in air-sand mixture with different density levels and use it as a basis for mathematical dependences.

In the context of hydromechanical extraction, amber dispacement within sand formation is characterized by the resistance of a dry friction type. Particles of amber, immersed into the environment, performs horizontal translatory vibration with ω frequency and trajectory radius r.

2 Mathematical representation of hydromechanical amber extraction

Consider a model of action of friction forces effecting amber particle (Fig. 1). Relying upon research by I.I. Blekhman for water-air mixture, specify resistance force in terms of any relative horizontal displacement of the particle through F_h; if the displacement is vertical, then it is specified through F_v. Specify a particle mass through m_1 taking into consideration associated environmental mass and the environmental mass in terms of volume being equal to the particle volume through m_0; specify average particle density-environment ratio through $\Delta = \rho / \rho_0$. To do that, assume that x, y, and z are components of rectangular coordinate system XYZ of relative particle velocity within the environment.

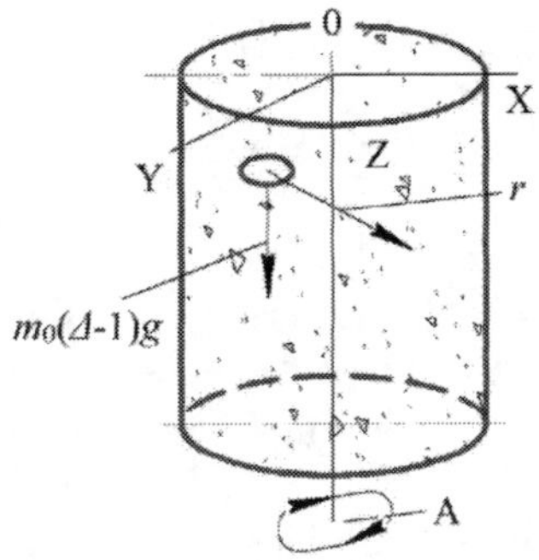

Fig. 1. Forces acting on the particle within the rock mass with the resistance of a dry friction type.

Differential equations of the particle motion as for the amber-bearing environment can be expressed as follows:

$$m_1 \ddot{x} = m_0 \left(\Delta - 1\right) r\omega^2 \cos\, \omega t - F_h \frac{\dot{x}}{\sqrt{\dot{x}^2 + \dot{y}^2 + \dot{z}^2}}; \tag{1}$$

$$m_1 \ddot{y} = m_0 \left(\Delta - 1\right) r\omega^2 \cos\, \omega t - F_h \frac{\dot{y}}{\sqrt{\dot{x}^2 + \dot{y}^2 + \dot{z}^2}}; \tag{2}$$

$$m_1 \ddot{z} = m_0 \left(\Delta - 1\right) g - F_v \frac{\dot{z}}{\sqrt{\dot{x}^2 + \dot{y}^2 + \dot{z}^2}}; \tag{3}$$

$$\sqrt{\dot{x}^2 + \dot{y}^2 + \dot{z}^2} \neq 0; \tag{3}$$

$$\sqrt{\dot{x}^2 + \dot{y}^2 + \dot{z}^2} \neq 0. \tag{4}$$

Consider a ratio:

$$\frac{F_v}{m_0 g} > \left|\Delta - 1\right| > \left[\left(\frac{m_0 r\omega^2}{F_h}\right)^2 + \left(\frac{m_0 g}{F_v}\right)^2\right]^{-1/2}, \tag{5}$$

reflecting following condition: on the one hand, the particle mass within the environment is less than its resistance force of its vertical motion; on the other hand, inertia in terms of relative motion is sufficient to overcome resistance force of dry friction type. At the same time, dependences (1 – 4) suppose accurate solution:

$$\dot{x} = R\omega \cos\left(\omega t + \beta\right),\ \dot{y} = R\omega \sin\left(\omega t + \beta\right),\ \dot{z} = \dot{Z}_0, \tag{6}$$

corresponding to the particle motion as for the environment in terms of a spiral path. In this context, following expressions are derived for radius of a spiral line R and vertical motion of $\dot{Z}_0$ particle velocity:

$$R = r\sqrt{\left[\frac{m_0}{m_1}\left(\Delta - 1\right)\right]^2 - \left[\frac{F_h}{m_1 r\omega^2}\right]^2 \left(1 - \delta^2\right)}; \tag{7}$$

$$\delta = m_0 \left(\Delta - 1\right) g / F_v; \tag{8}$$

$$\dot{Z}_0 = \frac{\delta}{\sqrt{1 - \delta^2}} R\omega. \tag{9}$$

Henceforth, expression for phase displacement β is insignificant and the determined motion is stable [17 – 19]. It follows from dependence (9) that the amber particles which density is more than the environment density ($\Delta > 1$) immerse; the particles which density is less than the environment density ($\Delta < 1$) float. In this context, an effect when dry friction is transforming into viscous friction takes place.

If we compare dependence (9) with the known expression for the velocity of free fall of a spherical particle in viscous liquid in terms of small numbers of Reynolds, we have:

$$\dot{Z}^* = \frac{m_0(\Delta-1)g}{3\pi\mu^* d}, \tag{10}$$

where d is the amber particle diameter, and μ^* is viscosity coefficient of a liquid.

Determine true viscosity of amber-bearing environment:

$$\mu^* = \frac{F_v\sqrt{1-\delta^2}}{3\pi dR\omega} = \frac{F_v}{3\pi dr\omega}\left\{\left[\frac{m_0(\Delta-1)}{m_1\sqrt{1-\delta^2}}\right]^2 - \left[\frac{F_h}{m_1 r\omega^2}\right]^2\right\}^{-1/2}, \tag{11}$$

The considered system is typical in the fact that the established motions are divided naturally: vertical particle motion is rapid and horizontal motion is slow. Moreover, in terms of $\dot{Z} = const$, dependences of rapid motions (1 – 4) accept accurate periodical solution of the type:

$$x^* = R_1\omega\cos(\omega t + \beta_1),\ \dot{y}^* = R_1\omega\sin(\omega t + \beta_1). \tag{12}$$

Substituting the solution for dependence (3), we obtain a dependence of a slow vertical motion of the particle (fundamental equation of vibration mechanics):

$$m_1\ddot{Z} = m_0(\Delta-1)g + V(\dot{Z}); \tag{13}$$

$$V(\dot{Z}) = -F_v\dot{Z}\left(R_1^2\omega^2 + \dot{Z}^2\right)^{-1/2}, \tag{14}$$

is vibration force which obtaining does not involve averaging. As we can see, the force is of nonlinear viscous resistance nature $[V(0)=0]$. In the context of full inertial approximation when it is possible to consider that $F_h \ll m_0(\Delta-1)r\omega^2$, we obtain $\beta \approx -\pi/2$ and $R_1 = r(\Delta-1)m_0/m_1$ from the dependences (1 – 4); according to formula (14) we obtain vibration force acting on the amber particle and making it move:

$$V(\dot{Z}) \approx -F_v\dot{Z}\left\{\left[\frac{m_0}{m_1}(\Delta-1)r\omega\right]^2 + \dot{Z}^2\right\}^{-1/2}. \tag{15}$$

To determine a period for different amber particles surfacing in terms of different operation modes while comparing data of natural experiment and theoretical studies, simulate distribution of amber concentration at the surface of the deposit. To do that it is required to apply nonlinear differential equations within partial derivatives describing one-dimensional slow process of changes in volumetric concentrations $c_1,\ldots,c_n$ of loose mixture components, i.e. corresponding equations of vibration mechanics:

$$\frac{\partial c_i}{\partial t} = \frac{\partial}{\partial t}\left[\frac{\partial c_i}{\partial z}\sum_{j-1}^{n} a_{ij}c_j\right] - \frac{\partial}{\partial z}\left[c_i\left(\sum_{j-1}^{n} b_{ij}c_j + \sum_{j-1}^{n} a_{ij}\frac{\partial c_j}{\partial z}\right)\right]\ (i=1,\ldots,n), \tag{16}$$

where Z is spatial coordinate; a_{ij} and b_{ij} are functions of z being determined either basing upon studies of "rapid" process or experimentally. They depend upon properties and characteristics of particles as well as upon vibration parameters in the neighborhood of the point [20 – 22].

Experimental methods in laboratory conditions are the most efficient, such as mining and laboratory investigations and modelling on equivalent materials [23, 24]. If the experimental research in real conditions is impossible, in connection with evaluation of the initial investigated task state or presence of constraints, mathematical modelling and computer experiment are interrelated [25, 26]. The process was modeled within software environment Matlab with the use of M-function "*pdepe*" meant to solve one-dimensional boundary problems for the systems of parabolic and elliptic differential equations within first-order partial differential derivatives in terms of one spatial variable and time.

3 Results and discussion

While using computer modeling according to the input data ($c|_{t=0} = 100$ g/m^3, $c|_{x=0} = 0$, $\left.\frac{\partial C}{\partial x}\right|_{x=L} = 0$, $D = 0.0001$ m^2/s, $L = 1$ m) following results were obtained for different amber fractions (Table 1).

Table 1. Classification of the extracted amber.

No.	Amber sizing according to weight, g	Distribution of the extracted amber, %
1	5	5
2	3	22
3	1	12
4	5	15
5	2	30
6	Others	16
	Total	100

Table 2 demonstrates the results of the research concerning distribution of amber concentration at the deposit surface for different fractions as well as in the form of dependence graphs resulting from computer modeling in terms of different operation modes of the facilities (i.e. combination of vibration, water, and air – v-w-a; supply of water and air – w-a; and water supply without air and vibration – w).

Table 2. Calculation data concerning distribution of amber concentration at the deposit surface, if mining depth is 1 m.

No.		Distribution of amber concentration at the deposit surface while mining, g/m^3									
Fraction 2 mm	v-w-a	0.826	22.86	29.19	30.54	30.84	30.9	30.91	30.92	30.92	30.92
	w-a	0.09	16.51	25.37	28.71	30.03	30.57	30.78	30.86	30.9	30.91
	w	0.004	10.46	20.06	24.9	27.54	29.02	29.85	30.31	30.58	30.73
Fraction 5 mm	v-w-a	0.533	12.73	15.07	15.39	15.43	15.44	15.44	15.44	15.44	15.44
	w-a	0.059	9.667	13.77	14.94	15.29	15.39	15.42	15.43	15.44	15.44
	w	0.002	6.36	11.44	13.59	14.57	15.03	15.24	15.35	15.39	15.42
Fraction 10 mm	v-w-a	0.476	10.84	12.58	12.78	12.81	12.81	12.81	12.81	12.81	12.81
	w-a	0.053	8.361	11.65	12.49	12.72	12.78	12.8	12.81	12.81	12.81
	w	0.053	8.361	11.65	12.49	12.72	12.78	12.8	12.81	12.81	12.81
Fraction 30 mm	v-w-a	7.124	21.91	22.13	22.13	22.13	22.13	22.13	22.13	22.13	22.13
	w-a	0.217	14.83	20.19	21.6	21.98	22.09	22.12	22.13	22.13	22.13
	w	0.001	7.124	14.25	17.8	19.72	20.78	21.37	21.7	21.89	21.99
Fraction 50 mm	v-w-a	0.803	4.776	4.849	4.85	4.85	4.85	4.85	4.85	4.85	4.85
	w-a	0.03	3.809	4.707	4.829	4.847	4.849	4.85	4.85	4.85	4.85
	w	4E-04	2.424	4.093	4.598	4.764	4.82	4.84	4.846	4.848	4.849

Fig. 2 and 3 have been developed relying upon the data represented in Table 2.

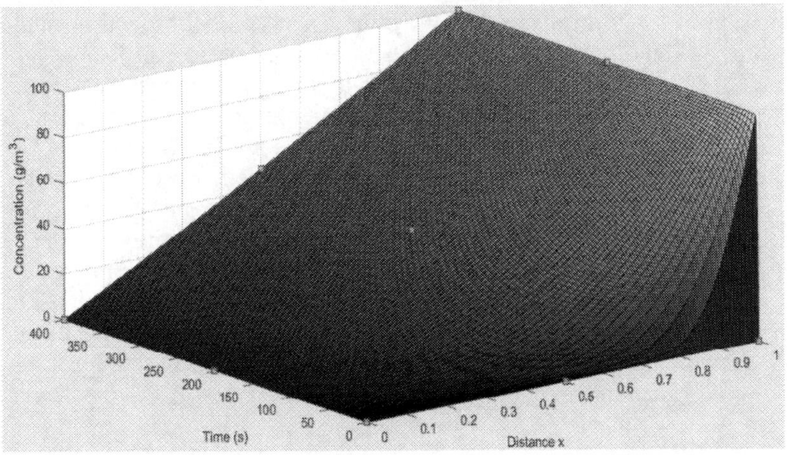

Fig. 2. Spatial distribution of amber concentration within amber-bearing deposit.

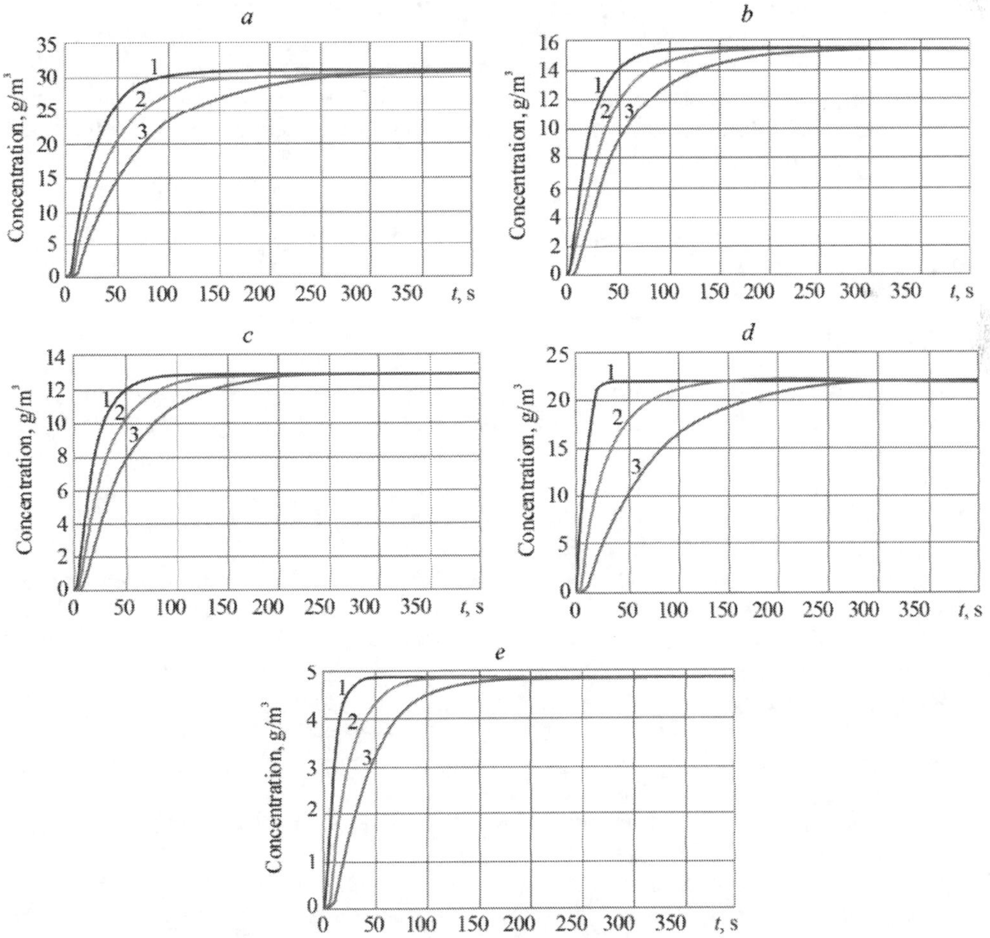

Fig. 3. Distribution of amber concentration at the deposit surface while mining: a – 2 mm; b – 5 mm; c – 10 mm; d – 30 mm; e – 50 mm; 1 – v-w-a; 2 – w-a; and 3 – w (2 mm fraction).

Therefore, it is obvious that improvement of technology and equipment is possible at the expense of amber extraction intensification when sand deposit is saturated with water, air, and vibration in the form of mechanical effect. Amber floating within sand deposit is characterized by the environmental resistance and effect of prevailing factors providing the amber surfacing. The modeling makes it possible to develop dependence as for distribution of amber concentration over the deposit surface when the time of different amber fractions flotation is determined in terms of different operation modes. Integrated action of vibration, water and air supply intensify the process of amber floating while determining surfacing velocity. Hence, the velocity is a technological parameter used to determine productivity of equipment. Complete floating of the finest amber fractions (i.e. 2 mm) from 1 m depth to a surface takes 50 seconds. Floating process duration helps determine productivity of equipment and simulate processes of amber flotation from different depths. The obtained adequate mathematical model to solve one-dimensional boundary problems describes accurately behaviour of different amber particles within amber-bearing deposits when vibration acts, and water and air are supplied.

4 Conclusions

Mathematical model to solve one-dimensional boundary problems for the systems of parabolic and elliptic differential equations within first-order partial derivatives on one spatial variable and time has been developed as a result of amber surfacing process. The model describes accurately the behaviour of different amber fractions within amber-bearing deposit in terms of vibration, and water and air supply. The research has determined that larger amber fractions surface faster.

This work was supported by the Ministry of Education and Science of Ukraine (State registration No. 0102U001910).

References

1. Mykhailov, V., & Kurylo, M. (2011). Estimation of flux reserve and resource base of Ukraine. *International Multidisciplinary Scientific GeoConference Surveying Geology and Mining Ecology Management, SGEM*, 135-140. https://doi.org/10.5593/sgem2015/b11/s1.018
2. Bondarenko, V., Cherniak, V., Cawood, F., & Chervatiuk, V. (2017). Technological safety of sustainable development of coal enterprises. *Mining of Mineral Deposits*, *11*(2), 1-11. https://doi.org/10.15407/mining11.02.001
3. Petlovanyi, M.V., Lozynskyi, V.H., Saik, P.B., & Sai, K.S. (2018). Modern experience of low-coal seams underground mining in Ukraine. *International Journal of Mining Science and Technology*. Article in press. https://doi.org/10.1016/j.ijmst.2018.05.014
4. Khomenko, O.Ye., Sudakov, A.K., Malanchuk, Z.R., & Malanchuk, Ye.Z. (2017). Principles of rock pressure energy usage during underground mining of deposits. *Naukovyi Visnyk Natsionalnoho Hirnychoho Universytetu*, (2), 34-43.
5. Bondarenko, V., Lozynskyi, V., Sai, K., & Anikushyna, K. (2015). An overview and prospectives of practical application of the biomass gasification technology in Ukraine. *New Developments in Mining Engineering 2015: Theoretical and Practical Solutions of Mineral Resources Mining*, 27-32. https://doi.org/10.1201/b19901-6
6. Kovalevskii, S.B., & Lehkyi, V.V. (2017). The integral estimate of the anthropogenic transformation of forest landscapes of Dubrovytsia region as a result of unsecured extraction of amber. *Scientific Bulletin of UNFU*, *27*(9), 52-55.
7. Masley, V.M., Mozgovoy, D.K., Bilousov, K.G., Horoshilov, V.S., Bushanska, O.S., & Galich, N.G. (2016). Methods of the impact evaluation of amber mining by multispectral satellite images.

Kosmicna Nauka i Tehnologia, 22(6), 26-36.

8. Alekseev, V.I. (2013). The beetles (Insecta: Coleoptera) of Baltic amber: the checklist of described species and preliminary analysis of biodiversity. *Zoology and Ecology, 23*(1), 5-12. https://doi.org/10.1080/21658005.2013.769717
9. Wang, S., Shi, C., Zhang, Y.-j., Hu, G.-x., & Gao, L.-z. (2016). Trading away ancient amber's secrets. *Science*, 351 (6276), 926. https://doi.org/10.1126/science.351.6276.926-a
10. Yablokov, A., Levchenko, V., & Kerzhentsev, A. (2017). The Biosphere as a Living System. On the Harmonization of Human and Biosphere. *Philosophy and Cosmology*, (18), 52-83.
11. Bulat, A.F. (2016). *Promyslovi tehnolohii vydobytku burshtynu*. Dnipro: Instytut heotehnichnoi mehaniky.
12. Krynytska, M.V., & Korniienko, V.Ya. (2018). Obgruntuvannia heolohichnykh umov ta tekhnolohichni osnovy vydobutku poliskoho burshtynu. In *Poturaivski Chytannia*, 32-39.
13. Macuj, V.M. (2013). Basic Amber Problems at the Present Stage. *Sbornik Nauchnyh Trudov, Posvyashchennyy 150-letiyu so Dnya Rozhdeniya Akademika P.A. Tutkovskogo*, 339-346.
14. Haletskyi, L.S., & Remezova, O.O. (2012). Outlooks of Prospecting for New Amber Deposits in Ukraine. *Identykatsiia Vykopnykh Smol. Zbornik Materialiv Naukovoho Seminaru*, 63.
15. Macuj, V.M. (2013). Marine Stage of Fossilization of Resin Extraction from Conifers in Transforming Them to Amber-Succinite. *Geologiya i poleznye iskopaemyeh.* (2), 101-108.
16. Belichenko, O., & Ladzhun, J. (2016). Complex gemological research of new types of treated amber. *Visnyk of Taras Shevchenko National University of Kyiv. Geology, 4*(75), 30-34. https://doi.org/10.17721/1728-2713.75.04
17. Malanchuk, Z., Korniyenko, V., Malanchuk, Y., & Khrystyuk, A. (2016). Results of experimental studies of amber extraction by hydromechanical method in Ukraine. *Eastern-European Journal of Enterprise Technologies, 3*(10(81)), 24. https://doi.org/10.15587/1729-4061.2016.72404
18. Nadutyy, V.P., & Korniyenko V.Ya. (2016). Research of process of extraction amber from amber containing deposits vibro classifier. *Vibratsiia v Tekhnitsi ta Tekhnolohiiakh, 3*(83), 117-121.
19. Nadutyi, V.P. (2016). Doslidzhennia vplyvu pokaznykiv vibratsii na vydobutok burshtynu z pishchanykh rodovyshch. *Heotekhnichna Mekhanika*, (126), 27-35.
20. Malanchuk, Z., Malanchuk, Ye., & Khrystiuk, A. (2016). Mathematical modeling of hydraulic mining from placer deposits of minerals. *Mining of Mineral Deposits, 10*(2), 18-24. https://doi.org/10.15407/mining10.02.018
21. Malanchuk, Z., Korniienko, V., & Malanchuk, Ye. (2017). Results of research into amber mining by hydromechanical method. *Mining of Mineral Deposits, 11*(1), 93-99. https://doi.org/10.15407/mining11.01.093
22. Malanchuk, Z., Korniienko, V., Malanchuk, Ye., Soroka, V., & Vasylchuk, O. (2018). Modeling the formation of high metal concentration zones in man-made deposits. *Mining of Mineral Deposits, 12*(2), 76-84. https://doi.org/10.15407/mining12.02.076
23. Lozynskyi, V.G., Dychkovskyi, R.O., Falshtynskyi, V.S., Saik, P.B., & Malanchuk, Ye.Z. (2016). Experimental study of the influence of crossing the disjunctive geological faults on thermal regime of underground gasifier. *Naukovyi Visnyk Natsionalnoho Hirnychoho Universytetu*, (5), 21-29.
24. Lozynskyi, V., Saik, P., Petlovanyi, M., Sai, K., & Malanchyk, Ye. (2018). Analytical Research of the Stress-Deformed State in the Rock Massif Around Faulting. *International Journal of Engineering Research in Africa*, (35), 77-88. https://doi.org/10.4028/www.scientific.net/JERA.35.77
25. Cabana, E.C. (2017). Formation of thermal fields by the energy-chemical complex of coal gasification. *Naukovyi Visnyk Natsionalnoho Hirnychoho Universytetu*, (5), 36-42.
26. Dychkovskyi, R.O., Lozynskyi, V.H., Saik, P.B., Petlovanyi, M.V., Malanchuk, Ye.Z., & Malanchuk, Z.R. (2018). Modeling of the disjunctive geological fault influence on the exploitation wells stability during underground coal gasification. *Archives of Civil and Mechanical Engineering, 18*(4). https://doi.org/10.1016/j.acme.2018.01.012

Resource evaluation of geothermal power plant under the conditions of carboniferous deposits usage in the Dnipro-Donetsk depression

Mykhailo Fyk[1*], *Volodymyr* Biletskyi[2], and *Mokhammed* Abbud[2]

[1]V.N. Karazin Kharkiv National University, School of Geology, Geography, Recreation and Tourism, 4 Svobody Sq., 61022 Kharkiv, Ukraine
[2]National Technical University "Kharkiv Polytechnic Institute", Institute of Chemical Technology and Engineering, 2 Kyrpychova Str., 61002 Kharkiv, Ukraine

Abstract. The objective of the work is to substantiate the actualization of the problem of obtaining geothermal energy from oil and gas wells of oil and gas production facilities in the central-eastern part of the Dnieper-Donetsk Depression (DDD). The research methodology is based on the collection and statistical processing geophysical data of the DDD oil and gas deposits, the use of balance equations for energy and substance matter. The main result of the work is that the principal technological scheme of the geothermal system has been developed and the geothermal potential of oil and gas wells in the experimental zone has been analyzed. There have been considered the technological and ecological aspects of geothermal heat usage from depleted deposits of the DDD wells, which were disclosed into carboniferous deposits productive horizons.

1 Introduction

Geothermal power plants (GeoPP) with a binary cycle are based on two closed cycles: one for geothermal liquid or fluid, the second for working fluid or gas with low boiling point. At these power plants, the working fluid heated by geothermal water is converted to steam, then a heat exchanger is used, and finally the steam enters the turbine, ensuring its rotation (Fig. 1). The peculiarity of the binary geothermal power plants work is that a working fluid in them, for example, a two-component water-ammonia mixture or a propane-butane mixture, evaporates at a lower temperature than water. Taking into account that in nature the sources of geothermal water with a temperature below 190 °C are commonly available at depths up to 3 – 4 km, then in the future one should expect the predominant construction of geothermal thermal power plants with a binary cycle. For geothermal power plants (GeoPP), even lower wellhead temperatures (40 – 70 °C) are sufficient. Multistage heat supply systems are built on the principles of cognitive technologies [1] in terms of simultaneous heat utilization and its distribution among consumers, extraction and intensification of hydrocarbon inflows, other extracting fluids [2]. The greatest development of these technologies was in the USA, Germany, Canada. More than 6 mln m^3

* Corresponding author: mfyk@ukr.net

of geothermal water per year are mined in Dagestan. In the Northern Caucasus, about 500 thousand people use the geothermal water-supply [3]. Capital costs for the GeoPP construction range currently from $ 1.500 to $ 2.500 per kilowatt of installed electric power, compared to a similar economic indicator for nuclear power plants and thermal power plants. The work [2] shows that the use of geothermal heat as a combination of heat and mass transfer processes with the production of hydrocarbons (cognitive-technological approach) is technically and economically profitable. It is shown in [4] that the temperatures of the coolant outlet flow from the Grobla deposit near Krakow reach more than 40 °C when developing depths of only 900 m. The oil and gas wells survey in the south of Texas [5] have shown the possibility of obtaining up to 3.5 MW of thermal power for wells with a depth of about 3 km, with a circulation velocity of the secondary coolant 3 – 5 m/s and pressures in the wells of 30 – 40 bars. The thermal insulation thickness of the well column, as evidenced by the authors of [5], should be not less than 1 inch, which will reduce the loss of the temperature potential up to 3 – 5 °C. Assessment of the use of abandoned deep hydrocarbon reservoirs and dry wells in the Croatian area of the Pannonian basin (Croatia) as a source of geothermal energy was made in [6]. It is proposed to receive geothermal energy according to the principle of a coaxial heat exchanger. The calculated capacity of the considered GeoPP is 400 kW (peak is 1.5 MW) for 20 years of operation, the temperature of the heat transfer fluid reaches a stable state at 50 °C. The model-based analysis [7] of Vilfortune-Trekate, the large European oil field, shows that one well can produce 4 MW of thermal and 400 kW of electric power at an estimated cost of electricity of 0.05 euro/kWh (Italian consumers). In [8] the heat-flux density in wells of 75 – 100 mW/m^2 for geological formations of California (Santa Clara, Monterey and Santa Barbara) were investigated.

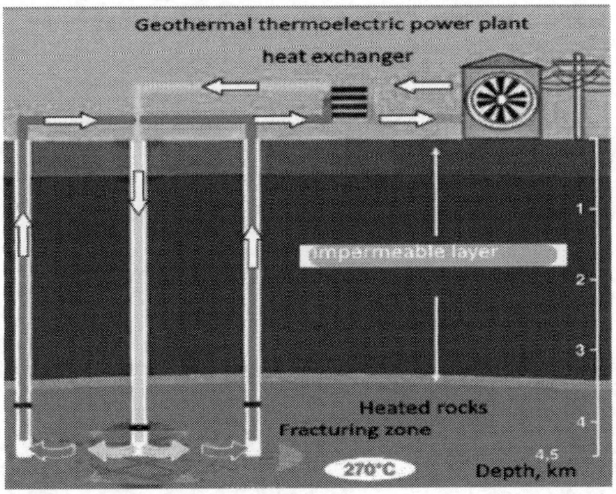

Fig. 1. Binary GeoPP Fractured Reservoirs in Australia.

Work [10] shows the significant potential of depleted and abandoned oil wells in Pakistan from the point of view of geothermal energy of oil and gas reservoirs with significant water-flood operation in the end.

This actualizes the research task of the fluid production thermobaric regimes from the horizons of the DDD carboniferous deposits, the development of multilevel utilization complexes for geothermal heat.

The circulation technology for the development of geothermal deposits with natural reservoirs (Fig. 1) has been successfully applied in France, has industrial distribution in

Germany, Denmark, Switzerland, the USA, Poland, Russia (Chechnia, Dagestan), Australia, etc. Ukraine also has a binary GeoPP in Autonomous Republic of Crimea. Research GeoPP is also built in Ukraine near the city of Uzhhorod for heat supply of the greenhouse complex and livestock farm; the depth of this system is 2.3 km, the water temperature is 124 °C. In general, there are two main types of geothermal energy sources in the form of dry rocks heat concentrated mainly in the northern and central regions of the country, occurring at depths of up to 5 km and having relatively low temperatures (up to 100 °C), and in the form of thermal waters laying at a depth of up to 3000 m and heated to 120 °C that are concentrated in the Autonomous Republic of Crimea, Zakarpattia, Kharkiv and Poltava regions, as well as in the Carpathian region.

2 Results and discussion

The research methodology includes the collection and statistical processing of geophysical data of oil and gas wells of the DDD oil and gas deposits. In this case, the balance equations for the energy and substance matter fluxes of the primary and intermediate coolant are used.

The analysis of works [1 – 11] proves the possibility of using lithologically and tectonically limited reservoirs of the experimental oil and gas deposits for the purposes of space heating and multi-level heat supply. This actualizes the research task of the fluid production thermobaric regimes from the horizons of the DDD carboniferous deposits, the development of multilevel utilization complexes for geothermal heat based on cognitive technologies, and the assessment of the main environmental aspects of the GeoPP implementation.

On the basis of UkrNDIgaz Institute projects for the oil and gas deposits development and the results of works [11 – 13], Table 1 shows the data of 32 oil-gas and gas condensate deposits in the DDD central and eastern area. The calculation of the thermal power is made according to the formulas from [2, 13].

It can be seen from Table 1 that most deposits have reservoir temperatures above 50 – 60 °C, can be used for local complex heat supply and space heating. In this sample, only three deposits (Leventsivske, Vergunske, Hnidyntsivske – less than 10%) show a temperature regime of fluids in the carbon less than 50 °C.

By analogy with the operating GeoPP in Germany (see Fig. 2, with a monocycle), a GeoPP scheme has been developed for DDD conditions with three-level heat supply (see Fig. 3, with a full binary cycle).

The schematic diagram (Fig. 3) is completely isolated between the formation circulating coolant (underground circuit) and the water in the heat supply system (the coolant-water circuit in the surface part) due to the heat exchangers of units 4 and 13. The difference between the developed GeoPP scheme and the prototype [12] is in the application of the binary GeoPP principle, as well as in the introduction of the third heat exchanger after leaving the heat pump 13. This allows to work not with three, but with four thermal potentials, indicated in Fig. 2 with different 4 colors.

The underground circuit may contain natural hydrocarbon mixtures. For the deposits considered in Table 1, the extracting hydrocarbon mixtures are oil, gas and gas condensate.

It is submitted the comparative analysis of the Geotech circuitry descriptive variants. The simplest type of GeoPP use: one well is made thermally insulated and lifts hot water or heat coolant (see Fig. 1 – 3). After passing the heat pump and turbine, cold water is discharged into the same (see Fig. 2) or another (Fig. 3) well at the same depth, the system is hermetically sealed, that is, the pump only overcomes the resistance of the underground filtration and the resistance of the tubing pipes. The second well is made on the contrary as heat-exchanging, so that along the path of discharged water downstream, it begins to be heated along the path from the mouth to the productive horizon. With a sufficient flow rate of water or other liquid coolant, a drop of even a few tens of degrees will give economically significant amounts of energy. Instead of water, light oil product can be used.

Table 1. Thermobaric conditions for production of formation fluids from 32 oil and gas fields in the DDD central and eastern area.

№	The oil deposit, (area of DDD)	Depth H, m	Actual temperature at depth H, °C	Geothermal degree at depths H, m/°C	Heat power of the well, MW
1	Leventsivske	750	30	33.90	0.53
2	Verhunske	1165	35	43.00	0.79
3	Zachepylivske	1350	36	48.00	0.84
4	Borivske	1510	55	35.00	1.84
5	Prylutske	1602	63	29.10	2.26
6	Vilkhivske	1669	57	33.90	1.94
7	Hnidyntsivske	1730	45	46.60	1.31
8	Leliakivske	1850	48	46.10	1.47
9	Skhidno-Novoselivske	1922	53	42.60	1.73
10	Novohryhorivske	2000	70	32.00	2.63
11	Kremenivske	2215	64	39.40	2.31
12	Malosorochynske	2257	79	31.70	3.10
13	Krasnopopivske	2311	92	27.40	3.78
14	Shebelynske	2430	64	43.30	2.31
15	Bohdanivske	2495	85	31.90	3.41
16	Pereschepynske	2640	75	39.30	2.89
17	Chornukhinske	2957	84	38.80	3.36
18	Milkivske	3000	89	36.90	3.62
19	Volokhivske	3060	83	40.70	3.31
20	Melykhivske	3100	72	45.50	2.73
21	Velykobubnivske	3105	95	35.60	3.94
22	Solokhivske	3120	88	38.90	3.57
23	Zakhidno-Vedmedivske	3190	75	47.50	2.89
24	Monastyryshchenske	3355	97	37.60	4.04
25	Rybalske	3384	91	40.70	3.73
26	Zakhidno- Khrestyschenske	3400	76	49.90	2.94
27	Verkhnolannivske	3400	74	51.40	2,84
28	Novotroitske	3400	87	42.90	3.52
29	Zakhidno-Sosnivske	3676	78	52.40	3.05
30	Opishnianske	3695	101	39.60	4,25
31	Hlynsko-Rozbyshivske	3748	101	40.20	4,25
32	Mashivske	4060	108	40.50	4.62

Next, we concentrate on the energy calculation of GeoPP geothermal power. Geothermal energy by mass can be obtained from the expression:

$$W = C(T1 - T2) = 4200(80 - 30) = 210000 \text{ J/kg}, \qquad (1)$$

where $T1$ is the inlet temperature of the coolant (fluid), 80 °C, $T2$ is the outlet temperature of the coolant out of the system, 30 °C, C is the thermal power of the fluid.

For the thermal power of the wells (Fig. 3) at water flow rate $Q = 20$ kg/sec we have:

$$P = Q \cdot W = 210000 \cdot 20 = 4200000 \text{ W}. \qquad (2)$$

Thus, we obtain ~ 4 MW from two wells with a depth of 3 – 4 km (with a difference in temperature on the choke – 50 °C). For the above example of implementation in Germany, the temperature at the mouth is 65 °C, and the coolant returns with a temperature of up to 5 – 15 °C. So, the temperature difference on the choke of the heat pump is more than 50 °C.

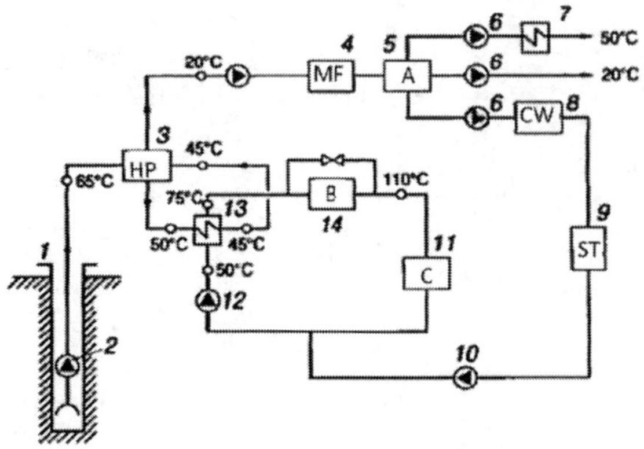

Fig. 2. The technological scheme of GeoPP operating in Germany: 1 – geothermal well with depth of 2350 m with a temperature of 65 °C; 2 – submerged-type well pump, which provides geothermal water consumption from 24 to 55 l/s; 3 – absorption heat pump (HP) with a thermal power of 6.8 MW; 4 – membrane filters that purify geothermal water to the norm of cold water supply (TP); 5 – cold water accumulator (A); 6 – cold water pumps; 7 – heat exchanger for water heating for the water park; 8 – chemical water purification of the heat supply system feeding (CW); 9 – storage tank for make-up water (B); 10 – feed pump; 11 – consumers of the heat supply system (C); 12 – network pump; 13 – heat pump heat exchanger; 14 – boiler (B).

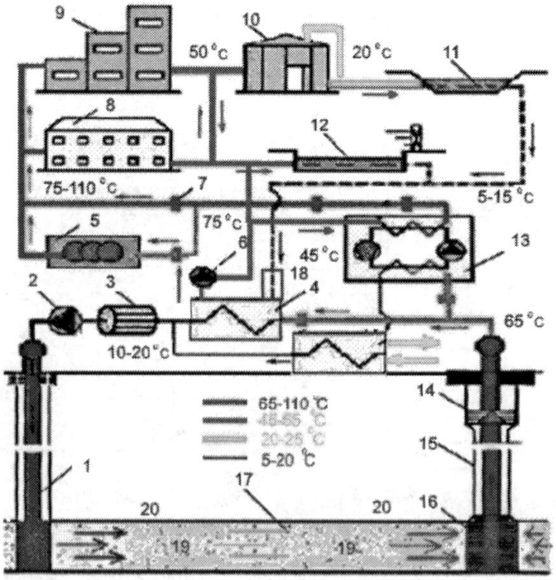

Fig. 3. Schematic diagram of the proposed GeoPP for the development of carboniferous deposits collectors in Ukrainian DDD: 1 – injection well; 2 – surface pump unit; 3 – gas cleaning and coolant preparation system; 4 – heat exchangers; 5 – boiler room for finish heating up (at the outlet to 115 °C); 6 – network pump; 7 – main heat transport systems; 8 – residential area; 9 – industrial facilities; 10 – greenhouse complex; 11 – fisheries; 12 – balneal and sports complex; 13 – thermal heat pumps (capacity to 6 – 7 MW); 14 – submerged-type well pumps; 15 – extracting (water-lifting) heat-insulated well with depth of 750 – 4000 m; 16 – bottomhole filter system, submersible well pump (geothermal coolant flow rate from 25 to 50 l/s); 17 – channel of heat exchange between the permeable collector and recycled water; 18 – chemical water purification of the heat supply system feeding; 19 – natural permeable collector; 20 – natural geological impermeable layer.

Thus, the studies [1 – 12] and the above-mentioned data prove that two wells disclosed at the depth of carboniferous deposits can confidently provide 0.4 – 4.5 MW of heat output per local heat supply. The lower and upper limits of the power range comply with experimental studies [4 – 9, 14]. The generalization of the data array of the thermal loggings of 130 oil and gas fields wells in the DDD central and eastern area has allowed to synthesize the curve of the trend line $P = f(H)$ – the dependence of the average thermal power P of the well on its depth H (Fig. 4):

$$P = 0.6046H^{0.5611}, \tag{3}$$

with that the approximation error R^2, which is greater than 0.8 when using the function (3).

Fig. 4 shows the regression field of the potential thermal power of a well with a flow rate of 25 kg/s according to equations (1 – 3) and data given in Table 1.

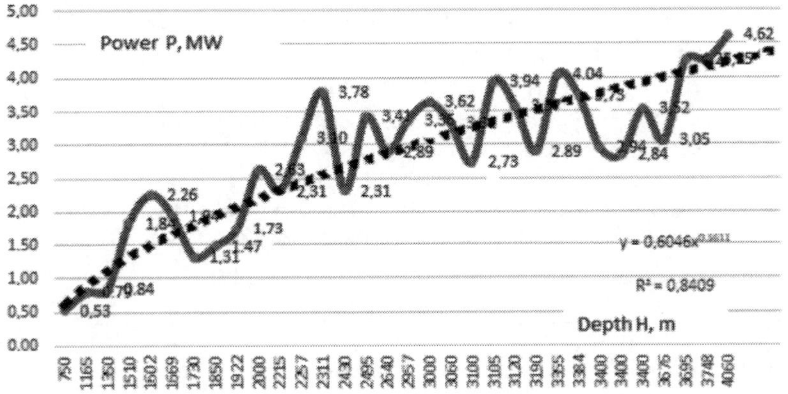

Fig. 4. Dependence of the potential thermal power of the DDD oil and gas wells on the depth.

The calculation of the thermal power (axis of ordinates in Fig. 4, is also shown in the last column of Table 1) is made according to formulas (1 – 2) [2] and with account of COP of heat pumps in geothermal systems (COP = 6 – 6.2), as well as with account of the power efficiency of heat exchangers (49%) [14]. From Fig. 4 it can be seen that the trend is observed of close to linear increase in the thermal power of the well, and at depths of 2000 – 3500 m (where carboniferous deposits are very common in the DDD zone), the theoretically achievable thermal power of a geothermal well is kept at 2.5 – 4 MW. The use of geothermal energy is not associated with the use of significant areas on the Earth's surface and is not accompanied by significant emissions into the atmosphere and waste products. The impact on the environment when using geothermal resources is the least negative among all energy sources. The emissions of carbon dioxide at the modern geothermal power plants are minimal – they are either completely absent or insignificant and amount to about 0.45 kg per 1 MWh of the volume of electricity produced at the power plant, whereas at a natural gas-fired power plant they amount to 464 kg, on oil – 720 kg, on coal – 819 kg. The process geothermal water or other selected coolant is reliably isolated from groundwater through a pipeline installed into the well. The terrain around the geothermal installation is not spoiled by any mines, nor tunnels, nor piles of waste. Geothermal power plants need very small areas of land, much less than for other types of power plants, and these plants can be located on almost any land, in particular, on agricultural land. For Ukraine with a large area of valuable chernozems, energy technology of this type is indisputably an interesting prospect.

3 Conclusions

The technical and economic analysis carried out by the authors showed that on the basis of the DDD oil and gas wells it is possible to construct geothermal power plants with depths of drilling or disclosing wells up to 3 – 4.5 km. At such depths, the 90% thermal potential of geothermal waters in productive oil and gas horizons of carboniferous deposits does not exceed 108 °C. In this case, the replacement of organic fuel and electricity by the heat of geothermal waters and rocks is much more profitable for providing heat and heating (by 3 – 5 times). Two wells from the depth of carboniferous deposits can provide 0.4 – 4.5 MW of thermal energy.

The improvement of the technological scheme of the heat supply system from the horizons of DDD oil and gas deposits will lead to an increase by 30 – 40% of sampled heat for heating and heat supply purposes. To do this, you need to use a three-stage scheme, which will allow obtaining heat for three different categories of consumers, if it is necessary for the maximum inlet temperature. The calculation carried out by the authors shows that 90% of the productive horizons of the oil and gas fields of the DDD eastern and central parts are suitable for the construction of GeoPP.

Geothermal heat supply systems are much more environmentally friendly than other heat-generating systems with hydrocarbon fuels. In agriculture, fertile lands will not be disrupted by the secondary use of oil and gas wells.

The authors express their gratitude to the branches of "Ukrhaspromheofizyka" Central Administration and JSC "Ukrhasvydobuvannia" for the provided thermal logging data and professional assistance.

References

1. Biletskyy, V., Sergeyev, P., & Krut, O. (2013). Fundamentals of highly loaded coal water slurries. *Mining of Mineral Deposits*, 105-114. https://doi.org/10.1201/b16354-20
2. Kutia, M., Fyk, M., Kravchenko, O., Palis, S., & Fyk, I. (2016). Improvement of technological-mathematical model for the medium-term prediction of the work of a gas condensate field. *Eastern-European Journal of Enterprise Technologies*, 5(8(83)), 40-48. https://doi.org/10.15587/1729-4061.2016.80073
3. Dobrohotov, V.I. (2003). Ispol'zovaniye geotermal'nykh resursov v energetike Rossii. *Teploenergetika*, (1), 2-11.
4. Barbacki, A. (2000). The use of abandoned oil and gas wells in poland for recovering geothermal heat. *Poland Proceedings World Geothermal Congress* (pp. 3361-3365). Kyushu-Tohoku, Japan.
5. Davis, A., & Michaelides, E. (2009). Geothermal power production from abandoned oil wells. *Journal Energy*, (34), 866-872.
6. Maceni,' M., Kurevija, T. (2018). Revitalization of abandoned oil and gas wells for a geothermal heat exploitation by means of closed circulation – case study of deep dry well "Pčelić-1". *Interpretation*, *6*(1). https://doi.org/10.1190/INT-2017-0079.1
7. Soldo, E., & Alimonti, C. (2015). From an Oilfield to a Geothermal One: Use of a Selection Matrix to Choose Between Two Extraction Technologies. *Proceedings World Geothermal Congress*. Melbourne, Australia, 1-10.
8. Caulk, R., & Tomac, I. (2017). Reuse of Abandoned Oil and Gas Wells for Geothermal Energy Production. *Renewable Energy*, (112), 388-397. http://dx.doi.org/10.1016/j.renene.2017.05.042
9. Conser, A. (2013). Double Dipping: Utilizing Oil Wells for Geothermal Energy. *Wm. & Mary Envtl. L. & Pol'y Rev*, 37(3), 813-843. http://scholarship.law.wm.edu/wmelpr/vol37/iss3/7
10. Mehmood, A., Yao, J., Fun, DY., & Zafar, A. (2017). Geothermal Energy Potential of Pakistan on the Basis of Abandoned Oil and Gas wells. *Journal of Petroleum & Environmental Biotechnology*, (7), 332. https://doi.org/10.4172/2157-7463.1000332

11. Orlov, O., Fyk, I., Bodnarchuk, V., & Mazur, A. (2015). *Bitumonaftohazoheolohichne raionuvannia, naftovi i hazovi rodovyshcha ta pidzemni skhovyshcha hazu v Ukraini*. Ivano-Frankivsk: Symfoniia forte.
12. Butuzov, V. (2008). Geotermalnoye teplosnabzheniye g. Erdinga v Germanii. *Al'ternativnaya energetika i ekologiya*, (9), 59-60.
13. Jozsef Nyers (2016). COP and Economic Analysis of the Heat Recovery from Waste Water using Heat Pumps. *Acta Polytechnica Hungarica, 13*(5), 135-154.
14. Nian, Y-L., & Cheng, W-L. (2018). Evaluation of geothermal heating from abandoned oil wells. *Energy*, (142), 592-607. https://doi.org/10.1016/j.energy.2017.10.062

E3S Web of Conferences **60**, 00007 (2018) https://doi.org/10.1051/e3sconf/20186000007
Ukrainian School of Mining Engineering

Investigation of the influence of polyelectrolytes hydrodynamic properties on the hydrate formation process

Volodymyr Bondarenko[1], *Olena* Svietkina[2], *Kateryna* Sai[1*], and *Vasyl* Klymenko[3]

[1]National Mining University, Department of Underground Mining, 19 Yavornytskoho Ave., 49005 Dnipro, Ukraine
[2]National Mining University, Department of Chemistry, 19 Yavornytskoho Ave., 49005 Dnipro, Ukraine
[3]Central Ukrainian National Technical University, Department of Electrotechnical Systems and Energy Management, 8 Universytetskyi Ave., 25006 Kropyvnytskyi, Ukraine

Abstract. One of the most perspective non-traditional sources of hydrocarbon raw materials and energy is gas hydrates (GH), which stimulates their large-scale study in many laboratories and scientific centers around the world. The interest in the GH is related to the possibility of their industrial application. The use of technologies for storage and transportation of natural gas in the form of GH requires fundamental kinetic studies of the hydration process at atmospheric pressure in the presence of chemical impurities (catalysts, surfactants, etc.). During the research, an increase in the rate of gas hydrates formation was discovered at the expense of activated impurities and composite compounds obtained on their basis, which lead to changes in the hydrochemical and hydrodynamical modes of systems. It was established that the process of GH formation occurs due to the binding of methane macromolecules and other chemical impurities, which contribute to increase the number of moles of gaseous methane, due to the flow of exchange processes between the polymer matrix and water soluble salts, as well as a result of the change in the hydrodynamics of the water system. The purpose of this work is to investigate the effect of polyelectrolytes solutions hydrodynamic properties on the GH formation. We used a number of ionenes based on polymeric quaternary ammonium salts, side aliphatic radicals of different lengths contained in the acyl fragment.

1 Introduction

The unconventional methods of recovering and obtaining various energy sources are of great interest in recent years, and are directly related to the annual increase in consumption of fuel and energy resources. For this very reason the intensification of geological and prospecting works of all types of energy raw materials is certainly of high priority in Ukraine. One of the alternatives to increasing energy consumption is the search and

[*] Corresponding author: kateryna.sai@gmail.com

evaluation for unconventional hydrocarbon deposits [1 – 3].

The coal industry of Ukraine providing the extraction and primary processing of hard and brown coal, is one of the main branches of the Ukrainian fuel industry. However, the majority of coal reserves are in difficult mining and geological conditions (low thickness of coal seams, great depths of extracting, low rocks stability, increased water-inflow coefficient, seams high gas content) [4]. It forces one to create new strategies for energy development of the country.

The problem of rational use of the existing potential of hydrocarbon gases, in particular methane, remains relevant both as for coal, oil and gas industries, so for petrochemical industry [5 – 7]. This is particularly true for the transportation and storage technologies of natural gas resources [8], coalmine methane [9], methane of drainage boreholes [10], gas of marginal field and marine field [11], and oil-associated gas.

The introduction of gas hydrate technologies based on the ability of gas and water molecules to form stable clathrate structures, which will provide the possibility of obtaining an additional energy source, is a promising direction in solution of this problem. However, the hydrate formation process, which is the key one, requires intensive binding of significant gas volumes into the gas-hydrate form. Its efficiency determines the economic performance of the technology and the quality of the final product.

Gas hydrate technologies, in comparison with the existing ones, also provide an opportunity with greater energy efficiency to separate mixtures of gases and liquids, compress gases to high pressures, concentrate aquatic solutions, produce and accumulate cold, utilize and store up the carbon dioxide. To study them is also important for solving the problems of methane gas hydrates deposits development, in particular, in the Black Sea water area [12, 13].

From the fundamental point of view, gas hydrates are interesting for that they are objects of supramolecular (permolecular) chemistry – a relatively new science that studies the other, subtler principle of chemical matter organization, based not so much on a specific chemical binding as on the favourable spatial complementarity of the reactants, even in the case of only weak van der Waals interactions between them. The physical, physical and mechanical, thermodynamic and electromagnetic properties of gas hydrates, as a solid, are similar to ice, and the study of cryogenic systems has always been an important aspect both for the conditions of life advent and existence, and in order to control the speed and mechanisms of biological processes [14]. The study of the molecules interaction patterns in gas hydrate systems and the structure of the compounds formed, their thermodynamic stability zones (phase diagrams), the kinetics of formation and decomposition is of great importance for many branches of industry and science.

Generally, all hydrophobous gases and highly volatile liquids with molecular sizes of 3.8 – 9.2 Å, and also some hydrophilic compounds having a sufficiently weak interaction with water, which does not prevent clathrate formation, are capable of GH formation [15]. The existing and developed Heriot-Watt Hydrate model (HWHYD) does not consider the mechanism of stable GH compounds formation with almost any hydrophobous gases and highly volatile liquids having molecular dimensions indicated above. Without solving these issues, the technological problems related to gas hydrate systems cannot be solved successfully.

From a practical point of view, the research results are necessary to improve modern technologies for gas production, transportation and storage, as well as to improve the safety of mining operations, and develop new technologies for the integrated development of coal deposits [16]. It can be stated that the study of the GH formation mechanism seems to be one of the most important and urgent problems.

Based on research conducted in the innovative technologies laboratory of National Mining University (National Technical University "Dnipro Polytechnic"), it was found out that all the solutions of surface-active materials (surfactants) reduce the surface tension of the solution by 79 – 96%, although the hydrate formation rate is significantly changed. Thus, it was concluded that the solubilization, passing at the nanoscale level, is the mechanism of hydrate formation. For these nonionic surface-active materials (nonionic surfactants), a feature has been found related to the existence of two zones available for the solubilizate – a hydrophobous hydrocarbon core and a thick (polymolecular) hydrophilic polyoxyethylene capsule, wherein the solubilizate can dissolve both in the core and in the hydrophilic outer micelles capsule. Consequently, it is distributed in equilibrium between these two zones, while the extremely hydrophobous paraffin hydrocarbons, in particular methane, are localized when solubilizing only in the hydrocarbon core of the micelles. The studies of X-ray diffraction in concentrated solutions of surfactants containing solubilizate have shown that when hydrocarbons are dissolved, there occurs the introduction of solubilizate molecules into the intramicellar space between the hydrocarbon-terminated molecules of surfactants.

The peculiarity of this mechanism is that as a result of the surfactants application, the inverse microgas emulsions are formed [17]. This process is shown schematically in Fig. 1.

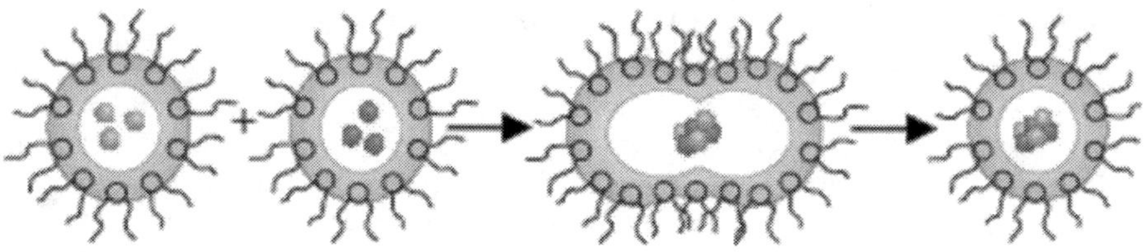

Fig. 1. Scheme of formation of surfactants microgas emulsions: green – water; red – gaseous methane.

Such microgas emulsions, according to Plato, arise when foaming the system "methane + surfactant + water". In the zone of three films contact that belong to three or more contiguous bubbles, a Plato channel is formed. Herein, the methane bubbles in GH have the shape of pentagonal dodecahedron. It is necessary to consider the fact that, at concentrations of $10^{-4} – 10^{-2}$ mol/l at a temperature of 274 K, gaseous methane is concentrated inside the pentagonal dodecahedron. This occurs as a result of cryoprotectant properties of surfactants at the indicated temperatures. The resulting micelles are actually the inverse, concentrating the gaseous methane.

2 Literature data analysis and formulation of the problem

The process of the GH formation is in the formation of a new phase (solid). The uncontrolled processes proceeding are caused by the fact that the new state of the system under the changed conditions is more stable and has a lower energy reserve. The formation of interphase surfaces and associated surface phenomena is due to excess surface energy. This is due to the fact that the molecules of the gaseous phase (methane) are in the interphase surface, have excess energy in comparison with the molecules in the volume because of their uncompensated intermolecular interactions. The resulting uncompensated intermolecular interactions are caused by the difference in the composition and structure of the contacting phases, and this leads to the occurrence of surface forces and an excess of energy – surface energy, on the interphase surface.

Transportation and use of GH consists of three phases: hydrate formation; transportation by itself; gas release in its decomposition. The first stage is the mixing of gas with water under the necessary conditions. One of the serious problems is the low rate of hydrate formation, which slows down the industrial application of this method. To solve this problem, surfactants and hydrotropes are often used [18], which makes it possible to find out the fine details of the mechanism of the hydrate formation process and its kinetics. Nevertheless, when adding surfactants, a side effect is often observed, which leads to a decrease in the methane hydrate stability. As it is shown in the work of [19], this can be avoided by adding a mixture of sodium dodecyl sulfate, xanthan, or starch.

The problem of the hydrate formation process acceleration is being studied by scientists from India, Iran, Canada, the USA, and Beijing. Thus, Iranian researchers studied the effect of copper oxide nanoparticles usage in order to increase the volume of gas absorbed by water to better convert water to a hydrate state and to increase the kinetics of the hydrate formation process [20]. The United States scientists have developed a technology for spraying water through the gas phase to increase the contact area of the hydrogen and gas phases, which leads to the hydrate formation process acceleration [21]. Indian scientists have proposed the use of surfactants to accelerate the process of hydrates nucleation, to provide for more active heat transfer and to increase the water and gas contact surface [22].

The first group in the field of studying the processes of hydrate formation in electrolytes solutions under laboratory conditions was the Canadian group of Professor P.R. Bishnoi [23]. These studies were carried out in a stirred-tank reactor in solutions of NaCl and KCl with different mineralization (up to 8 wt.%), in a narrow temperature range (270 – 274 K) at a fixed stirrer speed of 400 rpm. In this work, the diffusive mechanism of the hydrate formation process is considered. It is assumed that on the surface of the hydrate particles formed there is a water film through which the gas diffuses to the hydrate surface.

In works of [24, 25], the natural gas hydrates synthesized in sodium and calcium chlorides solutions were the objects of the study. The concentrations of sodium chloride solutions were 5, 10 and 15%, which corresponds to the total mineralization of salt brines 51.8; 107.1 and 166.3 grams per litre. It was identified that the chemical nature of salts affects the formation and decomposition processes, as well as the texture of natural gas hydrates [26, 27]. The hydrates formation in sodium chloride solutions proceeds more slowly than in calcium chloride solutions. With increase in mineralization of solutions, the rate of gas hydrates decomposition increases.

The kinetics of hydrate formation of carbon dioxide and gas mixtures containing carbon dioxide in sodium chloride solutions, was studied by the Beijing group of Professor T.M. Guo [28]. The experiments were also carried out in a stirred-tank reactor.

The work of [29] describes the kinetic curves of methane hydrate formation in distilled water and water containing inhibitors (concentration of salts – up to 5 wt.%). In accordance with the traditional concepts of formal kinetics, a process proceeding mechanism has been proposed, which includes the stages of gas dissolution, a metastable hydrate cluster formation and a gas hydrate particle growth. The potential reversibility of each process stage was taken into account, and a general kinetic equation was suggested. The pressure of the gas-hydrating agent was not maintained during the process. The absorption rate of methane was calculated according the pressure drop in the experimental cell.

Thus, the results of fundamental studies of the GH formation mechanism, as well as data on the chemical additives and surfactants which influence on the hydrate formation rate, can significantly affect the technological and economic attractiveness of natural gas storage in the gas hydrate form.

A new direction in the GH formation is their production in the presence of polyelectrolytes (PE), which are widely used in various industries as flocculants and coagulants of disperse systems, for example, for clarification of waste and turbid waters, for structuring of soils and grounds from radioactive elements [30], as well as for the colloids stabilization, in particular, for emulsions and foams, which is especially important in the obtaining GH. Besides, it is known that PE make an effect on the hydrodynamic properties of ecosystems [31]. The polymer dissolution in the liquid is accompanied not only by an increase in the viscosity, but also by an increase in the rate of the turbulent flow of the polymer solution in comparison with the flow rate of the initial low-viscous dissolvent. This paradox (Toms effect), which was discovered in the middle of the last century, up to the present time does not have a generally accepted theoretical basis. The first who engaged in a detailed study of the effect of the hydrodynamic drag reduce were hydromechanics who considered macromolecules of different chemical nature only as mathematical objects devoid of individual physical and chemical properties. This made possible to present a formal quantitative description of the phenomenon, but such a mechanistic approach is one of the reasons for the lack of the Toms effect explanation [32].

3 Materials and methods of research of hydrodynamic regimes influence on hydrate formation process

3.1 The essence of conducted research

The stalagmometric method with automatic photoelectron count of drops (measurement error 0.1%) was used. To determine the critical concentration of micelle formation (CMC), a conductometric method was used (conductivity meter No. 5721, Wroclaw: ElVRO). The electrical conductivity was measured with a Wheatstone bridge (measurement error 0.05 – 0.1%).

Interphase electric potential was measured by the potentiometric method using the PPTV1 potentiometer.

The number average molecular weight of the polyelectrolyte samples was determined by spectrophotometric titration of amine-terminated groups 0.01 with HCl alcoholic solution in triphenyl-verdazil presence, and was also evaluated from the reduced viscosity of aqueous solutions of ionenes at a concentration of 0.1 g/dl. The measurements of the PE solutions viscosity were carried out on a VPJ-1 viscometer with a capillary diameter of 0.54 mm after temperature equilibrium was set at 25 + 0.1 °C.

3.2 Experimental studies of the rate of gas hydrates formation

The research on the study of the GH formation mechanism in the presence of surfactants was carried out in an installation created in the innovative technologies laboratory of National Mining University (National Technical University "Dnipro Polytechnic") [33, 17]. The laboratory facility makes it possible to conduct the research of high-speed methane GH production under mild conditions (at pressures up to 7 MPa) and temperatures from 274 to 281 K.

The methods of experiments for obtaining methane GH in the presence of surfactants in laboratory conditions is described in the work of [34]. The rate of methane hydrate formation was determined by the kinetic method from the initial rate of the ice crust formation, and then the amount of methane per unit volume (C_m) was determined at different surfactant contents.

The methodology for determining the amount of methane per unit GH volume obtained at different surfactant content, as described in detail in work [34], include the following main stages:

1. Temperature decrease in the climatic thermal chamber and, accordingly, in the hydrate formation reactor from 274 to 263 – 258 K after the end of the GH production process, when all the water reacted with the gas;
2. Closing of valves on the cylinder with gas and on the reducer;
3. Disconnecting the high-pressure hose from the reducer and connecting it to the gas meter;

4. Opening the reactor valve and temperature rise to room temperature in the chamber and in the reactor;

5. Decomposition of methane hydrate and passing the released gas through the gas meter to determine its amount.

Knowing the value of the extracted methane and the volume of water poured into the reactor, the value of C_m was determined. Then the gas pressure in the reactor was reduced by 1 MPa and the determinations were repeated.

4 The results of experimental studies of the gas hydrates formation mechanism in the presence of polyelectrolytes

The hydrodynamic properties of PE are largely determined by intra- and intermolecular electrostatic interaction of charged sections of the chain, which leads to the formation of unfolded conformations [35]. On the other hand, the presence of hydrophobous groups, the possibility of forming intramolecular salt bridges and hydrogen bonds leads to the compact structures formation. In general, the conformational state of the PE macromolecule is determined by the ratio of forces of electrostatic and non-electrostatic interactions, which can be estimated from the values of the characteristic viscosities. A sharp increase in the reduced viscosity in the area of low polymer concentrations does not make it possible to determine the characteristic viscosity by extrapolation to zero concentration, and therefore the viscosimetrical data were linearized in the coordinates of the equations in the works of [36, 37].

The values of the characteristic viscosities for the fully-ionized $[\eta]_{fi}$ and fully nonionized $[\eta]_{fn}$ states are calculated by extrapolation for zero abscissa. It is obvious that the ratio of these quantities will characterize the change in the volume of the coil macromolecule at the changes of polyion dissociation degree from 0 to 1. Then the quantity of:

$$\alpha = \left(\frac{[\eta]_{fi}}{[\eta]_{fn}} \right)^{\frac{1}{3}}, \qquad (1)$$

will characterize the increase in the radius of the coil in the process of the polyelectrolyte swelling. All the calculated parameters for aqueous and mixed solvents show that the characteristic viscosity for a fully-ionized polymer decreases substantially with the elongation of the side alkyls. This is obviously connected with the hydrophobous interactions stabilizing a compact structure of the macromolecule. The consequence of this is the reduction of swelling parameter α.

The value of the true characteristic viscosity at an infinitely large ionic strength of the solution has been obtained by isoionic dilutions method. The constancy of the ionic strength of the solution in the process of dilution results in a linear dependence of the reduced viscosity on the polymer concentration (Fig. 2). This makes it possible to determine the characteristic viscosity at a given ionic strength of the solution $[\eta]_{Ii}$ as well as characteristic viscosity at the infinite ionic strength $[\eta]_{\infty}$ using the Fuoss-Strauss equation.

Based on the above experimental data, we attempted to determine some macromolecular parameters of synthesized ionenes, and also to estimate the change in nonelectrostatic interactions during PE solutions dilution with the use of empirical constants. The structure of the polymers obtained makes it possible to consider them as semirigid macromolecules and to determine the length of the statistical segment and the hydrodynamic radius of the macroion.

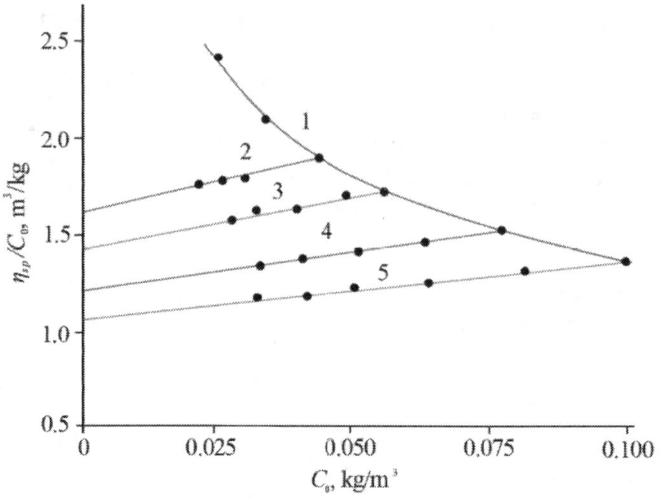

Fig. 2. Dependence of the reduced viscosity on the concentration for polyelectrolyte in water and in a salt brine: 1 – water; 2 – 5 – in KCl solution with ionic strength, mol/m^3; 2 – $1.2\cdot10^{-3}$; 3 – $1.3\cdot10^{-3}$; 4 – $1.6\cdot10^{-3}$; 5 – $1.7\cdot10^{-3}$.

The values $[\eta]_c$ and $[\eta]_\infty$ are the characteristic viscosities at the infinite ionic strength. However, the value of $[\eta]_c$ characterizes the state of the polymer at its concentration in a solution tending to infinity, that is, under conditions of maximum non-electrostatic interactions, while $[\eta]_\infty$ value is determined by extrapolation to zero concentration. Thus, the ratio:

$$\varphi_\infty = \frac{\eta_c}{\eta_\infty}, \tag{2}$$

will characterize the change in the volume of the coil macromolecule due to the hydrophobous bonds formation at infinite ionic strength. Then the ratio:

$$\varphi_{Ii} = \frac{\eta_c}{[\eta]_{Ii}}, \tag{3}$$

will characterize non-electrostatic bonds destruction in the process of polyelectrolyte swelling in passing from an infinitely great to a fixed value of ionic strength. The calculated macromolecular parameters, as well as the values of φ_∞ and φ_{Ii} show that an increase in the length of the side hydrocarbon radicals leads to intensification of non-electrostatic interactions and formation of more compact conformations.

This became a prerequisite for the use of watersoluble nitrogenated polymers, which include into the main chain the fragments of various organic acids, as the regulators of hydrochemical and hydrodynamic regimes in the process of GH production.

When adding this type of polymers in the concentration range of 10^{-3} – 10^{-6} g/l, the reduced viscosity η_r/C becomes negative, that is, the viscosity of the system is lower than the standard viscosity of water. The abnormal character of the aqueous solutions viscosity of the polymers presented leads to a change in the rate of hydrate formation processes. The degree of this influence is determined by the chemical structure and the molecular weight of the polymers.

The studies have been carried out in the laboratory conditions on the influence of five polymers on the hydrochemical and hydrodynamic regimes of aqueous systems (hereinafter polymers are represented by sequence numbers from 1 to 5).

The experiment was carried out according to the methodology described in the work of [34]. Each of five polymers were added at concentrations of: 0.01; 0.05; 0.1 mol/m^3. Analysis of polymers influencing the process of methane hydrate formation was determined by the moles number of methane in the GH as well as according to the rate of formation. The analysis of obtained data shows that at all concentrations for polymers No. 1, 2, 3 and at concentrations of 0.05 mol/m^3 for polymers No. 4 and 5, the concentration of gaseous methane is above the control point. The rate of the hydrate formation process compared to a control point at concentrations of 0.1 mol/m^3 for all polymers, and for polymer No. 4 at concentrations of 0.05 mol/m^3 – by 120 – 150%. The dependence is presented in Fig. 3.

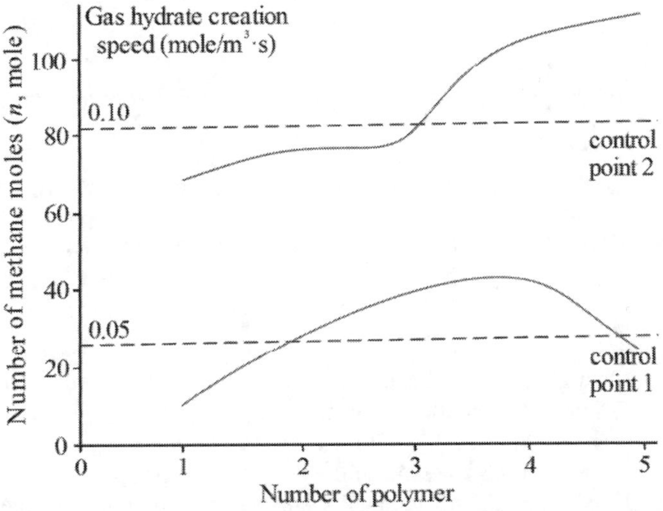

Fig. 3. Dependence of the amount of methane retention (*n*, mole) in GH (1); dependence of the rate of GH formation in the PE presence at a polymer concentration of 0.01 kg/m^3 (2).

It was found out in experimental industrial installations that the rate of methane gas hydrate formation during the experiment ranged from 0.02 to 0.16 mol/(m^3·s). The average values of this parameter indicate that for polymer No. 2 at its concentration of 0.1 mol/m^3 and No. 4 at a concentration of 0.01 mol/m^3, they exceed the control point value by 1.4 – 1.5 times. The concentration of hydrogen ions during the experiment varied within the range of pH = 8 – 9.

It should be noted that the coefficient of this parameter variation ranges from 1.7 for polymer No. 4 at a concentration of 0.1 mol/m^3 to 4 at a concentration of 0.01 mol/m^3. It was 2.5 in the control series. It is of interest that the coefficient for methane variation is much higher, it ranges from 12.5 for control to 36.8 for polymer No. 4 at a concentration of 0.1 mol/m^3.

The rate of GH formation at the beginning of the experiment was low (1.9 – 5.0 mol/(m^3·s)), and after a certain time it increased by 2 – 17 times, and, what is more, in the control point the increase was by 6 times. After one day, an increase in the rate of GH formation was observed for all polymers, except for control where it has decreased by 30%.

The dynamics of the rate of methane hydrate formation in the presence of PE is presented in Fig. 4.

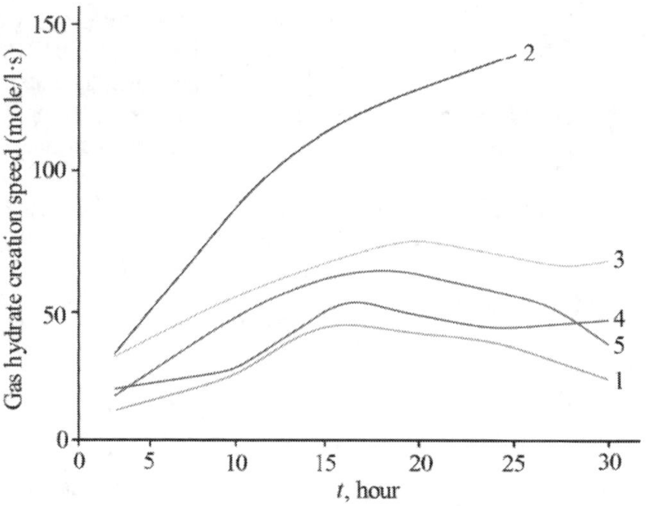

Fig. 4. Dynamics of the rate change in the methane hydrates formation: 1 – control; 2 – polymer No. 2 with a concentration of 0.05 mol/m^3; 3 – polymer No. 1 with a concentration 0.1 mol/m^3; 4 – polymer No. 5 with a concentration of 0.1 mol/m^3; 5 – polymer No. 3 with a concentration of 0.05 mol/m^3.

5 Conclusions

It has been established on the basis of conducted experiments, that the studied polyelectrolytes (PEs) have a "retention" ability with respect to methane in water during the hydrate formation process, and also the pH media stabilization occurs.

The formation of GH occurs as a result of the macromolecules binding under the influence of gaseous methane and some other chemical additives that contribute to the increase in the number of gaseous methane moles, due to the exchange processes between the polymer matrix and watersoluble salts, and also in the change of the water system hydrodynamics.

Taking into account the peculiarities of PE solutions under the condition of temperature or pH change, and adding of chemical additives (salts), the coacervation phenomenon takes place, as a result of which the system is divided into two phases, which leads to the formation of a spatial grid. As a result, an increase in the rate of GH formation was observed from 2 to 17 times, depending on the PE structure, its concentration and the reaction time.

The mechanism of this process is associated with a spontaneous, uncontrolled change in the PE solutions viscosity. As a result of changes in the hydrodynamic properties of water, an increase in the amount of gaseous methane in the GH is observed up to 36.8 mol per a mole of water at a PE concentration of 0.1 mol/m^3.

The presented results were obtained in the framework of the complex implementation of research projects "Development of scientific principles of phase transformations of technogenic and natural gas hydrates and creation of the newest technologies of their extraction" (State registration No. 0115U002294) and "Scientific substantiation and development of energy saving and low waste technologies of hydrocarbon and mineral raw materials extraction" (State registration No. 0116U008041).

References

1. Mykhailov, V. (2016). Prospection and estimation of unconventional hydrocarbon deposits in Ukraine. *Visnyk of Taras Shevchenko National University of Kyiv. Geology, 2*(73), 38-45. https://doi.org/10.17721/1728-2713.73.06

2. Samoilov, V.V. (2017). Planning of the industrial and hydrogeological research at the final stage of hydrocarbon deposits development. *Visnyk of V.N Karazin Kharkiv National University – Series Geology Geography Ecology*, (46), 45-49.
3. Naumko, I., Kurovets', I., Zubyk, M., Batsevych, N., Sakhno, B., & Chepusenko, P. (2017). Hydrocarbon compounds and plausible mechanism of gas generation in "shale" gas prospective Silurian deposits of Lviv paleozoic depression. *Geodynamics*, *1*(22), 26-41. https://doi.org/10.23939/jgd2017.01.026
4. Petlovanyi, M.V., Lozynskyi, V.H., Saik, P.B., & Sai, K.S. (2018). Modern experience of low-coal seams underground mining in Ukraine. *International Journal of Mining Science and Technology*, 1-7. https://doi.org/10.1016/j.ijmst.2018.05.014
5. Bondarenko, V., Tabachenko, M., & Wachowicz, J. (2010). Possibility of production complex of sufficient gasses in Ukraine. *New Techniques and Technologies in Mining – Proceedings of the School of Underground Mining*, 113-119. https://doi.org/10.1201/b11329-19
6. Saik, P., Petlovanyi, M., Lozynskyi, V., Sai, K., & Merzlikin, A. (2018). Innovative approach to the integrated use of energy resources of underground coal gasification. *Solid State Phenomena*, (277), 221-231. https://doi.org/10.4028/www.scientific.net/ssp.277.221
7. Bondarenko, V., Lozynskyi, V., Sai, K., & Anikushyna, K. (2015). An overview and prospectives of practical application of the biomass gasification technology in Ukraine. *New Developments in Mining Engineering 2015: Theoretical and Practical Solutions of Mineral Resources Mining*, 27-32. https://doi.org/10.1201/b19901-6
8. Pedchenko, M., Pedchenko, L., Nesterenko, T., & Dyczko, A. (2018). Technological solutions for the realization of NGH-technology for gas transportation and storage in gas hydrate form. *Solid State Phenomena*, (277), 123-136. https://doi.org/10.4028/www.scientific.net/ssp.277.123
9. Bondarenko, V.I., Kharin, Ye.N., Antoshchenko, N.I., & Gasyuk, R.L. (2013). Basic scientific positions of forecast of the dynamics of methane release when mining the gas bearing coal seams. *Naukovyi Visnyk Natsionalnoho Hirnychoho Universytetu*, (5), 24-30.
10. Bondarenko, V., Kovalevska, I., Astafiev, D., & Malova, O. (2018). Examination of phase transition of mine methane to gas hydrates and their sudden failure – Percy Bridgman's effect. *Solid State Phenomena*, (277), 137-146. https://doi.org/10.4028/www.scientific.net/ssp.277.137
11. Pedchenko, L., Niemchenko, K., Pedchenko, N., & Pedchenko, M. (2018). Use of alternative energy sources to improve the efficiency of natural gas hydrate technology for gas offshore deposits transportation. *Mining of Mineral Deposits*, *12*(2), 122-131. https://doi.org/10.15407/mining12.02.122
12. Bondarenko, V., Sai, K., Prokopenko, K., & Zhuravlov, D. (2018). Thermodynamic and geomechanical processes research in the development of gas hydrate deposits in the conditions of the Black Sea. *Mining of Mineral Deposits*, *12*(2), 104-115. https://doi.org/10.15407/mining12.02.104
13. Bondarenko, V.I., & Sai, K.S. (2018). Process pattern of heterogeneous gas hydrate deposits dissociation. *Naukovyi Visnyk Natsionalnoho Hirnychoho Universytetu*, (2), 21-28. https://doi.org/10.29202/nvngu/2018-2/4
14. Melnikov, V., & Gennadinik, V. (2018). Cryodiversity: the world of cold on the Earth and in the solar system. *Philosophy and Cosmology*, (20), 43-54. https://doi.org/10.29202/phil-cosm/20/4
15. Mohebbi, V., & Behbahani, R.M. (2014). Experimental study on gas hydrate formation from natural gas mixture. *Journal of Natural Gas Science and Engineering*, (18), 47-52. https://doi.org/10.1016/j.jngse.2014.01.016
16. Hanushevych, K., & Srivastava, V. (2017). Coalbed methane: places of origin, perspectives of extraction, alternative methods of transportation with the use of gas hydrate and nanotechnologies. *Mining of Mineral Deposits*, *11*(3), 23-34. https://doi.org/10.15407/mining11.03.023
17. Bondarenko, V., Svietkina, O., & Sai, K. (2017). Study of the formation mechanism of gas hydrates of methane in the presence of surface-active substances. *Eastern-European Journal of Enterprise Technologies*, *5*(6(89)), 48-55. https://doi.org/10.15587/1729-4061.2017.112313
18. Farhang, F. (2014). *Kinetics of the formation of CO_2 hydrates in the presence of sodium halides and hydrophobic fumed silica nanoparticles*. PhD Thesis. Queensland: The University of Queensland. https://doi.org/10.14264/uql.2014.385

19. Ganji, H., Manteghian, M., & Rahimi Mofrad, H. (2007). Effect of mixed compounds on methane hydrate formation and dissociation rates and storage capacity. *Fuel Processing Technology, 88*(9), 891-895. https://doi.org/10.1016/j.fuproc.2007.04.010
20. Najibi, H., Mirzaee Shayegan, M., & Heidary, H. (2015). Experimental investigation of methane hydrate formation in the presence of copper oxide nanoparticles and SDS. *Journal of Natural Gas Science and Engineering*, (23), 315-323. https://doi.org/10.1016/j.jngse.2015.02.009
21. Brown, T. D., & Taylor, C. E. (2013). Rapid gas hydrate formation process. Patent No. 8354565, USA.
22. Kumar, A., Bhattacharjee, G., Kulkarni, B. D., & Kumar, R. (2015). Role of surfactants in promoting gas hydrate formation. *Industrial & Engineering Chemistry Research, 54*(49), 12217-12232. https://doi.org/10.1021/acs.iecr.5b03476
23. Dholabhai, P.D., Kalogerakis, N., & Bishnoi, P.R. (1993). Kinetics of methane hydrate formation in aqueous electrolyte solutions. *The Canadian Journal of Chemical Engineering, 71*(1), 68-74. https://doi.org/10.1002/cjce.5450710110
24. Kalacheva, L.P., Rozhin, I.I., & Fedorova, A.F. (2016). Izuchenie zavisimosti processov obrazovaniya i razlozheniya gidratov prirodnogo gaza ot himicheskoy prirody rastvorov elektrolitov, imitiruyushchih plastovye flyuidy. *Mezhdunarodniy Zhurnal Prikladnyh i Fundamental'nyh Issledovaniy*, (8-4), 565-569.
25. Kalacheva, L.P., & Rozhin, I.I. (2017). The influence of the chloride-calcium-type water composition on the properties of natural gas hydrates. *Neftegazovaya Geologiya. Teoriya i Praktika, 12*(3). https://doi.org/10.17353/2070-5379/25_2017
26. Lang, X., Fan, S., & Wang, Y. (2010). Intensification of methane and hydrogen storage in clathrate hydrate and future prospect. *Journal of Natural Gas Chemistry, 19*(3), 203-209. https://doi.org/10.1016/s1003-9953(09)60079-7
27. Holzammer, C., Finckenstein, A., Will, S., & Braeuer, A.S. (2016). How sodium chloride salt inhibits the formation of CO_2 gas hydrates. *The Journal of Physical Chemistry B, 120*(9), 2452-2459. https://doi.org/10.1021/acs.jpcb.5b12487
28. Fan, S.-S., & Guo, T.-M. (1999). Hydrate formation of CO_2-rich binary and quaternary gas mixtures in aqueous sodium chloride solutions. *Journal of Chemical & Engineering Data, 44*(4), 829-832. https://doi.org/10.1021/je990011b
29. Guo T.-M., & Qiu, J.-H. (2002). Kinetic of methane hydrate formation in pure water and ingibitor containing systems. *Chinese Journal of Chemical Engineering, 10*(3), 490-497.
30. Svietkina, O. (2013). Receipt of coagulant of water treatment from radio-active elements. *Annual Scientific-Technical Collection – Mining of Mineral Deposits 2013*, 227-230. https://doi.org/10.1201/b16354-42
31. Svietkin, Yu.V., Riabenko, V.V., Volkov, M.F., & Varlan, K.E. (1984). *Sposob stymulyrovanyia rosta zooplanktona*. Patent No. 1137603, USSR.
32. Manzhai, V.N., Konovalov, K.B., & Kazarian, M.A. (2017). Model povedeniya makromolekul v turbulentnom potoke i eye analiticheskiye sledstviya. *Kratkiye Soobshcheniya po Fizike FIAN*, (12), 3-6.
33. Ovchynnikov, M., Ganushevych, K., & Sai, K. (2013). Methodology of gas hydrates formation from gaseous mixtures of various compositions. *Annual Scientific-Technical Collection – Mining of Mineral Deposits 2013*, 203-205. https://doi.org/10.1201/b16354-37
34. Bondarenko, V., Svietkina, O., & Sai, K. (2018). Effect of mechanoactivated hemical additives on the process of gas hydrate formation. *Eastern-European Journal of Enterprise Technologies, 1*(6(91)), 17-26. https://doi.org/10.15587/1729-4061.2018.123885
35. Svetkina, Ye.Yu. (2013). Intensification of concentration process through minerals vibroactivation. *Naukovyi Visnyk Natsionalnoho Hirnychoho Universytetu*, (2), 38-43.
36. Fuoss, R.M. (1948). Viscosity function for polyelectrolytes. *Journal of Polymer Science, 3*(4), 603-604. https://doi.org/10.1002/pol.1948.120030414
37. Yuan, L., Dougherty, T.J., & Stivala, S.S. (1972). A semi-empirical approach to the viscosities of polyelectrolyte solutions. *Journal of Polymer Science Part A-2: Polymer Physics, 10*(1), 171-189. https://doi.org/10.1002/pol.1972.160100113

Productive flows control in coal mines under the condition of diversification of production

Oleksandr Mamaikin[1*], *Vadym* Sotskov[1], *Yurii* Demchenko[1], and *Oleksandr* Prykhorchuk[2]

[1]National Mining University, Department of Underground Mining, 19 Yavornytskoho Ave., 49005 Dnipro, Ukraine
[2]PJSC "DTEK Pavlogradvuhillia", MA "Pershotravenske", 76 Soborna Ave., 51400 Pavlohrad, Ukraine

Abstract. The objective of the work consists in methodical bases of concept development for control by productive flows in coal mines with the use of cost-effective mechanisms. These mechanisms would provide their stable functioning in conditions of uncertainty and risks, formed by internal and external factors. The theoretic views have been used for the supply system structuring, when assessing the opportunities and reasonability to contract out the diversified production operations. The correlation of industrial factors has been defined, representing the mine as a multi-level system, and a logical interpretation of the obtained dependences. The system of efficiency evaluation has been proposed, which is intended to supplement the existing system of risk management with a kind of control element, an indicator of the positive processes in processing technogenic wastes of mines and dressing plants.

1 Setting the problem in general terms and its connection with important scientific and practical tasks

The result of the production activity of the underground complex of mining operations is the production on the surface of four productive components, which from the viewpoint of rational resource and nature management can be considered on the one hand, as factors that can have a negative impact on the environment. These components are: coal, rock, mine water and pit gas [1, 2]. Considering their dynamic nature, it is expedient for the mathematical description of moving processes of coal (C), rock (R), pit gas (G) and water (W) to present them as flows in some technological system of converters. In such a case, the aim of conversion is to obtain basic and additional resources of raw materials, as well as environmental damage minimization. At the same time, the degree of the impact of coal mining enterprise on the environment, on the one hand, is determined by the efficiency of each production unit included in it. On the other hand, it is determined by the mutual coordination of the objectives of all the production units that contribute to achieving the overall objectives of the enterprise. That is, for coal mines, which differ in the specific of the natural component effect, the possibilities of achieving the common goals represent the sum of local opportunities to achieve goals by all production units with the condition of

* Corresponding author: mamaykin@yahoo.com

minimizing the environmental footprint (F).

$$F = C + R + G + W \rightarrow min\,. \tag{1}$$

As can be seen in the expression (1), all components are positive, then minimization of the sum means minimization of each summand. The minimization of the first summand C is equivalent to the decrease in ash content and damage from surface underworking to the minimum permissible value. The minimization of the second summand R means the maximum volume of rock left in the mine and its use as an object of production diversification. In particular, if the entire rock is not released to the surface, R takes on its minimum value equal to zero. The minimization of the third summand G leads to reduction of methane emissions into the atmosphere, and minimization of the parameter W values – leads to a decrease in the volume of release into the environment of highly mineralized mine water.

All the summands in the expression (1) are unrelated and their minimization represents, in a number of cases, contradictory tendencies. Therefore, the comparative value of the coefficients in the objective function should reflect the balance of these contradictory tendencies impact [3].

For any coal mining enterprise, the most desirable is the economic situation, at which the full predictability of its development is achieved in all directions, both external and internal. Under these conditions, the planned volume of marketable products, sold in the energy market is easyly produced. Reserves are prepared for extraction without any obstacles, productivity increases within a predetermined range, and as a major result of its productive and economic activity, the mine receives a profit in that amount, which satisfies its interests to the fullest extent [4 – 6].

A review of the theoretical views of individual scientists has shown that the structuring of the supply system becomes a very important factor in assessing the opportunities and reasonability to contract out the diversified production operations [7]. For each enterprise, the consequences of making a decision to implement a contract form of work organization cannot be the same, and if so, they should be different in weight and value. So, it is necessary to weigh carefully all the pros and cons, as well as to consider other possibilities for the use of rock, gas or mine water.

2 Presentation of the main research material with full substantiation of scientific results obtained

Obviously, it is desirable to allocate capital investments for the development of mines, where the greatest effect is expected. Difficulty in determining such mines (and their sequence) is due to the complexity of the mine as a natural-technological system [8]. The results of the mine operation depend on a wide range of factors, which act strictly individually in the specific conditions of each mine and often have the opposite direction. For example, the mine may have a relatively shallow depth of development, but also a low thickness of a seam [9, 10]. Natural factors exist from time eternal, they are not controlled and their effect is largely predictable. Industrial factors – machinery, technology, organization of production – in principle are controlled, but often influenced by natural factors and their changes, and are usually associated with high costs [11].

Industrial factors, which describe the mine as a multi-level system, are quite varied and not always adequately describe the state of the system both qualitatively and quantitatively. The requirements for choice of the most specific factors and those, which make it possible to differentiate one enterprise from another, can be the following:

– indicators of the state of the mine are of significant importance – its prospects, the complexity of the production system, the situation (in terms of the enterprise capacity) among other mines, the ratio of mine production processes;

– it is desirable that the number of factors taken into account was not too large, they should be quite clear in their content and obtained in the simplest way from the materials of the current reporting.

In our opinion, such factors may be the annual production (D), level of concentration of work (K), producing cost (S), and labor productivity of a mining worker (P). These parameters describe the coal mining enterprise not only as technical and economic indicators, but they give the chance to estimate the unused internal reserves to increase or improve the production state, or for calculations to substantiate the need for diversification at the best case with the subsequent prospect of closure and total liquidation.

Identifying the relationship of these factors to each other and the logical interpretation of the dependencies obtained can provide an answer to one of the main issues of the future fate of the domestic coal industry in general and specific enterprises related to each other by identical geological and geological conditions, territorially and socially.

In the market interpretation, with an increase in the labor productivity and the level of concentration of work, the producing cost should decrease. By increasing the length of the face line, (and this is an indicator of the growth in the deconcentration of work), the producing cost should increase. These assumptions have been tested in the analysis of the work of 10 mines in the Selidovo and Mirgorodsk districts. Linear relationships between producing costs and parameters P, K, and D have been constructed with the aim of subsequent construction of linear optimization models. Static totality was quite indicative, as confirmed by the relevant fitting criterion (a period of 10 years was considered – from 2007 to 2016).

The results of calculations of multiple regressions are given in Table 1, and in Table 2 – the corresponding equations of net regression. Noteworthy is the fact that in some mines, with an increase in labor productivity, the producing cost rises [12, 13]. This can be interpreted as nonsense, although this phenomenon can be explained, as well as the growth of producing costs while increasing production in mine.

In connection with an improper tendency of the interrelation between the producing costs and the argument factors, it follows that the data are discrepant and can not be the basis as coefficients for unknown variables in constructing a complex model of the mines parameters optimization.

The main reason for the inadequacy of the obtained results is in the inconsistency of the applied technologies to modern trends or in the unreliable technical and economic performance indicators of enterprises over recent years, especially with the introduction of such term as "commercial confidentiality".

Table 1. The equation of multiple regression in the mines of Myrnohradskyi and Selidovskyi districts.

Mine	The dependence of the producing costs (S) from		
	labor productivity of a worker in production (P), t/month	the actual extraction of the mine (D), m/month	the level of work concentration in mine (K), m/1000 t
Stakhanova	$S = 2655.9 - 24.1P - 0.1D - 3.9K$		
Kotliarevska	$S = 989.2 + 70.9P - 5.2D + 2.6K$		
Kurakhivska	$S = 5461.2 - 239.9P + 221.6D + 0.87K$		
Rodinska	$S = 1930.8 - 4.7P - 6.6D + 4.1K$		
Yuzhnodonbaska No. 1	$S = 1095.2 + 31.5P - 1.9D + 2.6K$		
Yuzhnodonbaska No. 3	$S = 4582.5 + 23.6P - 2.9D - 4.5K$		
1-3 Novohrodivska	$S = 5331.8 - 42P - 2.1D - 4.9K$		
Ukraine	$S = 4383.4 - 126.4P + 8.1D + 0.4K$		
Tsentralna	$S = 3169.4 + 17.8P - 9.6D - 0.5K$		
Mine 5/6	$S = -1428.7 + 86.8P - 11.8D + 5.3K$		

Table 2. The equation of net regression in the mines of Myrnohradskyi and Selidovskyi districts.

Mine	The dependence of the producing costs (S) from		
	labor productivity of a worker in production (P), t/month.	the actual extraction of the mine (D), m/month.	the level of work concentration in mine (K), m/1000 t
Stakhanova	$S = 2409.5 - 38.8P$	$S = 2343.7 - 2.3D$	$S = 1998.2 - 5.7K$
Kotliarevska	$S = 2275.1 - 30.8P$	$S = 2570.3 - 2.1D$	$S = 118.9 + 2.6K$
Kurakhivska	$S = 4577.8 - 90.1P$	$S = 3408.8 - 10.5D$	$S = -2328.5 - 1.7K$
Rodinska	$S = 3997.2 - 76.9P$	$S = 3728.9 - 18.3D$	$S = 742.0 + 6.5K$
Yuzhnodonbaska No. 1	$S = 2239.2 - 27.8P$	$S = 2559.9 - 1.9D$	$S = -703.8 + 8.1K$
Yuzhnodonbaska No. 3	$S = 1272.2 - 5.2P$	$S = 2194.6 - 1.3D$	$S = 2899.1 - 4.6K$
1-3 Novohrodivska	$S = 3081.5 - 41.8P$	$S = 2186.7 - 3.2D$	$S = 581.1 + 1.9K$
Ukraine	$S = 3403.8 - 26.5P$	$S = 2876.4 - 4.1D$	$S = 2357.1 + 0.2K$
Tsentralna	$S = 2557.6 - 30.5P$	$S = 1884.5 - 1.1D$	$S = 1745.4 - 1.2K$
Mine 5/6	$S = 1398.3 + 12.2P$	$S = 3115.3 - 10.6D$	$S = -452.8 + 4.9K$

But the limited size of mine fields (1000 – 1500 m) as compared to 2000 – 3000 m in other countries, provides evidenced both of the insufficient level of planning decisions and the insufficient energy saturation and reliability of extracting machines.

The amount of comprehensive faces, working with a low load (less than 500 tons/day), is 65%.

The most important is to solve the issues related not only to changes in coal mining technology in enterprises, which, ipso facto, have the internal reserves to increase labor productivity, but also involvement in this not simple process of low-potential coal-mining enterprises with low profitability or working with low productivity caused by objective circumstances.

Modern trends in the coal industry maintenance are intimately connected with the diversification of mining enterprises (see Fig. 1). One of the most promising trend is the maintenance of capacity of mines, primarily unprofitable ones, in the direction of advanced processing of coal, coalmine methane and coal waste at the place of their extraction through the production of thermal and electric energy, the introduction of highly efficient waste processing technologies, etc. [14, 15]. This will change the price of the final product structure, where the producing cost of coal will no longer play a decisive role, and will be one of the components.

There are more than 1.500 dumps in the Donbas territory, and in each of them there is an average of 1.144 cubic meters of rock. They extend for significant areas and have a negative impact on the atmosphere, soils, surface and underground water sources. Therefore, the problem of dumps recycling is urgent at the moment [16 – 18]. There are proposed more and more projects, according to which mini-factories for the extraction of coal and strategically important aluminum, germanium, scandium, gallium, yttrium and even zirconium, can be built on the basis of the closed mines. With the help of advanced technologies, a whole range of problems can be solved: environmental (in relation to mine communities), social (new jobs), land (the release of territories for construction). Moreover, the raw material received from one average volume of a dump costs at least $ 100 million dollars, since the presence in mines of industrial infrastructure will reduce the costs and implementation deadline of the planned production. The proposed projects emphasize the need for about 5 million dollars to equip the complex processing of refuse ore with smelting furnaces and electrohydraulic crushers, as well as about the same amount for the purchase of chemical reagents.

In a similar way it can be considered, from the perspective of the mine potential level (EVA), the reasonability of involving into mining the worst reserves, which, in the form of complex mining and geological conditions and the conditioning reserves depletion, are obviously unprofitable in the sale of coal at existing wholesale prices (2). Obviously, the question arises on the reasonability of involving into mining of complex reserves of deeply unprofitable mines. Despite the fact that the involvement for mining of these areas of the mine field leads to increased loss, the final decision depends on the adopted system of government subsidies to such enterprises. If the subsidy is payed in the fixed form, then it is not profitable for the enterprise to involve a complex area for mining, since in this case, without any effort on its part, the mine assumes a more favorable position when refusing complex reserves mining.

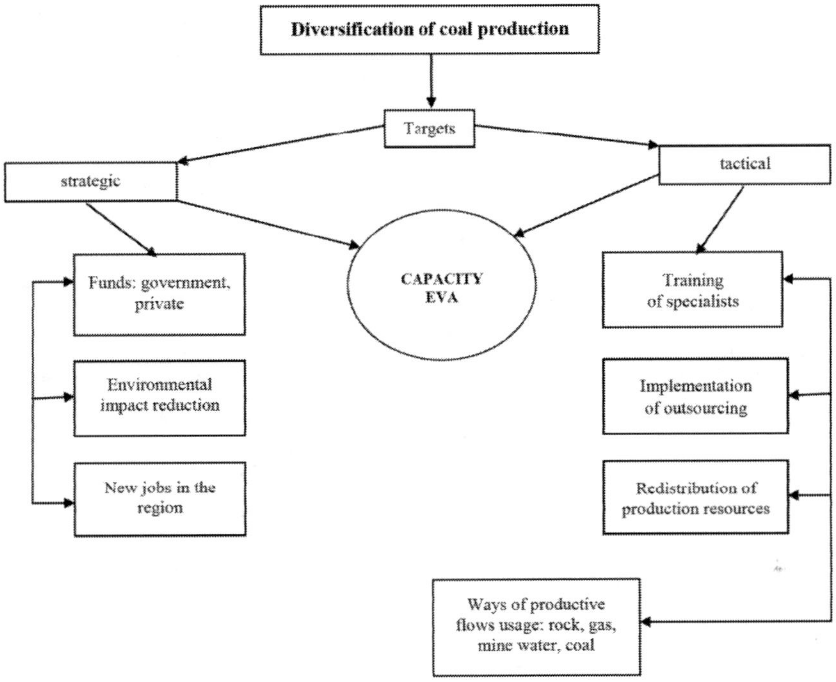

Fig. 1. Tactical and strategic features of production diversification.

With the help of relevant indicators, the capabilities of selected outsourcers are compared, and the feasibility of the contract form or the feasibility of a separate enterprise establishment for the production of diversified production on the basis of existing divisions, are analysed. Mathematically, the indicators are as follows: T_i – annual coal production from the dumps dismantling; W_i – energy, obtained from water use i-th mine in thermal pumps; B_l – volume of ceramic products from rocks l-th dump; R_l – volume of building materials after dismantling l-th dump for road construction.

$$EVA(max) = \begin{cases} P(T,W,B,R), \ if \ \lambda \geq 1; \\ P(T,W,B,R) + S_i, \ if \ 1 \geq \lambda \leq 0.5; \\ S_i, \ if \ \lambda \leq 0.5, \end{cases} \quad (2)$$

where $P(T,W,B,R)$ is the only outsourcing services; λ is the relative evaluation for each type of outsourcing services in the market of waste recycling; S_i is the services of the i-th type by own divisions.

The determination of the criteria weight in each of the cases is carried out on the basis of the expert assessments method. The overall assessment of the possibilities of the contract application $P(T,W,B,R)$, or independent performance of these works S_i, is calculated as the sum of all criteria values with regard to their weight. Since the capabilities of outsourcers are evaluated by several independent indicators, therefore, the choice can be made based on the principle of minimax (maximin). At the same time, the optimal alternative is the alternative, which is not worse by all indicators and better, at least, by one indicator.

When assessing the consequences of diversified enterprises creation, a clearly expressed problem of uncertainty is possible. Uncertainty is understood to be the incompleteness and (or) inaccuracy of information on the conditions for the diversified products realization, including the necessary costs and expected results. The uncertainty associated with the possibility of the adverse situations emergence and consequences is characterized by the concept of risk. When managing the processes of diversified coal enterprises support, both objective and forecast uncertainties must be considered. The probability of the latter one is greater, because it is associated with changes in market conditions, prices, subsidy policy, etc. Since all of these factors can not be completely overcome, hence the nature of the risks at the stage of new companies' privatization, the change of ownership, change of resource potential. Basically, the management by sustainable development and costs, anyhow, implies the use of two criteria – the maximum production with the given capabilities of the resource potential and the minimum production costs. The value of both functionals is important for the manager. If they are significantly different, it is necessary to apply mathematical techniques associated with decision-making under uncertainty. Unfortunately, there are no general rules for assessing the applicability of a criterion, since the behavior (often changing) of an investor who makes a decision due to the uncertainty of the situation, is the most important factor in choosing the appropriate criterion.

It is known that coal-mining enterprises (mainly unprofitable) are transformed to new conditions with difficulty, and to some extent, they restrain the course of market reforms in the country [19, 20]. But backlog, in particular, such as the diversification of coal production, can be explained, above all, by the fact that the economic mechanism for regulation of the enterprises capacity in a particular region has not been worked out, rather than by the lack of experience. The system of efficiency evaluation has been proposed, which is intended to supplement the existing system of risk management with a kind of control element, an indicator of the positive processes in processing technogenic wastes of mines and dressing plants.

3 Conclusions

Based on the definition, diversification can be characterized as subsidiary production with its own divisions. Therefore, an assumption can be made regarding performance by them of service functions, that is, services, provision of services, and thus, their involvement into the service sector.

The supply system structuring is relevant to the rendering of service functions of the coal company outside the enterprise, if they are organized in the form of subsidiary facilities. The main production regulates the demand and distributes it among the divisions. Even the service divisions themselves can act as economic participants and be closely interrelated, such as the introduction of thermal pumps for heating the housing or extracting coal from the dump.

The possibility of a significant reduction in costs by eliminating unprofitable sites and subdivisions engaged in secondary, periodically performed work on environmental protection, processing of rock, etc., facilitates the development of outsourcing in the coal-mining regions. An important aspect of the advantages of outsourcing is the use of special production, technological and laboratory equipment, storage facilities, vehicles and construction mechanisms.

The research is carried out within the framework of scientific topics “Resource-saving geotechnical and hydrodynamic parametrization of the extraction of low-capacity mineral raw materials in an technogenically loaded environment” (State registration No. 0117U006753), financed by the state budget of Ukraine.

References

1. Mamaikin, O., Kicki, J., Salli, S., & Horbatova, V. (2017). Coal industry in the context of Ukraine economic security. Mining of *Mineral Deposits, 11*(1), 17-22. https://doi.org/10.15407/mining11.01.017
2. Mamaikin, O.R., Salli, S.V., Pochepov, V.M. , & Ashcheulova, O.M. (2015). Complex assessment of the recovery ratio of unprofitable mines. *Mining of Mineral Deposits, 9*(1), 135-139. https://doi.org/10.15407/mining09.01.135
3. Coyle, E.D., & Simmons, R.A. (2014). *Understanding the global energy crisis.* Purdue University Press.
4. Vermeylen, K. (2015). *The Stochastic Growth Model.* BookBoon.
5. Opara, V., Buzina, I., & Vynohradenko, S. (2017). Environmental-economic efficiency of land use improvement reasoning. *Visnyk of V.N Karazin Kharkiv national university-series geology geography ecology*, (46), 152-157. https://doi.org/10.26565/2410-7360-2017-46-21
6. Minke, A.G. (2013). Conducting transatlantic business: Basic legal distinctions in the US and Europe. *BookBoon*, 59.
7. Mazzeo, N. (2013). *Chemistry, emission control, radioactive pollution and indoor air quality.* InTech.
8. Fomichov, V., Sotskov, V., & Malykhin, A. (2014). Determination and analysis of the acceptable benchmark changes of the stress strain state of frame and bolt fastening elements of dismantling drift when approaching a working face. *Naukovyi Visnyk Natsionalnoho Hirnychoho Universytetu*, (1), 22-26.
9. Sotskov, V.O., Demchenko, Yu.I., Salli, S.V. & Dereviahina, N.I (2017). Optimization parameters of overworked mining gallery’s support during carrying out long-wall face workings. *Naukovyi Visnyk Natsionalnoho Hirnychoho Universytetu*, (6), 34-40.
10. Kovalevska, I., Barabash, M., & Snihur, V. (2018). Development of a research methodology and analysis of the stress state of a parting under the joint and downward mining of coal seams. *Mining of Mineral Deposits, 12*(1), 76-84. https://doi.org/10.15407/mining12.01.076
11. Sotskov, V., & Saleev, I. (2013). Investigation of the rock massif stress strain state in conditions of the drainage drift overworking. *Mining of Mineral Deposits*, 197-202. https://doi.org/10.1201/b16354-36
12. Bai, J., Ghysels, E., & Wright, J. (2013). State space models and MIDAS regressions. *Econometric Reviews*, (32), 779-813.
13. Goncalves, S. & Perron, B. (2017). Tests of equal accuracy for nested models with estimated factors. *Journal of Econometrics, 198*(2), 231-252.
14. Clark, T.E., & McCracken, M.W. (2017). Tests of predictive ability for vector autoregressions used for conditional forecasting. *Journal of Applied Econometrics*, (32), 533-553.
15. Sadovenko, I.O., Zahrytsenko, A.M., Podvihina, O.O. & Dereviahina, N.I. (2017). Water balance control within rock mass using the capacity of water-bearing formations. *Scientific Bulletin of National Mining University*. (4), 19-26.
16. Demchenko, O., Evtekhov, V., Evtekhova, A., & Georgiieva, H. (2018). Grounding for effective technology of reprocessing coarse grained screenings from crushing and screening plants of Kryvyi Rih basin. *Visnyk of Taras Shevchenko National University of Kyiv. Geology, 1*, 89-92.

https://doi.org/10.17721/1728-2713.80.09

17. Hang, Q. (2016). A computationally efficient method for vector autoregression with mixed frequency data. *Journal of Econometrics*, (193), 433-437.
18. Fomichov, V., Pochepov, V., Sotskov, V., & Mamaikin, O. (2018).Formation of a calculation model determining optimal rate of stoping face movement with a large deformation of a rock massif. *Journal of Engineering and Applied Sciences*, *13*(7), 2381-2389.
19. Bondarenko, V., Cherniak, V., Cawood, F., & Chervatiuk, V. (2017). Technological safety of sustainable development of coal enterprises. *Mining of Mineral Deposits*, *11*(2), 1-11. https://doi.org/10.15407/mining11.02.001
20. Malashkevych, D., Sotskov V., Medyanyk V. & Prykhodchenko D. (2018). Integrated Evaluation of the Worked-Out Area Partial Backfill Effect of Stress-Strain State of Coal-Bearing Rock Mass. *Solid State Phenomena*, (277), 213-220. https://doi.org/10.4028/www.scientific.net/ssp.277.213

Self-regulating roof-bolting with the rock pressure energy use

Oleh Khomenko[1*], *Maksym* Kononenko[1], *Ihor* Kovalenko[2], and *Denys* Astafiev[1]

[1]National Mining University, Department of Underground Mining, 19 Yavornytskoho Ave., 49005 Dnipro, Ukraine
[2]Ukrainian State Chemical Technology University, Inorganic Chemistry Department, 8 Haharina Ave., 49000 Dnipro, Ukraine

Abstract. The perspective of discovery of zonal disintegration phenomenon of rocks around underground mine workings is analyzed. The methodological stages for research of this phenomenon are shown. The physics of zonal disintegration of rocks around underground mine workings is disclosed. There are described the possibilities of advanced entropy method and developed energy method that allow to investigate a phenomenon of zonal capsulation of underground mine workings. The sequence of research of this phenomenon parameters is presented. The order of choice of stable shape and resource-saving support in underground mine workings is substantiated, for the deep horizons of the Kryvorizkyi Iron-Ore Basin mines. The method of parameters calculation of self-regulating roof-bolting in underground mine workings is substantiated, which allows to use the rock pressure energy. The design of a self-regulating roof-bolting is developed, which allows applying metal, polymer and rope bolts. The economic efficiency of rock pressure energy usage is substantiated in case of support setting at great depths in underground mine workings.

1 Introduction

The problem of physics disclosure of phenomenon of zonal disintegration of rocks around mine workings, which first was discovered in 1972 in the South African mine and then in the USSR, Russia, Ukraine, Kazakhstan, the USA, Japan and other countries of the world, for a long time could not find a solution [1 – 3]. The discovery of the phenomenon itself has given only a comprehension about formation of the self-organized circular-type structures around underground mine workings. The first hypotheses rejected the physical reality of the discovery, but since the 1990s, the new methods which also did not describe the mechanisms for the energy zones formation, have been actively used in mining science [4]. Therefore, this phenomenon has been recognized as one of the most difficult problems in mining science, and cannot be described on the basis of the classical laws of physics and geomechanics. At the very stage of a scientific problem analysis, with the use of synergetic methodological search, we have managed to disclose the physics of the studied phenomenon and called it as zonal

* Corresponding author: rudana.in.ua@gmail.com

capsulation of mine working using a system of annular energy zones [5, 6].

Then, by means of the improved entropy method, the virgin massif state has been modeled. It was established that the processes of energy exchange in rocks proceed in mutually perpendicular directions corresponding to vertical and horizontal energy flows. It made possible to describe the processes, to determine the parameters and to establish the patterns of potential energy redistribution in rocks of the Ukrainian crystalline shield [7, 8]. Then, by means of the developed energy method, the parameters of mine workings capsulation were investigated using the 'convergence' and 'divergence' categories, which are phases of the centripetal self-organization of open systems. As a result of the convergence, the rock massif containing mine working is divided into integration subzones, and at divergence - disintegration subzones [9]. Due to the convergence and divergence, the massif resists to mining operations in it and restores the continuity, forming the protective capsule, which consists of the energy zones system [10]. The conducted research of a zonal capsulation phenomenon of mine workings with the help of new methods have made it possible to establish the exact quantity, the sizes and a shape of energy zones, and, most importantly to reveal the sinusoidal stresses and annular areas of deformation [11]. It is established that the energy of the safety capsule, formed around the mine working, can be used according to parabolic-and-hyperbolic dependence, the values on which correspond to the reasonable principles: resistance, equilibration, facilitating, elimination and transmission of concentrated energy of the massif on which the proposed technology solutions are based [12].

2 Methods

The prospect of Kryvorizkyi Iron-Ore Basin development on mining of thick steep deposits in the range of 1500 – 3000 m depths is connected with application of high-performance self-propelled and, accordingly, the large-sized drilling and loading transport equipment [13]. Ensuring the preparatory mine workings stability with a cross-section area up to 20sq.m and above is rational with the use of a stable shape. But, with an increase in depth to more than 1000 m, it is reasonable already to set the self-regulating roof-bolting with the use of the rock pressure energy [14].

The calculated values of a shape coefficient of energy zones λ entering the safety capsule of the mine working are defined as the ratio of vertical potential stress to horizontal one ($\lambda = \Delta\sigma_{z(y)} / \Delta\sigma_{x(y)}$), and are represented in the form of multiple exponential functions in Fig. 1.

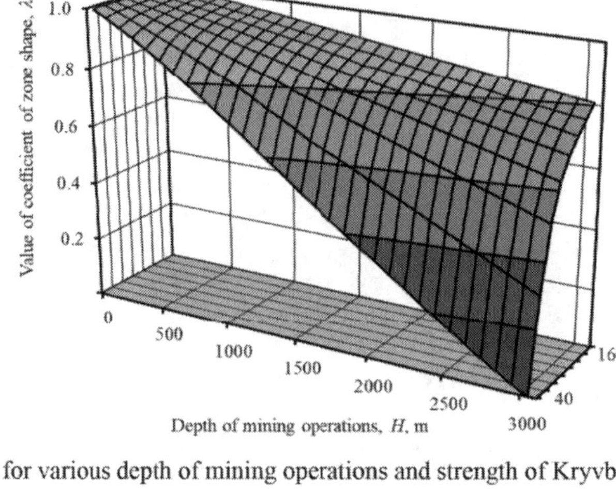

Fig. 1. Values of λ for various depth of mining operations and strength of Kryvbas rocks.

With the use of λ coefficient, the sizes of the vertical a and horizontal b semi-axes of the energy zone bordering on the mine working are determined as, m

$$a = 0.5\left(d\lambda^{-1} + h\right); \; b = a\lambda, \qquad (1)$$

where d is the width and h is the height of preparatory mine working, m [15].

When calculating a stable contour of mine working, the minimum value λ is accepted as equal to 0.5. If $\lambda < 0.5$, then the rock pressure on the support increases significantly. In this case, the rock swelling is possible, and this forces to form a complete ellipsoid with the outcrop of massif and to fasten it with pliable metal sets, framed through 0.2 – 0.5 m, or with monolithic concrete. The results of modeling the stable shape of a mine working and application of the traditional for ore mines and new way of support setting, innovative self-regulating roof-bolting, are set out in Table 1 [16].

Table 1. The stable cross-sectional shape and resource-saving type of support setting in the preparatory mine working, constructed in rocks with a strength of 80 MPa.

Laying depth of mine working, m	350	700	1000	1600	3000
Value λ	1.0 – 0.9	0.9 – 0.8	0.8 – 0.7	0.7 – 0.5	0.5 – 0.1
Traditional shape and support					
Shape	trapezium-shaped		arched		
Support	wooden		reduced-weight arched	improved arched	monolithic concrete
Resource-saving shape and support					
Shape	circular	arched	crescent-shaped	semielliptical	elliptical
Support	without support	shotcrete	bolting	reduced-weight combined	

The proposed approach to the choice of the new type of support is implemented according to the principle of resistance of support to the pressure of enclosing rocks. In other words, with an increase in depth of mine working laying, there is an increase in the volume of rocks, which must be supported [17, 18]. Partially, it is possible to lower the loading (up to 10%) by means of giving a stable contour to the outcrop of massif, and the rest part of loading is supported by setting the support itself. The higher the loading value, the more complex and expensive the constructions of applied supports. Until now, the complexity of mining-and-technical conditions caused an adequate support structure modification only for the purpose of minimizing costs. The problem of using the energy of triaxial compression of rocks as the main resource of efficiency increasing in the mine workings operation, the quality of setting and maintenance their support has been solved only in recent years [19].

Due to a system research of the phenomenon of zonal capsulation of mine workings, an analytical description of a pattern of self-organization of the massif disturbed by the mine working, the possibility is opened of using a new principle in the technology of support setting and maintenance of mine workings. The innovative principle practically excludes the influence of depth in mine workings drivage on the prime cost of their support setting [20, 21]. Its essence consists in that the most part of loading on the support, created by the pressure of the underworked thickness (convergence), is counterbalanced by the loading from the divergence processes development by means of application of self-regulating roof-bolts as a part of the combined support. The possible depths of the proposed technological solution application are closely connected with the process of zonal capsulation, i.e. formation of the second and the subsequent energy zones around mine workings. Based on the research results, the authors have developed a method for calculating the support setting passport of preparatory mine workings which uses the rock pressure energy [22].

1. Roof-bolts length:

– for a roof and a bottom of mine working at the small radius of contour corners, m

$$l_a = 2a - r + l_n ; \tag{2}$$

– for mine working sides at the large radius of contour corners, m

$$l_a = 2b - R + l_n , \tag{3}$$

where l_n is the length of the roof-bolt part protruding from the blast-hole, which depends on its design and the bearing-supporting elements thickness, equal to 0.05 – 0.2 m; radius of corner of arched parts of the mine face, i.e. small radius, m

$$r = \frac{\left(a^2 + c^2\right)^{0.5} - a + c}{2cos\psi} ; \tag{4}$$

radius of corner of the central part of the mine face i.e. large radius, m

$$R = c + (a - r)tg\theta , \tag{5}$$

where θ is the angle of normal stress action

$$\theta = \frac{\pi}{2} - \psi , \tag{6}$$

where ψ is the angle between the principal normal stress and the vertical plane, degree

$$tg\psi = \lambda . \tag{7}$$

2. The density of the roof-bolts setting is determined from the P_m – the perimeter of the mine working with an ellipsoidal shape, m

$$P_m = 4\frac{\pi ab + (a - b)^2}{a + b} , \tag{8}$$

where V is the volume of massif, enclosed into the border energy zone, which is determined as the average value of the sums of differences between the vertical semi-axis of the zone a and half of the mine working height $0.5h$, and between the horizontal semi-axis of the zone b and half of the mine working width $0.5d$, m

$$V = \frac{(a - 0.5h) + (b - 0.5d)}{2} . \tag{9}$$

3. The distance between the roof-bolts, m

$$a = \sqrt{\frac{P_a}{U\gamma k_o}} , \tag{10}$$

where P_m is the active perimeter of mine face, m; γ is the bulk density of mine rocks within the area of possible destruction, kN/cu.m; k_o is the roof-bolt overload coefficient; P_a is the load-bearing capacity of the roof-bolt determined from the Table [23].

Roof-bolt type	Compressive resistance of rocks, MPa	Load-bearing capacity of the roof-bolt, P_a, kN
Reinforced-concrete	> 20	100
Slot-and-wedge (joisting)	60 – 100	40 – 80
Tubular (with expanding joist)	> 40	60 – 120
Resin-grouted	Any	140 – 234
Rope (ampoule)		

After determination of key parameters, the support setting passports of preparatory mine working are developed in accordance to the computational scheme (Fig. 2).

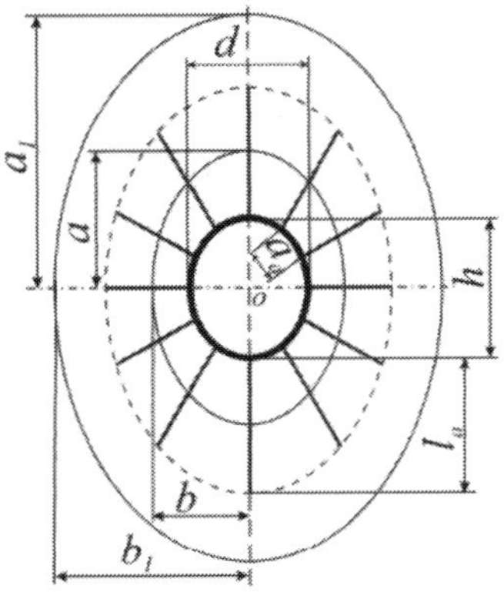

Fig. 2. The computational scheme to the drawing up of mine working support passport.

Since the actions of radial stresses are directed to the geometrical center of mine working, the drilling of blast-holes and, respectively, the roof-bolts setting are made from the center of mine working at the appropriate angles to the contour of massif outcrop [24, 25]. The mine working bottom is formed by means of ballast or concreting the lower part with laying in the center of drainage structure systems (flumes with covers, perforated pipes, shuttering). In order to avoid cleavage of small pieces of ore and rock during mine workings drivage in fractured rocks and ores, the combined support setting is used. For this purpose, the roof-bolts with Rabitz type steel-wire fabric, raised-perforated flooring, reinforced sheets etc., should be applied [26]. It is necessary to place the steel-wire fabric or other supporting structures along the mine working contour, in the roof and sides up to 80% of the fixed site length, and to fix with roof-bolts. In case, when the mine working intersects the ore and rocks strata, the roof-bolts should be set at an angle perpendicular to strata, which will lead to an increase in their length by 20 – 30% [27]. The required length of the roof-bolt set at an angle, perpendicular to strata, is the following, m

$$l_y = l_a \sin\alpha , \tag{11}$$

where α is the angle of rocks strata or deposit dip, degree.

For support setting of prestressed rocks, it is reasonable the application of new generation roof-bolts, i.e. elongated roof-bolts with a joist location in the center of the energy imbalance. It allows to satisfy a condition of active resistance to massif destruction by using the elastic energy of the massif [28]. At the same time, the tensile energy of rocks is counterbalanced with energy of compression in the depth of the massif on outcrop of stope face or preparatory mine workings [29]. This makes possible to increase the rational depths of application of the roof-bolting and combined roof-bolting in preparatory mine workings, and also to exclude the application of expensive frame or concrete supports in preparatory mine workings [30, 31]. At the same time, a basic element of such a support is the design of the roof-bolt and joists that are packed in the zone of active compression, but not just outside the area of rocks deformation [32]. The design and parameters of the separate elements of the self-regulating roof-bolt are represented in Fig. 3.

Fig. 3. Design of the self-regulating roof-bolt: 1 – roof-bolt (rope, polymer); 2 – hold-down strap and tightening nut; 3 – cased pipe; 4 – part of the blast-hole, free from casing; 5 – roof-bolt joist (concrete, wedge, chemical); 6 – free coal-face part of the blast-hole.

The length of cased pipe is determined for vertical and horizontal outcrops of the rocks massif and backfilling:

– roof-bolts set vertically, m

$$l_{ov} = a - 0.5h + \frac{0.5a_n}{3}; \tag{12}$$

– roof-bolts set horizontally, m

$$l_{oh} = b - 0.5d + \frac{0.5b_n}{3}, \tag{13}$$

where a_n is the number of an energy imbalance zone, having the highest stiffness of elastic energy. As a rule, in the safety capsule this is the second zone from the mine working which size equals to $2a$.

The length of a part of the blast-hole free from casing, the roof-bolt joist and free coal-face part 6 are determined by an identical formula:

– roof-bolts set vertically, m

$$l_{cv} = \frac{0.5a_n}{3}; \tag{14}$$

– roof-bolts set horizontally, m

$$l_{ch} = \frac{0.5b_n}{3}. \tag{15}$$

The roof-bolt design (Fig. 3) allows the bolt to move inside the cased pipe by means of setting the joist in the zone of maximum contraction stresses directed backwards from the mine working. This design allows to counterbalance the maximum volume of energy consumed by the massif for destruction of a mine working contour by means of elastic energy resistance of the massif divergence [33].

3 Results and discussion

In accordance with the substantiated interrelation of a cross sectional shape of preparatory mine workings and the type of their support, represented in Table 1, it has been determined the cost of 1 long metre of mine working drivage (Table 2) [34]. The prime cost was determined by main types of expenses for mining-and-geological conditions of preparatory mine working drivage: cross-sectional area of preparatory mine working S_{dr} =10.5 sq.m (h = 3 m, d = 3.5 m); depth of laying H = 1200 m; strength of martite ores is 80 MPa; sizes of semi-axes of border and subsequent energy zones $a = 3.98$ m, $a_1 = 7.96$ m, $a_2 = 11.93$ m, $a_3 = 15.92$ m and $b = 2.81$ m, $b_1 = 5.62$ m, $b_2 = 8.43$ m, $b_3 = 11.2$ m. Being the basic shapes of cross section in preparatory mine workings and types of their support setting, the accepted data have gained the greatest distribution in ore mines of Ukraine [35, 36] with the types of expenses and the prices operating since 2012 [37, 38].

Table 2. The prime cost of 1 long metre of preparatory mine working drivage depending on its cross sectional shape and support setting type.

Laying depth of mine working, m	Basic parameter			Proposed parameter		
	cross section shape	applied support	prime cost, UAH/m	cross section shape	applied support	prime cost, UAH/m
350	trapezium-shaped	wooden	4044.0	circular	without support	2643.0
700				arched	shotcrete	3164.0
1000	arched	reduced-weight arched	3882.0	crescent-shaped	bolting	3241.0
1600		improved arched	4624.0	semielliptical	reduced-weight combined	3464.0
3000		monolithic concrete	5493.0	elliptical		

From Table 2 it can be seen that application of a stable shape of cross section and of the new support of mine working allows reducing the cost of its support setting. Thus, the laying of preparatory mine working with a stable shape allows to save by 1 long meter: at a depth up to 350 m – 1401 UAH, up to 700 m – 880 UAH, up to 1000 m – 641 UAH, up to 1600 m – 1160 UAH and up to 3000 m – 2029 UAH [39]. There is a stable exponential dependence with an extremum point at a depth of 1000 m, at which a sharp increase in saved costs begins on a support setting in the preparatory mine workings [40].

Then, the specific efficiency of preparatory operations has been determined, which depends on the massif energy used (Fig. 4).

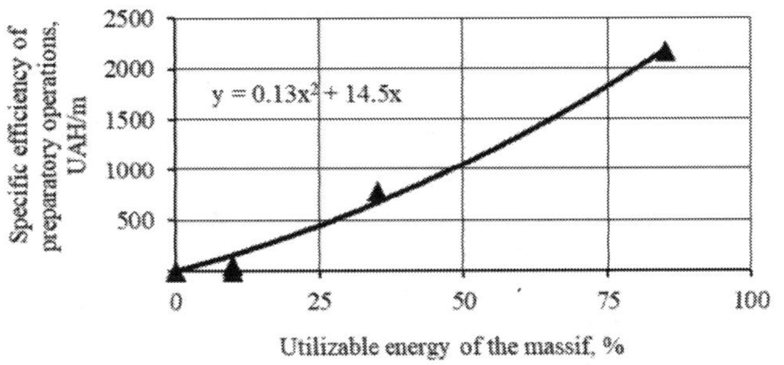

Fig. 4. The pattern of specific prime cost of preparatory operations with the massif energy use.

4 Conclusions

The energy of the safety capsule formed around the mine working is used according to the parabolic-and-hyperbolic dependence, the values on which correspond to the principles of resistance, equilibration, facilitating, elimination and transmission of concentrated energy of the massif. The support setting in mine workings with the designed principle of resistance is implemented with the energy costs equivalent to the energy of the massif (100%). When using the principle of transmission, up to 86% of the massif energy can be used.

The improvement of preparatory operations technology included giving the ellipsoidal contours both to the plane of the coal-face and to the cross section of preparatory mine workings of PJSC "Kryvorizkyi Iron-Ore Plant" (IOP) mines allows to counterbalance the stability of outcrops with the safety capsule in energy state without energy costs for maintenance. And the additional application of self-regulating roof-bolting which works by the principle of transmission of convergence energy, destroying the outcrop contour, into the zone with active divergence, and allows to use the maximum amount of the energy generated in the safety capsule of the mine working with the resources saving up to 37%.

The efficiency assessment of the developed resource-saving technologies implementation for mine workings support setting in Ukrainian ore mines is based on the method of local and general prime cost determination during ore deposits mining. The use of up to 86% of the relative amount of the energy concentrated in the massif around preparatory mine workings increases by exponential function the specific economic efficiency of preparatory operations. It allows to reduce the expenses up to 2029 UAH per 1 long meter during mine workings drivage with application of the self-regulating roof-bolting in mines of PJSC "Kryvorizkyi IOP".

The authors express their gratitude to Usatyi V.Yu., the production director of PJSC "Zaporizkyi IOP", Rychko V.S., the technical director of PJSC "Kryvorizkyi IOP", Pukhalskyi V.N., the chief engineer of SE 'Shidnyi Mining and Beneficiation Plant' for assistance in providing production data and implementation of the received results.

References

1. Shemyakin, E.I., Kurlenya, M.V., Oparin, V.N., Reva, V.N., Glushikhin, F.P., & Rozenbaum, M.A. (1992). *Yavlenie zonal'noy dezintegratsii gornykh porod vokrug podzemnykh vyrabotok.* Otkrytie No. 400, SSSR.
2. Kirichenko, V.Ya., Zvyagi'skiy, E.L., Lishin, A.V., Usachenko, B.M., & Khalimendik, Yu.M. (2002). *Yavlenie obrazovaniya peremeshchayushchikhsya narushennykh zon v napryazhennykh gornykh porodakh.* Otkrytie No. 188. Ukraine.
3. Baysarov, L.V., Il'yashov, M.A., Levit, V.V., Palamarchuk, T.A., Usachenko, V.B., & Yalanskiy, A.A. (2008). *Zakonomernost' samoorganizatsii gruntovykh i porodnykh massivov vokrug protyazhennykh podzemnykh vyrabotok.* Otkrytie No. 318. Ukraine.
4. Khomenko, O. (2012). Implementation of energy method in study of zonal disintegration of rocks. *Naukovyi Visnyk Natsionalnoho Hirnychoho Universytetu*, (4), 44-54.
5. Khomenko, O., Kononenko, M., & Bilegsaikhan, J., (2018). Classification of Theories About Rock Pressure. *Solid State Phenomena*, (277), 157-167. https://doi.org/10.4028/www.scientific.net/ssp.277.157
6. Kononenko, M., Khomenko, O., Sudakov, A., Drobot, S., & Lkhgva, Ts. (2016). Numerical modelling of massif zonal structuring around underground working. *Mining of Mineral Deposits, 10*(3), 101-106. http://dx.doi.org/10.15407/mining10.03.101
7. Sobolev, V., Ustimenko, Y., Nalisko, M., & Kovalenko, I. (2018). The macrokinetics parameters of the hydrocarbons combustion in the numerical calculation of accidental explosions in mines. *Naukovyi Visnyk Natsionalnoho Hirnychoho Universytetu*, (1), 89-98. https://doi.org/10.29202/nvngu/2018-1/8
8. Stupnik, N., Kalinichenko, V., & Pismennyi S. (2013). Pillars sizing at magnetite quartzites room-work. *Mining Of Mineral Deposits*, 11-15. https://doi.org/10.1201/b16354-4
9. Soboliev, V.V., Bilan, N.V., & Khalimendik, A.V. (2017). On formation of electrically conductive phases under electrothermal activation of ferruginous carbonate. *Naukovyi Visnyk Natsionalnoho Hirnychoho Universytetu*, (4), 53-60.
10. Lozynskyi, V., Saik, P., Petlovanyi, M., Sai, K., & Malanchyk, Ye. (2018). Analytical Research of the Stress-Deformed State in the Rock Massif Around Faulting. *International Journal of*

Engineering Research in Africa, (35), 77-88. https://doi.org/10.4028/www.scientific.net/JERA.35.77

11. Dychkovskyi, R.O., Lozynskyi, V.H., Saik, P.B., Petlovanyi, M.V., Malanchuk, Ye.Z., & Malanchuk, Z.R. (2018). Modeling of the disjunctive geological fault influence on the exploitation wells stability during underground coal gasification. *Archives of Civil and Mechanical Engineering*, *18*(4), 1183-1197. https://doi.org/10.1016/j.acme.2018.01.012
12. Khomenko, O.Ye., Sudakov, A.K., Malanchuk, Z.R., & Malanchuk, Ye.Z. (2017). Principles of rock pressure energy usage during underground mining of deposits. *Naukovyi Visnyk Natsionalnoho Hirnychoho Universytetu*, (2), 35-43.
13. Gorova, A., Pavlychenko, A., Borysovs'ka, O., & Krups'ka, L. (2013). The development of methodology for assessment of environmental risk degree in mining regions. *Annual Scientific-Technical Collection – Mining of Mineral Deposits*, 207-209. https://doi.org/10.1201/b16354-38
14. Astafiev, D., & Shapovalov, Y. (2013). On the question of implementation prospects of selective mining for exploitation unconditional coal seams. *Annual Scientific-Technical Collection – Mining of Mineral Deposits*, 313-316. https://doi.org/10.1201/b16354-58
15. Babets, D., Sdvyzhkova, O., Larionov, M., & Tereshchuk, R. (2017). Estimation of rock mass stability based on probability approach and rating systems. *Naukovyi Visnyk Natsionalnoho Hirnychoho Universytetu*, (2), 58-64.
16. Khomenko, O., & Maltsev, D. (2013). Laboratory research of influence of face area dimensions on the state of uranium ore layers being broken. *Naukovyi Visnyk Natsionalnoho Hirnychoho Universytetu*, (2), 31-37.
17. Stupnik, N.I., Kalinichenko, V.A., Kolosov, V.A., Pismenniy, S.V., & Fedko, M.B. (2014). Testing complex-structural magnetite quartzite deposits chamber system design theme. *Metallurgical and mining industry*, (2), 89-93.
18. Bondarenko, V., Symanovych, G., & Koval, O. (2012). The mechanism of over-coal thin-layered massif deformation of weak rocks in a longwall. *Geomechanical Processes During Underground Mining: School of Underground Mining 2012*, 41-44. https://doi.org/10.1201/b13157-8
19. Vladyko, O., Kononenko, M., & Khomenko, O. (2012). Imitating modeling stability of mine workings. *Geomechanical Processes During Underground Mining*, 147-150. https://doi.org/10.1201/b13157-26
20. Kononenko, M., & Khomenko, O. (2010). Technology of support of workings near to extraction chambers. *New Techniques And Technologies In Mining*, 193-197. https://doi.org/10.1201/b11329-32
21. Khomenko, O., Kononenko, M., & Myronova, I. (2013). Blasting works technology to decrease an emission of harmful matters into the mine atmosphere. *Annual Scientific-Technical Collection – Mining of Mineral Deposits*, 231-235. https://doi.org/10.1201/b16354-43
22. Khomenko, O., Sudakov, A., Malanchuk, Z., & Malanchuk, Ye. (2017). Principles of rock pressure energy usage during underground mining of deposits. *Naukovyi Visnyk Natsionalnoho Hirnychoho Universytetu*, (2), 35-43.
23. Belmas, I., Kolosov, D., & Onyshchenko, S. (2018). Stress-strain state of rubber-cable tractive element of tubular shape. *Naukovyi Visnyk Natsionalnoho Hirnychoho Universytetu*, (2), 60-69. https://doi.org/10.29202/nvngu/2018-2/5
24. Dreus, A.J., Sudakov, A.K., & Kozhevnikov, A.A. (2016). Investigation of heating of the drilling bits and definition of the energy efficient drilling modes. *Eastern-European Journal of Enterprise Technologies. Technologies*, *3*(7(81)), 41-46. https://doi.org/10.15587/1729-4061.2016.71995
25. Kolosov, D., Dolgov, O., & Kolosov, A. (2013). The stress-strain state of the belt on a drum under compression by flat plates. *Annual Scientific-Technical Collection – Mining of Mineral Deposits*, 351-357. https://doi.org/10.1201/b16354-65
26. Khomenko, O., Kononenko, M., & Myronova, I. (2017). Ecological and technological aspects of iron-ore underground mining. *Mining of Mineral Deposits*, *11*(2), 59-67. https://doi.org/10.15407/mining11.02.059

27. Dreus, A., Kozhevnikov, A., Lysenko, K., & Sudakov, A. (2016). Investigation of heating of the drilling bits and definition of the energy efficient drilling modes. *Eastern-European Journal of Enterprise Technologies*, *3*(7(81), 41-46. https://doi.org/10.15587/1729-4061.2016.71995
28. Dreus A.Yu., Sudakov, A.K., Kozhevnikov, A.A., & Vakhalin, Yu.N. (2016). Study on thermal strength reduction of rock formation in the diamond core drilling process using pulse flushing mode. *Naukovyi Visnyk Natsionalnoho Hirnychoho Universytetu*, (3), 5-10.
29. Sudakov, A., Dreus, A., Khomenko, O., & Sudakova, D. (2017). Analytical study of heat transfer in absorptive horizons of borehole at forming cryogenic protecting of the plugging material. *Naukovyi Visnyk Natsionalnoho Hirnychoho Universytetu*, (3), 38-42.
30. Khomenko, O., Kononenko, M., Myronova, I., Sudakov, A. (2018). Increasing ecological safety during underground mining of iron-ore deposits deposits. *Naukovyi Visnyk Natsionalnoho Hirnychoho Universytetu*, (2), 29-38. https://doi.org/10.29202/nvngu/2018-2/3
31. Bondarenko, V., Kovalevs'ka, I., & Fomychov, V. (2012). Features of carrying out experiment using finite-element methodat multivariate calculation of mine massif – combined support system. *Geomechanical Processes During Underground Mining*, 7-13. https://doi.org/10.1201/b13157-3
32. Kozhevnykov, A., Dreus, A., Baochang, L., & Sudakov, A. (2018). Drilling fluid circulation rate influence on the contact temperature during borehole drilling. *Naukovyi Visnyk Natsionalnoho Hirnychoho Universytetu*, (1), 35-42. https://doi.org/10.29202/nvngu/2018-1/14
33. Sudakov, A. Khomenko, O., Isakova, M., & Sudakova, D. (2016). Concept of numerical experimentof isolation of absorptive horizons by thermoplastic materials. *Naukovyi Visnyk Natsionalnoho Hirnychoho Universytetu*, (5), 12-16.
34. Fedko, M.B., Kolosov, V.A., Pismennyi, S.V., & Kalinichenko, Ye.V. (2014). Economic aspects of change-over to TNT-free explosives for the purposes of ore underground mining in Kryvyi Rih basin. *Naukovyi Visnyk Natsionalnoho Hirnychoho Universytetu*, (4), 79-84.
35. Khomenko, O., & Rudakov, D. (2010). The first Ukrainian corporative university. *New Techniques and Technologies in Mining*, 203-206. https://doi.org/10.1201/b11329-34
36. Gorova, A., Pavlychenko, A., Kulyna, S., & Shkremetko, O. (2012). Ecological problems of post-industrial mining areas. *Geomechanical Processes During Underground Mining*, 35-40. https://doi.org/10.1201/b13157-7
37. Stupnik, N.I., Fedko, M.B. Pismenniy, S.V., & Kolosov, V.A. (2014). Development of recommendations for choosing excavation support types and junctions for uranium mines of state-owned enterprise “SkhidHZK”. *Naukovyi Visnyk Natsionalnoho Hirnychoho Universytetu*, (5), 21-25.
38. Khomenko, O., Tsendjav, L., Kononenko, M., & Janchiv, B. (2017). Nuclear-and-fuel power industry of Ukraine: production, science, education. *Mining of Mineral Deposits*, *11*(4), 86-95. https://doi.org/10.15407/mining11.04.086
39. Fuławka, K., Mertuszka, P., & Pytel, W. (2018). Monitoring of the stability of underground workings in Polish copper mines conditions. *E3S Web Of Conferences*, (29), 00008. https://doi.org/10.1051/e3sconf/20182900008
40. Pivnyak, G., Dychkovskyi, R., Smirnov, A., & Cherednichenko, Y. (2013). Some aspects on the software simulation implementation in thin coal seams mining. *Energy Efficiency Improvement of Geotechnical Systems*, 1-10. https://doi.org/10.1201/b16355-2

Influence of man-made edaphotopes of the spoil heap on biota

Vasyl Popovych[1*], *Oleksandr* Kuzmenko[2], *Andrii* Voloshchyshyn[1], and *Mykhailo* Petlovanyi[2]

[1]Lviv State University of Life Safety, 35 Kleparyvska Str., 79007 Lviv, Ukraine
[2]National Mining University, Department of Underground Mining, 19 Yavornytskoho Ave., 49005 Dnipro, Ukraine

Abstract. The aim of the work is to investigate the influence of man-made edaphotopes of colliery spoil heap on biota by bioindication method. The physicochemical properties of edaphotopes were investigated: humidity, substrate temperature, humus content at different intervals of their depth and on different sections of the waste heap. The storage technologies of mineral waste in the dumps disregard its further use and transforming that occurs over time. It was revealed that on the northern slope exposures the greatest amount of moisture was concentrated and temperature regimes were acceptable for the activity of the soil biota, which was proved by the largest amount of biomass. New dependences of humus content in edaphotopes of the heap in different intervals of their depth from biomass of humus-forming organisms were established. Slight successional processes on the surface of the heap were observed, where young flora began to develop. This positive phenomenon will contribute to the humus of the genetic horizons and the development of tree and shrub vegetation. Favourable conditions for the development of neo-terrain and the formation of a new landscape in mining regions were confirmed.

1 Introduction

The development of mineral deposits is accompanied by mining works in rocks that doesn't contain commercial components, or their number is not profitable for different reasons. Mining technologies and technological parameters of development systems do not provide for the separate extraction and storage of mineral waste in heaps with their further use in other sectors of the economy over time. A lot of waste heaps are located in the Donetsk coal basin, where the extraction of coal beds is conducted in complicated geological conditions from ultrathin and thin (0.65 – 1.2 m) layers [1], which necessitates the continuous removal of rock and coal during development workings.

Thus, coal of the humite class, and sometimes sapropelitic-humic layers and rocks, represented by argillites, siltstones, sandstones, more rarely by limestones, fall into the waste heaps. These rocks are stored on the ground surface in significant volumes near or even close to mining enterprise, thus forming a specific relief. Waste heaps form a spoil tips

* Corresponding author: popovich2007@ukr.net

up to 60 – 80 m high ridge-like (92%), as well as flat (8%). The rocks are heterogeneous in grain size distribution – from clay particles to boulders – containing wood, metal objects (pipes, wires, etc.) [2]. Mass of colliery spoils and shale deposits exceeds 18.7 billion tons.

In [3, 4] the authors mention that besides the planned waste of mining production, significant additional volumes of rock and minerals that peel off as a result of the stress-deformed rock and artificial solid mass conditions, are added to them. All of them fall into mine roadway and are winded on the ground surface under mining operations. Furthermore waste heaps, sludge and sediment located near the coal-preparation plants should be added. These rocks are more broken than in the waste heaps of mining enterprises. Large areas of the ground surface are taken out of circulation for their storage. For example, in the waste heaps of the Lviv-Volyn region 40.0 million cubic meters of gas have been accumulated. The area occupied by them is the second largest (320 hectares) after the Pokrovskyi district of the Donetsk region. Coal-bearing strata occupy 15% of the continent's territory [2]. Distribution of waste heaps of coal mines of Lviv-Volyn region by their volume and area is shown in Fig. 1.

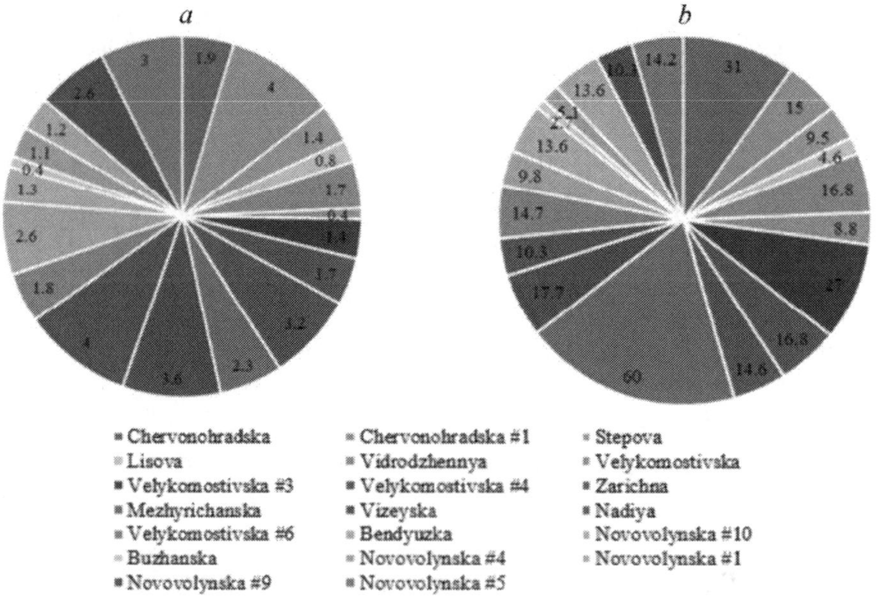

Fig. 1. Distribution of waste heaps of coal mines of Lviv-Volyn region by their volume in million m^3 (a) and area in hectares (b).

Thus, an enclave of environmental pollution is formed in a separate industrial area, which by its volume affects the general sanitary state of the region and the distribution of atmospheric flows. Donetsk and Lviv-Volyn coal basin with the developed infrastructure of mining enterprises create a threatening situation that may transform into the zone of ecological disaster. The chemical composition of the spoil tip rocks is close to the rocks of the coal-bearing series of the Lviv-Volyn Basin [5] (Table 1).

This chemical composition of rocks is also characteristic of other waste heaps of the mines of this region. The rock is broken and most of the chemical elements that form complex geological structures interact with the atmosphere and sediments. Besides to the formation of new chemical compounds in the waste heaps, mined gases of CH_4, CO_2, SiO_2, CO, TiO_2 and others, which rise to the ground surface through cracks and tectonic fractures, or during mining operations or carbonization of coal layers are added [6, 7]. In the initial, active phase of the waste heaps, living creatures are almost absent, but there are transitional processes of rocks oxidation (the temperature of spoil tips combustion is 900 – 1300 degrees), their transferring into a resting phase, which creates favorable conditions for the rebirth of invertebrate organisms.

Table 1. The chemical composition of the rocks in the waste heap of Chervonohradska mine [5].

Mine rocks	SiO_2	Al_2O_3	TiO_2	FeO	CaO	MgO	K_2O+ Na_2O	H_2O	SO_3	P_2O_5
Sandstone	75-85	5.7-9.7	0.2-0.6	2.7-4.0	0.1-0.5	0.4-1.3	0.9-1.5	1.8-4.0	0.1-0.2	0.0-0.1
Claystone	49-58	9.6-22.5	0.4-1.0	8.8-9.9	0.7-3.5	1.2-2.0	1.2-3.0	1.9-4.5	0.2-6.5	0.2-2.0
Siltite	49.8-70	16.2-23.0	0.8-1.5	1.2-1.5	0.7-1.4	0.8-1.4	1.5-3.9	0.5-1.5	0-0.3	0.1-0.5
Burnt rock	60	22.7	1.0	8.0	0.8	1.5	2.5	2.5-	0.95	0.1
Weathered rock	38	16.5	0.8	10.5	0.8	1.9	2.2	27.9	0.55	0.6
Unweathered rock	38.5	17.4	0.7	12.0	0.8	1.5	2.3	24.8	0.35	0.2

2 Aim, tasks and methods of research

The aim of the work is the bioindication of man-made edaphotopes of coal mines waste heaps by the activity of its biota. According to the stated aim, the following tasks were set:

– to investigate the environmental safety in the area of coal mines waste heaps;

– to identify the main factors of increase of environmental hazard caused by waste heaps;

– to establish the basic physical and chemical properties of the rock refuse;

– to investigate the conditions for the development of land biota in the area of waste heaps influence.

The object of the study is the bioindication of the edaphic conditions of the spoil heaps for the development of the land biota.

The subject of the study is the peculiarities of the biomass formation, depending on the physico-chemical changes of the edaphotopes on the surface of the spoil heaps.

Methods of the research. In this work bioindicative methods of research, system analysis and methods of logistic constructions are used. The study of the life activity of the main species of edaphotopes biota was carried out according to the methods of M.S. Gilyarov [8]. The species of earthworms, as a bioindicator of edaphotopes, were determined by T.S. Perel's method [9].

For soil-zoological studies direct methods are used to determine the number of soil animals in the soil (on 1 m^2). We have used the method of lit-by-lit digging and investigating of soil samples. The size of the samples was 0.25 m^2 (50×50 cm). Samples were taken up to the lower level of invertebrates in the soil.

The sampling process is as follows: first, the area of the sample is marked by hammering the pegs in the corners of this square, then a fencing tape is passed through them. Then from the edges of the measured area the mortality and litter are bent in different directions. Dense fabric is spread near sample, where the soil selected from the sample is displaced.

At first, mortality, leaves, plant residues are displaced from the sample. They are hand-picked and all fallen individuals are taken into account.

The research was carried out in four - one digs the soil, two hand-pick the sample, one collects and describes the identified biotypes.

Instruments and materials: humidity meter MG-44, pH-meter KC-300V, portable meteorological station "Kestrel-4000", dense fabric, fencing tape.

3 Main part

The research was carried out in the spoil heaps area of the Mezhyrichanska mine, that functioned in the Chervonohrad mining area (Lviv-Volyn coal basin).

Waste storage was carried out on two spoil heaps (1 – reclaimed, nonfunctioning). Currently there is 1 spoil heap. It should be noted that on the reclaimed heap natural reforestation processes take place. They are accompanied by the development of pioneer species – drooping birch, Scotch pine and bush grass. As the rock refuse is used for road construction frequent caving and shearing of the spoils are observed on the refuse dumps. An important negative environmental factor is that the mined rocks are weathered which causes the air polluting, rainwash and polluting of the soils.

Thus, the waste heaps of the Mezhyrichanska mine due to the violation of safe operating conditions are a powerful man-made factor of environmental pollution. On the spoil heaps, like on the devastated landscapes, specific edaphic and climatic conditions for the development of vegetation were established. The physicochemical and mechanical properties of edaphotopes in different parts of the waste heaps can vary significantly. The reason for such a phenomenon is the violation of the operation requirements of the waste heaps, the unevenness of spreading (or absence) of ground mixtures, chaotic disposal and accumulation of filtrate, etc. In such conditions, the successional processes proceed slowly, and in places where the land is aligned with bulldozers, they are absent at all. Detritophags – the representatives of the soil fauna – play a significant role in humus generation and the edaphotope formation. The ecological conditions of the degraded landscapes have a powerful effect on the activity of the representatives of the soil fauna [10].

One of the most common bioindicators of environmental conditions are the following criteria: the number of individuals, the survival capacity, the degree of covers damage, the intensity of metabolism [11, 12]. In some studies of technogenically polluted edaphotopes, a criterion of the rate of digging in the soil is used [13]. Investigation of the living organisms' activity will help to develop the processes of remediation and revegetation on the spoil heaps in order to reduce the harmful effects on the environment. Climatic and physico-chemical conditions are presented in scientific works [14, 15].

To study the biota's activity in the edaphotopes of the spoil heaps and adjoining area 5 sites were selected on the waste heap surface:

1 – top;

2 – northern slope exposition (middle tier);

3 – southern slope exposition (middle tier);

4 – western slope exposition (middle tier);

5 – east slope exposure (middle tier).

The chosen type of worm is *Lumbricus terrestris.*

It should be noted that according to the research methodology, the level of activity was investigated at 2 levels – in the depths of 0 – 10 cm and 10 – 20 cm. Background values were collected in the area of meadow soil, which is located 1.5 km southward of the investigated waste heap.

As a result of erosion processes on the slopes the capacity of the humus horizon is lower, the content of phosphorus and potassium is lower, the hydrolytic acidity is higher, and the soil saturation at the bottom is low. An important factor of the medium is the acidity of the soil, the optimal ratio of the various species on agricultural land is almost

 https://doi.org/10.1051/e3sconf/20186000010

neutral [16].

The pH reaction of the waste heap edaphotopes at a depth of 10 cm was slightly acidic and acidic. The highest humus index was found on the northern and western slope exposures (3.36 and 3.27%). In the background soils, the humus content was 7.25%. The highest humidity of the edaphotope was detected at the northern slope exposure – 86.7%. The lowest humidity is on the southern exposure (65.4%). The temperature of the substrate was the highest in the southern region – + 8.5 °C, and the lowest in the northern region – + 6.5 °C (Fig. 2).

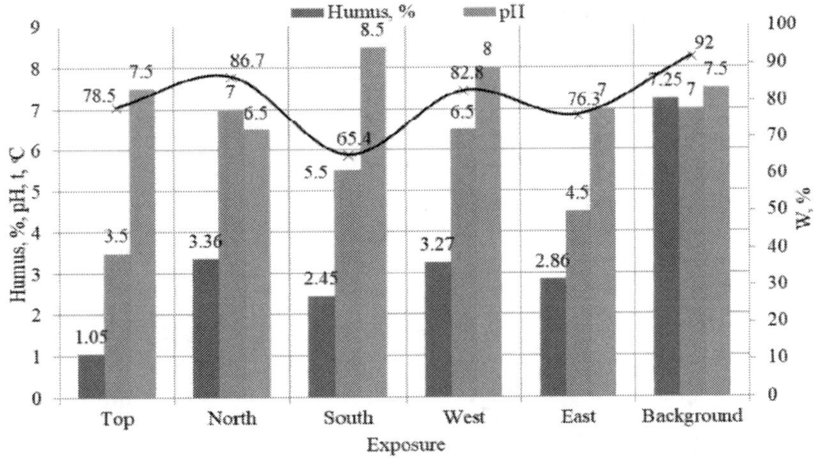

Fig. 2. Physicochemical properties of the waste heap edaphotope at a depth of 0 – 10 cm.

The pH reaction of the waste heap edaphotopes at a depth of 20 cm was acidic. The highest humus index was found in the northern and western slope exposures (1.15% and 1.1%). In the background soils (depth up to 20 cm) the humus content was 7.14%. The highest humidity of the edaphotope was detected on the northern slope exposure – 74.8%. The lowest humidity is in the eastern exposure (54.3%). The temperature of the substrate was the highest in the southern region – +6.5 °C, and the lowest in the eastern region was +5 °C (Fig. 3).

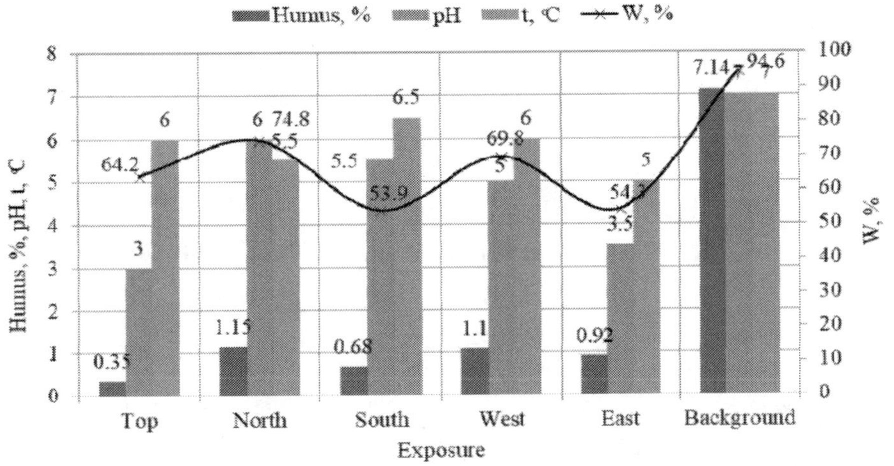

Fig. 3. Physicochemical properties of the waste heap edaphotope at a depth of 10 – 20 cm.

Taking into account the acidic reaction and the low temperature of the substrates of the earthworms colonies of in the waste heaps edaphotopes are inconsiderable. These data are supported by the following indicators. Since, on the northern slope exposure the highest humidity and acceptable temperature rates for the activity of earthworms are set, where their largest colonies are observed: at a depth of 10 cm – 18 pcs. with a biomass of 9.6 g/m^2; at a depth of 20 cm – 7 pcs. with a biomass of 4.8 g/m^2 (Fig. 4 and 5). In the southern and eastern exposures of the pine the investigated zoocenosis was not observed. The reason for this phenomenon is the drying of the substrate surface the by wind and solar radiation. In the background areas, the indicators of activity significantly exceed the conditions of artificial substrates and are as follows: at a depth of 10 cm – 22 pcs. with a biomass of 10.7 g/m^2; at a depth of 20 cm – 25 pcs. with a biomass of 12.1 g/m^2.

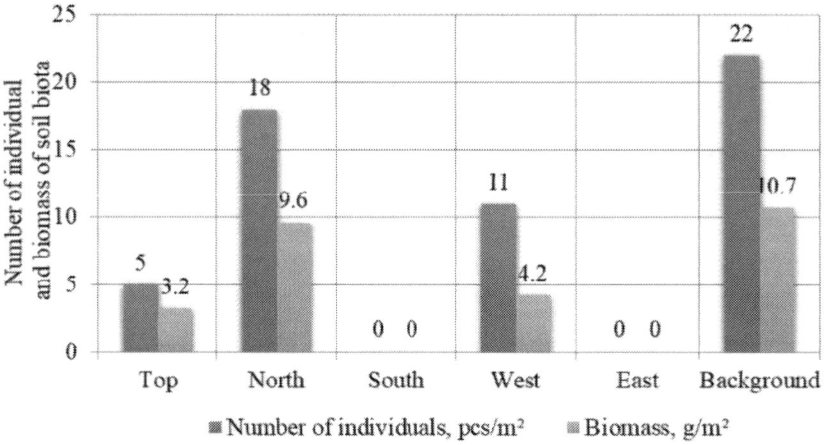

Fig. 4. The number of individuals of earthworms and their biomass at a depth of 0 – 10 cm.

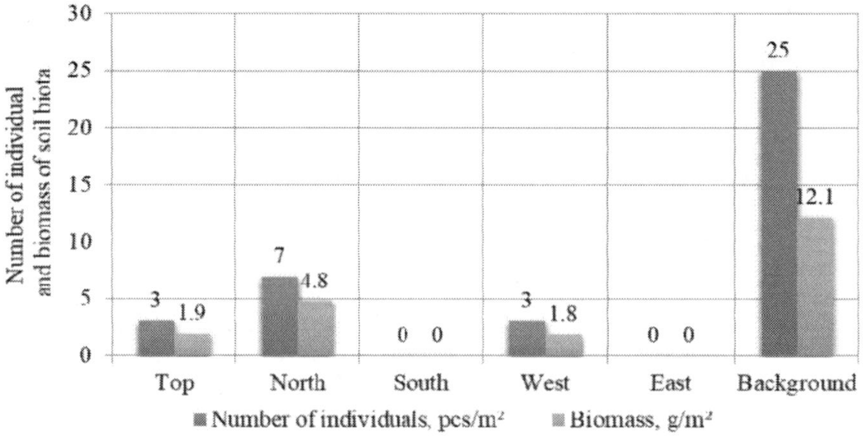

Fig. 5. The number of individuals of earthworms and their biomass at a depth of 10 – 20 cm.

The obtained data on the activity of the earthworms and the content of humus in of their substrate give an opportunity to determine the dependence of humus content on the number of individuals (Fig. 6).

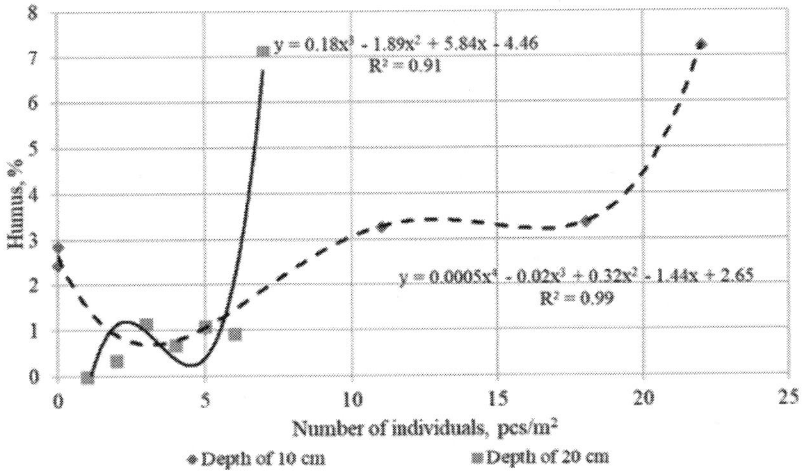

Fig. 6. Dependence of the humus content in the edaphotopes of the waste heap from biomass on the area of 1 m^2.

The most important reasons for the absence of populations of living organisms in edaphotopes at different sites of the spoil heap are:

– high acidity of the waste heap edaphotopes;

– toxicity of the waste heap edaphotopes;

– high wind speed;

– slow rise of the neo-terrain and the light successional processes on the surface of the waste heap;

– landfalls and plump holes on the spoil heaps.

So, it can be stated that the technologies of underground mining are conducted without selective separation of rocks during their storage and the formation of artificial deposits. Gross waste storage is carried out without taking into account the chemical processes and the environmental activity. The forms and technological parameters of the storage of colliery waste don't take into account the further use of mineral raw materials for getting new biological resources for other important sectors of the economy.

4 Conclusions

It is recommended the separate storage of rocks on the ground surface near the mines, taking into account further processes of physical and chemical transformation for the effective use of improvements in the economy and landscape formation, causing the reduce of the environmental impact.

The zonal formation of the edaphic-climatic conditions for the development of flora has been established. It gives an opportunity to determine the order and way of application of newly created products on burned spoil heaps in various fields of the economy and converts the waste storage into secondary biological resources of the ecosystem. On the spoil heaps, as well as on the degraded landscapes, specific edaphic and climatic conditions for the development of vegetation were formed, and the physicochemical and mechanical properties of edaphotopes in different parts of the heaps may differ significantly. The reason for such a phenomenon is the violation of operation requirements of the waste heaps, the unevenness of spreading (or absence) of ground mixtures, chaotic disposal and accumulation of filtrate, etc.

It was found that the highest humus index is in northern and western slope exposures (3.36 and 3.27%). In the background soils, the humus content was 7.25%. The highest humidity of the edaphotopes was detected on the northern slope exposure – 86.7%. The lowest humidity is on the southern exposure (65.4%). The temperature of the substrate was the highest on the southern section is +8.5 °C, and the lowest in the northern area is +6.5 °C. The presence of sedimentary rocks and humus coal in the mining waste during the oxidation process creates favorable conditions for the development of neo-relief and the new landscape formation in mining regions.

The formation conditions of the waste heaps edaphotopes are revealed. It is important for development of their mining and biological reclamation. Thus, during the mining-engineering recultivation is the possibility of adjusting the parameters of the fertile soil layer, which is being applied in the last thing, taking into account the properties of the already formed edaphotopes on the surface of the waste heap as a result of the complex physical and chemical transformations. At the stage of biological reclamation, while choosing of the artificial forest plantations, attention should be paid to certain species of tree and shrub vegetation adapted to the waste heap edaphotopes and have a high regenerative capacity.

The authors express their gratitude to Prof. Kucheriavyi V.P. for valuable advices and ideas expressed during the researches.

References

1. Kuz'menko, A., Pochepov, V., & Ryabychev, V. (2010). Dependence of effectiveness of development of mining operations on processibility of coal seams deposits with thickness of 1.2 m. *New Techniques and Technologies in Mining*, 51-55. https://doi.org/10.1201/b11329-10
2. Maksimovich, N.G., & Gorbunova, K.A. (1991). Geokhimicheskie izmeneniya geologicheskoy sredy pri razrabotke ugol'nykh mestorozhdeniy. *Geologiya i razvedka*, (5), 137-140.
3. Kuz'menko, A., Furman, A., & Usatyy, V. (2010). Improvement of mining methods with consolidating stowing of iron-ore deposits on big depths. *New Techniques and Technologies in Mining*, 131-136. https://doi.org/10.1201/b11329-22
4. Kuz'menko, O., Petlyovanyy, M., & Stupnik, M. (2013). The influence of fine particles of binding materials on the strength properties of hardening backfill. *Annual Scientific-Technical Collection – Mining of Mineral Deposits*, 45-48. https://doi.org/10.1201/b16354-10
5. Knysh, I.B. (2015). Heokhimichni doslidzhennia porid terykonu shakhty "Chervonohradska" Lvivsko-Volynskoho baseinu. V *Materialakh XI naukovo-praktychnoi konferentsii "Suchasna heolohichna nauka i praktyka v doslidzhenniakh studentiv i molodykh fakhivtsiv"* (pp. 82-86). Kryvyi Rih: Kryvorizkyi natsionalnyi universytet.
6. Falshtynskyi, V., Saik, P., Lozynskyi, V., Dychkovskyi, R., & Petlovanyi, M. (2018). Innovative Aspects of Underground Coal Gasification Technology in Mine Conditions. *Mining of Mineral Deposits*, *12*(2), 68-75. https://doi.org/10.15407/mining12.02.068
7. Khomenko, O., Kononenko, M., Myronova, I., & Sudakov, A. (2018). Increasing ecological safety during underground mining of iron-ore deposits deposits. *Naukovyi Visnyk Natsionalnoho Hirnychoho Universytetu*, (2), 29-38. doi:10.29202/nvngu/2018-2/3
8. Gilyarov, M.S. (1975). Uchet krupnykh pochvennykh bespozvonochnykh (mezofauny). *Metody pochvenno-zoologicheskikh issledovaniy*, 12-29.
9. Perel', T.S. (1976). Kriticheskiy analiz sistemy Lumbricidae (s opredelitel'noy tablitsey rodov fauny SSSR). *Zoologicheskiy zhurnal*, *36*(6), 823-836.
10. Kucheriavyi, V.P. (2010). *Zahalna ekolohiia*. Lviv: Svit.
11. Kibardin, V.M., & Zherebtsov, A.K. (1990). Vliyanie neftyanogo zagryazneniya na dozhdevykh chervey. V *Materialakh VI nauchnoy konferentsii molodykh uchenykh Kazanskogo instituta*

biologii (pp. 230-232). Kazan': Kazanskiy filial Akademii nauk SSSR.

12. Popovych, V.V. (2015). Bioindykatsiia edafichnykh umov smittiezvalyshch za dopomohoiu vyvchennia zhyttiediialnosti doshchovykh cherviakiv. V *Tezakh dopovidei III Mizhnarodnoi naukovo-praktychnoi konferentsii "Novitni dosiahnennia biotekhnolohii ta nanofarmakolohii"* (pp. 93-94). Lviv: Mehaprynt.
13. Kuz'min, Ye.V. (2008). Skorost' zakapyvaniya i vyzhivaemost' dozhdevykh chervey v usloviyakh neftyanogo zagryazneniya razlichnoy intensivnosti. V *Materialakh IV nauchno-prakticheskoy konferentsii "Aktual'nye problemy ekologii Yaroslavskoy oblasti"* (pp. 297-301). Izdanie VVO REA.
14. Popovych, V.V. (2008). Doslidzhennia fizyko-khimichnykh vlastyvostei hruntiv i perehorilykh porid na terykonakh Novovolynskoho hirnychopromyslovoho raionu. *Naukovyi Visnyk Natsionalnoho Lvivskoho Tekhnichnoho Universytetu Ukrainy, 18*(12), 258-264.
15. Popovych, V.V. (2009). Vplyv klimatychnykh umov na rozvytok roslynnosti tekhnohennykh landshaftiv Maloho Polissia u zymovyi period. *Naukovyi Visnyk Natsionalnoho Lvivskoho Tekhnichnoho Universytetu Ukrainy, 19*(3), 37-42.
16. Petrov, V.S. (1946). Aktivnaya reaktsiya pochvy (pH) kak faktor rasprostraneniya dozhdevykh chervey (Lumbricidae, Oligocheta). *Zoologicheskiy zhurnal, 25*(2), 107-110.

Explosive wave propagation in the presence of antiseismic protective curtain

Viktor Kravets[1*], *Nataliia* Remez[1], *Andrii* Kovtun[1], and *Azer* Shukiurov[1]

[1]National Technical University of Ukraine "Igor Sikorsky Kyiv Polytechnic Institute", Department of Geobuilding and Mining Technologies, 37 Peremohy Ave., 30056 Kyiv, Ukraine

Abstract. The objective of this work is to study theoretically the stress wave formation in the borehole-to-borehole space and on the boundary with the borehole contour set by creating a directed fracture system in it. The features of an interaction of the weak stress wave at a large-scale blasting in the short-delay blast mode (SDB) with the massif and the protective contour set of boreholes, are considered. It is noted the effect of the contour set parameters on wave processes. The constructive solutions are proposed in order to reduce the dynamic loadings in the peripheral rock massif. The proposals consist in applying a combined construction of garland charges to form the seismo protective structure that consists of a contour fracture and one-sided disrupted zone within the boundaries of the blasted block.

1 Introduction

The development of a seismically secure switching circuit for detonating system of large-scale blasting involves the primary blast of a contour set of boreholes [1], which reduces the loading from the last-line blast on the peripheral rock massif, comes out from the simplified representation about charge groups interaction. In this case, three consecutive mechanisms of interaction are considered – at the level of interference of the stress waves, at the level of the fracture system development and at the level of contact between the rocks blocks, bounded by adjacent groups. The first level provides the preliminary disturbance of microstructural bonds in the massif being destroyed, the development of microdefects, and the creation of a stress state in the massif. The following two levels practically solve the problem of separating the moving massif into standard sized parts. However, under real conditions of large-scale blasting, the processes of charge interaction and the resulting dynamic phenomena [2] become more complicated, acquiring additional features that should be studied.

2 Prerequisites for setting the research tasks

When implementing a large-scale blasting in the SDB mode, a relatively long time-extended oscillation field, formed in three stages, is actually created in the rock massif.

* Corresponding author: kravets@geobud.kiev.ua

These stages involve the blast of borehole charges in a certain sequence.

The first stage is realized during hundreds of microseconds in the process of sequential dynamic loading of the massif during the detonation decomposition of a separate elongated charge in the group and the radiation from shock wave at its contact with the rock massif. Taking into account that the propagation velocity of the detonation wave in the charge is of 4 – 5 thousand m/s for modern explosives, at a charge length of 10 m, its detonation decomposition will be about 500 microseconds. With the parameters of the system of borehole charges 5×5 m, during this time, the stress wave front at the level of the intermediate detonator (gunner) activation at its velocity in the hard rock also about 5 thousand m/s, will cross the 2 next groups of charges in the SDB scheme, which should be blasted much later – in tens of milliseconds. In this case, the stress wave front is located at an angle of 45° to the horizontal. Accordingly, there are changes in geometry of the rock masses displacement and interaction, driven by the charges in the set, as well as adjacent groups.

Thus, in the interval between two adjacent groups of charges due to non-simultaneous occurrence of the stress wave, the time difference in dynamic "pre-destruction" can be about 500 microseconds, and the time of the first arrival of stress front on an adjacent group will be 250 microseconds.

The second mode is related to the sequence of blasting the borehole charges in a group. Based on the features of the charge system assembly, the first and last charge in the group only theoretically should be blasted simultaneously. In fact, these charges are activated in sequence or along a common segment of the detonating cord, as well as in case of application of individual segments of the detonating waveguide of any modern initiation system of the "NONEL" type. In the latter variant, each subsequent charge in the group at a waveguide detonation velocity of 2 thousand m/s will be activated through 2.5 milliseconds relative to the previous one. Together, the whole group of 5 charges will be activated in 10 milliseconds.

The third mode assumes a consistent blasting of the charge groups with a deceleration within a few tens of milliseconds (from 20 to 70 – 100 ms).

3 Material and research results

The processes of blasting in mining and construction are accompanied by the propagation of stress waves outside and within the boundary of the massif being destroyed, which along with the massif disruption to the specified granulometric composition, have a harmful effect on the surrounding buildings, mine workings and the massif itself in the form of elastic-plastic (residual) and elastic deformation [3]. With account of the danger of seismic phenomena within the boundary of the bench which should be destroyed, it is considered appropriate, if possible, to protect the massif outside the blast block from damage, which in the future will lead to a weakening of the massif adjacent to disintegration of rock. These phenomena will either be a source of increased oversized output, or will need to increase the explosives consumption. Consequently, a decrease in the seismic effect outside the massif being blasted will reduce the explosive materials (EM) consumption.

In the specific conditions of production, the question arises of developing an integrated approach to managing the explosion energy – increasing the fraction of energy consumed for destruction, and reducing the seismic waves energy. One of such effective method of influencing the result of large-scale blasting is shielding [1].

The essence of the shielding method is in the formation on the boundary of the block, which should be destroyed, of the contour set of boreholes designed in two variants – either with the formation of a continuous mono gapping in the plane of contour charges system, or a plane zone of intense fracturing. The contour set being a curtain on the boundary of the surrounding massif from seismic oscillations, which reflects, scatters and refracts the waves generated by the blast. Outside the shield, the refracted wave carries only a part of the energy into the medium. It is important to note that the energy of the reflected wave, returning to the destroyed massif, increases the destructive effect of the blast.

Physical phenomena that contribute to the shielding process have not been studied sufficiently, therefore, it is proposed to apply the shields in a complex, that is, to apply the common and new types of shields with altered geometric, kinematic and dynamic parameters in order to increase the effectiveness of blasting operations. Solving problems on optimal parameters determination of such shields will ultimately increase the blast destruction effectiveness and reduce the seismic effect in the protected zone.

Such shielding consists in creating on the boundary of a blasted block, a zone of broken rock with a correspondingly reduced bandwidth throughput for elastic waves of sufficiently high intensity. In general, the effectiveness of destruction and the degree of shielding is estimated by the ratio of mass velocities in the massif with and without the shield at the equal distances from the charge.

To assess the capabilities of such a shield, let us consider in a simplified form the interaction of two adjacent boreholes filled with a gas-generating emulsion substance, wherein the first charge performs the role of active, that is detonating charge, the second performs the passive role.

In the first place, the patterns are investigated of stress wave propagation into a rock massif if to blast the cylindrical explosive charge, as well as its interaction with the material filling the adjacent borehole.

The movement of the detonation products, the mine rock and explosive materials are studied within the framework of continuum mechanics [4]. This motion is described by the laws of momentum and mass conservation. For the case of axial symmetry, they have the form [5]:

$$\frac{\partial \sigma_{rr}}{\partial z}+\frac{\partial \tau_{rz}}{\partial r}+\frac{\tau_{rz}}{r}=\rho\frac{du}{dt},\ u=\frac{dz}{dt}; \tag{1}$$

$$\frac{\partial \tau_{rz}}{\partial z}+\frac{\partial \sigma_{zz}}{\partial r}+\frac{\sigma_{zz}-\sigma_{\theta\theta}}{r}=\rho\frac{dw}{dt},\quad w=\frac{dr}{dt}; \tag{2}$$

$$\frac{1}{V}\frac{dV}{dt}=\frac{\partial u}{\partial z}+\frac{\partial w}{\partial r}+\frac{w}{r}; \tag{3}$$

$$\sigma_{zz}=S_{zz}-P,\ \sigma_{rr}=S_{rr}-P,\ \sigma_{\theta\theta}=S_{\theta\theta}-P; \tag{4}$$

$$P=\frac{1}{3}(\sigma_{rr}+\sigma_{\theta\theta}+\sigma_{zz});\ V=\frac{\rho_0}{\rho}, \tag{5}$$

where z, r are the coordinates; t is the time; $\sigma_{rr}, \sigma_{\theta\theta}, \sigma_{zz}$ are the normal stresses; τ_{zr} is tangential stress; $S_{zz}, S_{rr}, S_{\theta\theta}$ are the components of the stress tensor deviator; P is the average hydrostatic pressure; ρ is the density; u, w are velocity components, V is the specific volume.

The extension of the detonation products (DP) occurs by the binomial isentrope, the equation of which is proposed in [6]:

$$P^n=A\cdot\rho^n+B\rho^{\gamma+1}, \tag{6}$$

where A, B, n, γ are the constants for this explosive material.

For mine rocks, the ratios between stresses and deformations are written on the basis of the differential theory of plasticity [7]. The values of the deformation tensor components are represented as the sum of the elastic and plastic deformations:

$$\varepsilon_{rr} = \varepsilon'_{rr} + \varepsilon''_{rr},\ \varepsilon_{zz} = \varepsilon'_{zz} + \varepsilon''_{zz},\ \varepsilon_{rz} = \varepsilon'_{rz} + \varepsilon''_{rz}. \tag{7}$$

The connection between the stress and elastic deformations is established by the formulas of the generalized Hooke's law for isotropic material:

$$\sigma_{rr} = \frac{E}{1-\nu^2}\left(\varepsilon'_{rr} + \varepsilon''_{zz}\right),\ \sigma_{zz} = \frac{E}{1-\nu^2}\left(\varepsilon'_{zz} + \nu\varepsilon'_{rr}\right),\ \tau_{rz} = G\varepsilon'_{rz}, \tag{8}$$

where E, G, ν are the physical and mechanical characteristics of the mine rock.

In the case of plastic deformations, the ratio between the corresponding deformations and stresses is written as:

$$d\varepsilon''_{rr} = \lambda S_{rr},\ d\varepsilon''_{zz} = \lambda S_{zz},\ d\varepsilon''_{rz} = \lambda S_{rz} \tag{9}$$

when using the plasticity condition with linear kinematic displacement:

$$S_{rr}^2 + S_{zz}^2 + 2S_{rz}^2 \le \frac{2}{3}\sigma_T^2, \tag{10}$$

where λ is the coefficient of displacement; σ_T is the ultimate strength.

The emulsion explosive materials (EEM) of a cylindrical charge is modeled by two-component nonlinear-elastic medium, which consists of air and liquid components [8]. It is believed that under the action of dynamic loading, each component has the same pressure and moves at the same velocity. At atmospheric pressure, the density of EM is determined by the formula:

$$\rho_0 = \alpha_1\rho_{10} + \alpha_2\rho_{20},\ \alpha_1 + \alpha_2 = 1, \tag{11}$$

where a_1, a_2 and ρ_{10}, ρ_{20} is the volumetric content of components and their density.

The equation of volumetric compression of nonlinear-elastic two-component medium has the form:

$$\frac{\rho_0}{\rho} = \sum_{i=1}^{2} \alpha_i \left[\frac{\gamma_i(P-P_0)}{\rho_{i0}c^2_{i0}} + 1\right]^{-1/\gamma_i}, \tag{12}$$

where c_{i0}, γ_i is the sound velocuty and the index of isentrope of the i-th component.

The initial conditions of the problem are the following:

$$u = 0, w = 0, P = P_n, \rho = \rho_n \text{ at } z < l, r < r_0; \tag{13}$$

$$u = 0, w = 0, P = P_0,\ \rho = \rho_0 \text{ at } z > l, r > r_0. \tag{14}$$

The boundary conditions are:

$$u_{DP} = u_{RM}, P_{DP} = P_{RM}, \rho_{DP} \ne \rho_{RM} \text{ at } z = l; \tag{15}$$

$$w_{DP} = w_{RM}, P_{DP} = P_{RM}, \rho_{DP} \ne \rho_{RM} \text{ at } r = r. \tag{16}$$

The DP and RM are the indices of the detonation products and rock mass, respectively.

To solve the problem, the finite difference method is chosen in a moving Lagrangian coordinate system with a moving grid, which is automatically extended at the end of each computational cycle. The finite-difference scheme of the 'cross' type of the second order of accuracy [5] was used. The algorithm and the program for the PC were developed. In the numerical solution of the problem, a blast of EEM charge in granite was considered.

The detonation characteristics of the EEM are the following: $P_n = 3.9 \cdot 10^9$ Pa, $\rho_n = 1200$ kg/m^3, $D = 5000$ m/c, $n = 1.4$.

The parameters for the state equation of the DP (6) are as follow:

$$A = 56.68,\ B = 0.68 \cdot 10^5,\ n = 2.482,\ \gamma_0 = 0.242.$$

The cylindrical charge has a diameter of 0.2 m and a length of 15 m.

The physical and mechanical parameters of the mine rock are as follows:

$$\rho_0 = 2650 \text{ kg/m},\ c_p = 4170 \text{ m/c},\ c_s = 2800 \text{ m/c},\ E = 4.6 \cdot 10^{10} \text{ Pa},\ \nu = 0.09.$$

Fig. 1 shows the theoretical dependence of the initial pressure on the time in the mine rock at the boundary with the detonation products during the blast of the EEM cylindrical charge. As shown in Fig. 1, the pulse has a characteristic shape for the shock wave. According to the theory, the maximum pressure in the shock wave is achieved almost instantaneously (in fact, about 200 microseconds), but the time of the stress decrease to 0 reaches 100 milliseconds.

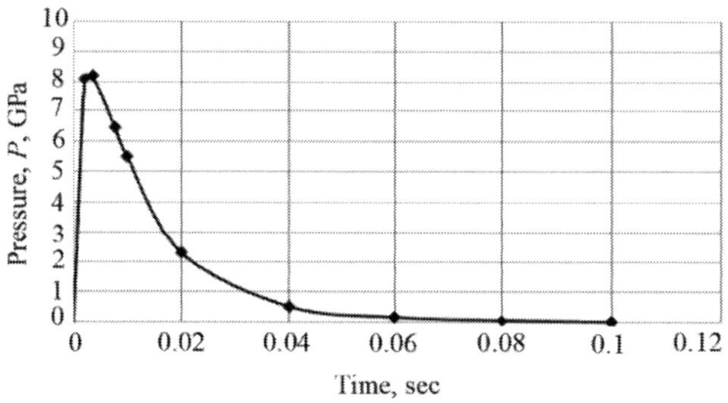

Fig.1 The curve of pressure in the mine rock at the boundary with the detonation products of the EEM cylindrical charge.

Since the energy of the explosive pulse is commensurate with the area of the curve shown in Fig. 1, the pulse existence time is extremely important. A known means for reducing the maximum pressure level in the initial pulse with the simultaneous lengthening of the pulse existence time while maintaining the total area of the pulse is the use of air gaps in the charge, that is, the dispersion of the elongated charge. This method leads to a decrease in P_{MAX}, reducing the size of the zone of rock overmilling and, consequently, the energy loss in the near area of the blast effect, and at the same time increases the pulse existence time. Due to energy losses in the zone of rock overmilling, or in the plastic deformations zone, the shock wave quickly transforms into a stress wave.

The further development of the process is shown in Fig. 2, where curve 1 corresponds to the dependence of the maximum pressure distribution (MPa) in the incident stress wave, which propagates into the borehole-to-borehole space $a = 5$ m in the mine rock, the curve 2 – in the reflected stress wave. The reflection phenomenon is a sign of the effectiveness of the obstacle in the path of the incident stress wave movement.

From Fig. 2 it is evident that in the borehole-to-borehole space at a distance of 25 – 35 radii of the active (detonating) charge, a significant jump in the pressure is fixed compared to the background one. The greatest pressure in the reflected wave is observed in the mine rock at a distance of 30 charge radii (3 m) from the active charge, accounting for more than 7 MPa. At the boundary with the adjacent borehole ($R_0 = 45$), the value of the average hydrostatic pressure reaches 3.8 MPa (Fig. 2).

Since the passive borehole in the task plays the role of the shielding cavity, it should be filled with an effective absorbing material. If such a shield has to exist for a certain time, it should simultaneously keep the borehole walls from being destructed under the influence of dynamic loadings. In the task we use the condition of filling the borehole with gas-generating emulsion material in two variants (Fig. 3).

The dependence of the density on the coordinate in the passive charge under the action of the stress wave of the active charge has a nonlinear character, since the emulsion is deformed in accordance with the law of the nonlinear-elastic medium. As it follows from the calculations (Fig. 3), this compliant material with an initial density of 1200 kg/m^3 under the action of $3.8-3.5$ MPa pressure acquires a new density. In the first variant, the emulsion density increases on the front wall of the borehole from 1200 to 1478.5 kg/m^3 , and when the wave passes through the borehole, it decreases to 1476 kg/m^3 on the opposite wall. In practice, one can consider such a change insignificant. This means that since in the first variant of the problem the emulsion is only 2% saturated with gas bubbles, its role as a means of absorbing the wave energy is insignificant.

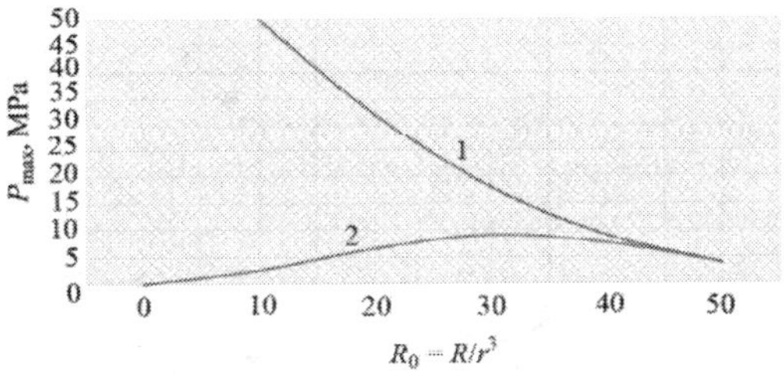

Fig. 2. The distribution of the maximum pressure P_{max} in incident (1) and reflected (2) stress waves between adjacent boreholes.

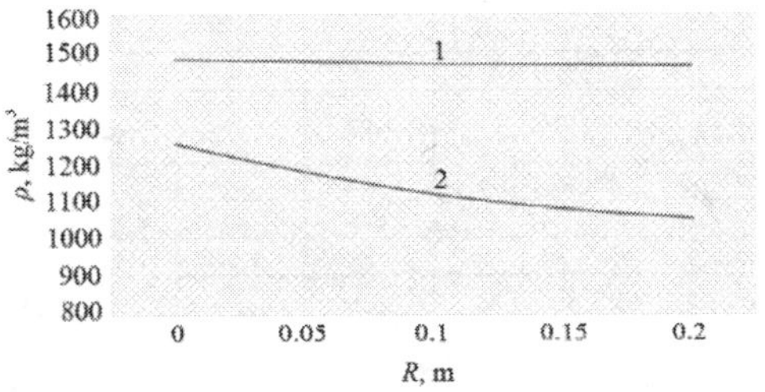

Fig. 3. The density distribution of the porous filler under the action of a stress wave in the section of the contour borehole with an emulsion, gasgenerated with sensitizer in an amount: $1-2\%$; $2-15\%$.

In the second variant of the problem, an emulsion with a density of 800 kg/m^3 has been used as absorbing material. Based on the character of curve 2 in Fig. 3, it follows that by virtue of the greater absorption capacity of the shield material, thus, the dissipation efficiency of the stress wave energy also increases significantly. With a significant compaction of the emulsion material (up to 1250 kg/m^3) on the front wall of the borehole, its density on the opposite wall is much less and is 1050 kg/m^3. Such a decrease in density within the borehole diameter indicates a significant increase in the effectiveness of the shield material.

Consequently, the shield effectiveness depends to a large extent on the filler material, as evidenced by graphs 1 and 2 in Fig. 3. However, the filtering performance of the protective structure relative to the stress wave is determined by its size in the direction of the wave motion, which should be commensurate with the wavelength. It is known [9] that when the relative density of the shield material is 0.9, the relative wave resistance is 0.63, whereas at a relative density of the shield material 0.6 this value is reduced to 0.12, that is, almost by 5 times. However, a change in the ratio of the shield width to the wavelength from 0.05 to 0.8, that is, by 16 times, changes the ratio of energy in the refracted and incident waves from 0.65 to 0.33, that is, only twice. So, if to consider the effect of these factors separately, the construction of a protective shield should begin with the selection of the filler material, and then to project the shield size.

Traditionally in mining, the contour set is arranged in such a way that the peripheral rock massif has got a minimum of damage. For this purpose, the diameter decreases of that boreholes, in which the garland charges of economical action are located, which are intended to create only fracture between the boreholes in the plane of the shield. Such a technique limits maximally the possibility of applying the main factor – shield density due to its inefficient width.

Based on the above, there is another technical solution. A variant of the complex method is possible, when the boreholes of increased diameter are drilled in the contour set to fill with the material absorbing the wave energy. To enhance the absorption capacity of the shield, the rock massif surrounding the borehole in the area wrapped up to the incident wave should be pre-saturated with a system of radial and slabbing fractures. In this case, the peripheral rock massif is minimally damaged.

To construct a contour shield, it is possible to use the properties of both factors in the complex by applying the garland charges of directional action. The peculiarities of the method is that the construction of such charges and the parameters of their location are based on the applying of systems of concentrated charges with a special shape, including the cumulative concavity. These charges are manufactured mainly at the blast site.

4 Conclusions

The analysis of the stress wave transformation in the conditions of shielding blast of the rock massif blocks by the system of borehole charges indicates the existence of a significant number of physical and technological factors determining the effectiveness of the contour method for protecting the rock massif adjacent to the place of large-scale blasting.

Theoretical studies and existing experience allow to develop and apply a new solution for the creation of a protective plane zone on the boundary of the large-scale blasting place with the adjacent rock massif. It is considered expedient to use the possibility of controlled formation of a protective fractured zone using the features of the mechanical effect development if to blast the concentrated charges with conic shape, which are manufactured at the site of blast operations.

Authors express their gratitude to the head of the Underground Mining Department at the National Mining University for help and consultations during the work.

References

1. Kravets, V.H., Korobiychuk, V.V., &. Boiko, V.V. (2015). *Fizychni protsesy prykladnoi heodynamiky vybukhu*. Zhytomyr: Zhytomyrskyi derzhavnyi tekhnichnyi universytet.
2. Khomenko, O.Ye., & Maltsev, D.V. (2005). K obosnovaniyu ratsional'noy tekhnologii burovzryvnykh rabot v usloviyakh shakht GP "VostGOK". *Naukovyi Visnyk Natsionalnoho Hirnychoho Universytetu*, (10), 3-6.
3. Bondarenko, V. Kovalevs'ka, I., & Cherednychenko, Y. (2010). Substantiation of design and installation technology of tubular rock bolts by explosive method. *New Techniques and Technologies in Mining*, 9-14. https://doi.org/10.1201/b11329-3
4. Kovalevs'ka, I., Symanovych, G., & Fomychov, V. (2013). Research of stress-strain state of cracked coal-containing massif near-the-working area using finite elements technique. *Annual Scientific-Technical Collection – Mining of Mineral Deposits 2013*, 159-163. https://doi.org/10.1201/b16354-28

5. Luchko, I.A., Plaksiy, V.A., & Remez, N.S. (1989). Mekhanicheskiy effekt vzryva v gruntakh. Kyiv: Naukova dumka.
6. *Fizika vzryva.* (2002). Pod redaktsiyey L. Orlenko. Moskva: Fizmatlit.
7. Bondarenko, V., Kovalevs'ka, I., & Fomychov, V. (2012). Features of carrying out experiment using finite-element methodat multivariate calculation of mine massif – combined support system. *Geomechanical Processes During Underground Mining,* 7-13. https://doi.org/10.1201/b13157-3
8. Lyakhov, G.M. (1982). *Volny v gruntakh i poristykh mnogokomponentnykh sredakh.* Moskva: Nauka.
9. Vovk, O.O. (mol.), Kravets, V.H., Isaienko, V.M., Vovk, O.O., & Remez, N.S. (2014). *Vplyv tekhnohennykh dynamichnykh protsesiv na stan pryrodnykh i inzhenernykh obiektiv.* Kyiv: Vydavnytstvo Natsionalnoho Pedahohichnoho Universytetu im. M.P. Drahomanova.

Analysis of pressure difference changes in respirator filters while dusting

Serhii Cheberiachko[1*], *Olena* Yavorska[1], *Yurii* Cheberiachko[1], and *Andrii* Yavorskyi[2]

[1]National Mining University, Department of Aerology and Labour Protection, 19 Yavornytskoho Ave., 49005 Dnipro, Ukraine
[2]National Mining University, Department of Underground Mining, 19 Yavornytskoho Ave., 49005 Dnipro, Ukraine

Abstract. Objective is to determine the relation between pressure difference of a filter and parameters of a filtering layer to lengthen the protection period. The study involved elements of system analysis and mathematical modeling. Basic statements of the theory of nonstationary filtration and aerohydrodynamics were applied to develop a dust-loaded filter model. Dependence of pressure difference in a filter upon certain changes in packaging density of dust-loaded fibers has been determined; the dependence makes it possible to define minimum fiber packaging density to provide maximum dust capacity. To provide maximum dust capacity and high protective efficiency, the number of filtering layers in multilayered filters depends upon dust concentration within the air of the working zone, specified protection level of a respirator, air loss, and maximum dust volume which may be accumulated in the finishing filtering layer in terms of fiber packaging density being determined according to the minimum pressure difference in a filter at final time of respirator operation. A model of changes in pressure difference in filtering respirators in the process of aerosol particles depositing on filters has been improved; contrary to other available models, that one takes into consideration changes in fiber packaging density while dusting, filtration coefficient, and the amount of aerosol particles.

1 Introduction

Coal dust does not only contaminate atmospheric air and environment [1], it is also dangerous for those peoples working immediately at mining enterprises. Efficiency of mining enterprises is based upon sustainable development of economic constituents of mineral extraction during both underground [2 – 5] and open-cast [6] mining with the provision of environmental issues of the process. Today's world practice demonstrates the tendency of broad-scale implementation of clean coal technologies being formed at the territory of mining enterprises. The technological solutions are the basis for the operation of energy and chemical complexes for coal gasification [7, 8]. Gasification process takes place within a closed system: underground generator – ground complex that practically restricts the emission of harmful generator gases into the environment [9]. However, it is impossible

* Corresponding author: sicheb@ukr.net

to eliminate completely negative effect of dust upon people. Thus, use of dust respirator under such conditions is the only way to protect respiratory organs of miners.

To provide high-efficiency dust protection, it is required to use low breathing resistance respirators which are characterized by minor effect upon functional state of workers. Otherwise, they will not be used during the whole shift [15]. Thus, the studies aimed at the improvement of ergonomic parameters of respiratory protective devices are of special topicality [10, 11, 16].

2 Statement of the problem

There are following basic structural, functional, and operational characteristics of filters [11]: protection coefficient – K_z; initial pressure difference – Δp; protection period – t_k; filtration area – F; shape and size of a filter – H_f, length – L_f; properties of a filtering layer which effect penetration coefficient: fiber radius – a, fiber packaging density – β; thickness of a filtering layer – H; and airflow resistance – R_0.

The first three characteristics are selected basing upon the purpose and conditions of filter application. The last three ones are determined on the basis of the preset conditions. There are numerous variants which can meet the initial data. Their functional relations are defined by following system of equations [13, 17]:

$$K_z = \frac{100}{\exp\left(-\dfrac{2 \cdot \beta \cdot H \cdot \eta(a)}{\pi a}\right)},$$

$$E = 1 - \exp\left(-\frac{2 \cdot \beta \cdot H \cdot \eta(a)}{\pi a}\right),$$

$$\Delta p_f = \mu \cdot v \cdot R_0 \cdot L .$$

where $\eta(a)$ is total coefficient of aerosol particles catching; μ is dynamic air viscosity, Pa·s; v is filtering rate, m/s; R_0 is non-dimensional force of fiber airflow resistance; L is total length of fibers in a filter, m^{-1}:

$$L = \frac{\beta H}{\pi a^2}.$$

However, to determine filtering layer parameters. It is necessary to define their effect upon pressure difference on a filter upon its dusting time; that will make it possible to predict the protection period. The latter is limited by the boundary admissible pressure difference that is determined by the corresponding normative documents. For instance, according to EN149:2003, in terms of filtering half-masks it should be not more than 400 Pa for class P1; not more than 500 Pa for class P2, and not more than 700 Pa for class P3, if air loss in 95 l/min. Determination of the mentioned relation is of special importance to calculate the parameters of multilayered filters when there is the necessity in defining packaging density of previous filtering layers.

3 Theoretical studies

It is required to determine changes in pressure differences during dust deposit accumulation on fibers within time t in terms of filtering layer thickness (x) at known parameters of filtering material (radius of fibers, density of filtering layer thickness packaging), and the preset conditions of a filtering process (airflow rate, particles concentration, and their average radius).

Use two cross sections to highlight elementary filtering layer with width Δx at the distance x from the filter surface. Suppose that the area of cross section is equal to a unit. Air with the dust amount n_1 goes to section 1-1; dust amount n_2 goes out through section 2-2 (Fig. 1).

Decrease in the amount of particles in the elementary filtering layer is:

$$\Delta n = -(n_2 - n_1) = -\frac{\partial n}{\partial x}\Delta x . \quad (1)$$

Derivative $\frac{\partial n}{\partial x}$ demonstrates certain change in the amount of deposited aerosol particles per unit of filtering layer thickness. The value is expressed by means of a derivative since the amount of particles in each section depends upon two derivatives: x – upon the distance from filter surface and t – from filtering time. The sign of minus in equation (1) shows that the amount of particles within the flow decreases along with the increase in filtering layer thickness.

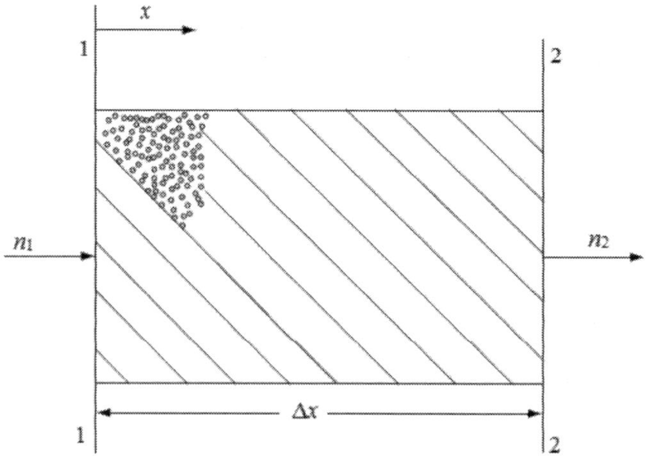

Fig. 1. Scheme of dust deposit accumulation on a filter during volumetric filtration.

Certain amount of air with rate v goes through a unit of cross section area of the highlighted layer. Respectively, the amount of substance caught on a filtering layer per time t is:

$$\Delta G = v(t)\Delta n = -v(t)\frac{\partial n}{\partial x}\Delta x . \quad (2)$$

Particles caught by a filtering layer form deposit on filtering fibers that is accumulated while filtering being the reason of further increase in filter density. Its growth in thickness Δx per time moment t will be Δx while the rate of accumulations deposited in a filtering layer per time unit is:

$$\Delta G = \frac{\partial \beta_1}{\partial t} \Delta x, \tag{3}$$

where β_1 is the density of deposit packaging in a filtering layer related to a volume unit of the elementary filtering layer.

While equalizing expressions (2) and (3), we obtain:

$$\frac{\partial \beta_1(x,t)}{\partial t} = -v(t)\frac{\partial n(x,t)}{\partial x}. \tag{4}$$

That differential equation is a well-known mass balance equation. It shows that the amount of the substance passing through a filtering layer per time unit is equal to the amount of the deposited substance in that layer per similar period of time. Equation (4) includes two dependent values: $n(x,t)$ and $\beta_1(x,t)$. Thus, to describe the process, we need other equation demonstrating specificity of a filtering process.

Consider the process of particles deposition on an obstacle as a total result of two contrary processes: particles catching by means of their sticking to filter fibers and tearing off the stuck particles while exhaling. In that case, decrease in the amount of the accumulated particles within the area Δx may be determined according to formula

$$\Delta n = \Delta n_1 - \Delta n_2, \tag{5}$$

where Δn_1 is the reduction in the particles amount at the expense of their sticking; Δn_2 is the increase of the particles amount at the expense of their tearing off.

Decrease in the particles amount due to their sticking may be taken as the one to be proportional to average amount of particles within the airflow passing through a filter as Ivasaki proposed

$$\Delta n_1 = \gamma n \Delta x. \tag{6}$$

Assume the increase in the amount in the air entering a filter as the one to be proportional to the amount of the accumulated deposit before time t, then equation (6) may be represented as:

$$\Delta n_2 = a_1 \beta_1 \Delta x, \tag{7}$$

where a_1 is the coefficient depending upon filter parameters, mechanism of particles tearing off, and size of the particles.

Equation (7) expresses the assumption that the rate of particles tearing off follows such a law:

$$\frac{\partial \beta_1}{\partial t} = -a_1 \beta_1. \tag{8}$$

It is quite possible to have a balance equation for the process of dust deposit destruction between filtering layer fibers which will be similar to equation (4).

Having substituted Δn_1 from (2) and Δn_2 from (7) into equation (1), we obtain:

$$-\frac{\partial n(x,t)}{\partial x} = \gamma(x,t)n(x,t) - a_1\beta_1, \tag{9}$$

where $\gamma(x,t)$ is the filtration coefficient demonstrating a mechanism of particles deposition on fibers.

Basing upon equation (4) and (9), it is possible to determine changes in dust deposit density while filtering:

$$\frac{\partial \beta_1(x,t)}{\partial t} = v(t)\gamma(x,t)n(x,t) - a_1\beta_1 . \quad (10)$$

The phenomenon of particles tearing off while filtering in elastomeric half-masks may be neglected as the exhale is performed through the exhalation valve; thus, dust deposit structure is not broken:

$$-\frac{\partial \beta_1(x,t)}{\partial t} = v(t)\gamma(x,t)n(x,t) . \quad (11)$$

If we represent expressions (9) and (4) as a system of equations with the initial conditions $\beta_1(x,0) = \beta$; $n(0,t) = n_0$, where β is the density of clean filter fiber packaging; n_0 is the initial concentration of aerosol particles within the air mixture (mg/m^3), entering a filter, then its solution is as follows:

$$n(x,t) = n_0 \exp\left(-\int_0^H \gamma(x,t)dx\right); \quad (12)$$

$$\beta_1(x,t) = \beta + n(x,t)\theta(t)A_1(x); \quad (13)$$

where $$\theta(t) = \int_0^t v(t)dt ; \; A_1(x) = \gamma(x,t)\exp\left(-\int_0^H \gamma(x,t)dx\right).$$

To calculate dust-loaded fiber packaging density, it is required to set the change in filtration coefficient in the process of aerosol particles deposition $\gamma(x,t)$.

It is usually represented as linear:

$$\gamma(x,t) \approx \gamma_0 + bxt ,$$

where γ_0 is the initial filtration coefficient of a clean filter.

Filtration coefficient for a dust-free fibrous filter may be determined according to a known formula:

$$\gamma_0 = \frac{2 \cdot \eta(a) \cdot \beta \cdot H}{\pi a^2} .$$

Coefficient a is determined experimentally; it depends upon the mechanisms of aerosol particles catching as well as their size and properties. Paper [14] indicates that its value for dolomite dust particles of 1-5 mcm may be within the range of 0.25-0.5.

In other cases it is possible to use the formula proposed by L.V. Radushkevych:

$$b = \frac{k}{r_c \cdot n \cdot t} ,$$

where k is the coefficient of test-aerosol penetration through a filtering layer; r_c is the aerosol particle radius, m.

Fig. 2 represents the curves of distribution of dust particles deposition and dust deposit packaging density in terms of thickness of a filtering polypropylene layer for which $\gamma_0 = 0.5$; $a = 2.5$ mcm; $H = 4$ mm; $\beta = 0.05$. The amount of dust particles of 5 mcm in diameter was determined upon the condition that their concentration was 100 mg/m^3 within the volume of 1 m^3. Filtration rate was 0.01 m/s.

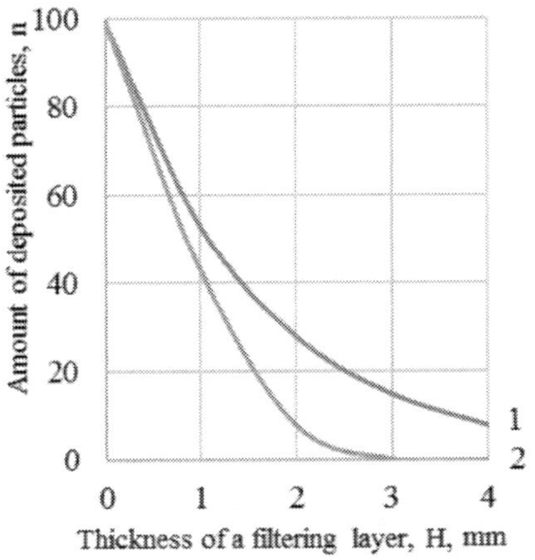

Fig. 2. Distribution of particles amount in a filtering layer in terms of different filtration coefficients: 0.25 (1) and 0.5 (2).

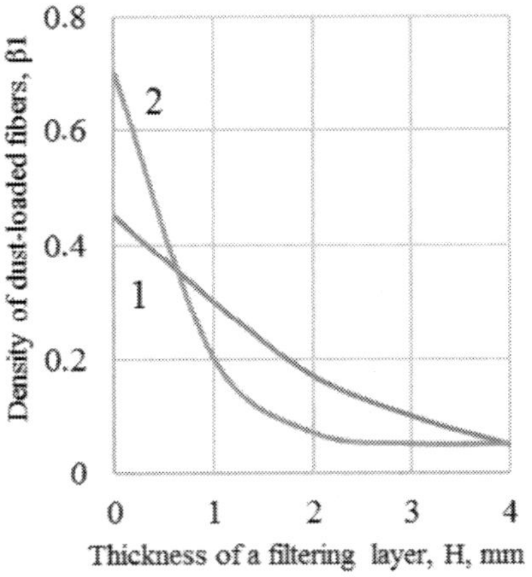

Fig. 3. Distribution of dust-loaded fiber packaging density in a filtering layer in terms of different filtration coefficients: 0.25 (1) and 0.5 (2).

Along with the increase of the filtration coefficient, we can observe that the basic mass of aerosol particles deposits within the upper shares of a filtering layer, hence reducing the period of a volumetric filtration phase. For instance, in case of multilayered filters, it is essential to provide uniform dust saturation for the whole filter thickness. Whereas in case of one-layer filter the efficiency of aerosol particle catching is the first thing to consider. Thus, the first case requires selecting a filter with low filtration coefficient while the second one requires a filter with high filtration coefficient.

If we imagine that particles are accumulated on a filter according to a dendritic model based upon the assumption on the formation of additional chains-dendrites, then they can be represented as fibers with the radius similar to the radius of the deposited particles. In such case, pressure difference of a dust-loaded filter is expressed according to formula:

$$\Delta p(x,t) = \frac{4\mu v\beta_1}{\pi a^2}\int_0^H F\left[\beta_1(x,t)\right]dx =$$

$$= \frac{2\mu vH\beta_1}{\pi a^2}\int_0^H \left[2\beta + a\beta_1 \exp\left(n(x,t)\theta(t)A_1(x)\right) + 3a^2\beta_1 \exp\left(2n(x,t)a\theta(t)A_1(x)\right)\right]dx. \quad (14)$$

Expression (14) makes it possible to analyze the effect of various parameters upon filter indices (Fig. 4 and 5) and detect the most influential ones.

Fiber density packaging also effects the value of pressure difference and mechanism of deposit distribution in terms of its thickness. The less it is, the deeper particles may penetrate into the middle of a filtering layer and effect the time of reaching critical breathing resistance (Fig. 6 and 7).

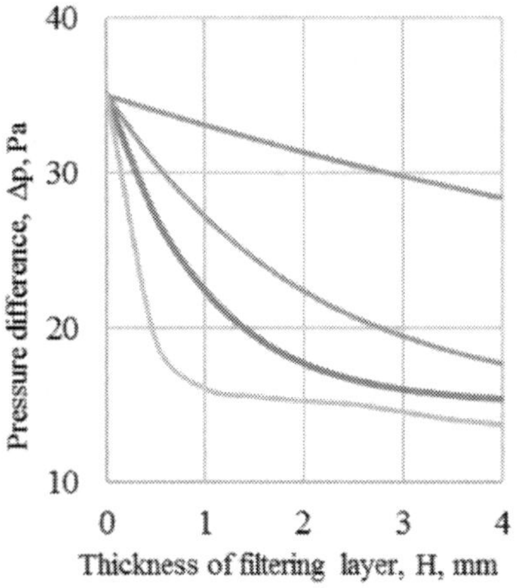

Fig. 4. Curve of changes in pressure difference in terms of different initial fiber packaging density of a filter: 0.5 (1); 0.1 (2); 0.07 (3); 0.05 (1).

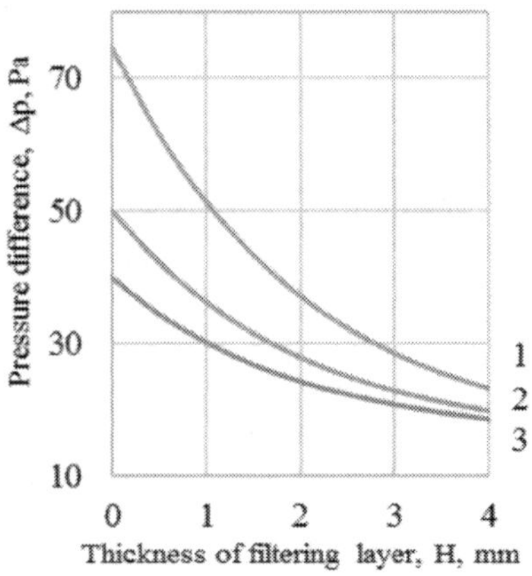

Fig. 5. Curve of changes in pressure difference in terms of filter thickness at different initial concentration of dust particles, mg/m^3: 100 (1); 50 (2); 25 (3).

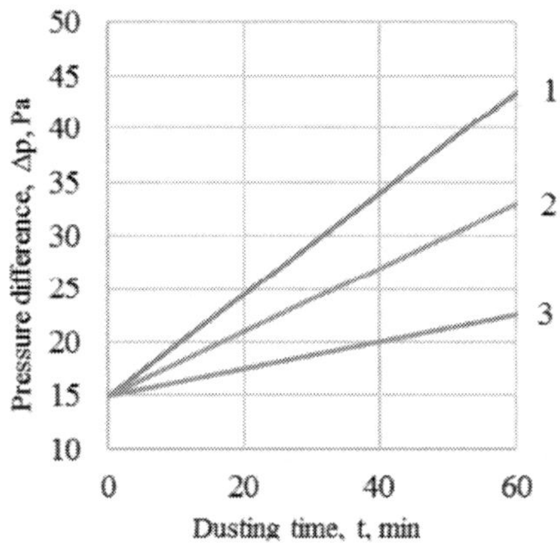

Fig. 6. Change in pressure difference during time in terms of different initial fiber packaging density: 0.05 (1); 0.1 (2); (3); 0.5 (1).

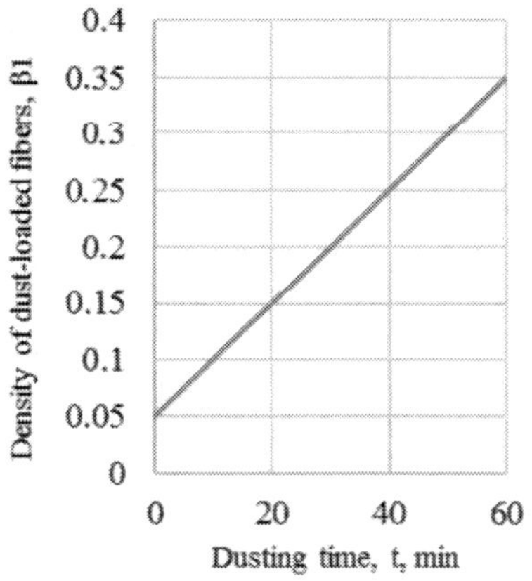

Fig. 7. Change in density of dust-loaded fiber packaging during time in terms of the initial fiber packaging density being 0.05.

Formula (14) (on the condition of substituting β_1 for β and assuming that filtration coefficient depends only upon fiber packaging density $\gamma(x) = B \cdot H \cdot \beta(x)$ allows determining distribution pattern of fiber packaging density in terms of dustless filter as for the thickness to provide maximum accumulated amount of dust per specified time:

$$\beta(x) = \frac{G}{B \cdot H}\left(1 + \tau_1 n(x,t)\right)^{-\frac{q}{(q+1)}}, \tag{15}$$

where $B = \dfrac{2\eta(a)}{\pi a^2}$; $G = \int_0^H \left(1 + \tau_1 n(x,t)\right)^{\frac{q}{q+1}} dx$ is the function determining changes in aerosol particles concentration in terms of filter thickness; $\tau_1 = \frac{\theta(t)B}{A_1}$; q is the coefficient which takes into account changes in fiber packaging density in terms of dust deposit accumulation (0.5 – 2.8).

Determine minimum fiber packaging density in terms of thickness for polypropylene filtering material, if protection coefficient is at the level of 100 units. Calculation results are represented on Fig. 8 and 9. It can be observed that to expand the protection period it is required to decrease fiber packaging density within the upper share of a filter. In this context, lower layers should be made less dense to provide sufficient protection level.

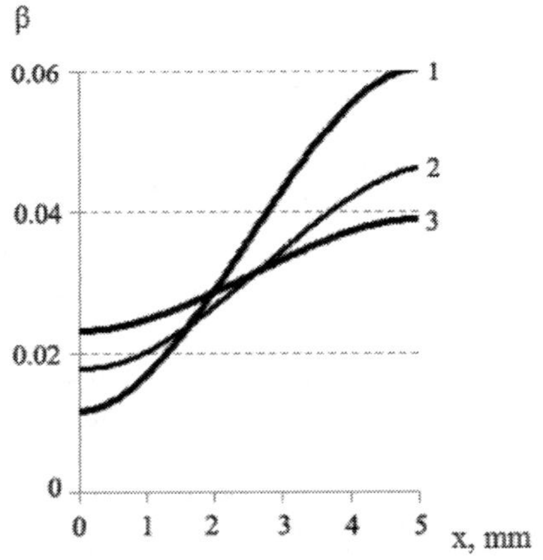

Fig. 8. Distribution of fiber packaging density (β) in terms of filtering layer thickness (x) for the final time of dusting t_k = 30 min; if q = 0.5; τ_1 = 1.5 (1); q = 0.5; τ_1 = 1 (2); q = 0.5; τ_1 = 0.5 (3).

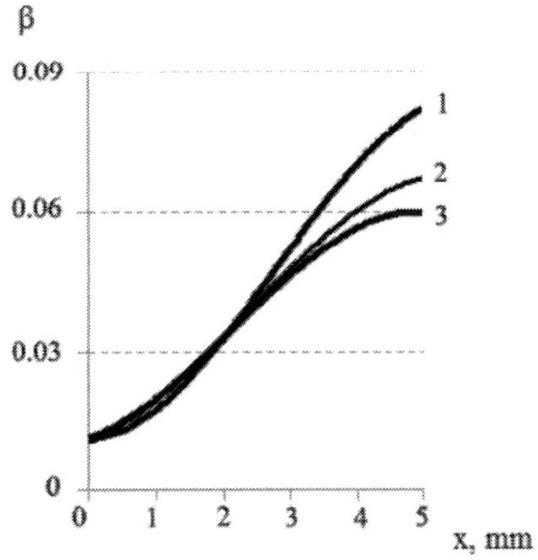

Fig. 9. Distribution of fiber packaging density (β) in terms of filtering layer thickness (x) for the final time of dusting t_k = 60 min; if q = 0.5; τ_1 = 1.5 (1); q = 0.5; τ_1 = 1 (2); q = 0.5; τ_1 = 0.5 (3).

Since it is quite a complex task to manufacture filters with changeable packaging density, it is proposed to make them of several various-density layers. The number of the required layers (k_s) will depend upon fiber packaging density of the last layer calculated on the basis of specified efficiency and maximum dust volume which will be accumulated on a filtering layer in terms of minimum pressure difference at final time:

$$k_s = \frac{C \cdot Q \cdot t_k}{\rho_c \cdot V_{max}},$$

where $V_{max} = \pi \cdot \beta_1 \cdot \eta(a) \cdot r_c^3$ is maximum dust volume which may be accumulated on one filtering layer at minimum pressure difference in a filter at final time; Q is air loss, l/min; t_k is the completion time of volumetric filtration phase, min.

To determine completion time of a volumetric accumulation phase, imagine that generally the character of time changes of various-fraction particles amount on the filter surface is uniform. In this context, the amount of deposited particles on fibers increases monotonously: in the beginning, the growth is quite fast; then it becomes slower being as follows within the specified time:

$$n(x) = \frac{1}{k} \cdot \frac{\eta(a)}{a},$$

while the time of dust deposit accumulation may be determined as:

$$t_k = \frac{2.3}{a \cdot v \cdot n(x)} \lg \left(1 - 2^{-\frac{1}{k}}\right)^{-1}.$$

Process of volumetric filtering of aerosol particles is capable of increasing protective period of filters. It is especially required for multilayered filters. Since it is the time of pore plugging of a previous layer that is the determining one to calculate the parameters of a finishing layer, then the analysis of the obtained dependences makes it possible to state that fiber packaging density has the most considerable effect upon the increase in pressure difference at minor filtration rates being up to 0.01 m/s (that is characteristic for respirator filters) while dusting. It is the sharp rise of filter breathing resistance (Fig. 6) in the beginning of its operation that tells about the fact that the main dust mass deposits within the upper layers not penetrating lower due to certain fiber density. Dendritic model proposed to calculate changes in pressure difference provides that, having entered the filter thickness, aerosol particles deposit on clean fibers first. Then they began depositing on each other forming certain branching similar to fibers, i.e. the formed chains increases the density of filter packaging. Thus, there is the necessity to determine distribution of fiber density in terms of filter thickness to provide uniform distribution of dust deposit. It is quite difficult to take into consideration that a filter should also have a specified coefficient of protection. Along with the lengthening of filter operating time, there is a necessity to reduce fiber packaging density; if there is a necessity to provide the required protection efficiency, fiber packaging density should be increased. The solution is right in the production of multilayered filters with different packaging density. To do that, it is recommended to define the amount of layers and to select other filter parameters basing upon the completed volumetric filtration phase that will provide maximum dust accumulation (maximum use of a filtering layer).

4 Conclusions

Dependence of pressure difference in a filter upon changes in density of dust-loaded fiber packaging making it possible to define minimum density of fiber packaging to provide maximum dust capability has been determined.

The number of filtering layers in multilayered filters to provide their maximum dust capability and high protective efficiency depends upon dust concentration within the air of a working zone, specified level of respirator protection, air loss, and maximum dust volume that may be accumulated in the finishing filtering layer in terms of fiber packaging determined according to minimum pressure difference in a filter at finite time of respirator operation.

Model of pressure difference changes in filtering respirators in the process of aerosol particles deposition on filters has been improved. Contrary to many other models, that one takes into consideration changes in fiber packaging density, filtration coefficient, and the amount of aerosol particles.

The study is carried out within the framework of the research subject "Scientific basics to form unified system to preserve and generate power of the objects of fuel and energy complex of Ukraine" (State registration No. 0117U001127).

References

1. Kolesnik, V.Ye, Pavlichenko, A.V., & Buchavy, Yu.V. (2016). Determination of dynamic parameters of dust emission from a coal mine fang. *Naukovyi Visnyk Natsionalnoho Hirnychoho Universytetu*, (2), 81-87.
2. Buzylo, V., Yavorsk'yy, A., & Yavorsk'yy, V. (2012). Analysis of stress-strain state of rock mass while mining chain pillars by chambers. *Geomechanical Processes During Underground Mining*, 95-98. https://doi.org/10.1201/b13157-16
3. Buzilo, V., Serdyuk, V., & Yavorsky, A. (2010). Research of influence of support resistance of the stope in the immediate roof condition. *New Techniques and Technologies in Mining*, 127-130. https://doi.org/10.1201/b11329-21
4. Pivnyak, G.G., & Shashenko, O.M. (2015). Innovations and safety for coal mines in Ukraine. *Naukovyi Visnyk Natsionalnoho Hirnychoho Universytetu*, (6), 118-121.
5. Golinko, V.I., Yavorskiy, A.V., Lebedev, Ya.Ya., & Yavorskaya, Ye.A. (2014). Estimation of frictional sparking effect on firedamp inflammation during fragmentation of gas-saturated rock massif. *Naukovyi Visnyk Natsionalnoho Hirnychoho Universytetu*, (6), 31-37.
6. Dryzhenko, A., Shustov, A., & Moldabayev, S. (2017). Justification of parameters of building inclined trenches using belt conveyors. *17th International Multidisciplinary Scientific GeoConference SGEM 2017*, 471-478. DOI:10.5593/sgem2017/13/S03.060
7. Falshtynskyi, V.S., Dychkovskyi, R.O., Saik, P.B., Lozynskyi, V.G., & E. Cáceres Cabana. (2017). Formation of thermal fields by the energy-chemical complex of coal gasification. *Naukovyi Visnyk Natsionalnoho Hirnychoho Universytetu*, (5), 36-42.
8. Lozynskyi, V., Dychkovskyi, R., Saik, P., Falshtynskyi, V. (2018). Coal Seam Gasification in Faulting Zones (Heat and Mass Balance Study). *Solid State Phenomena*, (277), 66-79. https://doi.org/10.4028/www.scientific.net/SSP.277.66
9. Bondarenko, V., Tabachenko, M., & Wachowicz, J. (2010). Possibility of production complex of sufficient gasses in Ukraine. *New Techniques and Technologies in Mining*, 113-119.
10. Golinko, V.I., Cheberiachko, S.I., Yavors'ka, O.O., & Tykhonenko, V.V. (2017). Assessment of protection level of dust respirators at coal mines. *Naukovyi Visnyk Natsionalnoho Hirnychoho Universytetu*, (3), 102-107.
11. Cheberiachko, S.I., Yavorska, O.O., & Morozova, T.I. (2013). Study of mechanical half-mask pressure along obturation bar. *Technical and Geoinformational Systems in Mining*, 317-323.
12. Newnum, J.D. (2010) The effect sofrelative humidity on respirator performance. *Master's thesis, University of Iowa*, http://ir.uiowa.edu/etd/861
13. Holinko, V., Cheberiachko, S., Yavorska, O., & Radchuk, D. (2016). Issledovanie zashchitnykh svoystv respiratorov-polumasok, ispol'zuemykh shakhterami. *Min. Miner. Depos.*, *10*(4), 29-36.
14. Fabbro, L.D., Brun, P., Laborde, J.C., Lacan, J., Renoux, A., & Ricciardi L. (2012). Contribution to the modelling of industrial pleated filters by solid particles. *American Journal of Environmental Science*, *8*(4), 385-395.
15. Tikhonov, M.M., Parmon, V.V., Boinichev, O.N., & Ryabtsev, V.N. (2015). Assessment of protective efficiency of respiratory protection devices to provide safe operation in the zone of radioactive contamination. *Vestnik Komandno-inzhenernogo instituta MChS Respubliki Belarus'*, *1*(21), 58-67.
16. Haruta, H., Honda, T., & Eninger, R.M. (2009). Experimental and theoretical investigation of the performance of N95 respirator filters against ultrafine aerosol particles tested at constant and cyclic flows. *Journal Int. Soc. Respir. Prot.*, (25), 75-88.
17. Feng, Z., Long, Z., & Chen, Q. (2014). Assessment of Various CFD Models for Predicting Airflow and Pressure Drop Through Pleated Filter System. *Build. Environ.*, (75), 132-141.

Ecological aspects of managing the stressed-deformed state of the mountain massif during the development of multiple coal layers

Volodymyr Buzylo[1], *Artem* Pavlychenko[2*], *Tamara* Savelieva[3], and *Olena* Borysovska[2]

[1]National Mining University, Department of Underground Mining, 19 Yavornytskoho Ave., 49005 Dnipro, Ukraine
[2]National Mining University, Department of Ecology and Environmental Protection Technologies, 19 Yavornytskoho Ave., 49005 Dnipro, Ukraine
[3]National Mining University, Department of Machinery Design Fundamentals, 19 Yavornytskoho Ave., 49005 Dnipro, Ukraine

Abstract. The ecological consequences of long-term development of coal deposits for the components of the environment, buildings, communications and infrastructure facilities have been analysed. The stress-deformed state of the rock massif around the preparatory working in the development of multiple coal layers is determined. The parameters of the development of the contiguous layers in the conditions of the mines of the Lviv-Volyn Coal Basin are substantiated. The complex of nature protection measures allowing to reduce the negative consequences of disturbance of the earth's surface in the development of multiple coal layers is substantiated.

1 Introduction

Coal has an important value in the global economy among natural energy resources and occupies the third position in terms of use after the oil and natural gas. The coal industry is an important component of the industrial potential of Ukraine and ensures the functioning of the leading sectors of the economy. The efficiency and stability of its functioning determines the sustainable development of the state, its energy and environmental security [1 – 3].

Exhaustion of balance reserves, unprofitability of most mines, complex mining and geological conditions and an unstable economic situation have led to a deterioration of the ecological condition of the mining regions. This led to the emergence of a complex of social, economic and environmental problems, the main of them are: subsidence of the earth's surface, deformation and destruction of buildings and infrastructure [4 – 6]; formation of waste heaps (dumps) of rock accumulated over many years of coal mining [7]; flooding of land and living quarters, deterioration of the quality of underground and surface water [8 11]; the formation of explosive methane-air mixtures, etc. As a result of this situation, most of the coal-mining regions have become zones of ecological and social disaster [12].

* Corresponding author: pavlichenko.a.v@nmu.one

The main tasks of managing the environmental safety of coal mining processes are to study the nature of the transformation of environmental components at all stages of field development [12 – 14]. In addition, it is necessary to introduce effective environmental measures timely, which will improve the quality of drinking water supply in coal mining regions [11, 15 – 17], as well as receive additional energy resources [18, 19].

The existing system of management of the ecological condition of coal mines doesn't allow proper solving environmental safety issues, both in the process of constant operation and further liquidation of coal mines [20 – 23]. Therefore, there is a need to establish patterns that govern the process of rock movement under the influence of underground mining of coal seams under natural objects, buildings, and settlements.

In this connection, the search for an analytical solution to the problem of rock mechanics is becoming topical in order to obtain a qualitative picture and quantitative estimates of the stressed deformed state (SDS) of the entire rock strata in conjunction with the parameters of the coal seam extraction at any stage of development. In this case, it is recommended to take into account the properties of the work-in-progress massif, the nature of the roof collapse, the intensity of the earth's surface displacement process, the depth of development, and specific mining and geological conditions [24, 25].

2 The research problem setting

The analysis of the literature sources revealed that the degree of negative influence of the reference pressure in the development of coal seams increases with the depth of development. So at a depth of 100 to 200 m, the noticeable influence of the reference pressure in the case of the imposition of the boundaries of clearing works was observed for interbeds of 15 – 25 m. At a depth of 300 – 340 m, debris in cleaning and preparatory workings took place even at interbed thicknesses of 50 – 60 m. This suggests that, in justifying the parameters for the development of the multiple contiguous layers, in each specific case, the SDS of the rock mass around the breakage face should be determined necessarily taking into account the overworking and (or) underworking, as well as taking into account the possible entry of the breakage face into the zone of pillars and layers selvages [24 – 27].

The speed of lowering the roof, hanging it behind the lining, the height of the inrush and the squeezing of coal are taken as the controlled parameters. It has been established that in the zone of increased rock pressure, the rate of lowering of the roof during the period of absence of the influence of production processes increases noticeably in comparison with the rate of lowering outside this zone [28, 29]. In the proposed unloading zone, which is adjacent to the zone of influence, it is 1.5 – 2.0 times smaller. In the zone of increased rock pressure, there was also an increase in the formation of the inrushes, with the height of the individual inrush reaching 0.8 m, while outside the zone of influence it did not exceed 0.1 – 0.2 m. The actual angle of influence of the pillar (≈ 84°) is obtained. The largest hangings of the roof behind the lining (not counting the end sections of the lava), equal to one step of the lining movement, were 0.5 – 0.7, and the largest values of the squeezing depth in the discharge zone were 0.4 – 0.5 m [26 – 29].

Obviously, there are not all the determining factors. For example, the formation may contain more than two closely spaced layers, an important role is played also by the physic and mechanical properties of coal and surrounding rocks that are not taken into account in the above-mentioned works, and so on. It is necessary to develop a methodology for calculating the values of subsidence of the earth's surface and recommendations for reducing the negative impact of surface underworking during the mining extraction of coal seams and rational management of coal reserves in safety pillars. But for this it is necessary to know the actual parameters of the demonstrations of rock pressure during the mining of coal seams using various development systems [28 – 31].

However, because of the consideration of the limited massif area in the vicinity of the breakage faces in the tasks of the mountain pressure with any method of solution, analytical or numerical, the determined stresses, deformations and deformation rates are not correlated with deformations of the earth's surface [24, 25, 29 – 31].

Most researchers associate the deformation of the earth's surface with angular parameters, consider, as a rule, the conditions of the completed process of displacement, and the main influencing factors are the depth of the coal layer and the dimensions of the breakage face. In this case, the data of mine surveying measurements are used, and therefore the obtained results do not possess sufficient generality [24, 25, 31, 32].

It should be noted that with underground coal mining irreversible changes occur in the rock mass that surrounds the working seam. These changes are a consequence of the collapse and subsidence of rocks over the breakage face, the redistribution of stresses around it. Changes in the mountain massif cause subsidence of the day surface over the mine workings. In addition to deformations from the influence of breakage heading, additional deformations and deformations of the mountain massif, the working seam and the earth's surface, are caused by the compression of the coal seam and host rocks under the influence of the reference pressure [32 – 35].

2 Determination of the stressed-deformed state around the preparatory working in the mining of multiple layers

The most important conditions for the safe operation of coal mines is to ensure the working condition of the entire network of mine workings from the surface to the breakage faces with minimal costs for their development and maintenance. The urgency of ensuring the operational stability of preparatory workings in the conditions of contiguous layers mining is constantly increasing due to the increase in the depth of mining operations [31 – 33]. At the same time, the nature of the distribution of rock pressure is significantly affected by overworking or underworking. Therefore, when choosing the method of protection and support of workings, especially in the case of weak enclosing rocks, it is necessary to proceed from the analysis of the SDS of the massif taking into account specific mining and technical situations as well as mining and geological conditions.

The layered heterogeneous environment in which the stress state of the roof and the soil of the contiguous layers undergoes significant changes under the influence of mining operations is typical for the mines of the Lviv-Volyn Basin. However, in the available industry documents regulating the location, protection and maintenance of mine workings, this change is accounted approximately by means of empirical coefficients established primarily for mining and geological conditions of other coal deposits. A similar approach is used in describing the structure of the rock massif.

To obtain quantitative estimates regarding the influence of interbed thickness and the depth of development, calculations were performed for a number of determining parameters (Fig. 1).

The analysis of the obtained data revealed that with a decrease in interbed thickness, the degree of stress growth in the reference pressure zone of the developing layer is increased. Thus, the maximum coefficient of stress concentration σ_y in the roof of the layer in the cross-section $x = 29$ m in relation to the level of γH is $k_y = 1.51$ at $h_2 = 6$ m and $k_y = 1.46$ at $h_2 = 20$ m. To a lesser extent the concentration of k_y is affected by the development depth H and at $H = 300$ m and at $H = 600$ m in the cross-section $x = 29$ m in the roof of the developing layer, k_y has the same value.

In the arch of preparatory workings at a depth of $H = 300$ m at $h_2 = 6$ m, the coefficient of equivalent stress concentration, calculated by the criterion of P.P. Balandin, reaches a value of 2.1 and remains the same at a depth of $H = 600$ m. The absolute value of the maximum equivalent tension at a depth $H = 500$ m becomes equal to the compressive strength of the rock, and at $H = 600$ m it is 1.2 times greater than it.

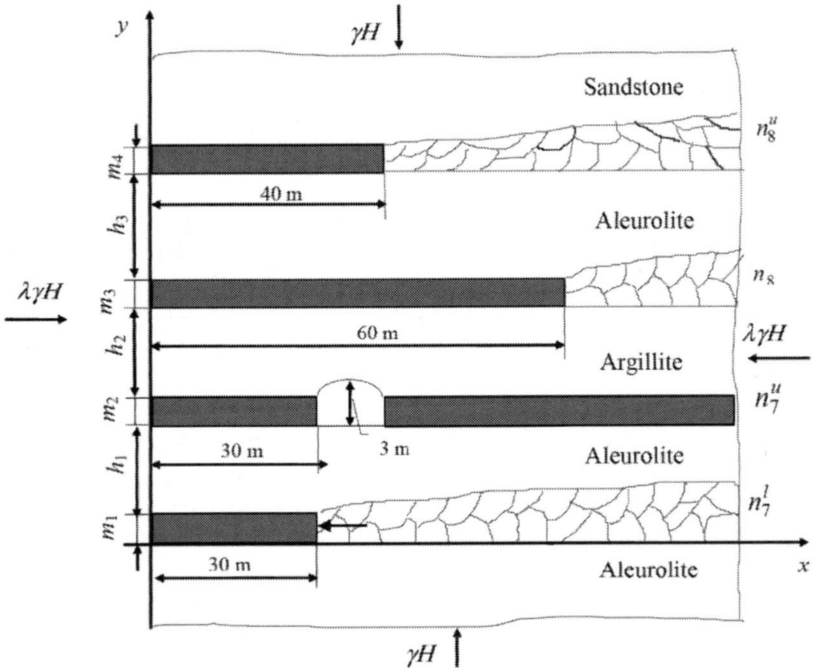

Fig. 1. The calculation scheme for determining the stress-deformed state of the massif around the working.

The growth of stresses in the massif surrounding the working, leads to an increase in the displacements on its contour. The performed studies showed that with increasing depth of development from 300 to 600 m, the maximum vertical displacements u_y in the arch grow by 1.79 times. Insufficient accounting of such SDS changes around preparatory workings that fall under the influence of clearing works on contiguous layers, when selecting lining, is one of the causes of deformation of preparatory workings.

3 Justification of the parameters for the development of contiguous coal layers in the mines of the Lviv-Volyn basin

Depending on the size of the mined-out space and the distance to the edge parts of the contiguous layers, in the roof and soil of the layer under development, zones of increased rock pressure of greater or lesser extent occur with different stress levels in them.

The most dangerous situation is the position of the line of the breakage face, in which overlay of reference pressure diagrams from the breakage face and the edge parts of the above- or lower-lying layer occurs in the roof and soil of the layer under development. The resulting total stresses, as a rule, exceed the admissible values. If, at the same time, a zone of extreme stress with considerable length is formed in the roof over the mined-out space, the mechanized lining is under great rock pressure [35 – 39]. The situation is aggravated by the fact that, a considerable additional load caused by the squeezing of coal is created by the tension of front moving breakage face $\sigma_c(V)$ under significant concentration of vertical normal stresses in the area of the reference pressure. As a result, an emergency situation may occur.

Based on the possible limits of the variation of the variable values, the ratio $\gamma_n H/(\sigma_c)_c$ in the calculation assumed the values [0.25; 0.27; 0.29; 0.31; 0.33; 0.35], and the ratio E_r/E_c belonged to the area [1.8; 2.8; 4.0] (E_r, E_c – are the elastic moduli of the roof rock and coal, γ_r – is the rock density of the roof, $(\sigma_c)_c$ – is the ultimate compressive strength of coal).

The geometric parameter of the investigated area l/l_n assumed the values [0.033; 0.166; 0.33; 0.66]. In this case, the length of the overworking zone was unchanged and amounted to $l_n = 150$ m, and the size of the mined-out space behind the breakage face $l\in$ was [5; 25; 50; 100] m. The width of the edge part of the layer was 70 m, the thickness of the interbed h_2 varied within 5 – 25 m. In total, 152 calculation options were considered for overworking conditions.

The values of the maximum tension concentration coefficients $(K_y)_{max}$ in zones of increased rock pressure in front of the breakage face, maximal approaches $(\Delta u)_{max}$ of the soil and roof in the mined-out space and loads P_0, were obtained using the boundary elements method for different values of the distance L between the breakage face and the edge parts of adjacent layers. On the basis of multiple correlation analysis of the obtained data, for the case of long face movement in the direction to the edge part of the adjacent layer under conditions of overworking, the ratios for the maximum values of the roof and soil approaches in long face in fractions of the thickness of the layer are established:

$$\frac{\Delta u_c}{m} = 0.107 + 0.097\frac{l}{l_n} - 0.065\frac{E_r}{E_c} + 0.530\frac{\gamma_r H}{(\sigma_c)_c} \qquad (1)$$

as well as the maximum relative tensions in the roof of the developed layer in front of the breakage face:

$$(K_y)_{\max} = \frac{(\sigma_{yy})_{\max}}{\gamma_n H} = 2.42 + 1.81\frac{l}{l_n} - 0.46\frac{E_c}{E_r}. \qquad (2)$$

Similar dependencies with regard to the conditions of underworking are:

$$\frac{\Delta u_y}{m} = -0.311 + 0.115\frac{l}{l_n} + 0.364\frac{E_c}{E_r} + 0.541\frac{\gamma_n H}{(\sigma_c)_y}; \qquad (3)$$

$$(K_y)_{\max} = \frac{(\sigma_{yy})_{\max}}{\gamma_n H} = 2.17 + 1.82\frac{l}{l_n} + 0.001\frac{E_c}{E_r}. \qquad (4)$$

In addition, correlation dependences are established for the case of long face outflow from under the edge part, which in the conditions of overworking have the form:

$$\frac{\Delta u_y}{m} = -0.314 + 0.350\frac{l}{l_n} + 0.310\frac{E_c}{E_r} + 0.446\frac{\gamma_n H}{(\sigma_c)_y}; \qquad (5)$$

$$(K_y)_{\max} = \frac{(\sigma_{yy})_{\max}}{\gamma_n H} = 2.38 + 2.468\frac{l}{l_n} - 0.196\frac{E_c}{E_r}. \qquad (6)$$

For the case of long face outflow from under the edge part of the adjacent layer under the conditions of underworking, the following dependences were obtained:

$$\frac{\Delta u_y}{m} = -0.280 + 0.301\frac{l}{l_n} + 0{,}285\frac{E_c}{E_r} + 0.433\frac{\gamma_n H}{(\sigma_c)_y}; \qquad (7)$$

$$\left(K_y\right)_{\max} = \frac{\left(\sigma_{yy}\right)_{\max}}{\gamma_n H} = 2.40 + 2.50\frac{l}{l_n} - 0{,}25\frac{E_c}{E_r}. \quad (8)$$

The correlation coefficients of dependences (1) and (2) are 0.88 and 0.82, respectively. The correlation coefficients of dependences (3) and (4) are equal to 0.97. For dependencies (5), (6), (7) and (8), the correlation coefficients are 0.97, 0.88, 0.97 and 0.89, respectively.

Similar studies were carried out for the case when the extraction is performed simultaneously in the conditions of underworking and overworking. The results of calculating one of the variants ($L = 65$ m) when the long face moves in the direction toward the edge part of the adjacent layer in the case of underworking in the form of diagrams of the normal tension σ_y and vertical displacements u_y are shown in Fig. 2 and 3. Displacements in the long face in this case $(\Delta u_y)_{max}$ do not exceed permissible for the mechanized support KM87.

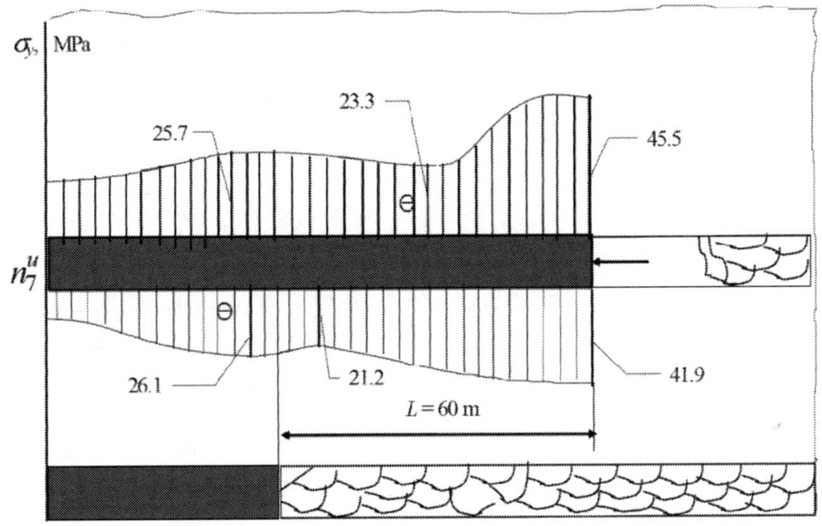

Fig. 2. Diagrams of normal stresses σ_y in the roof and soil of the layer n_7^u (underworking, movement of long face to the edge part of the adjacent layer).

In this sense, for underworking conditions, the $L = 65$ m value is acceptable. A similar analysis of the SDS pattern in the case of overworking has shown that $L = 60$ m is permissible.

Let us find out how the coefficient of tension concentration and the speed of moving the long face V affect the tension values $\sigma_y(V)$, appearing at the edge of the breakage face (in the cross-section $x = 0$, Fig. 1) and creating an additional load on the mechanized support.

Suppose that the speed of the breakage face movement is $V = 2$ m/day = 0.083 m/hour. Then a strip of coal $S = 1$ m will be worked out for $t = 6$ hours. The value of the tension concentration coefficient $(K_y)_{max}$ at $L = 65$ m under conditions of underworking is 3.6 and, therefore, $K = (K_y)_{max} - 1 = 2.6$ (if $L < 65$ m, then $(K_y)_{max} > 3.6$ and the strength condition is violated in the roof of the extractable layer in the zone of increased rock pressure in front of the breakage face). As a result of the calculations, we find that at a speed $V = 4$ m/day, $\sigma_y(V) = 19.4$ MPa. The corresponding calculations for the conditions, of overworking when $L = 60$ m, give us: $\sigma_y(V) = 7.56$ MPa at $V = 2$ m/day and $\sigma_y(V) = 16.63$ MPa at $V = 4$ m/day.

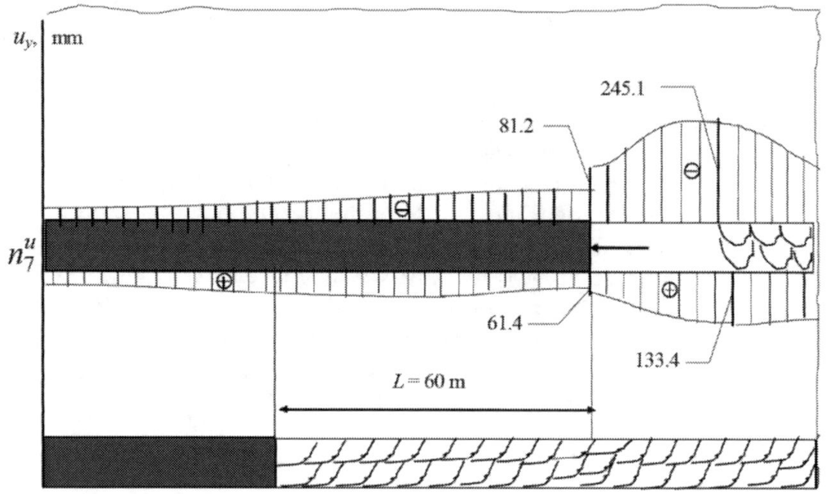

Fig. 3. Diagrams of vertical displacements u_y in the roof and soil of the layer n_7^u (underworking, movement of long face to the edge part of the adjacent layer).

Consequently, in the cross section $x = 0$ tensions are with increasing of rate $\sigma_y(V)$, and the faster the higher the tension concentration coefficient σ_y. However, this reduces the "removed" pressure, so the total load on the mechanized support decreases with increasing speed.

Of practical interest is the dependence of the value of the additional load $q(V) = 0.03P(V)$ on the tension concentration factor $(K_y)_{max}$. Such dependences for values of V, equal to 1.5; 2.0; 3.0 and 4.2 m/day, are shown in Fig. 4.

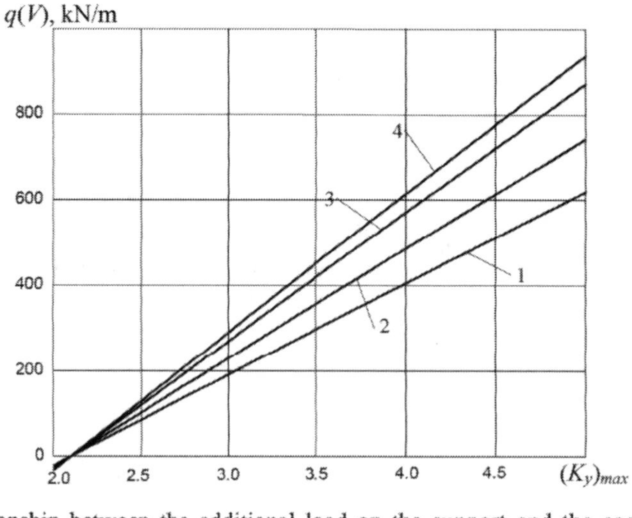

Fig. 4. The relationship between the additional load on the support and the coefficient of tension concentration: 1 – $V = 4.2$ m/day; 2 – $V = 3.0$ m/day; 3 – $V = 2.0$ m/day; 4 – $V = 1.5$ m/day.

Thus, given the values of L, l/l_n and $E_r/\gamma_n H$ and using on one of the formulas (2), (4), (6), or (8), depending on the direction of the long face movement and the development conditions under consideration, one can find the corresponding value of maximum tension concentration factor in front of the breakage face $(K_y)_{max}$, as well as the main load on the section of the mechanized support P_0 from the weight of the rocks in the zone of the ultimate stress state over the long face. Further, from the found value $(K_y)_{max}$ from the graph in Fig. 4, we determine the additional load on the section of the mechanized support $q(V) = 0.03P(V)$ at a given speed of long face movement.

The disadvantage of this approach is the fact that the reference pressure zone is considered only in front of the breakage face, and increased rock pressure behind the breakage face affects the processes in the breakage face too. In addition, the angle δ within the interbed thickness, on which, first of all, the SDS of the massif area of interest depends (the vicinity of the breakage face), can significantly differ from the average value δ of the entire superincumbent rocks. Finally, the picture of the SDS of the soil and the roof of the layers being developed is also determined by the speed at which the breakage face is moved.

4 Substantiation of ways to reduce the negative consequences of deformation and subsidence of the earth's surface in coal-mining regions

When breakage faces are working, dynamic demonstrations of rock pressure are possible, expressed in the instantaneous destruction of the lower layers of the roof or a large part of the rock massif in the vicinity of the breakage face up to the cutting of the entire interbed, which can lead to a catastrophic increase in load on the support. In this case, there is a sharp decrease in the stability of the immediate roof, often there are cases of blockages of long faces, "strict" clamping of mechanized supports sections, and also squeezing of coal and soil swelling.

Rock displacements and deformations occur during underground mining of coal layers. These displacements cause significant changes in the field of gravitational forces with the formation of support zones, in which the tensions are several times higher than the tension of the untouched massif [30 – 33]. Changes in the rock massif, depending on the natural, mining, technical and production factors, lead to significant difficulties in cleaning works conduction. The harmful influence of the reference pressure is evident not only in the plane of the layer, but also in rocks at a considerable distance from it up to the surface [35-38]. Significant values of subsidence on the earth's surface contribute to the disruption of the ecological balance of the coal-mining regions.

Zones of compression of rocks arise as a result of the demonstration of the reference pressure, and are the main cause of subsidence of rocks and the earth's surface beyond the boundaries of clearing works. And, consequently, they determine the position in shift trough of the points with critical values of deformations, along which the initial angular parameters used to calculate the displacements and deformations of rocks and the earth's surface are determined.

Deformations of the surface, appearing under the influence of mining operations, along with the disturbance of the soil layer make it difficult to process agricultural land. It is established that the subsidence of the earth's surface to a depth of 3 m reduces soil fertility by 10%, to 6 m – by 50%. When the surface subsidence is more than 8 m, the fertile soil layer is destroyed. Subsidence troughs are often filled with water, which causes flooding and waterlogging of land.

Protection of the surface from the negative effects of mining operations is possible in the following ways (or a combination thereof) [27, 31]:

– the abandonment of such coal pillars, when the disturbance of the earth's surface is prevented (thus a significant part of the coal reserves will be lost);

– implementation of a system of mining and technical measures to prevent or reduce subsidence of the earth's surface (laying the mined-out space with rock or a mixture of sand

and clay, the use of chamber development systems, etc.);

– implementation of a system of measures on the surface to neutralize the negative consequences of mining operations (advanced and subsequent reclamation, the construction of artificial water reservoirs, etc.). With this method, in addition to large material and financial costs, the time factor also plays a role (there is much more time to reclaim the surface than to subjugate these territories).

One of the most realistic ways to significantly reduce the negative impact of mining operations and reduce the cost of restoring the lands and structures that have been worked on is the use of coal mining technology that provides for the full or partial laying of the mined out space. The goaf stowing of the mined-out space, with its massive application, makes it possible to change the nature of the displacement of the underworked mountain massif. The angles of displacement are changing; the process of displacement is stretching in time by a factor of 3 – 5 compared with the total rock caving. The effectiveness of the goaf stowing is estimated by reducing the deformations of the underworked massif and the values of the surface lowering.

With the construction of filling massifs designed to protect underworked areas, the demand for high load-bearing capacity of the filling massif is in the foreground, which will minimize the displacement of the mined rocks and the harmful influence of underworking. However, the presence of high-quality filling material may not be sufficient if, for example, the goaf stowing of the mined-out space is made at a great distance from the breakage face of the long face, where significant subsidence of the roof rocks has already occurred. Therefore, the parameters of the technology used for development of the layer play not the least role in providing surface protection. It should be noted that the task of selecting the parameters of the filling massif and the influence of the technological parameters of layer development on the efficiency of goaf stowing, for the conditions of the main coal-mining basins, has not been completely solved.

Thus, the experience of applying a goaf stowing to the mined-out space indicates that prerequisites can be created for solving a whole complex of engineering and technical problems [39] in addition to reducing the number of rocks emitted from mines and improving due to this the ecological situation in coal-mining regions, the successful implementation of the protection of surface objects and the maintenance-free support of preparatory workings.

For example, due to full goaf stowing, methane flow into the breakage face from the mined-out space is sharply reduced and the probability of endogenous fires is completely eliminated, the intensity of the inrush formation process is reduced, and the state of the water resources of coal mining regions is improved [14, 40]. However, this experience was not continued due to increased labour intensity of work, reduced load on long face and the lack of incentives to improve the quality of extracted coal.

In view of the foregoing, in the projects for the preparation of breakage faces and in the passport for managing the roof and lining, it will be necessary to envisage the implementation of additional measures for managing the roof and of lining of the long face during the transition of zones of increased rock pressure that prevent or neutralize dangerous demonstrations of rock pressure. It is also necessary to continue developing methods to reduce subsidence of the earth's surface during the mining of coal layers, as well as to develop reserves under protected objects.

5 Conclusions

As a result of the studies, the following conclusions can be drawn:

1. On the basis of the method of boundary elements, an effective computational algorithm for determining the SDS of an inhomogeneous massif was developed with

reference to the development of a multiple coal flat-lying layers.

2. Using the developed algorithm for the conditions of the mines of the Lviv-Volyn Coal Basin, the boundaries of zones of increased rock pressure in the roof and soil of the extracting layer are established, the values of the maximum tension concentration coefficients σ_y relative to the γH level are determined, the load acting on the section of the mechanized support from the weight of rocks, located in the area of the ultimate tension state in the roof over the long face is found.

3. It is established that in the conditions of the mines of the state enterprise "Lvivvuhillia" during the work of breakage faces in the zones of influence of the edge parts of adjacent layers, the most dangerous from the point of view of the demonstration of rock pressure is the period of long face release from under the edge part of the underlying layer. Therefore, the rational power parameters of the mechanized support must be determined by the characteristics of the SDS of this particular area of the massif.

4. It is established that with the increase in the development depth from 300 to 600 m, the maximum displacements u_y in the arch of the preparatory working in the conditions of underworking increase 1.79 times.

5. One of the most realistic ways to significantly reduce the negative impact of mining operations and reduce the cost of restoring damaged lands and facilities is the use of coal mining technology, which provides for the full or partial goaf stowing of the mined-out space.

This work was conducted within the project "Development of technological bases of ecologically safe extraction of minerals in technogenic-loaded mining regions of Ukraine" (State registration No. 0117U001134).

References

1. Pivnyak, G.G., & Shashenko, O.M. (2015). Innovations and safety for coal mines in Ukraine. *Naukovyi Visnyk Natsionalnoho Hirnychoho Universytetu*, (6), 118-121.
2. Petlovanyi, M., Lozynskyi, V., Saik, P., & Sai, K. (2018). Modern experience of low-coal seams underground mining in Ukraine. *International Journal Of Mining Science And Technology*. https://doi.org/10.1016/j.ijmst.2018.05.014
3. Savchuk, V., Prykhodchenko, V., Buzylo, V., Prykhodchenko, D., & Tykhonenko, V. (2013). Complex use of coal of Northern part of Donbass. *Annual Scientific-Technical Collection – Mining of Mineral Deposits 2013*, 185-191. https://doi.org/10.1201/b16354-34
4. Belov, O., Shustov, O., Adamchuk, A. & Hladun, O. (2018). Complex processing of brown coal in Ukraine: history, experience, practice, prospects. *Solid State Phenomena*, (277), 251-268. https://doi.org/10.4028/www.scientific.net/SSP.277.251
5. Mormul, T.M., & Terekhov, Ye.V. (2017). Environmental and economic estimation of technological solutions in terms of land resource conservation in the process of open-cast mining. *Naukovyi Visnyk Natsionalnoho Hirnychoho Universytetu*, (3), 122-128.
6. Prokopenko, V. I., & Litvinov, Yu. I. (2017). Environmental orientable imperative of developing the technology and excavation of horizontal fields. *Naukovyy visnyk Natsionalnoho Hirnychoho Universytetu*, (2), 51-57.
7. Gumenik, I.L., Lozhnikov, O.V., & Panasenko, A.I. (2013). Deliberate dumping technology for mining reclamation effectiveness improvement. *Naukovyi Visnyk Natsionalnoho Hirnychoho Universytetu*, (5), 48-53.
8. Khomenko, O., Kononenko, M., Myronova, I., & Sudakov, A. (2018). Increasing ecological safety during underground mining of iron-ore deposits. *Naukovyi Visnyk Natsionalnoho Hirnychoho Universytetu*, (2), 29-38. https://doi.org/10.29202/nvngu/2018-2/3
9. Cherniaiev, O.V. (2017). Systematization of the hard rock non-metallic mineral deposits for improvement of their mining technologies. *Naukovyi Visnyk Natsionalnoho Hirnychoho Universytetu*, (5), 11-17.

10. Kolesnik, V.Ye., Fedotov, V.V., & Buchavy, Yu.V. (2012). Generalized algorithm of diversification of waste rock dump handling technologies in coal mines. *Naukovyi Visnyk Natsionalnoho Hirnychoho Universytetu*, (4), 138-142.
11. Gomelya, M.D., Trus, I.M., & Radovenchyk, I.V. (2014). Influence of stabilizing water treatment on weak acid cation exchange resin in acidic form on quality of mine water nanofiltration desalination. *Naukovyi Visnyk Natsionalnoho Hirnychoho Universytetu*, (5), 100-105.
12. Pavlichenko, A.V., & Kroik, A.A. (2013). Geochemical assessment of the role of aeration zone rocks in pollution of ground waters by heavy metals. *Naukovyi Visnyk Natsionalnoho Hirnychoho Universytetu*, (5), 93-99.
13. Kolesnyk, V., Kulikova, D. & Kovrov, S. (2013). In-stream settling tank for effective mine water clarification. *Annual Scientific-Technical Collection – Mining of Mineral Deposits 2013*, 285-289. https://doi.org/10.1201/b16354-53
14. Gorova, A., Pavlychenko, A., Kulyna, S., & Shkremetko, O. (2012). Ecological problems of post-industrial mining areas. *Geomechanical Processes During Underground Mining 2012*, 35-40. https://doi.org/10.1201/b13157-7
15. Vambol, S., Vambol, V., Kondratenko, O., Suchikova, Y., & Hurenko, O. (2017). Assessment of improvement of ecological safety of power plants by arranging the system of pollutant neutralization. *Eastern-European Journal of Enterprise Technologies*, *3*(10(87), 63-73. https://doi.org/10.15587/1729-4061.2017.102314
16. Gorova, A., & Klimkina, I. (2007). Methodology of Socio-Ecological Monitoring using Cytogenetic Methods. *Multiple Stressors: A Challenge For The Future*, 91-102. https://doi.org/10.1007/978-1-4020-6335-0_7
17. Malovanyy, A., Plaza, E., Trela, J., & Malovanyy, M. (2014). Combination of ion exchange and partial nitritation/Anammox process for ammonium removal from mainstream municipal wastewater. *Water Science & Technology*, *70*(1), 144-151. https://doi.org/10.2166/wst.2014.208
18. Shmandiy, V., Bezdeneznych, L., Kharlamova, O. at al. (2017). Methods of salt content stabilization in circulating water supply systems. *Chemistry & Chemical Technology*, *11*(2), 242-246. https://doi.org/10.23939/chcht11.02.242
19. Melnyk, L., Bessarab, O., Matko, S., & Malovanyy, M. (2015). Adsorption of Heavy Metals Ions from Liquid Media by Palygorskite. *Chemistry & Chemical Technology*, *9*(4), 467-470. https://doi.org/10.23939/chcht09.04.467
20. Khomenko, O., Sudakov, A., Malanchuk, Z. & Malanchuk, Ye. (2017). Principles of rock pressure energy usage during underground mining of deposits. *Naukovyi Visnyk Natsionalnoho Hirnychoho Universytetu*, (2), 35-43.
21. Malanchuk, Ye.Z., Lozynskyi, V.G., Dychkovskyi, R.O., Falshtynskyi, V.S., & Saik, P.B. (2016). Experimental study of the influence of crossing the disjunctive geological fault on thermal regime of underground gasifier. *Naukovyi Visnyk Natsionalnoho Hirnychoho Universytetu*, (5), 21-29.
22. Bondarenko, V., Symanovych, G., & Koval, O. (2012). The mechanism of over-coal thin-layered massif deformation of weak rocks in a longwall. *Geomechanical Processes During Underground Mining 2012*, 41-44. https://doi.org/10.1201/b13157-8
23. Khomenko, O. (2012). Implementation of energy method in study of zonal disintegration of rocks. *Naukovyi Visnyk Natsionalnoho Hirnychoho Universytetu*, (4), 44-54.
24. Bondarenko, V.I., Simanovich, G.A., Kovalevska, I.A., Fomichov, V.V., & Serdiuk, V.P. (2007). Research of Rock Stresses and Deformations Around Mining Workings. *Technical, Technological And Economical Aspects Of Thin-Seams Coal Mining, International Mining Forum 2007*, 47-56.
25. Kovalevs'ka, I., Symanovych, G., & Fomychov, V. (2013). Research of stress-strain state of cracked coal-containing massif near-the-working area using finite elements technique. *Annual Scientific-Technical Collection – Mining of Mineral Deposits 2013*, 159-163. https://doi.org/10.1201/b16354-28
26. Lozynskyi, V., Saik, P., Petlovanyi, M., Sai, K., & Malanchuk, Y. (2018). Analytical Research of the Stress-Deformed State in the Rock Massif around Faulting. *International Journal Of*

Engineering Research In Africa, (*35*), 77-88. https://doi.org/10.4028/www.scientific.net/jera.35.77

27. Buzylo V., Yavorsk'yy, A., & Yavorsk'yy. V. (2012). Analysis of stress-strain state of rock mass while mining chain pillars by chambers. *Geomechanical Processes during Underground Mining 2012*, 95-98. https://doi.org/10.1201/b13157-16
28. Khomenko, O., Kononenko, M., & Bilegsaikhan, J. (2018). Classification of Theories about Rock Pressure. *Solid State Phenomena*, (277), 157-167. https://doi.org/10.4028/www.scientific.net/ssp.277.157
29. Shashenko, O.M., & Kovrov, O.S. (2016). Comparative analysis of two failure criteria for rocks and massifs. *Naukovyi Visnyk Natsionalnoho Hirnychoho Universytetu*, (6), 54-59.
30. Soboliev, V., Bilan, N., & Samovik, D. (2013). Magnetic stimulation of transformations in coal. *Annual Scientific-Technical Collection – Mining of Mineral Deposits 2013*, 221-225. https://doi.org/10.1201/b16354-41
31. Buzylo, V., Savelieva, T., & Saveliev, V. (2012). Study of rock displacement with the help of equivalent materials using room-and-pillar mining method. *Geomechanical Processes During Underground Mining 2012*, 29-34. https://doi.org/10.1201/b13157-6
32. Kovalevska, I.A., Fomichov, V.V., & Skitenko, A.I. (2012). Influence of stratification characteristics and inclination angle of rock massif on mine workings stability computational test results. *Naukovyi Visnyk Natsionalnoho Hirnychoho Universytetu*, (3), 31-36.
33. Babets, D.V., Sdvyzhkova, O.O., Larionov, M.H., & Tereshchuk, R.M. (2017). Estimation of rock mass stability based on probability approach and rating systems. *Naukovyi Visnyk Natsionalnoho Hirnychoho Universytetu*. (2), 58-64.
34. Prusek, S. (2018). Perspektywy stosowania samodzielnej obudowy kotwowej w kopalniach JSW S.A. In *XXVII Szkola eksploatacji podziemnej*. Krakow.
35. Sdvizhkova, Ye.A., Babets, D.V., & Smirnov, A.V. (2014). Support loading of assembly chamber in terms of Western Donbas plough longwall. *Naukovyi Visnyk Natsionalnoho Hirnychoho Universytetu*, (5), 26-32.
36. Fuławka, K., Mertuszka, P., & Pytel, W. (2018). Monitoring of the stability of underground workings in Polish copper mines conditions. *E3S Web Of Conferences*, *29*, 00008. https://doi.org/10.1051/e3sconf/20182900008
37. Pytel, W., & Mertuszka, P. (2017). Blasting parameters alternate selection as a tool for elastic wave effect amplification at potentially instable locations within main roof strata. In *17th International Multidisciplinary Scientific Geoconference SGEM 2017. Science And Technologies In Geology, Exploration And Mining*. https://doi.org/10.5593/sgem2017/13/s03.015
38. Vladyko, O., Kononenko, M., & Khomenko, O. (2012). Imitating modeling stability of mine workings. *Geomechanical Processes During Underground Mining 2012*, 147-150. https://doi.org/10.1201/b13157-26
39. Kolosov, D., Bilous, O., Tantsura, H., & Onyshchenko, S. (2018). Stress-strain state of a flat tractive-bearing element of a lifting and transporting machine at operational changes of its parameters. *Solid State Phenomena*, (277), 188-201. https://doi.org/10.4028/www.scientific.net/ssp.277.188
40. Adamenko, Y.O., Arkhypova, L.M., & Mandryk, O.M. (2016). Territorial standard quality hidroekosystem protected areas, *Hydrobiological Journal*, *52*(6), 51-59.

A modern vision of simulation modelling in mining and near mining activity

Roman Dychkovskyi[1*], *Volodymyr* Falshtynskyi[1], *Vladyslav* Ruskykh[1], *Edgar* Cabana[2], and *Oleksandr* Kosobokov[3]

[1]National Mining University, Department of Underground Mining, 19 Yavornytskoho Ave., 49005 Dnipro, Ukraine
[2]San Agustin National University Arequipa, Institute of the Center of Renewable Energy and Energy Efficiency, 117 Santa Catalina St., 04000 Arequipa, Peru
[3]Company DTEK ENERGY LLC, Operational Efficiency Department, 57 Lva Tolstoho Str., 01032 Kyiv, Ukraine

Abstract. The paper represents the creation of the software simulation system, which reproduce the basic processes of mining and near production. It presents the consideration of such systems for both traditional and non-traditional mineral extraction systems. The principles of using computer recognition of processes are also presented in other processes of carbon-containing raw materials transition, as well as power production and waste utilization of mining production. These systems considerably expand the manageability of a rather complicated mining enterprise. The main purpose of such research is the simulation reproduction of all technological processors associated with the activity of mining enterprises on the display of the dispatch center. For this purpose, is used so-called UML-diagrams, which allows to simulate mining and near mining processes. Results of this investigation were included to the Roman Dychkovskyi thesis of the scientific degree of the Doctor of the Technique Sciences "Scientific Principles of Technologies Combination for Coal Mining in Weakly Metamorphoses Rockmass".

1 Introduction

Mine technological systems represent an ordered set of objects, the functioning of which in time and space implements a complete technological cycle of underground mining method. One of the main parts of technological system is a net of underground mining works that provide the preparation of reserves for processing, transportation of extracted raw materials, ventilation and delivery of people and materials [1 – 4].

The dynamic development of mining operations, the presence of various geodynamic stressed fields in anisotropic environment, the variation of the geological and structural formation of an array in the way of moving of a treatment face leads to the necessity of making adequate changes in mining technology and rock pressure control methods. The efficiency of operation of mechanized complexes depends on timely predicted correction of

* Corresponding author: dichre@yahoo.com

technological parameters and technical composition, an adequate change in the geo-mechanical situation at mining site.

The assessment of mining and technical situation during implementation of new radical mining technologies [3 – 11], change of physical and chemical properties under the influence of external factors [12, 13] is especially important. Extraction of minerals in complicated mining and geological conditions also are presented on specialized literature [14 – 16]. Detailed description of the mechanism which is using the method of final elements for determine the stresses in rock mass, and is proper for simulating the process of underground coal gasification is presented at works [3, 17, 18].

On the way of formation of man-made space in the mineral extraction the geo-mechanical situation may change within one mining area. Mining works develop dynamically into the depth of a rock array, and man-made stress fields are layered in the same direction. The presence of various geodynamic stressed fields in anisotropic geological environment influences the direction of the resulting stress vector having a different value depending on its orientation in the array. The development of mining works can be correlated differently to the highest values of the stress vector and this affects the stability of mining works. The entire mining equipment operation under these conditions depends on timely predicted correction of technological parameters in accordance with changes in the geo-mechanical situation [19 – 21].

The same situation is observed when mining production is curtailed. Utilization of waste, sewage or mine water requires continuous monitoring of this process. Any malfunction leads to significant surface contamination and adversely affects the flora and fauna of the after-mining areas [22, 23, 24].

Analysis of mining and geological conditions, geo-mechanical, mining and technical factors provide the opportunity to obtain data to determine technological parameters of all processes associated with mining. Solution of this task is provided by simulation modelling of production processes in a mine, taking into account changes in mining and geological, technological, mining and technical factors.

2 Source simulation data

Essence of software simulation processes of mining is presented at [3, 25]. So it is mentioned that creation of simulating model of all mine in the specified volumes demands not only big efforts at the level of the developed structure of programmers, but also high system opportunities of the software. Therefore, at design of such difficult information system it is offered to break it into components, each of which is considered separately. Two various ways of such splitting into subsystems are possible: structural (or functional) splitting and object (component) decomposition.

The structure of simulation modeling is divided into three sectors [26]. "The perfect idea" corresponds real value of object nature. This sector is ideal reproduction of process of modeling. When modeling, actually, has no opportunity to reach 100% convergence to results of "perfect idea"; it is possible to approach the received results to real conditions as much as possible only. The "zone of variable data" is changeable. Its purpose is the object adaptation – model to real object by means of comparison of fictitious results of model to real results in the conditions of mine. It is necessary in the course of modeling. The "zone of accidents" is sector of modeling at which the validity of process of modeling isn't observed. It is considered that here has no compliance of results of modeling.

The direction of increase of reliability of results of modeling is indicated the need as much as possible precisely to correspond sector by "perfect idea" is real value of modeling. A new interpretation of evolution of technologies and prospects of research with using big data can be also applied [27].

Use of simulation modeling in terms of modern mine is based upon the simulation representation of all its qualitative and quantitative characteristics: type and structure of a mine; mining techniques; schemes of mine field opening; mining systems; sequence of seam mining; plan of reserves extraction; types of underground transport; schemes and types of mine ventilation; number of mine field blocks within a simultaneous operation; number of levels and sublevels within a mine field and sequence of their extraction; number of panel mine sections within a level; number of seams and stopes being developed simultaneously; number of working days per year and mining shifts per day; number of hours per shift; sizes of mine fields and blocks; sizes of levels, panels, mine sections; length of longwalls and extraction pillars; advance rate of stopes; loads upon stopes, levels, panels, blocks; mine capacity; section areas of mine workings etc. The same situation is when we are going to simulate near mining processes. It is necessary to mentioned that should be simple and understandable to the performer.

3 System of functional representation

So-called activity diagrams are used to model the processes of operation performance in UML language. Graphical component used in them is like the state diagrams as the activity diagrams also contain the symbols of stationary states and qualitative transitions of the system. The difference is in the semantics of states and in the nonavailability of event signatures within the transitions. Each action of the diagram corresponds to the performance of some elementary operation; transition to the following state is possible only after the certain goal reaching. Graphically, the activity diagram is represented in the form of activity graph which peaks are actions; arcs are qualitative transitions from one state to another [26, 28].

It can represent interrelations between the idea and visual results as well as describe internal structure of the represented objects and relation type. The diagram cannot indicate the data concerning temporary aspects of the whole system functioning. Class diagram is the further development of the conceptual model.

Class diagram is a certain graph which peaks are the components of "classificatory" type connected with different structural relations. Class diagram may also contain interfaces, packages, connections, and separate software packages. The diagram means static structural model of the system being designed. However, it depends upon the changes in space and time of all the processes being modeled. Class diagram consists of numerous components demonstrating in their totality the declarative knowledge on the area under study. The knowledge is interpreted in basic concepts of UML language. In this context, separate diagram components may form the packages to represent more general model. If class diagram is the part of some package, then its components should correspond to the package components including possible references to the components of other packages.

In general case, package of a static structural model may be represented in the form of one or several class diagrams. Certain representation is decomposed into separate diagrams as a matter of convenience as well as for graphical visualization of structural relations of a visual area. In this context, diagram components correspond to the components of a static semantic model. In turn, a system model should be conformed to the internal class structure being described in UML language.

Class in UML language is to determine certain number of objects having similar structure, behavior, and relations with the objects of other classes. Graphically, the class is shown in the form of a square which can be divided additionally into sections by horizontal lines. The sections may contain the name of class, attributes (changeable data), and operations (methods to achieve the result).

Process of the class diagram development is in the center of object-oriented modeling of complex systems. Ability to select the classes properly as well as to find relations between them is often crucial both for programming and productivity of the programs execution. As practice shows, each programmer wants to use personal experience while developing new projects. It is stipulated by the desire to reduce new task into the partially completed ones to have the possibility of applying the tested fragments of a program code and separate constituents of the component library.

Such an approach makes it possible to reduce considerably the period of project implementation; however, it is reasonable only in case when new project does not differ too much from the previous ones in terms of its concept and technology.

After the class diagram development, modeling process may continue in two ways. On the one hand, if system behavior is trivial, then one can start developing cooperation diagram and component diagram. In terms of complex dynamic systems, the behavior becomes the essential aspect of their functioning. Behavior is the specified consequently while developing diagrams of state, situation change, and activity during the implementation of technological solutions.

The author has made an attempt to systematize into a single system a computer-aided design for simulation representation of a mine as a single model consisting of various class interactions according to certain clear scheme within the class diagram of functional design system (Fig. 1).

Class program consists of the functionals determining the mathematical mechanism to be involved into the modeling and the sequence of the mechanism involvement.

CCadApplication is the collection of the required documents (cross sections of mine workings and lithological difference of rock which a user deals with) and their processing: adding the necessary clarifications, deleting, and saving the documents. The class provides the connection with graphic interface of the program: it complements the toolbar with graphic element objects (lines, texts etc.) and commands (switching on/off 3D object demonstration).

CCadDocument is the collection of CcadShape objects (additional graphic elements: lines, splines, texts, coordinate grid, contact planes, representation of mine workings) and the collection handling: adding, deleting, clear line-tracing, object searching etc.).

CCadShape is the basic class to describe the objects behavior in the document. The essence and the objective of the class are as follows:

– processing the events performed by means of the control systems (touchpad, mouse, keyboard);

– clear line-tracing of the object;

– initializing the basic parameters (colour, lint thickness; object orientation both in space and in Cartesian reference system);

– saving the parameters restoring with the help of CCadParameters class. It is used to save the parameters of all the objects in the unified format. Moreover, it helps save data in a file and organize incorrect action cancelling; and

– operations with the object: moving, testing the conditions (a point in the document belongs to the object; square area of the document contains the object itself).

CCadDrawer is the clear line-tracing of the corresponding detalization (set of functions to perform graphic drawing); it is used by CCdShape objects and represents following object types either in display or printed form:

– Line (thickness, colour, type (solid, dashed);

– Spline (thickness, colour, type);

– Point (mostly to indicate the area of object editing);

– Text (font, size, inclination angle, colour);

– Rectangle, triangle, circle etc. (object of the preset shape).

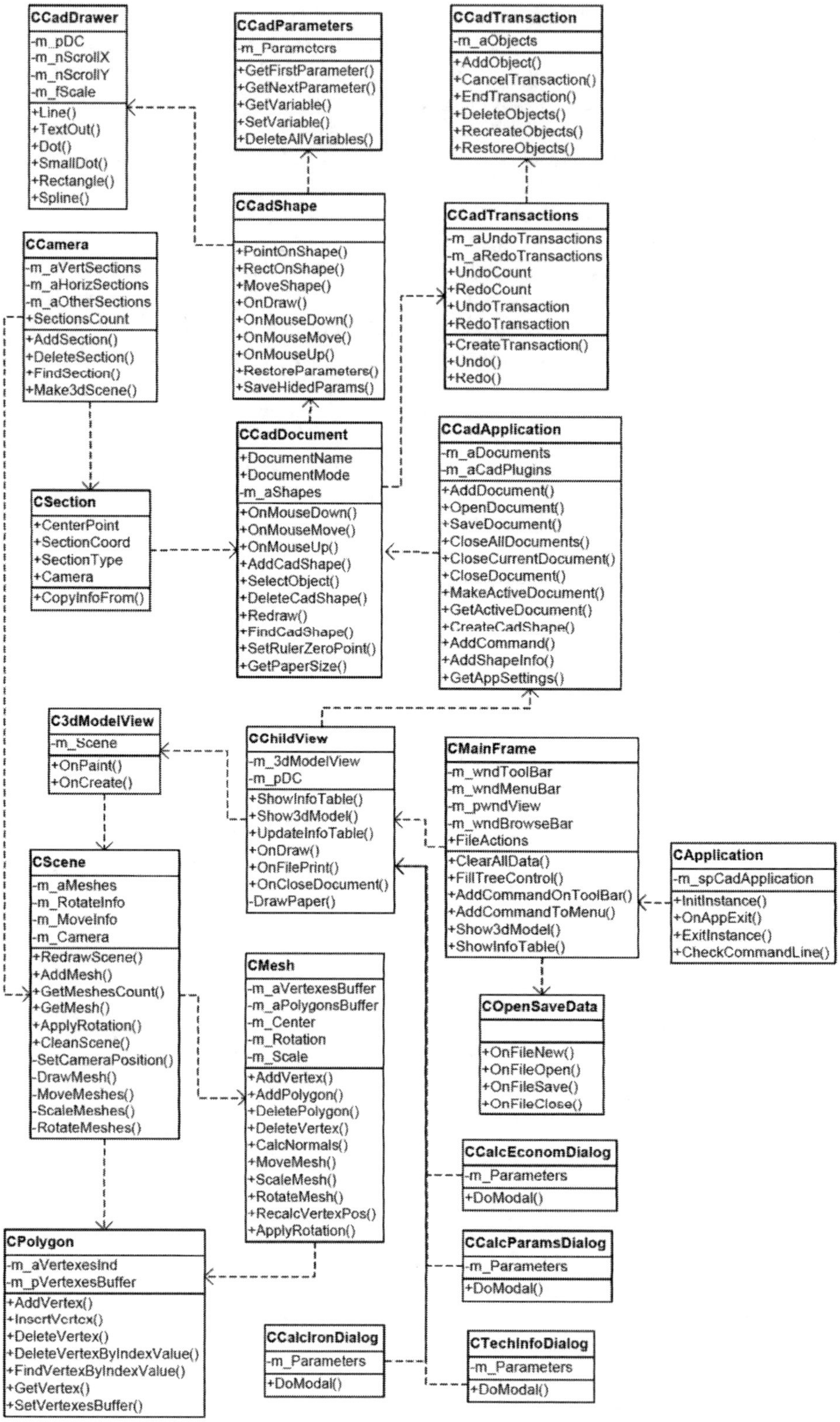

Fig. 1. Class diagram of the functional design system.

CCadParameters are the data storing the CCadShape objects. The data are stored in the form of pairs of rows called “variable name”, “value”, and “visible/invisible”. If the variable is “visible”, then its value may be changed in the property window.

CCamera is the collection of cross sections (a mine field site). Cross section objects may be deleted, added, corrected, and searched for. Similarly, 3D model of mine field parts is built in this class.

CSections is the cross section description (section types (vertical/horizontal), vertical coordinates, horizontal coordinates).

CCadTransaction, CCadTransactions store the object changes; they are also used to delete incorrect actions. Before any CcadShape graphic object changes, sample of CCadTransaction and copies of CCadShape object state in the form of CCadParameters sample are created.

C3dModelView is graphic elements for 3D mine representation. It initializes OPENGZ library, processes window messages of a mouse, and transmit them to the CScene sample.

SCsene is the class of the operation with 3D objects. It contains the collection of 3D objects (CMesh), position of a viewing point (a video-camera principle) within the “space - time” system. It implements clear line-tracing of all the objects, their selection, change in position and rotation as well as change in position of the point itself.

CMesh is the description of 3D object, movement operation, and rotation. It contains collection of polygons (CPolygon) and collection of points (CVertex). Together they gives the good tool for the visualization of processes.

CPolygon is the description of polygon (elementary 3D object). It contains the array of indices and points of spatial representation of different simulation objects located in CMesh.

CLine, CText, CSpline are basic objects of the document inherited from CCadShape. They are used to process and display both text and graphic information. In addition, their function is to visualize final information and demonstrate it to a user in the aesthetic and accessible form.

CSectionCoord is graphic representation of the document grid mesh. 3D model is used to determine the corresponding cross sections of all the modeled objects and to represent them in the form of lines.

CIronContour determines the design limits. In this case, there are two levels of representing the simulation representation limits: mineral contour within the whole mine and within the separate extraction site. The class is also responsible for the representation of rock difference within the vertical plane – limits for roof and floor rocks simulation.

CSurveyTraverse is the representation of man-made objects – all the underground and surface structures in reliance on their shapes and corresponding geometry.

COpenSaveData is operation and command class. It provides the opening and storing of the file containing the whole information. The class is responsible for the control of all the developed information.

CTechInfoDialog contains the information concerning databases. Calculation results obtained by other software products also act as the input information apart from the statistic data. In our case, these are the data from the assessment of geomechanical situation within a certain extraction site.

CCalcEconomDialog is the class of economic effect calculation; it is responsible for economic expediency of the implementation. The available principles to assess the efficiency of implementing new equipment and technology under certain mining and technical conditions are used as a mathematical mechanism.

CCalcIronDialog is responsible for rock pressure calculation; it represents geomechanical situation of the areas under study. In this case, Flac 5.01 software is applied; the software product uses finite difference mode in terms of elastic-plastic representation of the rock mass. The obtained results are compatible with CTechInfoDialog class.

CcalcParamsDialog is the calculation of technological parameters. It is a final stage of software operation and aimed at selecting rational use of high-load coal extraction relying upon technical provision of mechanical coal breaking from a seam, its transportation, technological support of mining operations, operational safety, economic and environmental expediency. The developed class diagram of the functional design system is not the end product in the form of separate software package in the computer-programming language. It contains all the required constituents for simulation mine representation in terms of high-productive mining and near mining processes. The author has managed to implement only certain parts and combine them into the consistent simulation system.

4 Conclusions

Simulation is widespread in all engineering studies, mainly because it implements the principles of decomposition, abstraction and hierarchy [3]. Each model efficiency is determined by the results obtained in ordinary and unordinary conditions. Comparison of simulation results with the field data gives an opportunity to carry out the relevant software updating and adjustments.

Simulation modeling is quite efficient, flexible, and dynamic method to represent complex systems; the method is the integral part of representation of complex mining production processes. It is based on a stable theoretical basis and at the same time gives the programmer a certain degree of self-expression will.

Development of a mine simulation model is not aimed at the development of a final product in the form of software package; its objective is to build up clear system of its implementation including all the mathematical algorithms used for calculations. Such models are also necessary in the secondary extraction of raw materials from dumps, recycling of mine wastes or purification of mine and wastewater.

This work was conducted within the projects "Scientific basics to form unified system to preserve and generate power of the objects of fuel and energy complex of Ukraine" (State registration No. 0117U001127) and "VirtualMine: as a Modeling Tool for Wider Society Learning" and Dubrovnik International ESEE Mining School (projects in the frame of EIT Raw Materials).

References

1. Avershyn, S.G. (1960). *Raschet deformatsiy massiva gornykh porod pod vliyaniem podzemnykh razrabotok*. Leningrad: Izdatel'stvo vsesoyuznogo nauchno-issledovatel'skogo marksheyderskogo instituta.
2. Zhanchiv, B., Rudakov, D., Khomenko, O., & Tsendzhav, L. (2013). Substantiation of mining parameters of Mongolia uranium deposits. *Naukovyi Visnyk Natsionalnoho Hirnychoho Universytetu*, (4), 10-18.
3. Dychkovskyi, R.O. (2013). *Naukovi zasady syntezu tekhnolohii vydobuvannia vuhillia iz tonkykh ta nadtonkykh plastiv u slabometamorfizovanykh porodakh pratsiuvannia*. Dnipropetrovsk: Natsionalnyi hirnychyi universytet.
4. Pivnyak, G.G., & Shashenko, O.M. (2015). Innovations and safety for coal mines in Ukraine. *Naukovyi Visnyk Natsionalnoho Hirnychoho Universytetu*, (6), 118-121.
5. Dychkovskyi, R.O., Lozynskyi, V.H., Saik, P.B., Petlovanyi, M.V., Malanchuk, Ye.Z., & Malanchuk, Z.R. (2018). Modeling of the disjunctive geological fault influence on the exploitation wells stability during underground coal gasification. *Archives of Civil and Mechanical Engineering*, *18*(4). https://doi.org/10.1016/j.acme.2018.01.012
6. Caceres, E., & Alca, J.J. (2016). Rural Electrification Using Gasification Technology: Experiences and Perspectives. IEEE *Latin America Transactions*, *14*(7), 3322-3328.

https://doi.org/10.1109/TLA.2016.7587637

7. Lavrov, N.V. (1957). Physical and chemical bases of combustion and gasification of fuel. Moscow: Metallizdat.
8. Cabana, E.C. (2017). Formation of thermal fields by the energy-chemical complex of coal gasification. *Naukovyi Visnyk Natsionalnoho Hirnychoho Universytetu*, (5), 36-42.
9. Xin, L., Wang, Z., Wang, G., Nie, W., Zhou, G., Cheng, W., & Xie, J. (2017). Technological aspects for underground coal gasification in steeply inclined thin coal seams at Zhongliangshan coal mine in China. *Fuel*, (191), 486-494. https://doi.org/10.1016/j.fuel.2016.11.102
10. Dychkovskyi, R.O. (2015). Forming the bilayer artificially shell of georeactor in underground coal gasification. *Naukovyi Visnyk Natsionalnoho Hirnychoho Universytetu*, (5), 37-42.
11. Dychkovskyi, R.O. (2015). Determination of the rock subsidence spacing in the well underground coal gasification. *Naukovyi Visnyk Natsionalnoho Hirnychoho Universytetu*, (6), 30-36.
12. Sobolev, V.V., & Usherenko, S.M. (2006). Shock-wave initiation of nuclear transmutation of chemical elements. *Journal de Physique IV (Proceedings)*, (134), 977-982.
13. Sdvizhkova, Ye.A., Babets, D.V., & Smirnov, A.V. (2014). Support loading of assembly chamber in terms of Western Donbas plough longwall. *Naukovyi Visnyk Natsionalnoho Hirnychoho Universytetu*, (5), 26-32.
14. Lozynskyi, V.G., Dychkovskyi, R.O., Falshtynskyi, V.S., Saik, P.B., & Malanchuk, Ye.Z. (2016). Experimental study of the influence of crossing the disjunctive geological faults on thermal regime of underground gasifier. *Naukovyi Visnyk Natsionalnoho Hirnychoho Universytetu*, (5), 21-29.
15. Khomenko, O.Ye., Sudakov, A.K., Malanchuk, Z.R., & Malanchuk, Ye.Z. (2017). Principles of rock pressure energy usage during underground mining of deposits. *Naukovyi Visnyk Natsionalnoho Hirnychoho Universytetu*, (2), 34-43.
16. Bondarenko, V., Lozynskyi, V., Sai, K., & Anikushyna, K. (2015). An overview and prospectives of practical application of the biomass gasification technology in Ukraine. *New Developments in Mining Engineering 2015: Theoretical and Practical Solutions of Mineral Resources Mining*, 27-32. https://doi.org/10.1201/b19901-6
17. Lozynskyi, V., Dychkovskyi, R., Saik, P., Falshtynskyi, V. (2018). Coal Seam Gasification in Faulting Zones (Heat and Mass Balance Study). *Solid State Phenomena*, (277), 66-79. https://doi.org/10.4028/www.scientific.net/SSP.277.66
18. Pivnyak, G., Dychkovskyi, R, Bobyliov, O., Cabana, C.E., & Smoliński, A. (2018). Mathematical and Geomechanical Model in Physical and Chemical Processes of Underground Coal Gasification. *Solid State Phenomena*, (277), 1-16. https://doi.org/10.4028/www.scientific.net/SSP.277.2
19. Gayko, H., Matviichuk, I., Biletskyi, V., & Saluga, P. (2018). Forecast assessment methods of geological environment promotion for the construction of urbanistics subjects. *Visnyk of V.N Karazin Kharkiv National University – Series Geology Geography Ecology*, (48), 39-51. https://doi.org/10.26565/2410-7360-2018-48-03
20. Stetsenko, V., & Zavgorodnyaya, V. (2017). Research of the aspects of folded structures of the north-western part at Skelyuvatske deposit (Kryvyi Rih basin). *Visnyk of Taras Shevchenko National University of Kyiv. Geology*, *2*(77), 89-92. https://doi.org/10.17721/1728-2713.77.11
21. Bezruchko, K.A., Diachenko, N., & Urazka, M. (2018). Influence of the Western Donbass share dislocation zone on the formation of gas accumulations in coal-bearing sediments. *Geodynamics*, *1*(24), 27-39. https://doi.org/doi:10.23939/jgd2018.01.027
22. Gorova, A., Pavlychenko, A., Kulyna, S., & Shkremetko, O. (2012). Ecological problems of post-industrial mining areas. *Geomechanical Processes During Underground Mining*, 35-40. https://doi.org/10.1201/b13157-7
23. Soboliev, V.V., Shiman, L.N., Nalisko, N.N., & Kirichenko, A.L. (2017). Computational modeling in research of ignition mechanism of explosives by laser radiation. *Naukovyi Visnyk Natsionalnoho Hirnychoho Universytetu*, (6), 53-60.
24. Caceres, E., & Alca, J.J. (2016). Potential For Energy Recovery From A Wastewater Treatment Plant. *IEEE Latin America Transactions*, *14*(7), 3316-3321.

https://doi.org/10.1109/TLA.2016.7587636

25. Pivnyak, G., Dychkovskyi, R., Smirnov, A., & Cherednichenko, Y. (2013). Some aspects on the software simulation implementation in thin coal seams mining. *Energy Efficiency Improvement of Geotechnical Systems*, 1-10.
26. Dychkovskyi, R.O., Avdiushchenko, A.S., Falshtynskyi, V.S., & Saik, P.B. (2013). On the issue of estimation of the coal mine extraction area economic efficiency. *Naukovyi Visnyk Natsionalnoho Hirnychoho Universytetu*, (4), 107-114.
27. Krichevsky, S. (2016). New model of evolution of technologies and prospects of research with using big data. *Philosophy and cosmology*, (17), 118-135.
28. Gryadushchiy, Yu.B., Dychkovskyi, R.O., & Ruskych, V.V. (2008). Software Simulation in Thin Coal Seams Mining. *21-st World Mining Congress & Expo 2008* Krakow – Katowice, Poland.

Rheology effect determination of the heterogeneous rock massif on the stress-strain state of the breakage face geomechanical system

Vadym Fomychov[1], *Viktor* Pochepov[1], *Liudmyla* Fomychova[1], and *Viktor* Lapko[1*]

[1]National Mining University, Department of Underground Mining, 19 Yavornytskoho Ave., 49005 Dnipro, Ukraine

Abstract. The purpose of this study is to determine the nature of influence of the rheological properties of the rocks on the results of computational experiments simulating the mine workings contour condition during the breakage face advance. To determine the deformations along the mine workings contour, the calculations have been made of the junction area between the preparatory and stope face mine workings by the finite element method for various parameters of the material models. These data provided the basis for analysis of the patterns development of the rheology modelling influence on the stress-strain state of the geomechanical system. The obtained patterns are described by a function of many variables, and this function determines the limit and degree of rheology effect on the results of calculations. For the first time this pattern takes into account the heterogeneity of the rheological characteristics of rocks and the specifics of the computational domain geometry. The use of the obtained patterns makes it possible to simplify significantly the process of modelling in time and space of the geomechanical systems, which have a complex structure describing the interaction of several objects.

1 Introduction

The decline in global underground coal production is primarily due to the low level of energy efficiency. In the absence of global industrial economy growth and an increase in the availability of other energy resources, underground coal mining becomes uncompetitive. Low indicators of self-recoupment and the high risks associated with ensuring the stable coal enterprises operation do not permit a policy of innovative development, which can provide a significant increase in energy efficiency [1, 2].

In such conditions, the most effective method for maintaining a sufficient level of coal mining is possible only due to the local optimizations of the production process with the design and service of coal mining enterprises. Most technological chains of the mine depend to a greater or lesser extent on the state of the rock massif within which the preparatory and stope face mine workings are operated [3, 4, 6, 11].

* Corresponding author: yvlapko@gmail.com

At the moment there are various methods for determining the state of the rock massif geomechanical systems and mine workings based on the full-scale, laboratory and computational experiments [8, 10]. But any method has its own accuracy limit of the obtained results and, thus, it has different levels of complexity in carrying out the experiments themselves. The recommendations received as a result of the conducted studies should ensure the effective operation of the mine. Therewith, the costs of the experiments themselves should be minimal.

The simplest way to reduce the cost for behaviour prediction of a structurally complex system with a wide range of possible description parameters is the segmentation of the description system of the studied objects in order to simplify the procedure for obtaining the results [5, 9]. Such an approach allows using different description types of physical object or process within the single methodology framework, depending on the requirements made during the research.

The investigation of the stress-strain state of breakage face geomechanical system, as a rule, is held in order to obtain technical solutions to minimize the deformations on the contours of the preparatory and stope face mine workings. It is a general practice here to perform estimative calculations by the simplified schemes. As a result, the adopted technological solutions are either excessive in their characteristics, or do not provide sufficient operational characteristics [7]. It turns out that the total costs for ensuring production simply do not justify engineering work.

The main objects, the parameters of which have a wide range of possible values, are the rock layers. These parameters are controlling ones for the determination of acceptable technological solutions on operation with a specific geomechanical system. Thus, the complexity of the simulation model and its adequacy is directly dependent on the method of the laminal rock massif description.

Over time, the growth in the mine working contour deformation causes a change in the mechanical characteristics of the limited volumes of rocks. This induces the formation of main cracks, which, in combination with water-cut, fracturing and tectonic stresses, modifies the initial conditions of mine working operation. Therefore, when carrying out a computational experiment, one must resort to the use of various methods for these characteristics description. An account in the rheology calculations allows to make a comprehensive description of the influence of such physical features of rocks on the deformation of the mine working contour. However, a rapid increase in computational costs during the computational experiments requires the determination of the effective range in the application of this modelling type.

2 Specifics of simulation models formation and stages of computational experiment realization

In this study, a computational experiment was based on the finite element method implemented in the ANSYS software package.

In general view, the geomechanical model of the breakage face has been implemented as a stack of parallelepipeds with different heights, but with the same width and length of the base. Each parallelepiped is used to describe the geometry of a certain rock layer. In this model of the rock massif, three cuttings are made, which represent the preparatory and stope face mine workings, and the space that arises after the collapse of the breakage face immediate roof. The geometric model of the computational domain was fixed along the lateral borders of the coal seam from the side of virgin rock massif. The loadings were applied on the upper and lower borders of the compound parallelepiped.

The impact on the rock massif of the preparatory mine working support and extracting equipment working bodies was modelled as a combination of forces. The mechanized support was modelled by a set of parallelepipeds in accordance with the real objects dimensions. When performing the calculations, a bench was modelled with a depth of 0.3 m on the breakage face plane. This bench indicates the location of the working body of the plough or cutter-loader.

When performing calculations, the parameters of five materials were used: steel, coal, argillite, siltstone and sandstone. These parameters were selected as averaged characteristics of rocks with different strength properties. The curves of rheological increments were constructed for each material (Fig. 1).

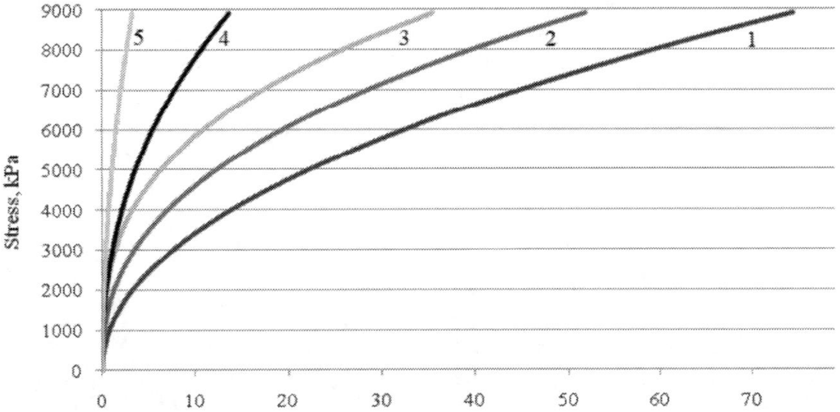

Fig. 1. Curves of rheological deformations increment for various materials of the simulation model: 1 – steel; 2 – coal; 3 – argillite; 4 – siltstone; 5 – sandstone.

These curves for coal and rocks were obtained in two stages. At the first stage, laboratory tests were carried out to determine the deformation characteristics of the test samples of materials. Conducting a study on the rheology effect on the stress-strain state of the breakage face geomechanical system suggests the calculation of deformations on a time interval of only 80 hours. Therefore, the tests of coal and rock samples were carried out in real time, and this greatly simplified the procedure of the experiment.

At the second stage, the laboratory test results were subjected to statistical processing to identify the averaged parameters of the deformation characteristics change of the selected materials. The obtained curves of total deformation had to be decomposed on the basis of the superposition principle into the elastic-plastic deformation curve and the curve of deformation under the rheology effect. The last one has already been used in the basic calculations of a computational experiment.

During the computational experiment, all calculations were carried out in an elastic-plastic setting without or with regard to rheology. The materials of rock layers' models, the velocity of the combined machine, which moved along the breakage face plane and the velocity of the breakage face displacement deep into the virgin rock massif varied in the calculations.

In total, there were performed 43 basic calculations out of 81 planned ones. The change in the number of calculations is due to the peculiarities of the obtained results. So, for example, from the analysis of the curves of already performed calculations it became obvious that changes in the rock layers' parameters, which were modelled second from above and from below, do not effect on the contour deformations of the studied mine workings. Thus, the calculations were cancelled, the performance of which could not give serious SSS deviations of the geomechanical system.

3 Analysis of the obtained data and formation of the identified patterns description

The basis for further research was to analyse the stresses intensity and linear deformations pattern in the area of junction of the preparatory and stope face mine workings. For most of calculation results by finite element method, the main differences were concentrated in the computational domain restricted to 30 m along the breakage face, 60 m along the axis of the preparatory mine working and 15 m in height. The centre of this area was 2 m behind the mechanized support model in the mined-out space, 6 m above the coal seam and 11 m from the longitudinal axis of the preparatory mine working from the side of the breakage face.

As the standard, four calculations were made in an elastic-plastic setting with different combinations of mechanical characteristics of the rock layers and with two different velocities of the breakage face plane displacements. An illustration of the results of one of the calculations is given for two cross-sections of three dimensional curves of stresses intensity (Fig. 2).

a *b*

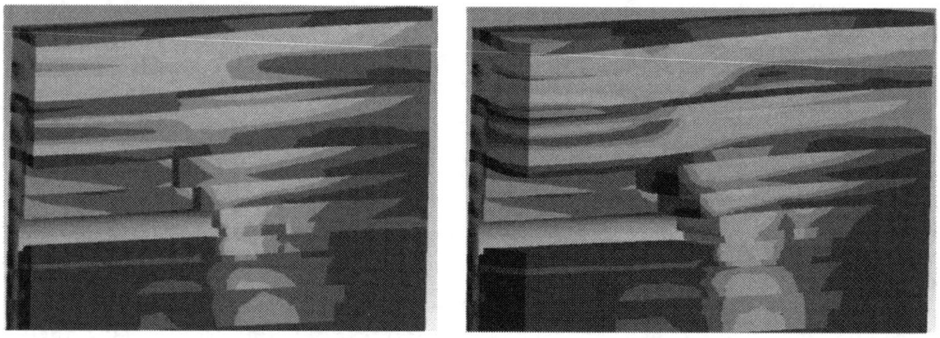

Fig. 2. Stresses intensity distribution in vertical planes directed along the axis of preparatory mine working and located in relation to it depthward into the stope face mine working at a distance of a – 8 m and b – 16 m.

The analysis of the presented figures shows that with this variant of calculation, above and below the mechanized support, a zone of increased rock pressure is formed, the value of which increases toward the breakage face. In this case, the velocity of displacement of the breakage face plane increases the geometric size of this zone, but does not change the maximum values of stresses intensity.

The stresses pattern presented in Fig. 2a shows that in the rock massif area, which is directly adjacent to both sides of the preparatory mine working, the stresses concentration has one order. For all variants of calculations this pattern does not change. It turns out that, for both soft and robust rocks, without taking rheology into account, the contour deformation forecast of the preparatory mine working has one kind and a relatively narrow range of absolute values variation. For the considered variants of calculations, with a variation of the deformation characteristics within 82%, the absolute deformations varied within 23%.

The stresses intensity distribution in Fig. 2b corresponds to the cross-section through which the bench has already passed, which models the contact surface with the working body of the cutter-loader. When comparing the stresses distribution in the virgin massif near the breakage face plane in Fig. 3a, b, we get a stresses drop in the bottom, an increase of the stresses intensity in the rock massif in front and above the breakage face plane. This peculiarity is manifested most intensively with a combination of high velocity of the cutter-loader moving and the occurrence of soft rocks in the roof and bottom of the stope face mine working.

Three specific cross-sections of stresses intensity curves in Fig. 3 show the SSS main features of the breakage face geomechanical system in calculations performed with account of the rheology curves presented in Fig. 1.

a

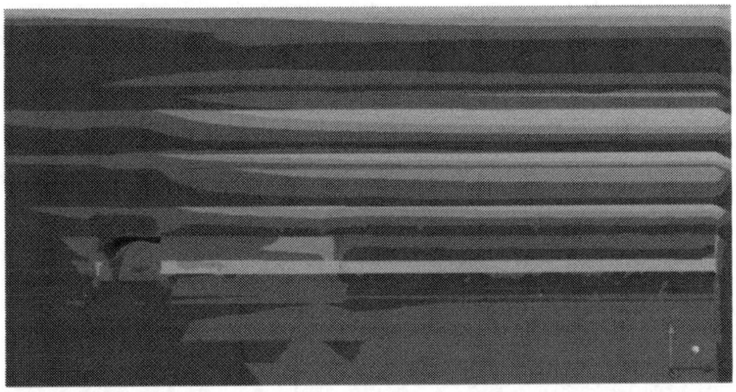

b

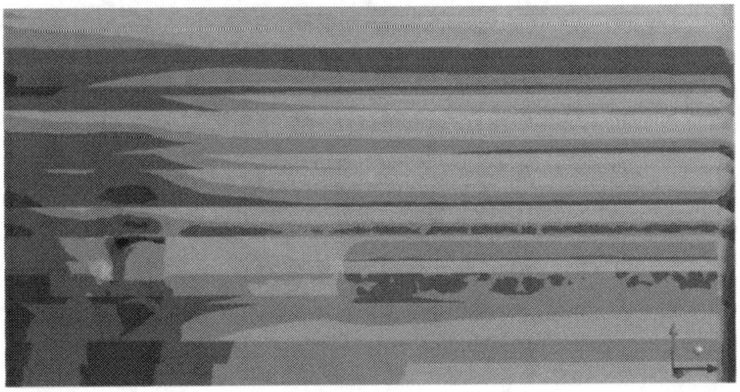

c

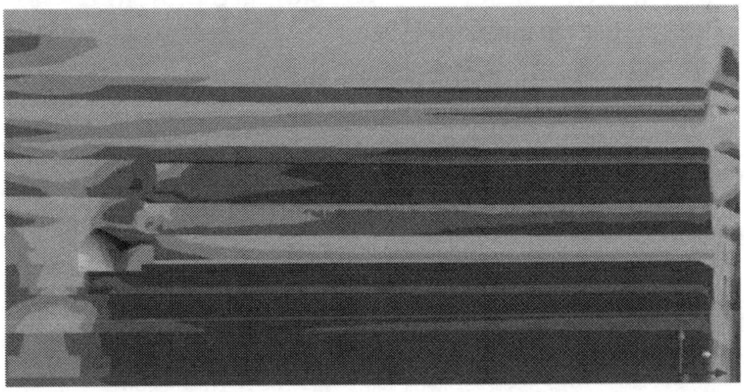

Fig. 3. The stresses intensity field in a geomechanical system along the cross-section: a – at a distance of 5 m in front of the breakage face plane; b – along the breakage face plane; c – at 60 m behind the breakage face.

When considering the cross-section located in the virgin rock massif in front of the breakage face plane (Fig. 3a), there are two main peculiarities of the obtained results. The first peculiarity – soft rocks are stressed by 1.5 – 2 times less than rocks with higher strength characteristics. The second peculiarity – with the approach of the bench formed by the working body of the cutter-loader, a 'wave' of stresses intensity is formed in the coal seam and the adjacent rock layers to the longitudinal axis of the extraction working for a distance of three widths (about 15 m). This effect is more manifested when the mechanical characteristics of the rock layers are alternated. The absolute values of the mechanical parameters of rock layers have only quantitative effect on the stresses intensity change.

In the cross-section, which coincides with the breakage face plane (Fig. 3b), the stresses intensity in the coal seam has the maximum values. The deformation of the preparatory working contour is still insignificant and does not exceed 0.3 m. In general, the evolution of the stresses pattern is observed, at which the rheological deformations growth is not yet observed. The drop in stresses in the rock layers adjacent to the coal seam indicates the beginning of formation of the unloading centres and the changeover of the simulation model part to the limiting state.

When analysing the velocity effect of the breakage face plane displacement and the bench formed by the working body of the cutter-loader we obtain the following: the first is that a high velocity of the bench displacement leads to significant (up to 37%) differences in stresses intensity in one rock layer; the second – an increase in the velocity of the breakage face plane displacement determines the velocity of total deformations of the mine working contour, but in a certain range and in case of a large spread in the values of the rock layers strength parameters; third – the considered combination of displacements works on the principle of mutual effect, that is, there is a stress intensity minimization in the coal seam and rock layers at certain velocity pairs.

Fig. 3c shows a development of rheological deformations in the rock massif in combination with the stresses intensity decrease after the collapse of the breakage face immediate roof. A zone of increased stresses is formed in the left side of the preparatory mine working. This peculiarity is manifested for all calculations using the rheology curves. Thus, when considering the rheology, the preparatory working support is under effect of a second wave of increased rock pressure over time after removing the breakage face plane from it. The stresses values in the zone of increased rock pressure depend on the combination of mechanical parameters of rock layers. The distance from the breakage face plane, on which this zone occurs, is directly proportional to the velocity of the breakage face displacement.

For the calculation, the results of which are presented above (Fig. 3), we analyse the deformation change in one of the selected points. In Fig. 4 the curves are shown of the displacement of the point located in the zenith of the preparatory mine working arch. At the initial time, this point was located in the plane of the breakage face. In the considered variant, the average velocity of the breakage face plane displacement was 1 m/hr.

The considered curves of total deformations in the point of the simulation model take into account the rheology of rocks and, at the same time, have one peculiarity, which is manifested only at the elastic-plastic calculation. If the deformation values along the axis of the preparatory mine working do not exceed 0.05 m and remain in the elastic zone, then the increase in vertical deformation is by an order greater and the pattern of its changes has an inflection point.

Such a peculiarity is observed not for all variants of computational experiment calculation. In addition, in one calculation for certain rock layers, two variants for changing vertical deformations can be realized. In the first variant, the deformation change corresponds to the characteristic presented in Fig. 4. In the second, the change in deformation occurs according to a power-law pattern without any peculiarities of the curve [12, 13].

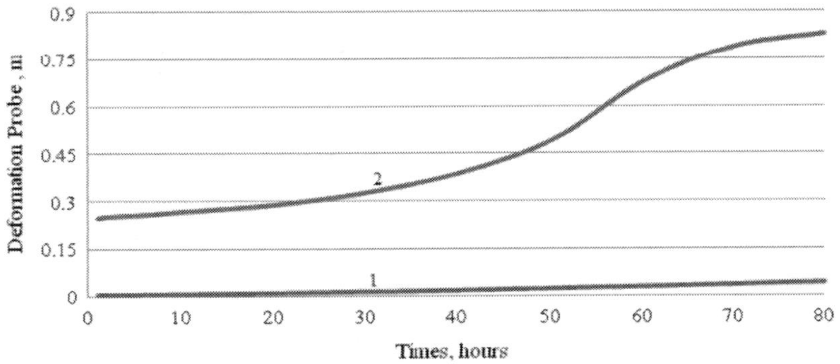

Fig. 4. Linear deformations in the selected point: 1 – directed along the axis of the preparatory mine working; 2 – vertical.

Thus, based on the data obtained by the deformations analysis in the calculation points of the model, an expression was obtained that describes the condition of necessity to take into account rheology depending on the number of layers and their deformation characteristics:

$$\frac{N \cdot \Delta\varepsilon_{max}}{11.41\left|\Delta\overline{\varepsilon}_i - \Delta\overline{\varepsilon}_j\right|} < 1, \quad (1)$$

where N is the number of modelled layers; $\Delta\varepsilon_{max}$ is the maximum increment of deformations in the computational domain; $\Delta\overline{\varepsilon}_i$ and $\Delta\overline{\varepsilon}_j$ are incremental matrices according to the layers of the model, where $i = \overline{1, N-1}$ and $j = \overline{2, N}$.

Using only the expression (1) to determine the range of calculations with account of rock layers' rheology does not allow to consider the dynamics of changes in time of the geometric characteristics of the simulation model. To account for these peculiarities of the computational experiment, it was decided to analyse the change in potential energy of deformation in the studied computational domain (Fig. 5).

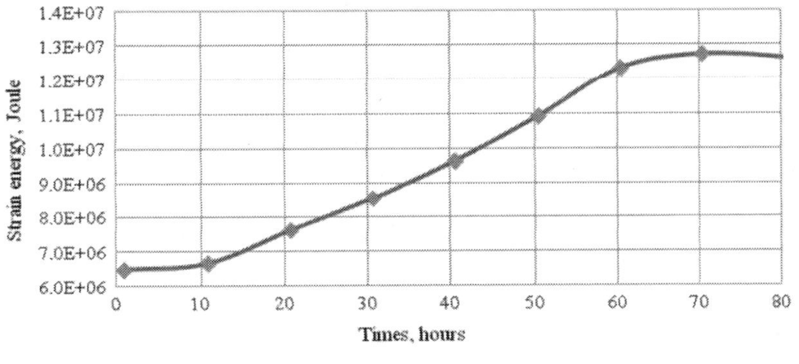

Fig. 5. The change of the deformation energy in calculations performed with account of rheology.

The curve of deformation energy change presented in Fig. 5, shows that taking into account rheology leads to energy dissipation when reaching a certain level of SSS calculated values in the chosen volume of the model. The analysis of similar curves for various combinations of the displacement velocities of the rock bench and the breakage face plane has shown that there is a limit beyond which the effect of rheology of rocks on the calculation result becomes minimal.

We obtained an expression relating the velocities of the rock bench and breakage face plane with deformations occurring in selected points of simulation model:

$$\frac{V_x}{14.8V_z} \cdot \frac{P_t \cdot \overline{\varepsilon}_{n \cdot t}}{\overline{U}_{n \cdot t}} > 5.12 , \quad (2)$$

where n is the number of a rock layer; V_x and V_z are velocities of displacement of the bench in the coal seam and the breakage face plane, respectively; P_t is the external pressure on the system at the selected time; $\overline{\varepsilon}_{n \cdot t}$ and $\overline{U}_{n \cdot t}$ are matrices of deformations and deformation energy, describing the state of the computational domain in the rock layer n and at the time of calculation t.

The systems of equations described in expressions (1) and (2) together determine the range of parameters of the simulation model of the breakage face geomechanical system, in which it is necessary to use the rheological characteristics of rocks in a computational experiment.

4 Conclusions

The obtained results of the computational experiment have shown that the effect of the rheology of rocks on the distribution of stresses and deformations in the breakage face geomechanical system has a different value and depends on two main factors - the structure of the rock massif and the velocity of the breakage face plane displacement.

There are upper and lower limits of the geometric and physical complexity of the simulation model description for which rheology accounting does not affect the quality of modelling or the conditions for a computational experiment performance. This makes it possible to neglect rheology indices even in calculations of structurally complex geomechanical systems.

The change in the stress-strain state of the breakage face geomechanical system under the effect of rheology of rocks increases in direct proportion with the increase in the number of rock layers which have different rheological parameters. It has a power-law dependence on the increase in the difference in time of deformations for different rock layers, and depends on the ratio of the averaged velocity of the breakage face plane displacement and the velocity of displacement of the rocks bench formed by the working body of the cutter-loader.

Thus, the results of these studies make it possible to provide acceptable indicators of the computational experiment quality in determining the optimal technological parameters of the geomechanical system in a context of minimizing the costs for calculations and subsequent mine workings operation. The resulting economic effect compared to traditional approaches can be up to 12%.

The authors express their gratitude to the staff of Department of Underground Mining of the National Mining University for the multifaceted assistance.

References

1. Barabash, M.V., & Cherednichenko, Yu.Ya. (2015). Transformation SHC "Pavlogradvugillia" in the world class coal-mining company – PJSC "DTEK Pavlogradvugillia". *Mining of Mineral Deposits*, *9*(1), 15-23. https://doi.org/10.15407/mining09.01.015

2. Fomychov, V. (2016). Efficiency of energy resource production while optimizing parameters of socio-economic balance. *Mining of Mineral Deposits*, *10*(1), 89-95. https://doi.org/10.15407/mining10.01.089
3. Bondarenko, V., Kovalevs'ka, I., & Fomychov, V. (2012). Features of carrying out experiment using finite-element method at multivariate calculation of mine massif – combined support system. *Geomechanical Processes During Underground Mining*, 7-13. https://doi.org/10.1201/b13157-3
4. Fomychov,V. (2012). Premise development of clearing models of the frame anchor support at nonlinear characteristics physics mediums behavior. *Naukovyi Visnyk Natsionalnoho Hirnychoho Universytetu*, (3): 45-52.
5. Fomychov, V., & Sotskov, V. (2018). Determination of parameters of non-uniform fractured rock massif in computing experiment. *Journal of Geology, Geography and Geoecology*, *26*(1), 26-32. https://doi.org/10.15421/111803
6. Małkowski, P., Niedbalski, Z., & Majcherczyk, T. (2016). Roadway design efficiency indices for hard coal mines. *Acta Geodynamicaet Geomaterialia*, *13*(2): 201-211.
7. Meng, L., Nan, Z., Jixiong, Z., & Zhicheng, L. (2017). Numerical Modelling of Mechanical Behavior of Coal Mining Hard Roofs in Different Backfill Ratios: A Case Study. *Energies*, *10*(7), 1-18. https://doi.org/10.3390/en10071005
8. Fomichov, V., Sotskov, V., & Malykhin, A. (2014). Determination and analysis of the acceptable benchmark changes of the stress strain state of frame and bolt fastening elements of dismantling drift when approaching a working face. *Naukovyi Visnyk Natsionalnoho Hirnychoho Universytetu*, (1), 22-26.
9. Kovalevska, I., Illiashov, M., Fomychov, V., & Chervatuk, V. (2012). The Formation of the Finite-Element Model of the System "Undermined Massif – Support of Stope". *Geomechanical Processes During Underground Mining: School Underground Mining 2012*, 73-79. https://doi.org/10.1201/b13157-13
10. Rosenblatt, J., & Bell, S. (1999). *Mathematical Analysis for Modeling*. Boca Raton.
11. Sotskov, V.O., Demchenko, Yu. I., Salli, S.V., & Dereviahina, N.I. (2017). Optimization of parameters of overwoked mining gallery support while carrying out long-wall face workings. *Naukovyi Visnyk Natsionalnoho Hirnychoho Universytetu*, (6), 34-40.
12. Zienkiewicz, O., Taylor, R. (2000). *The Finite Element Method. Vol. 1: The Basic*. Oxford: Butterworth-Heinemann.
13. Fomichov, V., Sotskov, V., Pochepov, V., & Mamaikin, O. (2018). Formation of a calculation model determining optimal rate of stoping face movement with a large deformation of a rock massif. *ARPN Journal of Engineering and Applied Sciences*, *13*(7), 2381-2389.

Formation of safety conditions for development of deposits by open mining

Oleh Anisimov[1*], *Volodymyr* Symonenko[1], *Oleksii* Cherniaiev[2], and *Oleksandr* Shustov[1]

[1]National Mining University, Department of Surface Mining, 19 Yavornytskoho Ave., 49005 Dnipro, Ukraine
[2]Institute for the Design of Mining Enterprises, National Mining University, 19 Yavornytskoho Ave., 49005 Dnipro, Ukraine

Abstract. The researched parameters of the catching device, which ensures the safe operation of equipment and working people in deep open pits, where are formed the benches by steep dipping layers. For obtaining the results, the following methods were used: analytical – in determining existing devices for the capture of rocks; geometric – in determining the direction of the flight of blocks, semigraphical – in plotting and analyzing them. The modern technological catching devices, their structure, construction are analyzed. These structures protective devices can be used in deep pits in the development of deposits. Analytical dependence of the height of the jump the block of rock after hitting about a horizontal surface is improved, which takes into account the rate of falling and the angle of rebound the block, the elasticity of the surface. Creating of catching designs which capture the blocks of rock in the falling from the upper horizons in open pits allows to operate safely the mining equipment and to save the lives of workers. Especially this is important in the development of deposits by steep dipping layers. Proposed solutions can improve the safe using of dump trucks and moving workers in the work zone in the mining ledges of height over 30 meters.

1 Introduction

In the process of working off the deposit, the working area of the open pit develops with the movement of some ledges on the boundary contours. There is a need to ensure the safe operation of equipment and people on the lower horizons of the open pit, especially in areas prone to destruction. The assessment of the local stability rockfalling slopes of the ledges should be made on the basis of quantitative and qualitative characteristics of crevice the rock, with the composition of the forecast of the intensity of precipitation of weathering products and the size of the rock blocks, taking into account the possible seismic effect as a result of blasting operations.

Permanent movement of working benches reduces the risk of violations of their integrity from slowly occurring deformations. But artificially made blocking out layers and blocks make a number of additional conditions that reduce resistance.

[*] Corresponding author: ipgpnmu@ukr.net

Rockfalls and inrushes are the most widely distributed types of violations of the steep benches and slopes of the ledges in rock and semi-rocky, which have substantial crevice (block structure). Caving and falling of individual blocks are usually prepared in advance by the breakdown of solid bundles of the rock in the form of gradually growing and deep cracks. The same rockfalls occur with catastrophic speed. Movement of the rock is accompanied by its overturning and rolling, sometimes with jumps on preventive berms and on other obstacles. Processes of weathering, penetration of water into the cracks, any dynamic loads and shakes contribute to the rockfalls. Among the latter, the greatest destruction of them is accompanied by seismic phenomena, which are caused, in contrast to the natural – organization of blasting operations.

The development of deformations of rock masses in benches of open pits takes a certain period of time, as and the occurrence of violations of stability. Any violation of the stability, even the caving of the mass, has a "hidden" stage of preparation, in the course of which there is a consistent destruction of numerous structural connections of rocks.

The weathering of rocks in the slope zones of the open pits is based on the development in time talus of the slopes formed mainly hard rocks. The following factors influence the development of the talus: angle of slope the bench, the intensity of crevice and the orientation of cracks to the top bench, the methods of preparation the benches, the technology of drilling and blasting operations, the climatic conditions of the area, the material composition of rocks.

There is a tendency to increasing the angle of slope of the non-working edges of the benches to 70 – 75°. This is due to the fact that the slight flattening out of the angles of the slopes the edges with further deepening of the bottom the open pit leads to a significant increasing in the volume of overburden removing, which is calculated in millions of cubic meters. There are the following types of deformations: talus, caving, rockfalls of individual plots in the shale deposits of iron-ore open pits [1, 2].

Taluses are observed almost on all the horizons of deep open pits. In case of their untimely liquidation the pit edge may have a kind of continuous slope of high altitude, which creates a danger for the equipment and people at the lower benches.

The above resulted the relevance of the problem and the need to study the parameters of catching devices in terms of working off deep open pits with the construction of heavy pitching benches.

Decisions on the formation of stable pit edges were discussed in the publications of G. Feesenko, A. Shaparya, E. Galustyana, A. Lomtadze, S. Polischuk, V. Simonenko, A. Halperin, Parham Ashaier, V. Kaiumova and others.

The width of preventive berms in open pits can be reduced significantly as a result of the application of special methods for protecting the lower horizons from rockfalls [3]. These methods can be divided into two groups: prevention of deformations slopes of benches; the formation of preventive berms of special construction and rock catching devices. This theme for iron-ore open pits is not completely considered and requires additional researches.

On the basis of the analyzed data systematization of non-ore deposits in the scientific work [3] was conducted to determine the rational technological scheme for the development of useful minerals in a steep digging layer on non-ore open pits. The results of the above researches are reflected in the established dependences between the parameters of the system of development [4]. On their basis are determined the efficiency of the application of a selected technological scheme of working off rock in a steep digging layer. This also allows to determine the basic parameters of the steep layers that influence the size of the working area.

The main indicators of the blocks that roll and fall from a significant height are determined in the research of V. Kaiumova. However, too high rates of their jump (6 m), calculated on a computer program are ambiguous.

In the work of Parham Ashaier an analysis and modeling of rockfall processes in different conditions were carried out [5]. Including enabled software complexes that reproduce the process of falling and the trajectory of the movement of blocks during the falling and rebound from the surface.

The analysis of various literary sources related with the studying of safe catching devices allows us to conclude that it is necessary to improve existing methods of their determination [6 – 11].

The formation of steep slopes layers during the development of deep open pits is accompanied by the positioning the areas of the benches in the temporary non-operating position and the formation of a single slope in the height more than 30 m. In this case, there is a need for the safe placement of equipment on the lower horizons. During mining, untimely preparation the slopes of benches leads to significant self-caving of the solid mass and the falling of individual blocks. The working area should be protected from such negative manifestations, which is achieved by the creation of catching devices near the slope. The question arises how far a device should be created and what height it should be to protect workers and equipment.

2 Method of carrying out the researches

Methods of strengthening the slopes of the benches prone to rockfalls and overbreaks of the individual pieces the rock can be used only on the areas of the pit edges, where the work is complete and the edge is placed in the limit position. In the conditions of open pit working areas, the formation of permanent catching and fortified constructions with the using metal elements is undesirable. This is due to that the minerals are subject to further processing at the enrichment plants. In the process of crushing and grinding there will be problems with removal of metal remnants of reinforcing constructions. Therefore, the strengthening of rocks in the working area of the open pit may be concrete blends.

According to the recommendations of the Union research institute of transport construction, it is possible to determine the rate of rock fragmentation and its trajectory when falling from the slope with very steep slopes, which corresponds to the proposed technology of working out the open pit and the construction of heavy pitching benches. The estimated velocity of the debris for a monotonous slope along the straight profile is determined by the formulas for a section with very steep slopes more than 60° and height of slope more than 10 m.

The angle of slope of the lower part is formed as a result of the falling of small rock pieces on horizontal platforms (shelves), and also the spillage on basis the bench, which is formed from the small fractions as a result of erosion processes, the conducting of blasting operations in an open pit, etc.

The air resistance can be disregarded for the calculation of a rockfalling from a conditionally small height in most cases. The falling of the rock fragment with its displacement-along the slopes of the benches has three types of fall rates. The first is determined by the height of the cut-off from the slope of the ledge (v_p), the second – occurs when the stone falls from the height and the impact on the intermediate horizontal plane with a slope formed by the spillages ($v_{t(0)}$). After reflection from the intermediate platform, the rubble strikes the horizontal area of the lower bench and reflects from it with the third type of velocity (U) (Fig. 1).

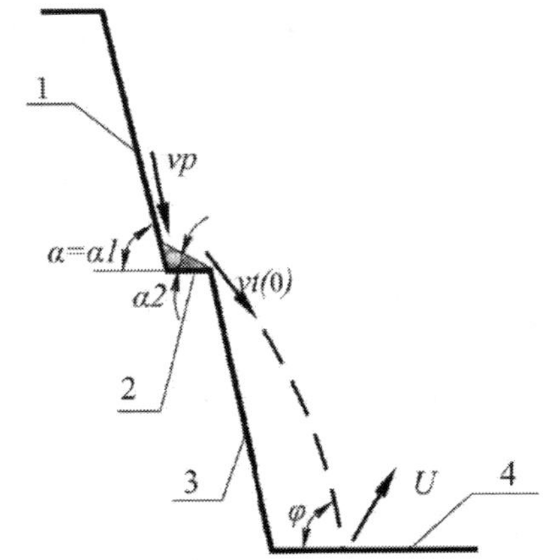

Fig. 1. Scheme of the movement of a rock rubble with the directions of speed at the points of change trajectory: 1 – the upper slope (the place of the chipping off a piece of rock; 2 – an intermediate area, covered by spillages; 3 – the lower slope; 4 – the lower area.

The overall dimensions of the in-depth catching shelves and the height of the soil shaft (m) for capturing blocks of rock should check the possibility of jumping the debris under their falling using the formula B. Kaiumova:

$$h_{\max} = \frac{U^2}{2g} \rho^2 \sin^2 \phi, \qquad (1)$$

where h_{max} is the calculated height of jumping the rubble, m; φ is the angle at which a rubble of rock falls on a horizontal surface, degrees.

Instead of the calculated rate of falling U it is proposed to substitute the speed resulting after the impact on the earth's surface. With absolutely elastic impact and the absence of impact friction, the speed of the material point before and after the impact is the same, and the incidence angle is equal to the angle of reflection. At the moment after impact on the vertical surface, the height of the jumping of a rock piece can be determined by the formula:

$$h_{\max} = \frac{\left(V_y \cdot \cos\alpha_3 \sqrt{tg^2\alpha_3 + K^2}\right)^2}{2g} \sin^2 \phi, \qquad (2)$$

where α_3 is the angle between the vertical component and the trajectory of the flight of the rubble at the time of impact, degrees ($\alpha_3 = 90 - \varphi$); φ is the angle between the horizontal surface of the trajectory of the rubble at the time of falling, degrees; K is the coefficient taking into account the elasticity of the surface (K =1 is the impact is absolutely elastic; K = 0 is the impact is absolutely not elastic; $0 < K < 1$ is the impact is partially elastic; V_y is the rate of falling of a stone at the moment of impact on the horizontal surface, m/s ($V_y = v_{t(0)}$).

3 Results

On the basis of the given method, calculations were made and schemes of the trajectories of the movement rock block after collision with the intermediate horizontal platform on different horizons, and also the distance of their landing from the lower platform were obtained. The width of the platform is 2 m. The angle of slope is 75°. The diameter of the boulders is 0.5 m. The example of the obtained graphic images at height of slope 45 m and the working off benches at height 15 m is presented in Fig. 2.

Similar calculations were performed for areas of height 40-meter (height of bench is 10 m), 48 m (the height of the bench is 12 m) and 40 m (the height of the bench is 20 m).

On the basis of received calculations and graphic images a table is prepared showing the distance of the flight of rock rubbles from the lower crest of bench and the possible height of their jumping on the lower platform at different angles of incidence. According to the calculations, in the formation of slopes with angles of 70 to 75 degrees, at the specified height of the slopes and temporarily inoperative their sections, the average indicators of the required width of the catching area in the basis of slopes and the height of the catching construction in the conditions of working out steep deeping extraction layers are established (Fig. 3) On average, a safe area from falling blocks is at a distance of 3.96 m from the lower crest of the bench, and the height of the catching construction should be not less than 2.37 m. Spillages, which are formed on intermediate horizontal platforms of open pits, in most cases, there is no possibility to clean, or need to hire rock climbers and manually clear the spillages. The work of climbers in such conditions is dangerous and costs considerable expenses for the mining enterprise.

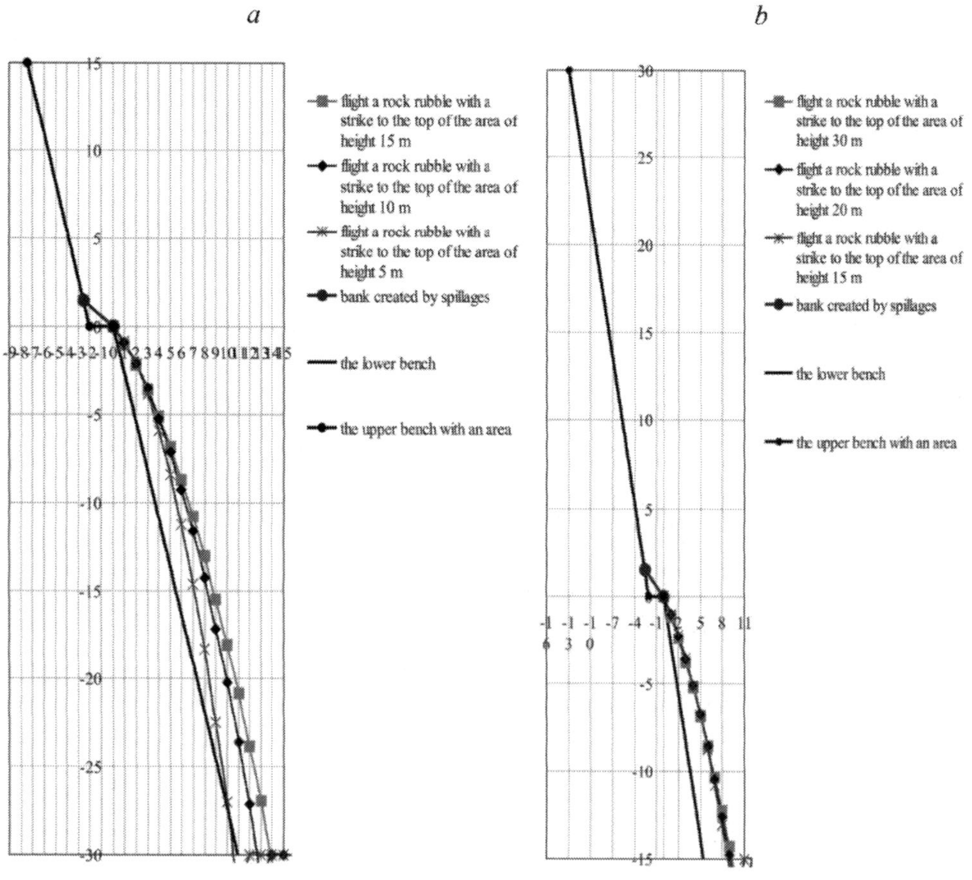

Fig. 2. Schemes of the trajectory the movement of rock rubble from a height 45, 40 and 35 m at different positions of the intermediate platform between the benches of height 15 m: a – intermediate platform at altitude of 30 m; b – intermediate platform at altitude of 15 m.

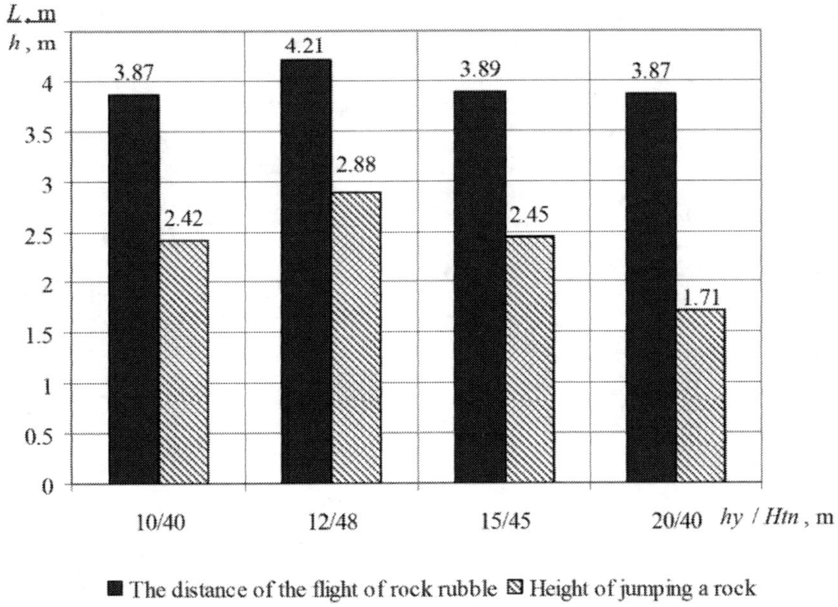

Fig. 3. Schedules of the distance the flight of rock rubbles and height of their rebound from the base of the face toe/bench bottom.

As can be seen from the table at working out individual benches of height 10 m and staging a section of the edge of height 40 m in a temporarily nonworking position, a rubble of rock that falls from the upper slopes of the bench, can land at a distance of 3.87 m and bounce from the ledger wall to height 2.42 m.

At working out the mountain masses by the benches of height 12 m with the possible building of a section of the edge of height till 48 m in a temporarily nonworking position, a rubble of rock that falls from the upper slopes of the bench, can land at a distance of 4.21 m from lower crest of bench and bounce from the ledger wall to height 2.88 m. The development of benches of 20 m high and their doubling till 40 m requires the establishment of safe boundaries from falling of rubbles at a distance of not less than 3.87 m from the lower crest of bench. At the same time, a possible ricochet can lead to a jumping of a piece to height 1.71 m. The benches of height 15 m with the setting in a temporarily non-operating position of a plot of 45 m height requires taking into account the safe location of the enclosing construction at a distance of 3.89 m from the lower crest of the lower bench. In this case, according to the above calculations there is a ricochet of a rubble rock with a jumping to height 2.45 m.

The obtained parameters allow, at given parameters of the working benches, to form catching and anti-caving constructions, which will protect industrial objects and workers in the working area of the open pit.

The retaining constructions should be foreseen to prevent shifts, rockfalls and inrushes of rubbles rock if it is impossible or economically inexpediency to change the size of the slope bench. In this case, retaining constructions are used in the form of:

– supporting walls – for strengthening the uplifting rock cornices;

– buttresses – separate supports, which are cut into stable layers of the rock for support of separate rock masses;

– supports – massive constructions to support unstable slope bench;

– facing walls – for prevention rocks from weathering and caving;

– fillings in the form of backfill cavities formed as a result of inrushes on slopes – for prevention rock from weathering and further destruction;

– anchoring – set as independent retaining constructions (with supporting slabs, beams, etc.) or in the form of fastening of separate rock blocks to a solid massif.

Table 1. Estimated values of distance of movement rock rubbles at falling from the upper benches of development the rock.

Height of slope of pit edge/height of bench, m	The height of the separation the rubble over the intermediate area, m	The placement of an intermediate platform with spillages in height, m	Angle of reflection of a rubble on the lower area, degrees*	The distance of falling a rubble from the lower edge of the lower bench, m*	Height of jumping a rubble on the lower area, m*
40/10	10	30	78/72	1.85/1.83	0.91/0.91
	8	30	79/73	1.36/1.14	0.74/0.74
	5	30	80/*slope*	0.32/*slope*	0.47/*slope*
	20	20	73/65	3.14/3.87	1.71/1.66
	10	20	76/69	2.29/2.65	0.88/0.87
	5	20	78/73	1.22/1.12	0.45/0.46
	30	10	69/61	2.42/3.16	2.42/2.34
	20	10	69/60	2.38/3.11	1.61/1.55
	10	10	72/65	2.12/2.66	0.85/0.82
40/20	20	20	73/66	3.12/3.87	1.71/1.66
	10	20	76/69	2.29/2.66	0.88/0.87
	5	20	79/73	1.22/1.12	0.46/0.46
45/15	15	30	77/70	2.68/2.98	1.33/1.32
	10	30	78/72	1.84/1.83	0.91/0.91
	5	30	81/*slope*	0.32/*slope*	0.47/*slope*
	30	15	70/62	3.04/3.89	2.45/2.38
	20	15	72/64	2.89/3.64	1.67/1.63
	15	15	74/66	2.69/3.34	1.28/1.25
48/12	12	36	78/73	1.87/1.74	1.09/1.1
	10	36	79/73	1.42/1.13	0.91/0.92
	5	36	*slope*	*slope*	*slope*
	24	24	74/67	3.42/4.21	2.07/2.03
	12	24	77/71	2.47/2.81	1.08/1.07
	6	24	79/73	1.23/1.09	0.55/0.55
	36	12	68/60	2.74/3.48	2.88/2.78
	24	12	70/62	2.69/3.43	1.96/1.89
	12	12	73/66	2.34/2.98	1.03/1.00

Notice: * – calculations are made at angles of slopes the benches 75 degrees/70 degrees; slope – means that the rubble fells not on the horizontal platform, but on the slope of the bench; the width of the intermediate platform between the benches is 2 m.

Retaining constructions in the conditions of a deep open pit can be arranged in the upper zone of open pit after the mine working on the design contours. Such constructions can be used on slopes near capital transport communications, near transshipment points, technological platforms of traffic, etc. The working area of the open pit changes constantly its position. In such conditions, the most reliable safety devices are small, movable catching devices that can be moved during moving the mining front on the bench.

Catching constructions and devices (walls, grids, shafts, trenches, shelves with retaining walls, spiles) should be provided for the protection of objects against the impact of talus, inrushes, falling of individual rock fragments, as well as rockfalls in that case if their warning or the arrangement of retaining constructions is impossible or economically inexpediently. Catching walls and grids are located at the ledger wall of slopes the benches, having angles more than 25 – 35°.

Catching trenches and shelves with a curb wall should be placed at the ledger wall of dangerous slopes the group of benches of height 60 m and steepness more than 35° for protection from the inrushes of individual rock debris volume till 1 m^3. The constructions of the main anti-caving devices and equipment are shown in Fig. 4 – 7.

Stationary protective devices (Fig. 5a, b) or anchoring are located only on areas that are no more involved in the development. This is due to the fact that the timbering consists of metal products. The anchors, as well as other timbering, remain in the slopes and can be adversely affected on the technological process of crushing in the event of their entry into the crusher. The placement of such timbering will be made on the finished slopes of the overburden rocks.

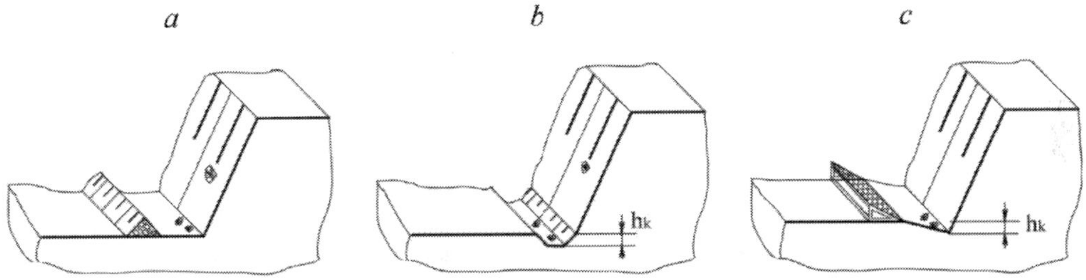

Fig. 4. Constructions of anti-caving catching constructions: a – ground shaft; b – catching ditch (drain); c – displaced mesh of the sloping construction: h_k – depth of the ditch, m.

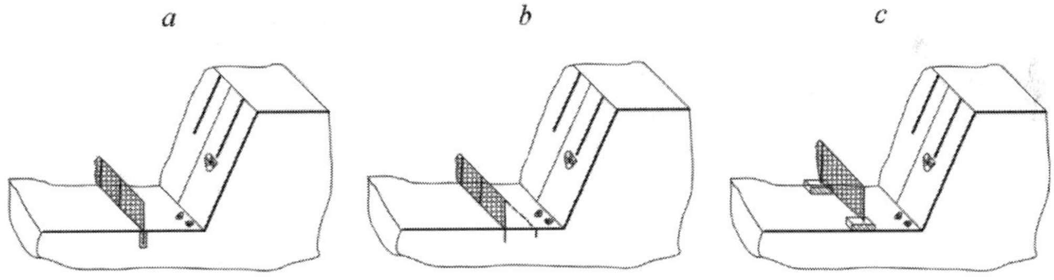

Fig. 5. Constructions of anti-caving catching meshes: a – stationary with concrete foundation; b – stationary with the strengthening of flexible traction; c – mobile (movable).

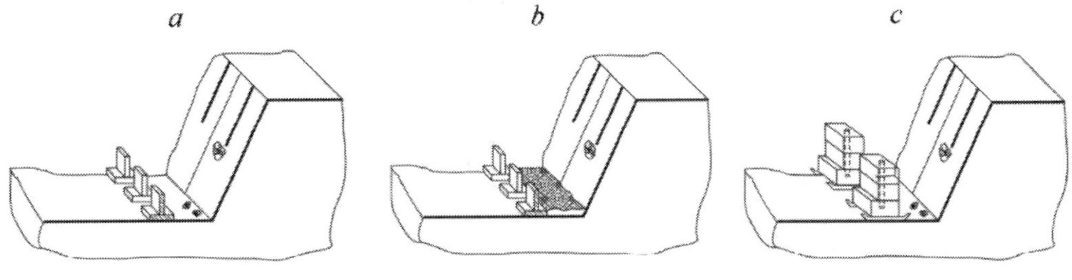

Fig. 6. Constructions of anti-caving catching, mobile concrete constructions: a – concrete blocks-catchers; b – concrete catchers with a soft “pillow”; c – mobile movable concrete blocks.

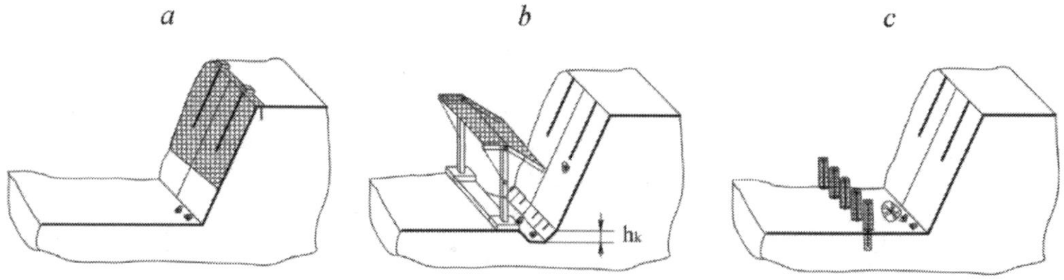

Fig. 7. Constructions of anti-caving devices: a – a mesh on the surface of slope the bench; b – mobile devices; c – rolls of a wound mesh on a concrete rod.

The most dangerous plots of the working areas are the places of traffic dumping trucks near the slopes above the bench. Under the action of dynamic loads and during drilling and blasting operations, the slope of the bench gradually loses its stable properties. Debris is cut off from the rock mass, which roll down. With significant dimensions, they have a great destructive power that can damage the equipment and are dangerous for workers in the open pits.

Widely accepted method of control with rockfalls of slopes the benches are shafts, ditches, timbering with a mesh and others. The disadvantage of the construction of retaining shafts is the need for their placement with considerable width and height at the loading platform near the lower crest of the upper bench. The height of the retaining shaft is 3 4 m or more, depending on the height and slope of the bench, and is constructed on a platform in the width of 4 – 6 m. The constructions of such shafts complicate the cleaning spillages.

For the purpose of improving the designs of devices that catch debris of rolling rock, a scheme is proposed for protection the plot on the non-working of the pit edge. Due to the improvement and introduction of new constructive solutions is attained significant reducing the width of the transport platform and protection the dumping trucks from falling large rock blocks, and, due to this, reducing the volume of mining capital operations on arrangement of temporarily non-working benches in the working zone. In addition, the placement of such constructions in the working zone allows the excavator to operate safely on the removal and loading into dumping trucks and other types of transport, used in deep open pits.

The new design of the protective device is mobile, which is very relevant in the working area of the open pit. As needed, this device can move around the working platform and be located near the lower crest of the bench. The moving of the construction is carried out using the stripes at the base of the device and the bulldozer. The bulldozer is used as auxiliary equipment. The principle of operation of the protective device is based on the accumulated rock mass in the stretched grid/mesh, with subsequent unloading it under the influence of rock weight and forces of gravity into the ditch. The metal grid/mesh allows to quench the energy of movement of individual blocks and catch them during the falling. The mobility of the catching device allows to move it following the mining work on the bench. As the rock masses accumulate in the ditch, the device moves toward the side and cleaning of the ditch is carried out with the using an excavator (Fig. 8). The excavator with the help of a bucket removes rock fragments from the ditch and places them on the plot or loads in dumping trucks. After cleaning the bottom part of the slope, the device returns to its place.

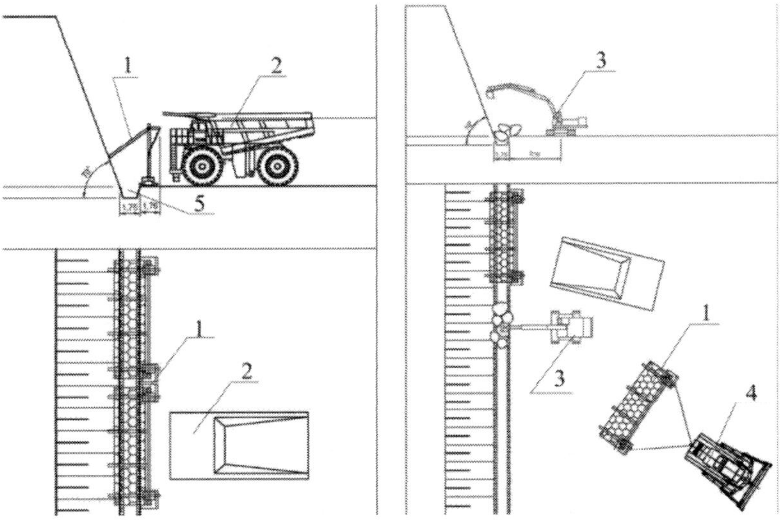

Fig. 8. Schemes for placing the protective device and cleaning the catching area from the rock blocks that fall down: 1 – a device for protecting the platform; 2 – an open pit dumping truck; 3 – excavator; 4 – bulldozer; 5 – catching ditch (drain).

The proposed design allows to reduce the width of the working platform and to protect dumping trucks from the falling of rock blocks. Dumping trucks carry out maneuvers of turning and movement on the working platform, and also people who may be in this zone. Compared to the fixing of the surface by a mesh throughout the plane of the slope, it allows to get a significant savings of material and monetary resources.

4 Conclusions

Thus, the formation of safe conditions for the formation of steep dipping excavation layers is possible due to mobile anti-caving devices, which allows to reduce the width of the working platform on 3 – 4 m, and according to reduce the volume of removal of the overburden rock. The proposed solutions allow to increase the safe using of dumping trucks and the movement of workers in the working area during working out the benches of the height more than 30 m. The safe area from falling blocks is at a distance of at least 4 m from the lower crest of bench where the trajectory of the movement reaches the highest point, and the height of the catching construction in this point must be at least 2.5 m from the horizontal plot of the bench.

These researches are carried out in the framework of the research work "Development of technological bases of ecologically safe mining in the technologically-loaded mining regions of Ukraine" (State registration No. 0117U001134).

References

1. Anisimov, O.A. (2015). *Tehnologiya stroitelstva i razrabotki glubokikh kar'yerov.* Dnipropetrovsk: Natsionalnyi hirnychyi universytet.
2. Abdellah, W.R., Beblawy, M.M., & Mohamed, M.T. (2018). Evaluation of open pit slope stability using various slope angles and element types. *Mining of Mineral Deposits, 12*(2), 47-57. https://doi.org/10.15407/mining12.02.047

3. Cherniaiev, O. (2017). Systematization non-metallic deposits of rocky minerals for improvements technologies of their development. *Naukovyi Visnyk Natsionalnoho Hirnychoho Universytetu*, (5), 11-17.
4. Symonenko, V., Cherniaiev, O., & Hrytsenko, L. (2016). Organization of non-metallic deposits development by steep excavation layers. *Mining of Mineral Deposits*, *10*(4), 68-73. https://doi.org/10.15407/mining10.04.068/
5. Parham Ashayer (2007). Application of rigid body impact mechanics and discrete element modeling to rockfall simulation. Department of Civil Engineering. University of Toronto.
6. Anisimov, O. (2018). Research on parameters of the working area on an internal dump for developing open pits. *Naukovyi Visnyk Natsionalnoho Hirnychoho Universytetu*, (1), 27-34. https://doi.org/10.29202/nvngu/2018-1/17/
7. Anisimov, O. (2015). The development of deep pits steep slope layers. *Theoretical and Practical Solutions of Mineral Resources Mining*, 243-246. ISBN:978-1-315-64877-4.
8. Cherniaiev, O. (2017). Systematization of the hard rock non-metallic mineral deposits for improvement of their mining technologies. *Naukovyi Visnyk Natsionalnoho Hirnychoho Universytetu*, (5), 11-17.
9. Gorova, A., Pavlychenko, A., Borysovs'ka, O., & Krups'ka, L. (2013). The development of methodology for assessment of environmental risk degree in mining regions. *Annual Scientific-Technical Collection – Mining of Mineral Deposits*, 207-209.
10. Saik, P., Petlevanyi, M., Lozynskyi, V., Sai, K., & Merzlikin, A. (2018). Innovative approach to the integrated use of energy resources of underground coal gasification. *Solid State Phenomena*, (277), 221-231. https://doi.org/10.4028/www.scientific.net/SSP.277.221
11. Shirin, L.N., Denishchenko, A.V., Yurchenko, O.O., & Mikhalev, D.V. (2012). Methodology for determination of rope vehicles energy consumption. *Naukovyi Visnyk Natsionalnoho Hirnychoho Universytetu*, (4), 464-469.
12. Pivnyak, G.G., & Shashenko, O.M. (2015). Innovations and safety for coal mines in Ukraine. *Naukovyi Visnyk Natsionalnoho Hirnychoho Universytetu*, (6), 118-121.

Improving the process of coal extraction based on the parameter optimization of mining equipment

Volodymyr Hrinov[1], and *Andrii* Khorolskyi[1*]

[1]Institute of Physics of the Mining Processes of National Academy of Sciences of Ukraine, 2a Simferopolska St., 49005 Dnipro

Abstract. The aim of this paper is to develop and validate methods of choosing the means of the mining face mechanization. This paper analyses existing methods of optimizing processes in mining. It was established that the effectiveness of the performance map of coal field is formed by a group of technological, operational, and economic parameters which can be represented as a vector of solutions. To find the optimal solution, it was suggested to use network models and graphs. The essence of the technique is to represent the input and output (production level, prime cost) resource flows in an organized structure. Regularities of forming technological schemes of coalfield operation with a given level of performance, taking into account the relationship between technological parameters of mining face, operational parameters of the stoping equipment, technical and economic performance are defined. We developed the system for decision-making support, which allows optimizing operational parameters, reducing the production prime cost, and selecting the structure of the mechanized complex of stoping equipment with a specified level of performance. This paper describes approaches that can be used at the design stage of mining face and in the process of operation.

1 Introduction

The coal industry is one of the most resource-intensive in the energy sector. For the effective operation of coalfields, it is necessary to solve a number of issues that are associated with the justification of rational operation of the stoping equipment, increase in the productivity of the mining face, decrease in the production prime cost, etc. Solving each task separately does not allow increasing the overall efficiency. Moreover, a group of technological, operational, and economic parameters forms performance indicators, so it is necessary to develop an integrated approach. Today, there is a number of methods and means of decision-making support in mining, which are aimed at improving the production process. One uses most frequently the multi-criteria decision analysis (MCDA), the essence of which is in priority setting, selecting the most important factors with further optimization. It is worth noting that, because of using MCDA, the solution is not optimal but efficient according to a specified function of "usefulness".

[*] Corresponding author: khorolskiyaa@ukr.net

In mining, one uses the analytic hierarchy process (AHP), designed by T. Saaty and adapted by scientists to operating condition of enterprises in the mining complex: M. Ataei [1], S. Alpay [2]. However, there is a number of difficulties: importance and priority of the criterion establish the decision-maker (DM) – for the accurate decision-making, there should be consistency between DMs (Pareto optimality); a great number of samples and a poll do not always guarantee the right choice; in addition, the selection criteria do not always correspond to the task [3]. Given the specificity of the mining industry, we developed modifications of the method taking into account deviations from the model of rational choice (TODIM) [4], as well as deviations from the "ideal" value with subsequent normalization [5] (Grey-AHP).

Also, the wide-spread is the method of preferences analysis (PROMETHEE), the essence of which is to set the preference function, comparison of options, and subsequent ranking of obtained preference functions. This method is used by Wang Chen [6] to solve tasks of transporting coal in the mining face, Zhang Jianping [7] to balance the supply of materials. The use of this method is difficult because nobody knows weighting coefficients for each parameter group (equipment, technological parameters, technical and economic indicators).

Analysis of the AHP method and its varieties (TODIM, Grey-AHP), as well as the method of preferences analysis PROMETHEE has found out that there is an issue of incommensurability and inconsistency in the selection of criteria. Therefore, the idea of adopting a "compromise solution" appeared [8]. This method was used at the stage of design transportation chains [9], to solve the tasks of production organization in the mining face [10]. However, there is the issue of "indistinct" uncertain objects that have different characteristics, for example: roof stability, soil rock category of opening, etc. To solve this issue, it was proposed to use fuzzy-set methods [11], while for the efficient design there should be the basis for "good" design decisions [12], which can be represented by graphs and network models.

The analysis of papers [1, 2, 4 – 12] allowed us to find out that the increase in the efficiency of technological schemes for the operation of coalfields is impossible without optimization of parameters of the stoping equipment. To achieve this task, one can use methods of the decision-making theory that are based on the comparison of possible solutions and priority setting. There are also factors that do not allow efficiently applying these approaches for the solution of tasks in the optimization of parameters of the stoping equipment:

– iincommensurability of efficiency criteria values; for example, the daily performance at mines of Ukraine is 500 to 3000 t/day, the level of significant preference $H_d = 0.5$ is adopted based on the condition of the break-even $Q \geq 1.000$ t/day, and the maximum preference $H_d = 1.0$ will correspond to the maximum capacity $Q \geq 3000$ t/day; in fact, 66% of performance indicators values will be preferred;

– when choosing means of the mining face mechanization in addition to quantitative criteria (performance, equipment life, prime cost, etc.) one also uses qualitative criteria – relationship with the pan line of armoured face conveyor, consistency of the advance increment of powered support and shearer, interrelation between the armoured face conveyor advance system and the shearer transfer system, etc.;

– as previously stated [11, 12], the search for the optimal solution to the task is to analyse good (rational for specific conditions) solutions, but this is impossible without the presence of the "bank of situational solutions";

– listed methods of decision-making [1 – 12] are based on the use of adjacency matrices with the dimensionality $n \times m$, where n is the number of alternatives, and m is the number of criteria that match the alternative n; for the final decision-making, we use methods of linear programming, special expensive software [13, 14]; at the same time, methods of discrete mathematics allow us to successfully use relevant approaches without additional resources.

Therefore, the process of improving the technology of mechanized coal mining is in support of rational (good) operating conditions with subsequent optimization of parameters of the stoping equipment. There is an area of "rational design" [15], and for our study, this is a set of conditions for the operation of the stoping equipment (layer thickness, face length, rocks length, roof length), which correspond to technical and economic, operational parameters.

To solve this task, we can use network models and graphs. In the paper [16], there is the methodology of selecting the stoping equipment using graphs and network models, and in the paper [17], there is its practical use. It is worth noting that this approach was used to solve transportation issues [18], tasks for the resource allocation at the mine [19], as well as for ventilation of mine workings [20], as well as surface mining of mineral deposits [21, 22].

Thus, establishing regularities of forming technological schemes of the stoping equipment with the set performance level, depending on technological, operational, economic parameters, is a topical scientific task. The aim of this paper is to develop a tool for optimizing parameters of the mining equipment. To do this, it is necessary to establish the rational area for the design of hardware, substantiate the system of principles for selecting the stopping equipment, develop the mathematical model, and offer the program implementation.

2 System principles for selecting the stoping equipment

In relation to the aim of the study, general system principles [23 – 25] are as follows:

– integrity, i.e. the mining equipment is considered as a single structure "mechanical powered support – shearer – face armoured face conveyor" without separation into individual components;

– hierarchy, i.e. one considers the relationship not only between the types of stoping equipment (for example: shearer-armoured face conveyor) but also between the assembly units of equipment (for example: the armoured face conveyor pan line and the shearer transfer system);

– structuring, i.e. the possibility of optimizing not the entire structure "powered support – shearer – armoured face armoured face conveyor", but only certain levels of "powered support – shearer", "shearer – armoured face conveyor";

– consistency, i.e. operating parameters of each type of equipment (layer thickness, mining face length, seam inclination) correspond to operating parameters of the structure "powered support – shearer – armoured face conveyor";

– multiplicity, i.e. the existence of standard models ("bank of good solutions" [12]) for a set of presented options.

Compliance with all principles allows obtaining the property of the "structural optimization" [26, 27] – the process of obtaining a series of system effects for the optimization of the objective function depending on given limitations.

As for the structure of the stoping equipment, we should consider the area of rational design, i.e. operating conditions in which indicators will be maximum and meet the break-even condition (for mines of Donbas $Q \geq 1000$ t/day).

As noted earlier, the efficiency of the production process P is affected by technological operational parameters P_T, means of mechanization P_E, and the production prime cost P_C. Then, the vector $p(P_T, P_E, P_C)$ will represent the rational option of integration of the stopping equipment. From the optimality condition:

$$p(P_T, P_E, P_C) = \min(P_T, P_E, P_C), \quad (1)$$

where $p = \{P_{1T}, P_{2T}, ..., P_{NT}; P_{1E}, P_{2E}, ..., P_{ET}; P_{1C}, P_{2C}, ..., P_{NC}\}$ – the vector and its components $P_{1T}, P_{2T}, ..., P_{NT}$ meet rational technological parameters of the working area, $P_{1E}, P_{2E}, ..., P_{NE}$ – types of the stoping equipment, $P_{1C}, P_{2C}, ..., P_{NC}$ – the production prime cost.

The group of technological parameters of the mining face $P_{1T}, P_{2T}, ..., P_{NT}$ includes the length of the working area l_n and the layer thickness m_n. The group of the "mining equipment" corresponds to the structure "powered support t_n, shearer c_n, armoured face conveyor k_n", which can provide the daily performance q_{nn} at the level of $Q \geq 1000$ t/day.

The level of costs for the purchase of machinery Z_n and equipment life R_n form the specific cost of production $P_{1C}, P_{2C}, ..., P_{NC}$.

Then, the mathematical meaning of the proposed approach can be formed as follows: the totality of types of the mining equipment $\langle t_n, c_n, k_n \rangle$, which is operated under given operating conditions $\langle t_n, m_n \rangle$, generates the level of performance q_{nn} at the prime cost s_n. It should be noted that any groups of parameters and performance criteria could be specified, as long as they meet the objectives of the study. In accordance with planned parameters, one can set the scope of rational design (Table 1).

Table 1. The scope of rational design for solving the task of optimizing the mining equipment.

Indicator	Criterion value	Limitation form
Layer thickness m_n, m	$M \subseteq H \Leftrightarrow \forall m_n \in M : m_n \in H$	$M \subseteq H$
Length of the mining face l_n, m	$L \subseteq W \Leftrightarrow \forall l_n \in L : l_n \in W$	$L \subseteq W$
Powered support t_n	$T \subseteq R \Leftrightarrow \forall t_n \in T : t_n \in R$	$T \subseteq R$
Shearer c_n	$C \subseteq S \Leftrightarrow \forall c_n \in C : c_n \in S$	$C \subseteq S$
Armoured face conveyor k_n	$K \subseteq A \Leftrightarrow \forall k_n \in K : k_n \in A$	$K \subseteq A$
Funds for the purchase of equipment Z_n, mln UAH	$Z_n \to \min$	$Z_{\max} \leq Z_n \leq Z_{\min}$
equipment life R_n, mln tonnes	$R_n \to \max$	$R_{\min} \leq R_n \leq R_{\max}$

where M, L – the subset of rational operational parameters, and H, W – the great number of possible technological parameters; T, C, K – the subset of rational types of the mining equipment, R, S, A – the great number of existing types of equipment.

From Table 1, it is seen that the given level of performance Q is generated only when technological parameters $\langle l_n, m_n \rangle$ are rational: $l_n \in L$, $m_n \in M$, i.e. they are at the subset of rational operational parameters M, L. In addition, the means of mechanization $\langle t_n, c_n, k_n \rangle$ have a high level of interrelation: $t_n \in T$, $c_n \in C$, $k_n \in K$, where T, C, K – the subset of rational types of equipment. The important condition is that the funds for the purchase of mechanization means should be kept at their minimum $Z_n \to \min$ and the equipment life – at the maximum $R_n \to \max$.

To optimize operational parameters, one should formalize them, i.e. present in the form of a network model. Then, the shortest route of the network model g_1 will correspond to the optimal solution. Conditions (2) and (3) shall be met:

$$(m_1 \cdots m_n) \to (l_1 \cdots l_n) \to (t_1 \cdots t_n) \to (c_1 \cdots c_n) \to \\ \to (k_1 \cdots k_n) \to (v_1 \cdots v_n) \to (s_{11} \cdots s_{1n}) \to g_1 \quad ; \tag{2}$$

$$(i; j) = s_n \text{ – the route exists, } (i; j) = \infty \text{ – the route is missing,} \tag{3}$$

where (2) – the condition for the existence of the alternative, i.e. the optimization requires that the route passes through vertices, which correspond to the layer thickness $(m_1 ... m_n)$, the length of the mining face $(l_1 ... l_n)$, powered support types $(t_1 .. t_n)$, shearer types $(c_1 ... c_n)$, armoured face conveyor types $(k_1 .. k_n)$, the level of performance $(v_1 ... v_n)$, the production prime cost $(s_{11} ... s_{1n})$; (3) – the condition for the existence of the link between vertices, i.e. the production prime cost for each type of equipment s_n can be expressed as the distance between vertices $(i; j)$ of the network model.

Then, the task of optimization by the parameter "prime cost" can be solved as follows:

1) define rational types of the mining equipment T, C, K and present them as vertices of the network model $(t_1 .. t_n)$, $(c_1 ... c_n)$, $(k_1 .. k_n)$;

2) define the production prime cost for each type of equipment $s_n = Z_n / R_n$ and present it as the length of the route (i, j) between vertices; if the interrelation between types of equipment is missing, then $(i, j) = \infty$;

3) find the best solution that corresponds to the shortest route from the vertice m_1 to the vertice g_1 using optimization algorithms or methods of linear programming.

The optimization parameter for the route length can be any characteristic of the process (maintenance costs, energy expenditure level, personnel number), and vertices can be represented by types of equipment, intermediate points, stages, etc.

We shall use mines "Krasnolymanska" and "Kotliarevska" as an example to consider the process of optimization and selection of the mining equipment according to the parameter of the "prime cost".

3 Optimization of operational parameters of the mining equipment

The task is in the optimum choice of the technological chain of the mining equipment for specified operational conditions (Table 2).

Table 2. Features of mining faces in mines "Krasnolymanska" and "Kotliarevska".

Mine	Layer thickness, m	Face length, m	Powered support type	Shearer type	Armoured face conveyor type
	m_n	l_n	t_n	c_n	k_n
Krasnolymanska	1.10 – 1.15	295	DM	RKU10	SP251
Kotliarevska	1.20 – 1.25	250	1KD90	1K101	SP26

To solve this task, we analyzed existing options for the integration of the mining face under given operational conditions. There are 57 such options for mines of Donbas. After the setting the minimum level of performance $Q \geq 1200$ t/day, the number of alternatives was reduced to 9.

However, 4 conditions remained for the mine "Krasnolymanska" and 3 conditions for the mine "Kotliarevska". Depending on the number of conditions (limitations), the result will be different, since this is the property of "structural optimization". Table 3 lists possible options available for the integration of mining faces for conditions of mines "Krasnolymanska" and "Kotliarevska".

Table 3. Possible options of completing for mining faces of mines "Krasnolymanska" and "Kotliarevska".

No.	Equipment types			Designation in Fig. 1
	powered support	shearer	armoured face conveyor	
1	KD80	KA80	SP251	1
2	DM	RKU10	SP26	2
3	KD80	KA80	SP250	3
4	1KD90	1K101	SP26	4
5	DM	UKD200	SP326	5
6	1KD90	1K101	SP326	6
7	KD80	1K101	SP250	7
8	DM	K103	SP250	8
9	1KD90	UKD400	SP326	9

For the mine "Krasnolymanska", requirements were as follows: low ash content of coal (excluding the alternative 2), series production of equipment (excluding the alternative 8), and screw type of the executive body (excluding alternatives 1, 3), interrelation of shearer and armoured face conveyor (excluding the alternative 4). In the end, 4 possible options were left (5, 6, 7, 9). Similarly, the choice was made for the mine "Kotliarevska". It is convenient to present the selection process in the form of a decision-making "tree" (Fig. 1).

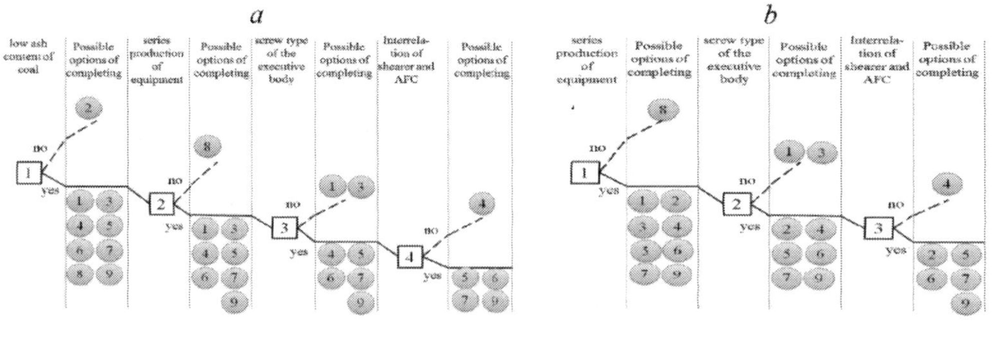

Fig. 1. Decision-making trees at different limitations: a – mine "Krasnolymanska", b – mine "Kotliarevska".

After the preliminary analysis of alternatives, one should choose a single alternative, optimal from the standpoint of minimization of the optimization parameter. In our case, this is the "production prime cost". We shall use the mine "Krasnolymanska" as an example to show the selection of the optimal chain of proposed options. To do this, let us present 4 options in the form of a network model. Vertices represent types of purification equipment, and the lengths between peaks are the values of specific cost (Table 4).

To automate the selection process, the personnel of the Institute of Physics of the Mining Processes of National Academy of Sciences of Ukraine developed the appropriate software [28 – 30]. The program will then automatically find the shortest route that will match the means of mechanization of the mining face, with a given level of performance $Q \geq 1200$ t/day and the minimum prime cost.

Table 4. Data required for optimization.

No.	Equipment type	Costs for the purchase, mln UAH	Equipment life, mln tonnes	Prime cost, UAH/t	Vertice of the network model (Fig. 2)
1	KD80	120.0	2.50	48	2
2	1KD90	193.5	5.00	39	3
3	DM	249.6	7.50	34	4
4	UKD200	35.0	0.60	58	5
5	1K101	15.0	0.29	52	6
6	UKD400	47.4	1.00	48	7
7	SP326	64.7	2.50	26	8
8	SP250	38.7	1.10	35	9

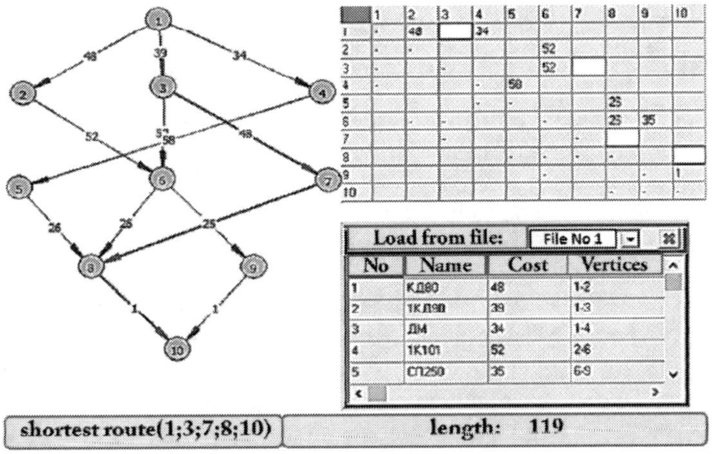

Fig. 2. Optimization of the process chain of the stopping equipment.

As can be seen from Figure 2, the shortest route is the one that passes through the following points: 1, 3, 7, 8, and 10. This corresponds to the equipment chain "powered support 1KD90, shearer UKD400, armoured face conveyor SP326". However, when one uses powered support DM the production prime cost is lower than with powered support 1KD90, so let us rearrange the structure of the production chain. For this purpose, we shall find powered support, shearer, and armoured face conveyor with the lowest production prime cost. In the end, the optimum chain of the mining equipment will consist of powered support DM, shearer UKD400, armoured face conveyor SP326.

Table 5. Technical and economic analysis of proposed solutions.

Equipment			Production volume, t/day	Increment, t/day	Prime cost, UAH/t	Reduction of prime cost, UAH/t
powered support	shearer	armoured face conveyor				
Mine "Krasnolymanska"						
DM	RKU10	SP251	1440	–	118	–
1KD90	UKD400	SP326	1480	40(2.8%)	113	5(4.2%)
DM	UKD400	SP326	1743	303(21.0%)	108	10(8.5%)
Mine "Kotliarevska"						
1KD90	1K101	SP26	848	–	124	–
DM	RKU10	SP326	1272	424(48.3%)	115	9(7.3%)

The program contains a database, provides various options for entering information and is integrated with office applications. For the purpose of optimization, we shall place points in the working area and set the length between points that correspond to vertices.

Because of the conducted selection, the prime cost will decrease by 10 UAH/t that will decrease the prime cost by 7.1 mln UAH (Table 5). For the mine "Kotliarevska", we also conducted the selection and optimization of operational parameters. Table 5 shows the analysis of proposed solutions.

Thus, the improvement of the mechanized coal mining is in the choice of means of mechanization of the mining face with a given level of performance. Subsequent application of network models and graphs allows optimizing the chain at one of the options. The described approach can be used as an addition to the already known ways of the planning mining works [31 – 32].

4 Conclusions

The study found that the performance of the mining face is an integral indicator, which is formed by a group of technological (layer thickness, length of the mining face), operational (means of mechanization), and economic factors. The effectiveness of the enterprise functioning is directly proportional to the level of interrelation between means of mechanization and inversely proportional to resources involved. This allowed us to draw the following conclusions:

– designing the technology for the mechanized coal mining is in the analysis of the set of rational solutions for specified operational conditions. Possible rational solutions can be formalized in the form of the "bank of decisions". This formalization allows to apply methods of structural optimization, analytic hierarchy process, linear and dynamic programming, and the graph interpretation allows automating the analysis process, increasing the dimension of solvable problems, which are found during the operation of coal fields;

– significant reserve of increasing technical and economic parameters of enterprises at the mining complex is the balanced flow of input and output resources; for this, it is necessary to present them in an organized structure as a network model. The subsequent application of optimization methods allows us to find the shortest route in the network, which corresponds to the balance of resources;

– scope of the described approach is not limited to coal fields and selection process of means of mechanization. Network models can be used to solve transportation problems, formulate emergency response plan, organize and manage resources, reduce the cost of equipment maintenance and repair.

One of the advantages of the proposed tool is its clearness, simplicity, ability to process and store large amounts of data. The characteristic feature of tools described in the paper is the ability to apply them together with MCDA methods at any stage of the study.

Authors express gratitude for the help and consultations during work performing for maintenance manager of "Krasnolymanska" mine Mykola Kostiev.

References

1. Ataci, M., Jamshidi, M., Sereshki, F., & Jalali I. (2008). S.M.E. Mining method selection by AHP approach. *Journal of the Southern African Institute of Mining and Metallurgy*, *108*(12), 741-749.
2. Mikaeil, R., Naghadehi, M., Ataei, M., & Khalokakaie, R. (2009). A decision support system using fuzzy analytical hierarchy process (FAHP) and TOPSIS approaches for selection of the optimum underground mining method. *Archives of Mining Sciences*, *54*(2), 349-368.

3. Saaty, T., & Shang, J. (2011). An innovative orders – of-magnitude approach to AHP-based Mutli-criteria decision making: Prioritizing divergent intangible humane acts. *European Journal of Operational Research*, *214*(3), 703-715. http://dx.doi.org/10.1016/j.ejor.2011.05.019
4. Dehghani, H., Siami, A., & Haghi, P. (2017). A new model for mining method selection based on grey and TODIM methods. *Journal of Mining & Environment*, *8*(1), 49-60. http://dx.doi.org/10.22044/JME.2016.626
5. Tzeng, G., & Huang, J. (2011). *Multiple attribute decision making: Methods and applications*. Boca Raton, FL, USA: Chapman and Hall/CRC Press.
6. Wang, Ch., Tu, Sh., Zhang, L., Yang, Q., & Tu, H. (2015). Auxiliary transportation mode in a fully-mechanized face in a nearly horizontal thin coal seam. *International Journal of Mining Science and Technology*, *25*(6), 963-968. https://doi.org/10.1016/j.ijmst.2015.09.013
7. Zhang, J. & Wang, Zh. (2012). The selection of coal strategic suppliers to the electric power enterprises based on cooperative game. China Coal, 3(4), 230-240.
8. Opricovic, S., & Tzeng, G.-H. (2007) Extended VIKOR method in comparison with outranking methods. *European Journal of Operational Research*, *178*(2), 514-529. https://doi.org/10.1016/j.ejor.2006.01.020
9. Grujic, M., & Tomasevic, A. (1996). Choice of outside transportation system in underground coal mines by multiple criteria analysis. *Underground Mining Engineering*, *4*(1), 62-70.
10. Vujić, S., Miljanović, I., Milutinović, A., Đorđević, D., Gojković, N., & Gajićet, G. (2011). Multiattributive prediction of terrain stability above underground mining operations. *Yugoslav Journal of Operations Research*, *21*(2), 275-291. https://doi.org/10.2298/YJOR1102275V
11. Nazymko, V.V. (2013). Pytannia pobudovy systemy avtomatyzovanoho upravlinnia proektom. *Upravlinnia rozvytkom sklatnykh system*, (14), 61-67.
12. Zakharova, L.N., & Nazymko, V.V. (2012). Doslidzhennia chutlyvosti prohramy rozvytku hirnychykh robit ta yii ryzykiv v umovakh vuhilnoi shakhty. *Radioelektronni i kompiuterni systemy*, *53*(1), 157-164.
13. Iphar, M., & Alpay, S. (2018) A mobile application based on multi-criteria decision-making methods for underground mining method selection. *International Journal of Mining, Reclamation and Environment*, *126*(3), 69-77. https://doi.org/10.1080/17480930.2018.1467655
14. Bogdanovic, D., Nikolic, D., & Ilic, I. (2012). Mining method selection by integrated AHP and PROMETHEE method. *Anais da Academia Brasileira de Ciências*, 84(1), 219-233. http://dx.doi.org/10.1590/S0001-37652012000100023.
15. Grinev, V., Cherepovskiy, P. & Deulenko, A. (2014). Tekhnologicheskiy aspekt formirovaniya ob'ema dobychi uglya s pozitsyi rynochnykh modeley. *Fiziko-tekhnicheskie problemy gornogo proizvodstva*, (17), 117-125.
16. Grinev, V. (2016). Grafy i seti dlya vybora gorno-shahtnogo oborudovaniya. Dnipro: Porogy.
17. Khorolskyi, A.A., Grinov, V.G., & Synkov, V.G. (2016). Vybir kompleksiv hirnycho-shakhtnoho obladnannia na osnovi teorii hrafiv. *Visnyk Natsionalnoho Tekhnicnoho Universytetu Ukrainy "Kyivskyi Politekhnicnyi Instytut"*, (31), 57-64. https://doi.org/10.20535/2079-5688.2016.31.69892
18. Fioroni, M., Santos, Letícia, C., Franzese, L., Seixas, J., Penna, B., & Alkmim, G. (2014). Logistic evaluation of an underground mine using simulation. *Rem: Revista Escola de Minas*, *67*(4), 447-454. https://dx.doi.org/10.1590/0370-44672014670181
19. Brazil, M., Thomas, D.A., & Weng, J.F. (2005). Cost Optimization for Underground Mining Networks. *Optimization and Engineering*, *6*(2), 241-256. https://doi.org/10.1007/s11081-005-6797-x
20. Guang. X., Jinxin, H., Baisheng, N., Chalmers, D. & Zhuoming, Y. (2017). Calibration of Mine Ventilation Network Models Using the Non-Linear Optimization Algorithm. *Energy*, *31*(11), 11-19. https://doi.org/10.3390/en11010031
21. Amankwah, H. (2011). Mathematical Optimization Models and Methods for Open-Pit Mining. Ph.D. Linköping University.

22. Shi, Q., & Erhan, K. (2016). New graph-based algorithms to efficiently solve large scale open pit mining optimization problems. *Expert Systems with Applications, 43*(1), 59-65. https://doi.org/10.1016/j.eswa.2015.08.044
23. Vladyko, O., Kononenko, M., & Khomenko, O. (2012). Imitating modeling stability of mine workings. *Geomechanical Processes During Underground Mining: School of Underground Mining 2012*, 147-150. https://doi.org/10.1201/b13157-26
24. Kirsch, U. (1981). *Optimum Structural Design: Concepts, Methods and Applications*. New York: McGraw-Hill.
25. Blauberg, I.V., Sadovskiy, V.N., & Yudin, E.G. (1978). Filosofskiy printsyp sistemnosti i sistemnyy podkhod. *Voprosy filosofii*, (8), 39-52.
26. O'Connor, J., & Mc. Dermott, I. (1997). *The art of systems thinking: essential skills for creativity and problem solving*. London: Thorsons.
27. Pidd, M. (2004). *Systems Modelling: Theory and Practic*. West Sussex: Wiley.
28. Grinev, V.G. & Khorolskiy, A.A. (2017). Obosnovanie parametrov vybora komplektatsii ochistnogo oborudovaniya s uchetom oblasti ratsional'noy ekspluatatsii. *Visti Donetskoho Hirnuchoho Instytutu, 40*(1), 139-144.
29. Khorolskiy, A.A. & Grinev, V.G. (2018). Setevye modeli kak instrument povysheniya organizatsionno-tekhnologicheskoy nadezhnosti proizvodstva. In *Innovative Technologies in Education, Science and Production*. Minsk: Belarusian National Technical University. https://rep.bntu.by/handle/data/36360
30. Khorolskyi, A.A., & Hrinov, V.G. (2017). Systemni pryntsypy ta otsinochni kryterii nadiinosti pry optymizatsii tekhnolohichnykh skhem vuhilnykh rodovyshch. *Visnyk Zhytomyrskoho Derzhavnoho Tekhnolohichnoho Universytetu, 80*(2), 225-233. https://doi.org/10.26642/tn-2017-2(80)-225-233
31. Mamaikin, O.R. (2014). *Obhruntuvannia parametriv tekhnolohichnykh skhem antratsytovykh shakht dlia ikh adaptatsii do innovatsii*. Ph.D. Natsionalnyi hirnychyi universytet.
32. Brazil, M., & Thomas, D. (2007). Network optimization for the design of underground mines. *Networks, 49*(1), 40-50. https://doi.org/10.1002/net.20140

Expanding of spheres the application of borehole hydro-production technology to develop deposits of non-traditional hydrocarbons

Mykhailo Pedchenko[1*], and *Larysa* Pedchenko[1]

[1]Poltava National Technical Yuri Kondratyuk University, Department of Oil and Gas Exploitation and Geotechnics, 24 Pershotravnevyi Ave., 36011 Poltava, Ukraine

Abstract. The reserves of oceanic hydrates of methane and natural bitumen in times exceed the stocks of traditional oil and gas. The purpose is the analysis of the adaptation possibility of borehole mining technology for the development of bituminous sands deposits and marine gas hydrates was the purpose of the study. The variants of the HBM technology adaptation for the development of bituminous sands and gas hydrates deposits are proposed. The method of extracting gas hydrates based on HBM technology involves them removal without the energy consumption for phase transition. The main advantages of the development of bituminous sands deposits on the basis of HBM technology are: the possibility them extraction in the range of depths 75 – 200 m, the potential high rate of bitumen extraction, the separation of bitumen from the rock in production, the minimum impact on the environment, the possibility of conducting extraction of natural bitumen in areas covered with water.

1 Introduction

The trend of long-term depletion of world stocks of traditional fuels has actualized for the world community the problem of further study of the resource potential of non-traditional sources of hydrocarbons [1, 2]. Oil of bituminous sands, coal methane, shale gas, shale oil and gas hydrates are unconventional sources of hydrocarbon raw materials. All these resources are explored or developed by companies all the world. However, based on explored stocks, oceanic methane hydrates and natural bitumen have particular importance for humanity [3 – 5]. Their stocks are more than the stocks of traditional oil and gas. All geological preconditions for the formation of large industrial clusters of high-viscosity oil, natural bitumen and marine gas hydrates are in Ukraine. According to preliminary estimates, total methane volumes accumulated in natural ocean hydrates are estimated at the level of $2.1\cdot10^{16}$ m^3 [6]. According to the French Institute of Petroleum, geological reserves of heavy oil and bitumen of all categories are estimated at 4.7 trillion barrels in oil equivalent. Among them, about 1 trillion are recoverable, which is roughly equal to the world resources of conventional oil.

* Corresponding author: pedchenkomm@ukr.net

2 Gas hydrates and methods for extracting of natural gas from gas hydrate deposits

Gas hydrates are clathrate compounds or combinations of inclusion. In appearance they are like thick snow. In 1 m^3 of natural gas hydrate contains about 160 m^3 of methane [7]. Underwater gas hydrates can form clusters on the seabed, a small pallet depth (first meters) or at a depth of hundred meters from the bottom.

The development of gas hydrate deposits can is carried out by a quarry method or free gas can is selected through wells after dissociation of the hydrate in the reservoir. Dissociation of hydrate to gas and water can be carried out by the following ways: lowering the pressure below equilibrium for hydrate formation, raising the temperature higher than equilibrium, injection of the substance-inhibitors for the process into the reservoir, or combination of the above methods. Significant energy costs are a disadvantage of thermal methods. In addition, the zone of thermal action in the reservoir will be limited to several meters.

In terms of energy costs, the way to reduce the formation pressure below equilibrium with the subsequent selection of free gas is most acceptable. However, in this way, secondary hydrate's formation will be observed in the prebottom of the well.

Gas hydrates can also serve as a "cement" for rock particles. In this case, the melting of the gas hydrate will lead to abnormally high porosity, the separation of large water masses and a significant reduction in the strength of the productive layer [6]. The removal of a significant amount of rock in the well will be observed as a result, which will interfere with its exploitation.

Methods of career development will be complicated by the fact that the bottom deposits of gas hydrate is mainly covered with a layer of deposits in several meters [8]. Also, this method will be effective only at significant content of gas hydrates in the rock. At present, industrial development of gas hydrates deposits is not conducted in the world due to lack of efficient and competitive technology.

3 Natural bitumens and methods of their extraction

Alternative to conventional crude oil is oil sands which (tar sands or bituminous sands). Oil sands are a loose sand deposit which contain a very viscous form of petroleum known as bitumen. These unconsolidated sandstone deposits comprise primarily of sand, clay and water saturated with bitumen. A typical oil sands deposits contains about 10% bitumen, 5% water and 85% solids. The largest world deposits are in Canada, Venezuela, Madagascar, USA and Russia [9]. In spite of the high energy consumption required in extraction of bitumen from tar sand, it is an economically viable enterprise and it is being undertaken in vast tar sand deposits in Canada. The majority of these reserves are found in Alberta's oil sands – over 174 billion barrels.

There are two basic methods of bitumen extraction used in the oil sands: surface mining and in-situ. The method used depends on the depth of the reservoir. The most effective way of extracting significant deposits of bituminous sands that lie at a shallow depth is the quarry. Deposits located at a depth of less than 75 m can technically be surface mined, although most deposits are located at a depth of less than 50 m below grade. It is widely distributed in Canada, where up to 20% of proven bitumen reserves lie in sandy collectors at a depth of 50 m.

The first stage mining by open method is the removal of a peat layer and rock 2 – 6 m thick. Preliminary the area is drained. The process can continue up to 3 years. The ore is dug up and mixed with warm water to separate and recover the bitumen from the sand. The separation takes place in separation cells. The combination of hot water and agitation releases bitumen from the oil sand, and causes tiny air bubbles to attach to the bitumen droplets, that float to the top of the separation vessel, where the bitumen can be skimmed off. Further processing removes residual water and solids. Bitumen normally requires dilution with lighter hydrocarbons to make it transportable by pipelines. The bitumen is then transported and eventually upgraded into synthetic crude oil.

This process recovers about 90% of the bitumen. The remaining 80% of the bitumen is too deep to be mined and can only be extracted in-situ. A number of methods for bitumen smelting in situ of bedding are developed in this regard. Most in-situ bitumen and heavy oil production comes from deposits buried more than 400 meters below the surface of the earth. Given the physical and chemical properties of natural bitumen, the application of methods that maximally reduce their viscosity, is the most rational. Among a such methods, thermal and thermochemicals occupy a special place. These involve different combinations of injecting steam or solvents through horizontal or vertical wells.

The first promyslovoyu tehnologiyehu becoming a method CSS (Cyclic Steam Stimulation). In this process, steam is added to the oil sands via vertical wells, and the liquefied bitumen is pumped to the surface using the same well. Steam at a temperature above 300 °C is pumped over several months. Then a well is heated for several weeks. After that, heated bitumen can is pumped during for several months. However, the process can be cost-effective only for the first few cycles (usually 3 – 5). This method allows to extract up to 25% bitumen, what contained in the sand.

The Steam-Assisted Gravity Drainage (SAGD) technology is believed one with the most effective and applicable. For implement SAGD, two horizontal wells are drilled (one over the other at a distance of 3 – 7 m). The steam is served during 90 days into the well and creates a warmed area, reducing the viscosity of bitumen (oil) and increasing mobility. The force of gravity directs condensed water and mobile hydrocarbons (bitumen) to the bottom of the mining well. In the second phase, the extraction of bitumen (high viscosity oil) is carried out from the extraction wells and steam is pumped into the inqection well. But at the same time the roof of the productive layer should be a reliable barrier to prevent the loss of steam. SAGD has demonstrated that its operations can recover as much as 60% of the bitumen inplace. SAGD wells operate at low vapor pressures. Steam Assisted Gravity Drainage (SAGD) is currently the most commonly used in situ recovery method.

"Cold" methods of bitumen extraction also exist and they predict instead of steam pumping bitumen is processed by the dissolvent in the layer (VAPEX method). This method is also implemented using two horizontal wells. A dissolvent (hydrocarbon dissolvents including ethane or propane) is pumped into the upper well instead steam. Bitumen is rarefied due of the dissolvent diffusion in it and flows to the mining well under the action of gravitational forces. This method requires no water and is 25% lower in capital costs than the SAGD process. Operating costs are half that of the SAGD process [10].

Also commercial efficiency of the extraction method using dissolvents in a vapor state (VAPor EXtraction – VAPEX) is justified. Hot ethanes and propane-butane are pumped into the well located at the top of the formation. Reduced viscosity of bitumen is achieved by heating and rarefing by the dissolvent, as a resulting in a mixture of bitumen-dissolvent gravity drains down into the mining well. Due to the using of hydrocarbon solvents it is possible to significantly reduce energy costs, as well as successfully develop collectors with high content of clay, where the using of the stea's method is unacceptable. The coefficient reaches 60%, however, the rate of extraction is extremely low.

The new technology, N-Solv, is a modification of the VAPEX method. The technology uses propane heated to 50 °C. It requires 80% less energy, and propane can be reused. The dissolvent dissolves bitumen, and the hardest components remain underground. The resulting product returns out less viscous and can be transported immediately along the oil pipeline. The efficiency of the process has been greatly increased by removing of the methane admixture from propane. Methane in the reservoir significantly degrades the heat exchange process at the boundary of the bitumen – dissolvent.

At present, the processes co-injection of vapor and dissolvent are also used: improved steam-gravity effect by the dissolvent – Expanding Solvent SAGD (ES SAGD), Process with addition of the dissolvent (SAP), alternating of reinqection dissolvent and vapor – Steam Alternating. However, the main problem with the injection of water vapor or hydrocarbon dissolvents is the heterogeneity of productive layers. This inevitably leads to their breakthrough to producing wells and the need to use flow-deflecting technologies (blocking gels, sediment-forming compositions).

The thermal methods for the development of bituminous sands also include the method of electric heating. This technology, as a rule, is used as an auxiliary in the implementation of SAGD and the development of deposits in a career way [11].

CHOPS technology [12] involves the "cold" extraction of heavy oil along with sand. For this purpose, depression is created in the layer. So-called "wormholes" are formed due to the destruction of weakly cemented rocks. Oil and sand by them come to the well. The CHOPS method is widely used for oil production in Canada (up to 15% of total production). It does not require large investments for development and has low operating costs. But the oil recovery ratio does not exceed 20%. Also, CHOPS requires constant monitoring, as conditions are constantly changing at the bottom.

Thus, the lower boundary of the career development of bituminous sands is a depth of 75 m. Methods for extracting bitumen in site of it occurrence are applied starting at a depth of 200 m due to the danger of a break through of the working agent on the surface. Therefore, the interval of bedding of the bituminous sands, for which there is not effective technology of their extraction, exists. In this regard, the important task is to develop an effective method for this interval of bedding of bituminous sands.

At first glance, the extraction technology of gas hydrates and natural bitumen must be significantly different. However, they are united by the fact that in the natural bedding the solid gas hydrate and viscous or hard bitumen do not exhibit fluid properties. Their mobility first of all affects extraction. For this, for example, a significant amount of thermal energy is introduced into the system. As a result, the gas hydrate dissociate into gas and water, and the viscosity in bitumen decreases. In terms of project effectiveness, it is better to do it in situ of their bedding.

4 Analysis of the features of hydraulic borehole mining the technology

Physical and chemical technology of hydraulic borehole mining (HBM) minerals is well known in mining. The cutting of a specific rock type is dependent on a multitude of factors involving both the physical properties of the rock and tool operating parameters. The primary cutting mechanism in a HBM system is the conversion of kinetic energy of the water jet into impact pressure, acting on the rock face. The technology involves the destruction of the rock in situ of it occurrence due to the hydraulic energy of the fluid jet, the formation of hydro-mixture (pulp) and it extraction on the surface [13]. Penetrating into an array due to filtration in pores and cracks, water reduces the strength of rocks, causing local hydrodynamic fractures and destruction [14 – 16]. In a flooded bottom, the technology allows to develop of non-bonded ore bodies in conditions of large water inflow, as well as under water and on the shelf of the seas.

However, the rapid dissipation of the energy of the jet occurs in condition a flooded bottoml [17]. Using of hoses or telescopic monitors lengthening as the bottom moves, is required. This complicates the design of the unit, requires the creation of special systems for monitoring and controlling the hydraulic monitor. At the same time, the amount of work for one well is limited in connection with the threat of collapse of the roof. After the collapse, the extended hydraulic monitor is difficult or impossible to remove. At the same time extraction of rock mass is difficult and exist high level of losses. All this limits the sphere of application HBM technological scheme.

However, the analysis of the characteristics of the HBM technology allows us to conclude that it is expedient to use for the development of bituminous sands deposits and marine gas hydrates. Therefore, the analysis of the adaptation possibility of borehole mining technology for the development of bituminous sands deposits and marine gas hydrates was the purpose of the study.

5 Features of the development of gas-hydrate deposits by the method of HBM

The above advantages and disadvantages of the HBM technology are taken into account in the method of developing marine deposits of gas hydrates (Fig. 1) [16].

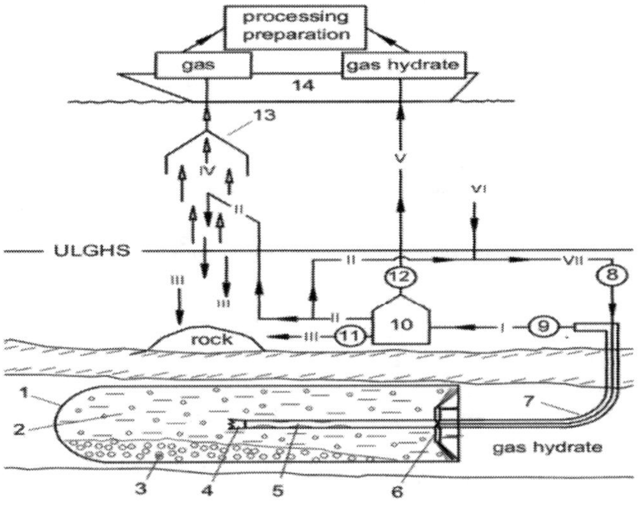

Fig. 1. Principal diagram of the method for marine gas hydrate deposits development: ULGHS – the upper limit of gas hydrate stability; 1 – excavation in the hydrate-saturated reservoir; 2 – space in the excavation site filled water-hydrate mineral pulp; 3 – sediment of rock solid inclusions; 4 – drill bit; 5 – pulp intake; 6 – jet gun device; 7 – well; 8, 9, 11, 12 – pump; 10 – gravity separator; 13 – gas collecting dome; 14 – production platforms; streams; I – water-hydrate mineral pulp; II – gas hydrates depleted pulp; III – empty rock; IV – gas released from the pulp as a result of gas hydrate dissociation; V – water gas hydrate mixture; VI – sea water; VII – working fluid for rock disintegration (sea water and pulp mixture).

The next sequence of technological operations is a feature of the proposed method of technical and technological decisions:

– drilling wells to maximum length;

– formation of a chamber on a bottom well hole for opening of a rod with nozzles of hydraulic monitors in working position;

– destruction of the rock with flooded jets and the formation of mining.

The decisive parameter in the energy transfer mechanism from the jet to the breed is the distance between them. Therefore, rods with hydromonitors rotate around the axis of the column, gradually approaching the front of destruction.

This method involves:

– destruction of the rock of the gas hydrate formation and it transfer into the composition of the pulp;

– preliminary enrichment of the pulp on the gas hydrate in the development as a result of gravitational separation of the fragments of rock and gas hydrate;

– separation on the seabed from a pulp of the gas hydrate in a gravity separator;

– dissociation of the residual gas hydrate in the impoverished pulp due to the thermal energy of sea water;

– supply of gas hydrate and gas to the extraction platform.

The dissipation of a large part of the jet energy will be observed in the hydrodynamic action in the process of HBM. In addition to increasing the flow temperature of the fluid in the pumps and the piping system, part of the energy will turn into heat at the moment of jet impact on the rock. However, the gas hydrate in the deposit is at a certain level of supercooling relative to equilibrium conditions. This is a condition for the existence of gas hydrates deposits. However, this overcooling is insignificant – it is several degrees. The local area heating will certainly be formed on the boundary of rock destruction by the jet. It is assumed that it will be limited by a few millimeters and will not go beyond the contour of the flow contact with the rock. Thermal energy in this zone will be spent on the increase of temperature including gas hydrate. Moreover, the allocated energy will be enough to disassociate its parts into gas and water. In addition, the gas hydrates in the rock can fulfill the function of it cement. Therefore, their melting on the boundary of the array and mining will positively affect the effectiveness of the technology. Thus, in the gas hydrates production by the method HBM, the heat energy of the dissipative processes will positively influence the efficiency of the disintegration of the rock saturated with gas hydrates. Also, according to the law of energy conservation, going beyond the boundaries of local zones heat energy, gas again enters to the conditions for the stability of the gas hydrate. Repeat formation of hydrate will occur as a result. Hence, hydraulic fracturing will be accompanied by recrystallization of a part of the gas hydrate. And melting occurs in the one situ, and the formation – in another (in the formed pulp). And most likely, the it overwhelming part will be separated from the rock particles.

In addition, the zone of pulp gravity separation will exist in the development of a certain distance from the zones of destruction and turbulent mixing of the pulp. Here, the fragments of the rock will fall into the precipitate, and the gas hydrate (natural and recrystallized) will fall to the cluster of excavation. (And it is here that it is supposed to place its pumping hopper to the extractive platform). Thus, the initial enrichment of the target product will occur even in the excavation. That is, there will be no need for the costs of lifting and subsequent operations with most of the breed ballast.

It also in mind that the method described in [18] Involves the use in the process of seawater (known mineralization). And the volume of the hydro-monitor flooded jet before the moment of reaching the destruction front will increase due to the addition of the pulp part together with particles of the breed [13]. In this case, the salts of sea water in the zone of the rock destruction will act as a inhibitor of the hydrate process, and grain of the rock – an abrasive material. All this will further increase the efficiency of this technology.

In addition, the proposed method for the development of gas hydrate deposits will allow the direct removal of gas hydrate without the energy consumption for phase transition. It is further assumed that on the extraction platform, the recovered gas hydrate will be recycled in accordance with the technology proposed in [19] for further transportation and storage. Given the enthalpy of hydrate formation (540 kJ/kg), this technology will have significant competitive advantages over methods that provide for it dissociation in the formation and extraction of gas.

6 Features of the development of bituminous sands deposits by the method of HBM

Taking into account the properties and features of the bituminous sands bedding, the technology described above for the development of gas hydrate deposits [18] can be adapted to extract bitumen from bituminous sands deposits. The technology based on HBM and adapted for bitumen mining is presented in Fig. 2. In essence, the main difference between this version of the technology is the heating of water before feeding it to the hydro-monitor and adding of the surface-active substances. That is, in addition to the rock destruction and the underground mining formation, the method involves bringing to it heat energy for reduce the viscosity of bitumen.

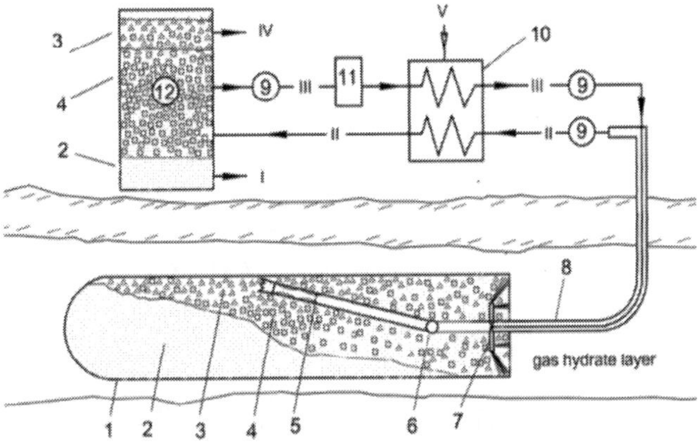

Fig. 2. Scheme of a method for the bitumen extraction from deposits of bituminous sands based on the technology of HBM: 1 – development; 2 – sand sediment; 3 – pulp enriched on bitumen; 4 – water with clay fraction; 5 – a shank with a drilling bit and a pulp picker; 6 – hinge; 7 – hydromonitoring device; 8 – well; 9 – pump; 10 – heating block; 11 – block of separation of solid phase; 12 – separator; streams: I – sand sediment; II – a pulp enriched with bitumen; III – clay solution; IV – bitumen; V – coolant.

The pressure of water for effective the rock crushing is within the range of 5 – 20 MPa. Given the phase equilibrium of the system "water – steam", its temperature can be increased to 400 – 500 K. After the jet out of the nozzle, part of the energy will be transferred to the attached pulp mass [13].

It is also assumed that outside the hydraulic system, where the pressure corresponds to the formation (reservoir), the level of liquid overheating in the destructe zone will be sufficient for it boiling. Bubbles and cracks of the rock will be forming and growing in the pores, destroying ones. Therefore, this process will also have a positive impact on the effectiveness of the technology. As a result, after the disintegration of the rock, the pulp will have a composition and temperature similar to maintained in the separators of known sets of the bitumen separation from bituminous sand. In this case, mine excavation will act as a separator itself. Similarly to the known technology, in these "separators", the active the pulp mixing will be observed near the destruction zone of the rock by hydro-motors. At some distance from this zone of movement will fade away. The pulp gravitational separation will occur in this place: the pulp concentrate with a high content of the bitumen and clay fraction rock ascends, and rock particles of a larger size – will settle down to the bottom of the excavation. Enriched pulp will be pumped from the excavation to the surface. The bitumen will be separated from the pulp after a slight additional heating. Warm water (uncleaned, even with certain rock content) returns to the process again. Judging by the processes that take place in the development, the rate of bitumen extraction by this technology can compete with the career method. Taking into account also the relatively low thermal conductivity of the rock, the possibility of heat energy recuperation, low water consumption and a potentially high bitumen recovery rate, this technology will have high competitiveness.

Variant of the deposit development on the land is shown in Fig. 2. However, the bitumen extraction in this way can be conducted and in areas covered with water and marshes. Also, this method will not require drainage at the place, and thus shorten the start date of extraction. The pressure in the production will be close to the hydrostatic, so the breakthrough of the pulp on the surface is excluded. There will also be no restrictions of the rock characteristics for the roof of the productive layer. The duration of the preparatory operations will be negligible. The product selection will begin almost immediately after the delivery of the coolant to the hydro-monitors. Therefore, the heat loss in the adjacent layers of the rock will be minimal.

Given the potentially high withdrawal ratio, the cost of bitumen extraction by this method will be low. The technology promises significantly lower capital and operating costs than conventional mining. Major oil sands projects require substantive environmental assessments before they are approved. In Canada oil sands development is subject to environmental standards that are among the most stringent in the world. An analysis of the proposed method shows that the negative impact on the environment will be minimal.

7 Conclusions

1. Analysis of the borehole hydraulic mining processes, technological operations and the sequence them implementation showed the prospect of the introduction such a method for the development of bituminous sands and gas hydrates deposits.

2. The main advantage of the method for the development of gas hydrate deposits based on HBM technology is remove the target product (gas hydrate) from the productive layer without the energy consumption for the phase transition.

3. The method of development of gas hydrate deposits based on HBM technology can be easily adapted for the bitumen extraction from bituminous sands.

4. The advantages of developing bituminous sands deposits on the basis of borehole hydraulic mining are:

– the possibility them extraction in the range of depths 75 – 200 m, where application other methods is problematic;

– potentially high coefficient of bitumen extraction, which will approach to the level of the career method;

– relatively low water consumption;

– minimal impact on the environment;

– similarly to the variant of marine gas hydrate deposits development, in this way it is possible to produce natural bitumen in areas covered with water (sea shelves, lakes), as well as in wetlands without drainage.

5. Hydraulic borehole mining can remove underground ore deposits tar sand that cannot be economically mined by the usual open pit or underground methods.

6. Selectivity of the borehole mining system also allows economical recovery of small or extraction ore deposits while minimizing the removal of host rock.

7. The development of technical solutions and refinement of the technology parameters requires research and industrial testing. Moreover, these decisions can differ significantly depending on the geological conditions of specific deposits.

This work has become possible due to financial and organizational support within the frames of the state budget research project under the auspices of the Ministry of Education and Science of Ukraine "Application of gas hydrate technology in the development of traditional and gas hydrate gas deposits" (State registration No. 0113U00857), "Research of influence of thermodynamic parameters of phase transitions in systems with gas hydrates on the efficiency of gas hydrate technology" (State registration No. 0115U002420).

References

1. Shunko, V. (2015). Monograph "Unconventional hydrocarbon resources of Ukraine". *Visnyk of Taras Shevchenko National University of Kyiv. Geology*, (3), 79-80.
2. Bondarenko, V., Lozynskyi, V., Sai, K., & Anikushyna, K. (2015). An overview and prospectives of practical application of the biomass gasification technology in Ukraine. *New Developments in Mining Engineering 2015: Theoretical and Practical Solutions of Mineral Resources Mining*, 27-32. https://doi.org/10.1201/b19901-6
3. Naumko, I., Kurovets', I., Zubyk, M., Batsevych, N., Sakhno, B., & Chepusenko, P. (2017). Hydrocarbon compounds and plausible mechanism of gas generation in "shale" gas prospective Silurian deposits of Lviv paleozoic depression. *Geodynamics*, *1*(22), 26-41. https://doi.org/10.23939/jgd2017.01.026
4. Samoylov, V. (2017). Planning of the industrial and hydrogeological research at the final stage of hydrocarbon deposits development. *Visnyk of V.N Karazin Kharkiv national university-series geology geography ecology*, (46), 45-49. https://doi.org/10.26565/2410-7360-2017-46-05
5. Melnikov, V., & Gennadinik, V. (2018). Cryodiversity: the world of cold on the Earth and in the solar system. Philosophy and Cosmology, (20), 43-54. https://doi.org/10.29202/phil-cosm/20/4
6. Kvenvolden, K.A. (1993). Gas hydrates – Geological Perspective and Global Change. *Reviews of Geophysics*, *31*(2), 173-187. http://dx.doi.org/10.1029/93rg00268
7. Carrol John (2009). *Natural Gas Hydrates*. A Guide for Engineers, 2nd Ed. Oxford: Gulf Publishing.
8. Shnyukov, E., Gozhik, P., Krayushkin, V., & Klochko, V. (2007). On the Eve of the World of Submarine Methane Hydrate Extraction. *Reports of National Academy of Sciences of Ukraine*, (6), 125-134.
9. Chilingar, G.V., & Yen, T.F. (1978). *Bitumens, asphalts, and tar sands*. New York, Elsevier Science.
10. Canada's Oil Sands, Opportunities and Challenges to 2015. *An Energy Market Assessment*, May 2004, National Energy Board, Canada, 108.
11. Briggs, P.J., Baron, P.R., & Fulleylove, R.J. (1988). Development of Heavy-Oil Reservoirs. *Journal of Petroleum Technology*, 206-214.
12. Sadler, K.W. (1995). An EUB Review of In Situ Oil Sands Bitumen Production. *SPE paper 30240- MS presented at SPE International Heavy Oil Symposium*, 19-21 June, Calgary, Alberta, Canada.
13. Arens, V., Babichev, A., Bashkatov, A., Gridin, O., Khrulev, A., & Khcheyan, G. (2007). *Borehole Hydro-Mining*. Proc. Manual: Mining Book.
14. Rehbinder, G. (1980). A Theory about Cutting Rock with Water Jet. *Rock Mechanics*, *12*(3-4), 247-257. http://dx.doi.org/10.1007/bf01251028
15. Bondarenko, V., Sai, K., Prokopenko, K., & Zhuravlov, D. (2018). Thermodynamic and Geomechanical Processes Research in the Development of Gas Hydrate Deposits in the Conditions of the Black Sea. *Mining of Mineral Deposits*, *12*(2), 104-115. https://doi.org/10.15407/mining12.02.104
16. Bondarenko, V., Kovalevska, I., Astafiev, D., & Malova, O. (2018). Examination of Phase Transition of Mine Methane to Gas Hydrates and their Sudden Failure – Percy Bridgman's Effect. *Solid State Phenomena*, (277), 137-146. https://doi.org/10.4028/www.scientific.net/SSP.277.137
17. Helgerud, M.B. (2001). *Wavespeeds in gas hydrate and sediments containing gashydrate: A laboratory and modeling study*. Ph.D. Stanford Univ. Press. Stanford, Calif.

18. Pedchenko, M., & Pedchenko, L. (2017). Analysis of gas hydrate deposits development by applying elements of hydraulic borehole mining technology. *Mining of Mineral Deposits, 11*(2), 52-58. https://doi.org/10.15407/mining11.02.052
19. Pedchenko, L., & Pedchenko, M. (2012). Substantiation of Method of Formation of Ice Hydrate Blocks with the Purpose of Transporting and Storage of Hydrate Gas. *Naukovyi Visnyk Natsionalnoho Hirnychoho Universytetu*, (1), 28-34.

Dependence of gas emission from the sources undermined by stope mine workings on the rocks displacement parameters

Mykhailo Filatiev[1*], *Elvira* Filatieva[2], and *Mykola* Antoshchenko[2]

[1]Donbas State Technical University, Mining Department, 84 Peremohy Ave., 93100 Lysychansk, Ukraine
[2]Donbas State Technical University, Department of Ecology, Life Safety and Labor Protection, 84 Peremohy Ave., 93100 Lysychansk, Ukraine

Abstract. The parameters are established in the work of the stope mine workings and rocks displacement, which correspond to the maxima of gas emission into the gas drainage boreholes. The angular and linear parameters of the crustal movement and rocks displacement have been calculated with the use of empirical dependences. These parameters correspond to the maxima of gas emission into the gas drainage boreholes. The obtained positive results for determining the possible areas of maximum gas emission allow us to recommend the empirical dependences for calculating the angles of maximum subsidence and complete displacements at the stage of stope works development, as well as the dimensions of the stope mine workings, at which the processes of rocks displacement reach the earth surface. The established relations between the gas drainage boreholes location and the rocks displacement zones with discontinuity make it possible to design the optimal schemes of draining-out of gases from the gas-and-coal deposits. For the first time, three independent methods have been used to determine the changes in dimensions of rocks displacement zones with discontinuity after the primary main roof squeezing until the formation of stable upper boundaries of these zones.

1 Introduction

The link between the level of gas emission from coal of the adjacent seams and its residual gas content with the intensity of the undermined rocks displacement is qualitatively confirmed by the results of mine experiments [1, 2]. The change in gas pressure and absolute methane emission from the undermined seams under the influence of the working stope face has been established by the studies [3]. Up to date, there are no works in which the parameters of rocks displacement and gas emission would have been considered together when removing the stope faces from the first cut workings in the initial period of the extraction sites operation.

Consideration of this issue is relevant to the creation of safe conditions during the stope works performance, as in this period the primary and subsequent squeezing of the main roof

* Corresponding author: mfilatev@gmail.com

occurs and the gas emission maximum is reached from the undermined coal and rock stratum. It has been established [4], that when removing a stope face from the face entry after the squeezing of the main roof, a periodic increase in methane emission into the gas drainage boreholes and into mine workings, is observed. Methane emission (I_m^0) into gas drainage boreholes begins since removal of the stope face from the first cut working (L_m^0) at a certain distance. And then there is an increase in gas emission and achieving of its maximum. After the first maximum (I_m^1), a decrease in methane emission is observed, and then there is an increase and the achievement of the next maximum of gas emission occurs. The alternation of gas emission maxima is determined through the entire time of mining the working area [5]. This is obviously caused by the periodic roof rocks squeezing as the stope face is removed from the face entry. On the basis of mine observations, it is possible to determine reliably the distances between the stope face and the first cut working (*L*), at which the gas emission maxima occur. According to the empirical dependences [6 – 10], it is possible to calculate the values of the angles of maximum subsidence of the earth surface (θ) and the complete displacement of the undermined rocks (ψ_1, ψ_2, ψ_3) with high accuracy (correlation ratios (*R*) more than 0.90) in relation of *L* to the depth (*H*) of stope works performance.

The maximum subsidence angle (θ) – from the side of the seam gradient, formed on the vertical section in the main cross section of the trough, transverse the course of seam by the horizontal line and the line connecting the middle of the mine working with the point of maximum subsidence at incomplete undermining of the earth surface [9]. According to the results of surveying observations [6], the empirical dependence has been obtained:

$$\theta = 86.3 - 56.5\exp\left[4\left(\frac{L}{H}\right)^{-1.9}\right], \text{ degree.} \quad (1)$$

According to the normative document [11], the angles of complete displacements are internal relative to the mined-out space formed on the vertical sections along the main cross sections of the trough. The flat bottom of the shift trough of the earth surface in the initial period of the extraction area operation is not formed. For this reason, the angles of complete displacements (ψ_1, ψ_2, ψ_3) are suggested to be calculated for flat seams according to the dependence [6]:

$$\psi = 93.7 \cdot \exp(-0.42 \cdot \frac{L}{H}), \text{ hail.} \quad (2)$$

In this case, the determination of the angles (ψ_1, ψ_2, ψ_3) was made by connection of the experimental point of maximum subsidence of the earth surface by means of one line segment with the fixed wall of the face entry, and by the second line segment – with the stope face position corresponding to this point.

The equation (2), with a high degree of probability, describes the change in the complete displacement angles at values of $L/H \le 1$ as from the side of the first cut working (ψ_3) so from the side of the stope face (ψ_p). At $L/H > 1$, the angles of complete displacements from the side of the stope face were in the range of 50 – 60° and averaged 55° [6]. Such values coincide with the recommendations of the normative document [10].

In addition to the angular parameters of the rocks displacement and the crustal movement, some characteristic linear dimensions of the stope mine workings are important. These include L_H and L_m. L_H is the removal of the stope face from the first cut working, at which the processes of rocks displacement reach the earth surface. According to [11], $L_H = (0.1 \div 0.3) \cdot H$. The parameter L_H depends on the seam thickness (m), the depth (*H*) of the stope works performance and the rate of the stope face advance (υ_{extr}). It is uniquely determined by the equation [6]:

$$L_H = H \cdot \left[(0.329 - 0.096 \cdot m) / 0.223 \right] \cdot (0.002 \cdot \upsilon_{extr} + 0.145) \text{, m.} \qquad (3)$$

The final formation of complete displacement area of the roof rocks occurs when the stope face is removed from the first cut working at a distance L, which is equal to the length of the longwall face or to the width of the mined-out space formed during the mining of several adjacent longwall faces L_l. In the mining extracted area of the longwall, a square is formed of the mined-out space, which is the zone of stope works influence. At further advance of the longwall, the zone of the mined-out space influence is formed, in which the parameter H_p of complete roof displacement is not changed until the longwall mining of the entire mining extracted area [12]. Based on this scientific position, the limit value of the height of the rocks displacement zone with discontinuity (H_p) is formed, when removing the stope face from the first cut working at a distance $L \geq L$.

Thus, the purpose of the work is to determine the parameters of the stope mine workings and the rocks displacements corresponding to the maxima of gas emission into the gas drainage boreholes. To do this, on the basis of mine experiments, it is necessary to determine the maxima of methane emission into the gas drainage boreholes drilled above the installation chambers, and corresponding removals of stope faces from the first cut workings. Having determined the dimensions of the stope mine workings and using empirical dependences, it is possible to calculate the angular and linear parameters of the crustal movement and rocks displacements corresponding to the maxima of gas emission into the gas drainage boreholes.

2 The research results

The studies were performed for the conditions of the anthratscite seam l_2^e mining by Gazeta Izviestiia Mine, SE Donbasantratsyt (Ukraine). The thickness of the mined seam was 0.9 m; the incidence angle was 2 – 3°. The longwalls (2nd bis and 3rd Western) were mined out by heading-and-stall method, their lengths were 185 and 205 m, (Fig. 1). The average rate of the stope face advance of the 2nd bis Western longwall for the entire observation period was 0.7 m/day, and of the 3rd Western longwall – 2.8 m/day. The depth of the stope works performance is 290 m.

In addition to the distances between the stope faces and the first cut workings, at which the maxima of gas emission were observed (Fig. 2), the distances corresponding to the beginning of the crustal movement (L_H) and the formation of the square of mined-out space ($L = L_l$), have been also taken into consideration during analysis.

For each characteristic dimension of the stope mine working (removal of the stope face) according to the equations (1, 2), the angles θ and ψ_p (Table 1) have been calculated, respectively. The parameter L_H is determined using equation (3). It shows that with such a dimension of the stope face, the displacement processes of the undermined rocks reach the earth surface, and the disturbance of the original natural state occurs in the entire coal-bearing strata. Upon availability of intersecting fractures, this creates conditions for the desorption of gas from the undermined seams and enclosing rocks.

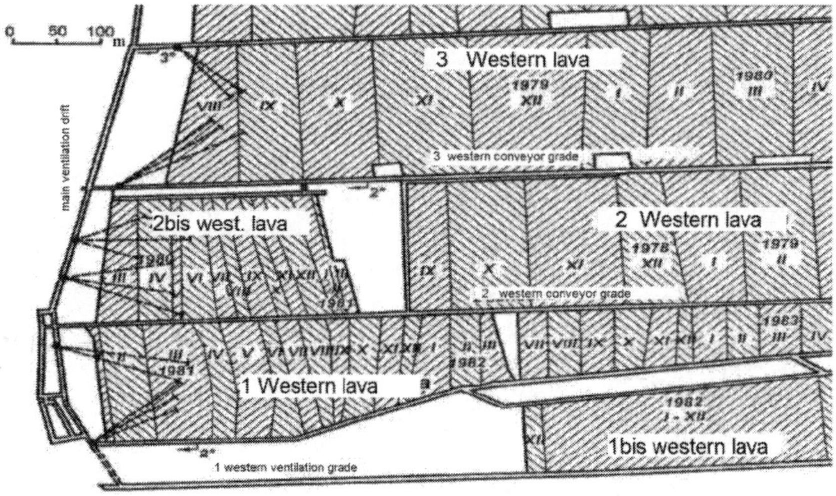

Fig. 1. A fragment of the plan for mining works through a stratum of Gazeta Izviestiia Mine, SE "Donbasantratsyt".

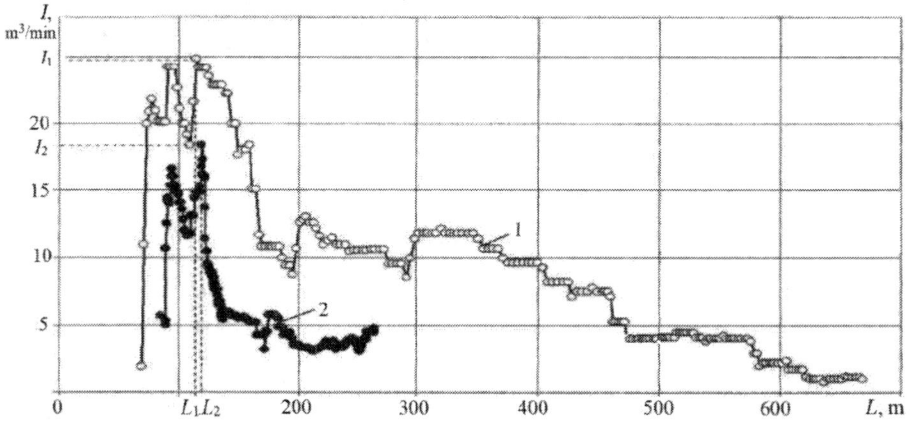

Fig. 2. Dependence of the flow rate I into the boreholes on distance L between the first cut workings and stope faces; 1 and 2 – change in gas emission into the boreholes of the 3-rd and 2-nd bis of the Western longwalls; ○, ● – experimental data.

The rocks displacement zone with discontinuity corresponds to these conditions. Its upper boundary H_p is determined in three ways using different initial parameters of rocks displacement. The values H'_p and H''_p (Table 1) are determined according to [6]:

$$H'_p = L \cdot \frac{tg\psi}{2}, \text{ m;} \quad (4)$$

$$H''_p = H \cdot \frac{tg\psi}{tg\psi_p}, \text{ m,} \quad (5)$$

where L is a removal of the stope face from the first cut working, m; $\psi = 55^o$ are the average values of complete displacement angles with sufficient stope works development (at $L/H > 1$); H is the depth of work performance, m; ψ_p is the angle corresponding to the maximum subsidence of the earth surface from the side of the first cut working when removing the stope face at a distance L.

To determine the parameter H_p''', a graphical method has been used. According to this method, the upper boundary of the rocks displacement zone with discontinuity has been determined by the point of line segment intersection, which determines the maximum subsidence of the earth surface, with a segment corresponding to the average value of the angle ψ. This method is designed to confirm the idea of compliance of the maximum gas emission with maximum rocks displacement.

Table 1. The initial experimental data on gas emission into the gas drainage boreholes and the calculated parameters of displacement of coal and rock stratum.

Characteristic indicators of gas emission and crustal movement	The distance between the stope face and the first cut working L, m	L/H	Parameters of the undermined rocks displacement				
			θ, degree	ψ_p, degree	H_p', m	H_p'', m	H_p''' m
2nd bis Western longwall							
Beginning of the crustal movement	59	0.20	86	86	42	29	40
Beginning of gas emission I_m^0	83	0.29	86	83	59	51	58
The first maximum of gas emission I_m^1	89	0.31	86	82	64	58	62
The second maximum of gas emission I_m^2	127	0.43	86	78	90	88	90
The third maximum of gas emission I_m^3	160	0.55	86	74	114	118	116
Removal of the stope face at a distance L_l	185	0.64	86	72	132	134	130
3rd Western longwall							
Beginning of the crustal movement	99	0.34	86	81	70	65	70
Beginning of gas emission I_m^0	75	0.26	86	84	53	43	48
The first maximum of gas emission I_m^1	80	0.28	86	83	57	51	52
The second maximum of gas emission I_m^2	104	0.36	86	81	74	65	68
The third maximum of gas emission I_m^3	114	0.39	86	80	81	73	80
The fourth maximum of gas emission I_m^4	202	0.70	86	70	143	150	136
Removal of the stope face at a distance L_l	205	0.71	86	70	146	150	140

As an example, a graphical interpretation is given of the change in the displacement parameters of the undermined coal and rock stratum, as well as the earth surface (Fig. 3).

The close values of the calculated parameters (H'_p, H''_p and H'''_p) for different distances between the stope face and the first cut working (Table 1) indicate approximately the same reliability of the results obtained by different methods. This is also an indirect confirmation of the compliance of the gas emission maxima with maximum rocks displacement.

The beginning of gas emission into the boreholes and its achievement of the first maximum value is caused by the primary main roof squeezing. In the cases under consideration (see Table 1), the beginning of the crustal movement and the achievement of the first gas emission maximum occurred approximately with the same removal of stope faces from the first cut workings. This is due to the low depth of the stope works ($H = 290$ m). It should be expected that, when performing the stope works at deeper horizons, the beginning of gas emission into gas drainage boreholes will occur much earlier than the beginning of the crustal movement.

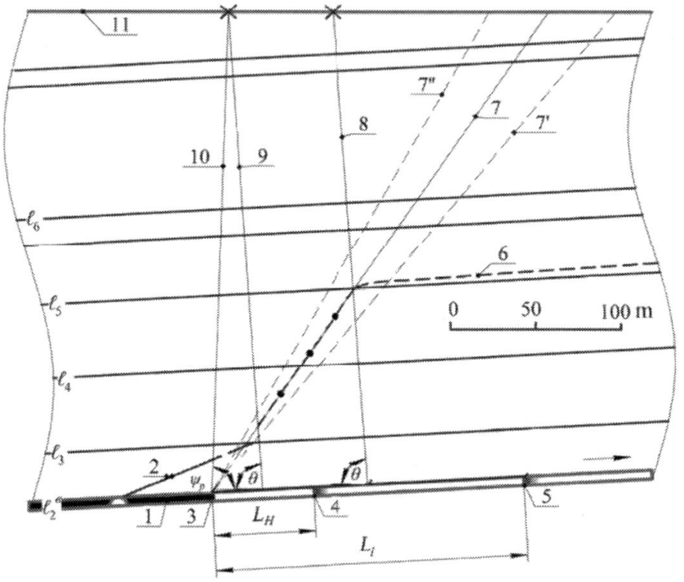

Fig. 3. The scheme of parameters relation of gas emission and rocks displacement in the operation of the 2-nd bis of the Western longwall: 1 – the mined seam; 2 – gas drainage boreholes; 3 – first cut working; 4, 5 – removals of the stope face from the first cut working, respectively, when the rocks displacement processes reach the earth surface L_H and formation of the square of the mined-out space with the side, which is equal to the length of the longwall (L_l); 6 – the upper boundary of the rocks displacement zones with discontinuity; 7 – the line segment, which determines the average angle $\psi = 55°$; 7′, 7″ – the line segments, corresponding to the values $\psi = 50°$and 60°; 8, 9 – the line segments, which determine the angles of maximum subsidence of the earth surface (θ), respectively, at $L = L_l$ and $L = L_H$; 10 – angle ψ_p, corresponding to L_H; 11 – the earth surface; × – the points of maximum rocks subsidence at $L = L_H$ and $L = L_l$; • – the points of maximum rocks subsidence, which correspond to the removals of the stope face from the first cut working, when reaching the maxima of gas emission; ⟶ – the direction of the stope face advance.

The first maxima of methane emission in the cases under consideration occurred after the beginning of gas emission with the further advance of stope faces at 5÷6 m. All the following maxima of gas emission were recorded when removing the stope faces from the face entries at distances less than the length of the longwall (see Fig. 3, Table 1), which corresponds to the scientific position [12]. When removing the stope faces from the face entries at distances more than L_l, a decrease in gas emission into the boreholes has been observed (see Fig. 2). This indicates the absence of additional methane inflow from the overlying coal and rock stratum and confirms the parameter H_p constancy in the subsequent stope face advance.

At $L > L_l$, the maximum rocks displacement, and, consequently, the maximum gas emission depend only on the position of the stope face relative to the undermined rocks. The efficiency of draining-out of gases, all other things being equal, will be determined by the location of the working sections of the boreholes in the undermined coal and rock stratum relative to the stope face.

The values close to each other H'_p, H''_p and H'''_p (see Table 1), obtained by different methods, indicate that the changes in the angles of maximum subsidence (θ) and complete displacements (ψ) in the stope works development can be accurately determined using the parameter L / H.

3 Conclusions

The maxima of methane emission into the boreholes drilled above the first cut workings are observed in the interval after the primary main roof squeezing till the removal of stope faces from the installation chambers at a distance not less than the length of the longwall.

With the development of the stope works, the position of the gas emission maxima is determined by the dimensions of the stope mine workings and the angle of maximum rocks subsidence.

For the first time, three independent methods have been used (including on the dynamics of gas emission) to determine changes in dimensions of rocks displacement zones with discontinuity after the primary main roof squeezing until the stable upper boundaries formation of these zones.

Into the gas drainage boreholes, the faces of which were located at a distance of 30 m from the mined seam, methane came from the complete displacement zones, the upper boundary of which was located at a distance of 130÷140 m.

The close results of the upper boundaries determinination of the rocks displacement zones with discontinuity by means of three independent methods, indicate about the reliability of the initial empirical data used in the calculations.

The established relations between the gas drainage boreholes location and the rocks displacement zones with discontinuity (see Fig. 3) make it possible to design the optimal schemes of draining-out of gases from the gas-and-coal deposits.

The obtained positive results for determining the possible areas of maximum gas emission allow us to recommend the empirical dependences for calculating the maximum subsidence angles and complete displacements at the stage of stope works development, as well as the dimensions of the stope mine workings, at which the processes of rocks displacement reach the earth surface

The authors express their gratitude to the management of the Gazeta Izviestiia Mine, SE "Donbasantratsyt" for providing the experimental data for the research work.

References

1. Morev, A.M., & Yevseyev, I.M. (1975). *Degazatsiya sblizhennykh plastov*. Moskva: Nedra.

2. Bondarenko, V.I., Kharin, Ye.N., Antoshchenko, N.I., & Gasyuk, R.L. (2013). Basic scientific positions of forecast of the dynamics of methane release when mining the gas bearing coal seams. *Naukovyi Visnyk Natsionalnoho Hirnychoho Universytetu*, (5), 24-30.
3. Ayruni, A.T. (1981). *Teoriya i praktika bor'by s rudnichnymi gazami na bol'shikh glubinakh*. Moskva: Nedra.
4. Antoshchenko, N.I., Okalelov, V.N., & Pavlov, V.I. (2013). *Formirovanie dinamiki metanovydeleniya iz podrabatyvayemogo massiva pri otrabotke gazonosnykh ugol'nykh plastov*. Alchevsk: Donbaskyi derzhavnyi tekhnichnyi universitet.
5. Antoshchenko, N.I., Koptikov, V.P., & Filimonov, P.Ye. (2014). *Bezopasnaya otrabotka gazonosnykh ugol'nykh plastov s uchetom geomekhanicheskikh protsessov sdvizheniya podrabotannykh porod*. Alchevsk: Donbaskyi derzhavnyi tekhnichnyi universitet.
6. Dubovik, A.I., Filat'ev M.V., & Filat'eva, E.N. (2017). *Inzhenernaya geomekhanika pri otrabotke ugol'nykh plastov*. Lysychansk: Donbaskyi derzhavnyi tekhnichnyi universitet.
7. Filat'ev, M.V., Antoshchenko, N.I., Gasyuk, R.L., & Pyzhov, S.V. (2015). Eksperimental'noye opredeleniye uglov maksimal'nykh osedaniy podrabotannykh ochistnymi vyrabotkami porod. *Ugol' Ukrainy*, (11), 3-6.
8. Filat'ev, M.V., Antoshchenko, N.I., Pyzhov, S.V., & Dubovik, A.I. (2016). Opredelenie zon sdvizheniya podrabotannykh porod s razryvom sploshnosti. *Ugol' Ukrainy*, (3), 9-16.
9. Filatiev, M. (2017). Determination of cross-correlation dependences between the parameters of swallies of the earth surface and the movement of underworked rocks. *Naukovyi Visnyk Natsionalnoho Hirnychoho Universytetu*, (3), 43-48.
10. Filatiev, M. (2017). Analytical determination of co-ordinates of d istinguished points of the Earth surface depression over broken workings. *Naukovyi Visnyk Natsionalnoho Hirnychoho Universytetu*, (1), 27-33.
11. GSTU 101.00159226.001-2003. (2004). *Pravyla pidrobky budivel, sporoud i pryrodnykh obiektiv pry vydobuvanni vuhillia pidzemnym sposobom*. Kyiv: Ministersto palyva ta enerhetyky Ukrainy.
12. Lobkov, M.I. (2012). *Rozvytok naukovykh osnov prohnozu obvalennia porid pokrivli pry vyimanni lavoiu polohoho plasta*. Ph.D. Instytut fizyky hirnychykh protsesiv.

Mathematical modeling of tight roof periodical falling

Vira Prykhodko[1*], *Nataliia* Ulanova[1], *Oleksandr* Haidai[2], and *Dina* Klymenko[1]

[1]National Mining University, Department of Higher Mathematics, 19 Yavornytskoho Ave., 49005 Dnipro, Ukraine
[2]National Mining University, Department of Underground Mining, 19 Yavornytskoho Ave., 49005 Dnipro, Ukraine

Abstract. The paper proposes a method to determine of a coal seam roof falling step basing upon the analysis of stress and strain state of the rock mass area with mine workings formed as a result of coal preparatory and extraction operations. A boundary element method has been applied to define stress and strain state (SSS). Fissuring of enclosing rocks was modeled by means of transversal-isotropic medium. Dependence of destructed rocks zone height within the roof of a seam being mined upon the weakening of the rock mass due to its fissuring and mine working geometry has been determined. Effect of fissility on the periodical roof falling step has been studied. Changes in support loads in the process of stope advance have been determined. A scheme of partial backfilling of the worked out area has been proposed to maintain the support in its working order.

1 Introduction

Efficiency of coal extraction in powered stopes depends considerably upon the character of rock roof falling. Major problems arise while developing seams with overlaying heavy roofs. In such cases, roof falling may result in deformation of powered support units, their displacement, and setting support stands in rigid position. In this context, stope operation is often accompanied by coal sloughing and rock outbursts into the working area. There are also certain difficulties with the maintenance of development workings in the neighbourhood of the stope.

In practice, to localize the described rock pressure manifestations, arbitrarily left support facilities in a worked out area are often applied i.e. complete or partial backfilling or coal pillars [1], [2], [3]. Another approach is in preliminary softening of tight roof rocks by means of drill and blast operations. Those measures are labour-consuming, dangerous, and not always efficient. There is an alternative method decreasing the intensity of rock pressure manifestation – stage-by-stage caving of tight roof. It is applied mostly in longwalls (longer than 200 m). The method provides caving of the rock mass within specific areas of its outcrop by turning the stope line [4].

Dangerous rock pressure manifestations result in accidents and injuries. According to

* Corresponding author: vprikhodko57@gmail.com

the data by the Ministry of Energy and Coal Mining, 68 rock fallings took place in Ukrainian mines within the period of 2015 – 2017, accounting for 35% of all the accident types; in 2017, total amount of injuries due to the factors made up 43%. It is also noted that mining and geological working conditions for Ukrainian miners are constantly deteriorating along with the deepening of mining operations where the amount of seams prone to intensive secondary roof caving increases.

There is a series of methods making it possible to evaluate with a certain degree of accuracy the risks of dangerous dynamic phenomena. In particular, there are papers containing recommendations for predicting roof caving and localizing their consequences. For instance, paper [5] proposes a method of roof control involving gradual turn of a longwall with roof caving by sections which length is not more than primary roof falling step; turning angle is connected with the direction of natural fissility of the rock mass. A value of the primary roof falling step was determined empirically.

To predict roof falling while coal seam extracting, it is important to evaluate the moment of the phenomenon beginning. Their periodicity is connected with the value of roof falling step. It is obvious that its determination should be based on the analysis of stress and strain state of rocks within potentially dangerous rock mass area.

Application of numerical methods has promoted considerably the process of rock mass modeling taking into consideration its structural heterogeneity and development of mining operations. Finite element method (FEM) is the one to be used more often. Thus, for instance, in paper [6] finite element method is aimed at studying rock mass-face support interaction; and papers [7], [8] consider stress and strain state of rock mass near a joint while coal seam mining. Paper [9] uses finite element method as the basis to model consequent movement of a longwall stope: going away from an installation chamber and approaching to a break-down chamber. The authors use an elastoplastic model of the medium; however, only plane or two-dimensional longwall cross-section is considered.

Paper [10] applied FEM in two-dimensional arrangement to study a change in stress and strain state of a border rock mass of a permanent anchor-supported inclined mine working in terms of its deepening.

As it is known, the highest intensity of rock pressure manifestation in mine workings is observed near the moving face. That is the result of changes in geometry of a worked out area as well as mutual effect of a stope and development mine workings. To take the facts into consideration, it is essential to consider spatial stress state of rock mass area nearby of conjugated workings. Moreover, to obtain solutions with the required accuracy within the framework of the mentioned FEM, the considered rock mass should be approximated with great number of elements that complicates the task significantly.

To represent mutual effect of a stope and development mine workings as well as such specificity of mining and geological conditions of Donbas mines as the presence of fissure system, it is required to develop more flexible algorithms making it possible to reduce problem order while preserving the advantages of a spatial model. Modification of a boundary element model [11] being quite efficient to model three-dimensional area of the rock mass containing thin coal seam helps do that.

The paper describes a method to determine roof falling step basing upon the analysis of spatial stress and strain state of rock mass area with mine workings formed as a result of flat coal seam preparatory and extraction operations. Numerical modeling is based upon one of the boundary element method modifications.

2 Methods and results of research

The paper has analyzed three-dimensional stress and strain state of the rock mass area containing a stope, development mine workings, coal pillar in front of the face, worked out area partially filled with the caved rocks, and areas of pillars or caved rocks bordering the development mine workings. Thus, computational area included all basic elements forming stress and strain state of roof rocks of the coal seam being mined.

The considered area of the rock mass was under the effect of the distributed load with γH components along the normal to the seam and $\lambda\gamma H$ along the strike and upwards. In this context, H is the seam occurrence depth, γ is the density of the enclosing rocks, and λ is the coefficient of lateral pressure. Computational algorithms are based on a method of boundary elements in the form of discontinuous displacements involving a seam element concept. As a result of the calculations, values of stresses and displacements both at the seam surface and in the analyzed area under the seam, i.e. in the roof, were determined.

A model of anisotropic medium was used to study the effect of system fissility of the enclosing rocks along with the model of elastic isotropic material. The applied method to model regular fissility is in its representation by means of transversal-isotropic medium [12]. In that case, it was assumed that along its fissures the medium has characteristics of undisturbed mass; and along the normal to the fissures plane, the medium characteristics are weakened. Similar representation does not preserve all the properties of the rock mass with regular fissility; however, it may be used to study the effects of anisotropy resulting from fissility.

The calculations involved modeling of averaged conditions of mine working construction being characteristic for Central Donbas mines. Properties of the enclosing rocks are: Young's modulus is $E = 3.5 \cdot 10^4$ MPa, Poison's ratio is $\nu = 0.17$, and volume weight is $\gamma = 2.5$ t/m^3 Coal properties are: $E_c = 3.2 \cdot 10^3$ MPa, $\nu_c = 0.13$. Occurrence depth and coal seam thickness are $H = 450$ m and $h = 1$ m. Characteristics of the caved rocks are: $E_{cr} = 3.0 \cdot 10^2$ MPa, $\nu_{cr} = 0.06$. Fissure characteristics are as follows: stiffness parameters along the fissures are $E_s = 3.5 \cdot 10^4$ MPa, $\nu_s = 0.17$; along the normal to fissures – $E_n = 3.5 \cdot 10^3$ MPa, $\nu_n = 0.02$. Distance between fissures is 0.1 m.

Boundaries of destructed rock zones in the mined seam roof are usually set according to one of the criteria. In terms of rock masses being in complex strain state, well-tested Parchevskiy-Shashenko, Balandin, and Hoek-Brown criteria [13], [14] are applied. The paper uses strength criterion by Balandin:

$$\sigma_{eq} = \frac{1}{2\psi}((1-\psi)(\sigma_1+\sigma_2+\sigma_3)+$$

$$+\sqrt{(1-\psi)^2(\sigma_1+\sigma_2+\sigma_3)^2+2\psi((\sigma_1-\sigma_2)^2+(\sigma_2-\sigma_3)^2+(\sigma_3-\sigma_1)^2)} \le \sigma^n,$$

where σ_1, σ_2, σ_3 are principal stresses; σ_{eq} is equivalent stress; $\psi = \sigma^t / \sigma^c$ is coefficient of rock brittleness; and σ^t and σ^c are tensile strength and compressive strength respectively. Value σ^n is used as ultimate stress; the value represents the stress calculated on the basis of elastic law corresponding to residual strength limit in three-link rock sample stress-strain diagram (Fig. 1).

Section of the diagram, characterizing the material breaking phase, corresponds to the values of elastic stresses exceeding σ^n. That is why the area where equivalent stresses σ_{eq} exceed σ^n is interpreted as the zone of destructed rocks. Stress lines $\sigma_{eq} = \sigma^n$ limit the destructed rocks zone in case when enclosing rocks are monolithic, i.e. there is no system fissility.

As a rule, geomechanical problems solving, rock mass fissility is taken into consideration by introducing a correction coefficient reducing calculated strength of the enclosing rocks – so-called coefficient of structural weakening. The latter is understood as rock mass strength - laboratory sample strength ratio. The coefficient is determined either empirically or on the basis of probabilistic and statistic approach [15]. The paper applies an analogue of the mentioned coefficient – rock weakening parameter $\tilde{k}$. It was determined as a ratio of maximum displacements of unsupported mine working contour within continuous and fissured rock masses $\tilde{k} = u / u^{fis}$. Fissured rocks were modeled by means of transversal isotropic medium. Thus, while studying fissure-weakened rocks, value $\tilde{k}\sigma^n$ was used as the ultimate stress. Values $\tilde{k}$, obtained for fissure angles to a stratification plane 0 to 90°, were 0.068 to 0.46. The results are well consistent with the studies [15] where fissility is characterized by means of geological strength index depending upon the fissures size and condition of their surfaces.

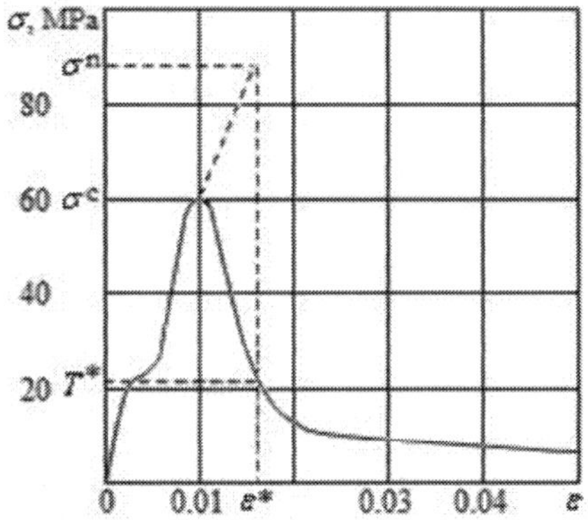

Fig.1. Clay shale compression diagram.

Maximum vertical extension of a rock destruction zone (destruction height) in the roof was determined for cases of continuous and fissured medium. In this context, values of geometrical parameters d (longwall length) and l (distance from a face entry to a longwall line) as well as weakening parameter $\tilde{k}$ were varied. As a result, the dependence of destruction height upon the indicated parameters for fissured rocks of the main roof was obtained [11]:

$$H_d^{fis} = 0.072 \cdot l^{0.75} \cdot d^{0.25} / \tilde{k} .$$

Previous dependence has helped evaluate roof falling step in the process of face advance. It was assumed that roof fall took place when the zone of broken rocks above the seam covered the whole main roof along the height. Value of distance from a face entry to a face line, in terms of which falling height became equal to the main roof thickness, was interpreted as a supposed roof falling step.

The described approach was used to determine primary roof falling step under following conditions. Tight roof thickness H_0 was specified as equal to 30 m. Face advance process was modeled in terms of step-by-step changes in distance l from a face entry to a face line, $0 < l < l_0$. Calculations were performed for control marks within the stope being $d_1 = 72$ m, $d_2 = 152$ m, and $d_3 = 216$ m, if $l_0 = 80$ m.

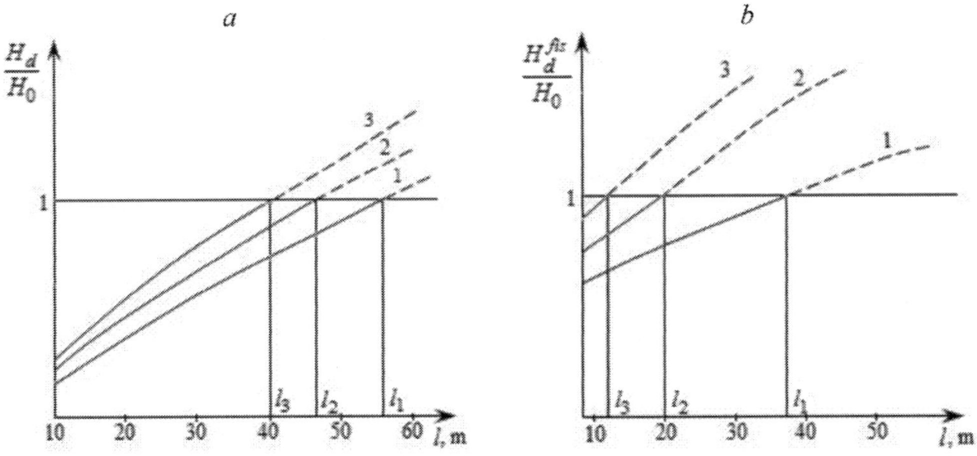

Fig. 2. Changes in destruction height while face advancing for longwall length values: *a* – unbroken medium; *b* – fissured medium; 1 – $d_1 = 72$ m; 2 – $d_2 = 152$ m; 3 – $d_3 = 216$ m.

Fig. 2a demonstrates curves of increase in destruction height without regard to the rock mass fissility. Destruction height H_d increases in the process of face advance until it reaches H_0 value. Horizontal line $H_d / H_0 = 1$ corresponds to the upper boundary of the main roof layer. Moments of primary falling correspond to the abscissas of intersections of curves 1, 2, and 3 with the mentioned line. They made up approximately $l_1 \approx 55$ m, $l_2 \approx 45$ m and $l_3 \approx 40$ m for values longwall length values d_1, d_2, and d_3 respectively. Fig. 2b shows similar results obtained considering the fact of fissility with fissure angle α to the stratification being 20°. In this context, destruction height H_p^{fis} rises faster, and value of primary falling step decreases respectively: $l_1 \approx 35$ m, $l_2 \approx 20$ m and $l_3 \approx 10$ m.

Thus, enclosing rocks fissility results in the reduction of primary roof falling step. The conclusion is consistent with the results of studies [16], where empiric dependences of falling step upon strength and geometry characteristics of the roof layers within 20 m thickness (with and without regard for fissility) are proposed. Data on 55 longwalls of thin gently sloping seams developed by Donbas mines were the basis for the studies. According to [16], ratio of roof falling step in fissured and non-fissured rocks varies from 0.35 up to 0.89. In the paper obtained value of the ratio as that being from 0.25 up to 0.65.

The obtained results were applied to determine support loads while stope advancing. It was assumed that the load acting on a powered support in a longwall is formed by the weight of that share of roof rocks which was within the destructed rocks zone. In case of tight roof hanging in the form of console, support load rises in the process of face advance up to another roof falling since the area of hanging share increases and destruction height grows.

Load, applied for powered support, was calculated as the weight of broken roof rocks related to the total area of canopies of all the support units. Unit sizes correspond to the support of 1KDD complex. Curves 1, 2, and 3 in Fig. 3 represent the load falling on the support unit for longwall length 152 m in three cases: fissured rocks with fissure angles to a stratification plane 20 and 70° and rocks without fissures. Horizontal line $F/F_{max} = 1$ corresponds to nominal value of the load on a support unit according to technical characteristics of 1KDD complex. The curves cross the indicated line at points with abscissas A, B, and C, i.e. value of support load exceeds the admissible one before the moment of the expected roof falling (points A_r, B_r and C_r). Moreover, if fissure angle is 20°, that happens with less than 20 m step.

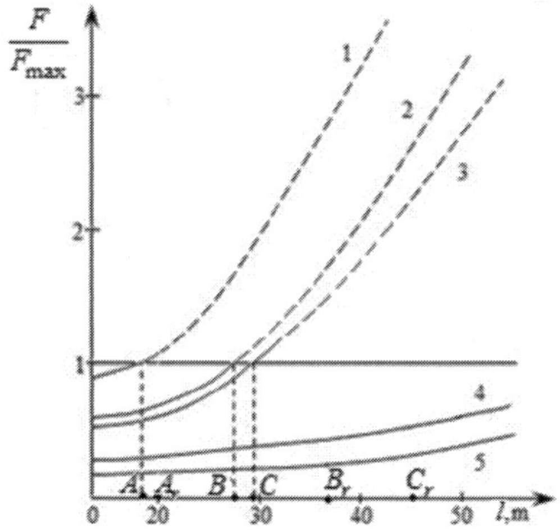

Fig. 3. Change in powered support load while face advancing. Complete caving: 1 – fissured rocks, 20° fissure angle; 2 – fissured rocks, 70° fissure angle; 3 – unbroken rocks. Partial backfilling with area width $a = 5$ m; 4 – distance between areas is $b = 8$ m; 5 – distance between areas is $b = 4$ m.

To avoid the described situation, it is recommended to backfill the worked out area by means of lines being perpendicular to a face line. Curves 4 and 5 in Fig. 3 demonstrate calculated values of support load for that variant of roof control, if backfill area width is $a = 5$ m and distance between areas is $b = 8$ m and $b = 4$ m respectively. In this context supported area is 35 and 50% of the total worked out area respectively.

Generalization of the research results has made it possible to obtain the dependence of support load upon a powered support unit while face advancing:

$$F = 0.66\frac{\gamma H_d l}{l_n},$$

where H_d is destruction height, m; l is distance from a face entry to a stope or to the nearest roof falling point, m; l_n is length of support canopies, m; and γ is volume weight of the rocks, t/m^3.

3 Conclusions

Geometry of the destructed rock zones in a coal seam roof has been obtained on the basis of the solution of spatial problem of the elasticity theory for the rock mass area near development mine workings and stope junction.

Values of step of roof falling, while face advancing, have been determined both in terms of rock masses without fissures and in terms of enclosing rocks with regular fissility (fissure angle to the stratification is 20°). It has been defined that the fissility effect is in the decrease in primary roof falling step by 1.57 – 4 times for a different values of longwall length.

Values of support loads, while face advancing, have been calculated in terms of various geometry of the study area as well as fissure angles to the stratification. Partial backfilling of the worked out area has been proposed to provide accident-free support operation; backfilling parameters for stability maintaining have been recommended.

The authors appreciate Liudmyli Novikovii, the Corresponding Member of the AMS of Ukraine, Doctor of Technical Sciences, Professor for her scientific guidance and support.

References

1. Hryhoriev, O.Ye. (2013). *Okhorona kapitalnykh vyrobok vugilnykh shakht zapobizhnymy cilykamy*. Dnipropetrovsk: Natsionalnyi hirnychyi universytet. http://ir.nmu.org.ua/handle/123456789/108279
2. Stupnik, N.I., & Gayday, A.A. (2013). Evaluation of the application of rectangular form of the coal pillars for the protection of the main drifts. *Coal of Ukraine*, (4), 8-10.
3. Hapieiev, S.M., Khoziaikina, N.V., Tereshchuk, R.M., & Kovalenko, V.V. (2016). *Resursozberigaiuchi tekhnolohii upravlinnia stiikistiu protiazhnykh vyrobok vugilnykh shakht*. Dnipropetrovsk: Natsionalnyi hirnychyi universytet.
4. Knyaz'kov, O.V., Averin, G.A., & Bolotov, A.P. (2007). Izmenenie polozheniya linii ochistnogo zaboya v prostranstve kak metod pervichnoy posadki trudnoobrushayemoy krovli. *Sbornik nauchnykh trudov Donetskogo gosudarstvennogo universiteta*, (25), 101-105.
5. Borzyh, A.Ph., Antjuhov, S.V., & Bolotov, A.P. (2008). *Sposib keruvannia pervynnoiu posadkoiu pokrivli v lavi*. Patent No. 34138, Ukraine.
6. Kovalevs'ka, I., Illiashov, M., Fomychov, V., & Chervatuk, V. (2012). The formation of the finite-element model of the system "undermined massif – support of stope". *Geomechanical Processes during Underground Mining*, 73-79. https://doi.org/10.1201/b13157-13
7. Kovalevs'ka, I., Symanovych, G., & Fomychov, V. (2013). Research of stress-strain state of cracked coal-containing massif near-the-working area using finite elements technique. *Annual Scientific-Technical Collection – Mining of Mineral Deposits*, 159-163. https://doi.org/10.1201/b16354-28
8. Bondarenko, V., Kovalevs'ka, I., & Fomychov, V. (2012). Features of carrying out experiment using finite-element method at multivariate calculation of mine massif – combined support system. *Geomechanical Processes During Underground Mining*, 7-13. https://doi.org/10.1201/b13157-3
9. Sdvyzhkova, O.O., Babets, D.V., & Smirnov, A.V. (2014). Support loading of assembly chamber in terms of Western Donbass plough longwall. *Naukovyi Visnyk Natsionalnoho Hirnychoho Universytetu*, (5), 26-32. http://www.nvngu.in.ua/index.php/en/component/jdownloads/finish/49-05/1454-2014-5-sdvizhkova/0
10. Tereshchuk, R.M., Khoziaikina, N.V., & Babets, D.V. (2018). Substantiation of rational roof-bolting parameters. *Naukovyi Visnyk Natsionalnoho Hirnychoho Universytetu*, (1), 19-26. http://nvngu.in.ua/jdownloads/pdf/2018/01/01_2018_Tereshchuk.pdf
11. Prykhodko, V.V., & Ulanova, N.P. (2018). Modeling of stress-strain state of fractured rock mass nearby of conjugated workings. *Naukovyi Visnyk Natsionalnoho Hirnychoho Universytetu*, (1), 5-11. https://doi.org/10.29202/nvngu/2018-1/15
12. Novikova, L.V., & Prikhod'ko, V.V. (1997). Otsenka vliyaniya sistemnoy treshchinovatosti porod na proyavleniya gornogo davleniya vokrug vyrabotok. *Gornyy informatsionno-analiticheskiy biulleten'*, (1), 148-150.
13. Sdvyzhkova, O., & Patynska, R. (2016). Effect of increasing mining on longwall coal mining – Western Donbass case study. *Studia Geotechnica et Mechanica, 38*(1), 91-99. https://doi.org/10.1515/sgem-2016-0010
14. Sdvyzhkova, O.O., Babets, D.V., Kravchenko K.V., & Smirnov, A.V. (2016). Determining the displacements of rock mass nearby the dismantling chamber under effect of plow longwall. *Naukovyi Visnyk Natsionalnoho Hirnychoho Universytetu*, (2), 34-42. http://www.nvngu.in.ua/index.php/en/component/jdownloads/finish/60-02/8483-2016-02-sdvyzhkova/0

15. Babets, D.V., Sdvyzhkova, O.O., Larionov, M.H., & Tereshchuk, R.M. (2017). Estimation of rock mass stability based on probability approach and rating systems. *Naukovyi Visnyk Natsionalnoho Hirnychoho Universytetu,* (2), 58-64. http://www.nvngu.in.ua/index.php/en/component/jdownloads/finish/67-02/8618-02-2017-babets/0
16. Borzyh, A.Ph., & Averin, G.A. (1990). Vliyanie treshchinovatosti na shag periodicheskikh osadok trudnoobrushayushcheysya krovli. *Ugol' Ukrainy*, (9), 10-11.

Investigation of muck drawing in blocks with trapezoid sills

Mykola Stupnik[1*], and *Olena* Kalinichenko[1]

[1]Kryvyi Rih National University, Department of Underground Mining of Mineral Deposit, 11 Matusevycha Ave., 50027 Kryvyi Rih, Ukraine

Abstract. The article is aimed at determining regularities of muck drawing in blocks and developing the technology of enhancing of muck ore parameters through shaping a trapezoid block sill. To confirm the suggested assumptions, study regularities of forming and determine contours of crests after full muck drawing from the second stage blocks, laboratory experiments have been conducted on drawing muck ore on the boundary with solidifying backfill massif of I stage stopes. For the first time there have been determined numerical dependencies of ore losses in crests on the contact with the solidifying backfill massif of the I stage stopes after ore drawing from the II stage blocks on the dead stope height and dips of side surfaces of trapezoid basis of sill pillars when forming the trapezoid sill with inclined side surfaces of the lower part of the I stage stopes. There are developed methods of calculating ore losses in crests on the contact with the solidifying backfill massif of I stage stopes after muck drawing from the II stage blocks. There are proposed methods of determining optimal parameters of mining ore deposits considering minimization of re losses in crests and increase of muck extraction from II stage blocks.

1 Introduction

Muck can be drawn in the contact with the muck ore waste rocks, on the contact with the solid ores or on the contact with the solidifying massif of artificial backfill. The character and regularities of drawing muck from a stope depending on characteristics of the contact. Considering these regularities, muck drawing on the contact with the solidifying artificial massif in mining the block with caving ore and enclosing rocks is similar to drawing ore on the contact with the solid ore massif [1 – 7, 11].

When drawing muck on the contact with the solid ore or solidifying artificial massif, muck loss in crests between drawpoints and on the contact with the solid massif becomes the main disadvantage.it should be noted that, according to G.M. Malakhov [1], muck losses in crests between drawpoints remain present regardless the drawn ore layer height.

According to investigations conducted, relative losses depend on the height of the layer of the drawn ore, its physical and mechanical characteristics and parameters of the drawing ellipsoid and the block geometry [8].

[*] Corresponding author: mstupnik2012@gmail.com

Currently, in Kryvbas when mining deposits with caving ores and enclosing rocks, ore losses in blocks reach 15 – 17% and sometimes more, including ore losses in crests accounting for up to 7 – 12% of general ore losses in blocks.

Analysis of the character of ore loss formation in crests on the contact with the solid massif after full caved drawing from the II stage blocks enables the figure of the muck crests that remain in the block as losses, Fig. 1.

To enhance indices of muck ore extraction from blocks, the authors propose the technology of mining and backfilling an area of the ore body enabling reduction of ore losses in crests to 4 – 8%.

The technology consists in the following.

First, I stage stopes are mined with forming the side surface of the lower part of the stope inclined towards II stage blocks (Fig. 2). The dip of the I stage stope lower part side surface is adequate to that of a drawing cone generatix of II stage blocks. In this case, from stopes of I stage blocks ore is drawn completely including the joined volume of the stope lower part (Fig. 2).

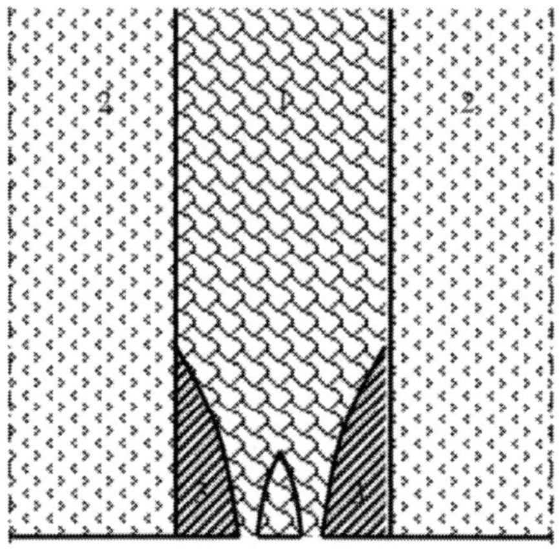

Fig. 1. Formation of ore losses in crests at the vertical contact of the drawn ore with the solidifying backfill massif of I stage stopes after full caved ore drawing from II stage blocks: 1 – waste rocks, 2 – solidifying backfill of I stage stopes, 3 – ore losses in crests on the contact with I stage stope backfill.

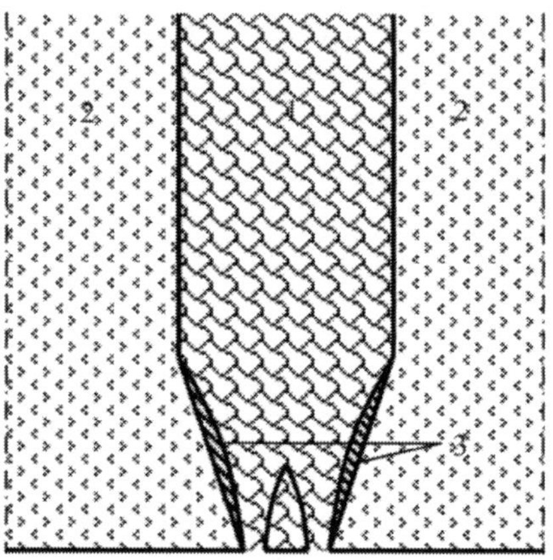

Fig. 2. Formation of ore losses in crests at the inclined contact of the drawn ore with the solidifying backfill massif of I stage stopes after full caved ore drawing from II stage blocks: 1 – waste rocks, 2 – solidifying backfill of I stage stopes, 3 – ore losses in crests on the contact with I stage stope backfill.

Then, a I stage stope is filled with the solidifying backfill. After the backfill is strengthened, ore and enclosing rocks of II stage blocks are caved.

Ore drawing under overlying rocks id performed according to the theory of muck drawing from volumes in the form of ellipsoid of revolution.

The proposed formation of the sill of a block with inclined side surfaces in the lower part of I stage stopes allows elimination of ore losses in crests on the contacts with solidifying backfill massif of I stage stopes after complete caved ore drawing from II stage blocks.

2 Methods

Laboratory experiments on caved ore drawing on the boundary with the solidifying backfill massif of I stage stopes were conducted to confirm the suggested assumptions, investigate regularities of formation and determine crest contours after complete caved ore drawing from II stage blocks.

The laboratory experiments were conducted at the ore drawing laboratory of the Department of underground mineral deposit mining, Kryvyi Rih National University, on standard static models of drawing for various mining and geological characteristics of rich iron ore deposits representative of Kryvyi Rih basin.

The problem of classifying drawing figures created when caved ore flows from drawpoints under overlying waste rocks is actual and one of the difficult in the caved ore drawing theory.

As a rule, the research findings to report up-to-date concern shapes and parameters of figures of drawing ore with various physical and mechanical properties of the ore and overlying rocks [1 – 10].

However, muck loss formation in crests on the contact with the solidifying backfill massif of I stage stopes after complete caved ore drawing from II stage blocks is still understudied.

The authors conducted laboratory experiments to fill this gap. The experiments were conducted as follows.

The first set of experiments (I option) models conditions of caved ore drawing from a II stage block located on the contact with the solidifying massif of I stage stopes.

Here, the muck ore–solid massif contact surface is assumed vertical according to the classical technologies of stope mining.

The basic laboratory model is given in Fig. 3.

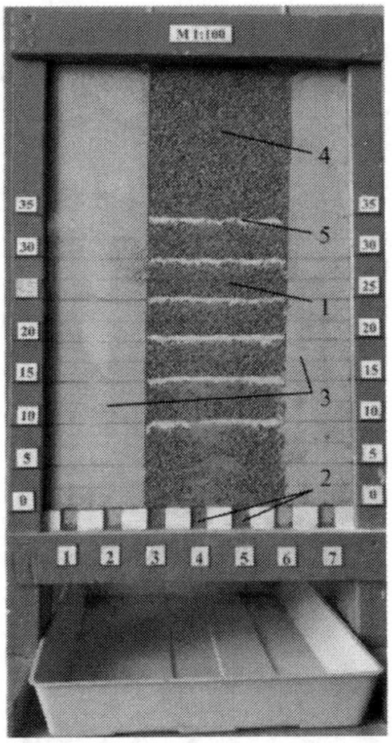

Fig. 3. The laboratory bench for drawing muck under overlying caved rocks on the contact with vertical side surfaces of the solidifying backfill of I stage stopes: 1 – II stage block muck, 2 – draw cones, 3 – the solidifying backfill massif of I stage stopes, 4 – waste rocks, 5 – the ore-rock contact line.

Fig. 4 presents the initial and final stages of muck drawing from a II stage block on the contact with vertical side surfaces of the solidifying backfill with muck loss formation in crests on the contact with the solidifying backfill massif.

The modeling technology consists in the following.

The model with the designed parameters is filled with the amount of ore according to the conditions of the experiment.

https://doi.org/10.1051/e3sconf/20186000021

Waste rocks are placed on above the ore.

Ore drawing is performed until its complete discharge from the block characterized by appearance of the white chalk line separating the ore from the overlying waste ores.

a *b*

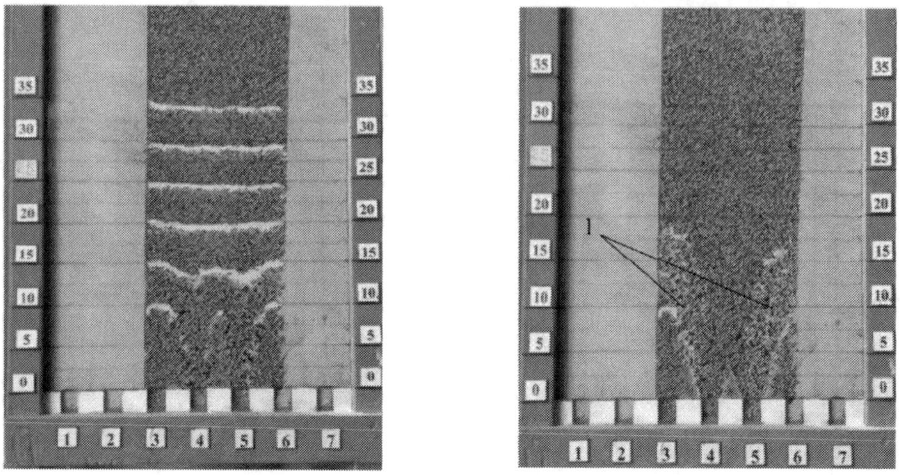

Fig. 4. The initial (a) and final (b) stages of muck drawing from a II stage block on the contact with vertical side surfaces of the solidifying backfill of I stage stopes: 1 – muck loss contours in crests.

The second set of experiments (II option) models conditions similar to those in the the previous set of tests with ore and wastes rocks.

Fig. 5 presents the laboratory bench for drawing muck under overlying caved rocks on the contour with inclined side surfaces of the solidifying backfill massif of I stage stopes.

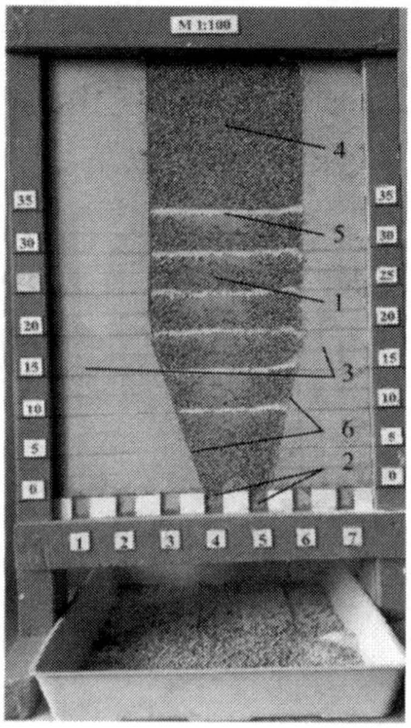

Fig. 5. The laboratory bench for drawing muck under overlying caved waste rocks on the contact with inclined side surfaces of the solidifying backfill of I stage stopes: 1 – II stage block muck, 2 – draw cones, 3 – the solidifying backfill massif of I stage stopes, 4 – waste rocks, 5 – the ore-rock contact line; 6 – the inclined side surfaces of the ore-solidifying backfill contact.

 https://doi.org/10.1051/e3sconf/20186000021

Unlike in the previous set of experiments, the second set provides for use of fragmented muck fraction and enclosing rocks, initial mining of I stage stopes with inclining the side surface of the lower part of the stope towards II stage blocks. The dip of the side surface of the lower part of the I stage stope equals to the generatix of the ellipsoid of drawing the muck from II stage blocks.

At the next stage the drawn ore is weighted, the ore remained in crests is weighted too. The obtained results are transformed into the drawn-lost ore percentage ratio.

Thus, the assumption that formation of the block sill with inclined side surfaces of the lower part of I stage stopes practically enables elimination of ore losses in crests on the contact with the solidifying backfill massif of I stage stopes after complete caved ore drawing from II stage blocks is confirmed.

Fig. 6 presents the initial and final stages of muck drawing from a II stage stope on the contact with inclined side surfaces of the solidifying backfill of I stage stopes and muck loss formation in crests on the contact with the solidifying backfill massif.

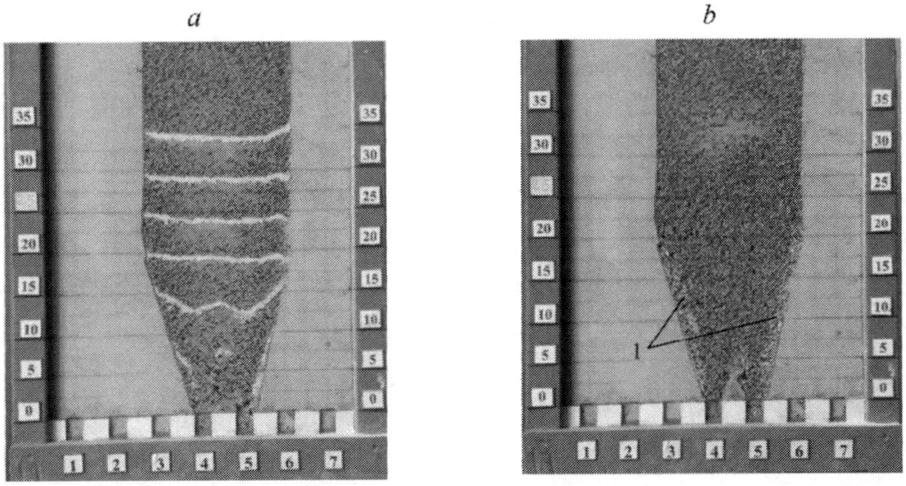

Fig. 6. The initial (a) and final (b) stages of muck drawing from a II stage block on the contact with inclined side surfaces of the solidifying backfill of I stage stopes: 1 – muck loss contours in crests.

3 Results and discussion

The investigations conducted result in the conclusion that ore losses in crests of dead II stage blocks on the contact with the backfill of I stage stopes depend on the dead stope height and the dip of the inclined side surfaces of the trapezoid basis of a sill pillar. The losses decrease to 0.7 – 1.2% if forming dips equal to the generatix of the draw cone which results in increase of extraction of ore from the block by 4.1 – 7.8%.

Besides, in the course of experiments the following dependencies have been determined.

Absolute ore losses in crests on the boundary with the I stage sill stopes do not practically depend on block parameters. So, academician G.M. Malakhov's conclusions [1] on ore losses in crests between drawpoints being steady regardless the drawn ore layer height are confirmed.

At the same time, relative ore losses in crests depend on the height and width of a block, a number of drawpoints and a distance between them.

Ore layers in the model are separated by chalk strips located parallel to each other every 5 cm for better visualization of the muck ore drawing from the stope. Fig. 5 shows results of muck drawing at the initial and final stages with formation of ore losses on the contact with inclined side surfaces of the solidifying backfill of I stage stopes.

Analysis of the experiments data results in the following regularities.

The functional of the variable value of ore losses in crests depending on the dip of the side surface of the trapezoid sill on the boundary with the I stage sill stope massif with the differentiated block height is well approximated with the second-order polynomial curve.

For 25-meter-high stopes (panels) the value of ore losses in crests can be determined from the expression:

$$L = 0.031\beta^2 - 4.473\beta + 163.42;\ R^2 = 0.9678,$$

where L is the value of ore losses in crests on the boundary with the I stage sill stope massif, %; β is the dip of the side surface of the trapezoid sill on the boundary with the I stage sill stope massif, degrees; R is validity of the approximation.

For 30-meter-high stopes (panels) the value of ore losses in crests can be determined from the expression:

$$L = 0.0269\beta^2 - 3.8965\beta + 143.02,\ R^2 = 0.9692.$$

For 35-meter-high panels the value of ore losses in crests is determined from the expression:

$$L = 0.0233\beta^2 - 3.3889\beta + 124.64,\ R^2 = 0.9695.$$

For 40-meter-high panels the value of ore losses in crests is determined from the expression:

$$L = 0.0201\beta^2 - 2.9114\beta + 106.99,\ R^2 = 0.9734.$$

For 80-meter-high blocks the value of ore losses in crests can be determined from the expression:

$$L = 0.0101\beta^2 - 1.4655\beta + 54.239,\ R^2 = 0.9663.$$

The generalized value of ore losses in crests depending on the dip of the side surface of the trapezoid sill on the boundary with the I stage sill stope massif with the differentiated block height is well approximated with the second-order polynomial curve and can be determined from the expression:

$$L = \left(0.0258\beta^2 - 3.7275\beta + 131.18\right) \cdot 1.5487e^{-0{,}0099hn},\ R^2 = 0.9993,$$

where h_n is the height of the dead panel, m.

4 Conclusions

For the first time there have been determined numerical dependencies of ore losses in crests on the contact with the solidifying backfill massif of the I stage stopes after ore drawing from the II stage blocks on the dead stope height and dips of side surfaces of the trapezoid basis of sill pillars when forming the trapezoid sill of the block with inclined side surfaces of the lower part of the I stage stopes. It was experimentally proved that ore losses in crests on the contact with the backfill massif of the I stage stopes depend on the dead stope height and the dip of the inclined side surfaces of the trapezoid basis of a sill pillar. The losses decrease to 0.7 – 1.2% if forming dips equal to the generatix of the draw cone which results in increase of extraction of ore from the block by 4.1 – 7.8%.

Considering the investigation findings and determined regularities of muck drawing on the contact with the solidifying backfill massif, the authors have developed a flowsheet of mining the area of the minefield with the trapezoid basis of blocks and combined backfill of the stope.

The proposed technology of forming the trapezoid block sill with inclined side surfaces of the lower part of I stage stopes enables elimination of ore losses in crests on the contacts with solidifying backfill massif of I stage stopes after complete caved ore drawing from II stage blocks.

The authors would like to express their sincere gratitude to the management of the PrJSC "Sukha Balka" for assistance in conducting the research and implementing its results.

References

1. Malakhov, G.M., Bezukh, V.P., & Petrenko, P.D. (1968). *Teoriya i praktika vypuska rudy.* Moskva: Nedra.
2. Kulikov, V.V. (1980). *Vypusk rudy.* Moskva: Nedra.
3. Dubynin, N.G. (1965). *Vypusk rudy pri podzevnoy razrabotke.* Moskva: Nedra.
4. Guy Lauret et Patrice Brasse. (1999). Explortation selective des amas uraniferez: evolution de la methode d'exploitation. *Mines carrierez – revue de la sosiete de l'industie minerale*, (81), 41-44.
5. Kunin, I.K. (1964). Vypusk i dostavka rudy pri *podzevnoy razrabotke.* Moskva: Nedra.
6. Azarian, A.A., Kolosov, V.A., Morgun, A.V., Popov, C.O., & Stupnik, N.I. (2012). *Instruktsiya po normirovaniyu pokazateley izvlecheniya rudy po tekhnicheskim, tekhnologicheskim i ekonomicheskim kriteriyam, ikh prognozirovaniyu i uchetu v protsesse podzemnoy razrabotki zhelezorudnykh mestorozhdeniy.* Kryvyi Rih: Mira.
7. Kulikov, V.V., & Deyneka, A.G. (1969). *Metodika prognozirovaniya pokazateley izvlecheniya rudy.* Moskva: Institut gornogo dela im. A.A. Skochinskogo.
8. Balkhavdarov, Kh.A., Sabatkoyev, M.M., & Gabarayev, O.Z. (1991). Dinamika poter' rudy na dnishche mezhdu voronkami. *Izvestiya Vuzov. Gornyy zhurnal*, (10), 24-26.
9. Stupnik, M.I., Kalinichenko, V.O., Khivrenko, O.Ya., Kalinichenko, O.V., Hrishchenko, M.A., & Teliapniov, V.O. (2016). Rozrobka ta doslidzhennia tekhnolohii vidpratsiuvannia zapasiv blokiv iz zastosuvanniam kombinovanoho vysokoefectyvnoho vypusku rudy. *Visnyk Kryvorizkoho Natsionalnoho Universytetu*, (43), 3-7.
10. Kalinichenko, O.V. (2017). Pidvyshchennia pokaznykiv vyluchennia zaliznykh rud pry vypusku obvalenoi rudnoi masy na kontakti z tverdiiuchym shtuchnym masyvom. *Visnyk Kryvorizkoho Natsionalnoho Universytetu*, (45), 118-122.
11. Stupnik, N., Kalinichenko, V., Kalinichenko, E., & Pismennij, S. (2015). Features of underlying levels opening at "ArcelorMittal Krivyi Rih" underground mine. *Theoretical and Practical Solutions of Mineral Resources Mining*, 39-45. https://doi.org/10.1201/b19901-8

Substantiating rational parameters of a method for shrinkage ore stoping while developing thin-vein steeply inclined deposits

Leonid Shyrin[1], *Yevhenii* Koroviaka[1*], *Valerii* Rastsvietaiev[1], and *Oleksandr* Denyshchenko[1]

[1]National Mining University, Department of Transport Systems and Technologies, 19 Yavornytskoho Ave., 49005 Dnipro, Ukraine

Abstract. Objective of the paper is to substantiate rational ore-stoping technique while using small wells in the context of thin-vein steeply inclined deposit mining. The technique is based upon the repeated field studies and simulation of ore drawing processes for shrinkage ore stoping in terms of the oriented drilling of periphery holes. A design of a blast-hole charge with low-density porous intermediate layer has been proposed as a result as well as a mechanism of shock-wave propagation within rock mass in the process of thin steeply inclined vein stoping. Scientific novelty is represented by means of analytical results of scientific sources, and dependences of ore losses on the vein wall hypsometry resulting from shrinkage stoping in the context of the technique being proposed. Practical relevance is to substantiate rational parameters of the ore-stoping technique being proposed. The technique involves designs of blast-hole charges with low-density porous intermediate layer in stemming. Moreover, the technique proposes to place the intermediate low-density stemming layer right after a blast hole was charged with explosives and live primers were inserted.

1 Introduction

Thin-vein steeply inclined deposits are something special for mining industry. The vein-type deposits have been formed when earth fissures were filled with mineral substance owing to hydrothermal processes and pneumatolysis. Their features, being typical for certain deposits, experience numerous variations in the context of a level and a block involving specific analysis and consideration while selecting and substantiating alternatives of mining which may differ greatly from traditional ones [1, 2].

As a rule, the extraction techniques with ore shrinkage, use of expansion-type support and worked-out area stowing applied in the process of thin-vein steeply inclined deposits mining, rely upon horizontal ore shrinkage to the rise.

Morphological characteristic and shape of ore body occurrence are important factors while selecting a technique for thin-vein deposit development. To decrease ore dilution degree, mining operations are performed along the vein only; besides, compulsory in-mine

* Corresponding author: koroviakaye@gmail.com

exploration is implemented constantly for each block to evaluate operatively vein occurrence and metal content in it. When the indices decrease, mining is ceased and due to the imperfection of extraction systems being applied the reserves are frozen.

2 Method of examination

The field studies which have been carried out in the mine workings of Zaporizhzhia Iron-Ore Integrated Works showed both mechanism and zones of active crushing taking place within rock mass when stress waves passed through it in the process of ore shrinkage. Moreover, they have helped forecast potential behaviour of rocks when ore bodies are mined horizontally to the rise [3].

Generally, veins occur within the mineralized rocks penetrated with folia nets. As a result, the rock mass represented by hard rocks cannot be considered as a continuum within mineralization area. Thus, when stress waves, resulting from explosion of blast-hole charges, pass through the rock mass, fissures occur and discontinuity takes place along the folia. Depth of the fissures, propagated beyond mineralization zones, can reach 1.0 m. In the context of the repeated stress waves, passing through the rock mass, fissures width as well as formation of partings being mutually fixed blocks of a spatial form takes place. When a contact between the fixed blocks is lost and external action on the rock mass happens, then rock exfoliation in the form of plates and blocks with 0.1 to 0.6 m width within 0.5 to 12.0 m^2 area is involved; in the context of ore drawing, it results in its additional dilution since it mixes with the broken-down rocks.

Fig. 1 demonstrates typical site of ore direction to the dip resulting from mine surveying while constructing rising and change in the dynamic axis of the face space within specific sections of I-I, II-II, III-III block (Fig. 1b).

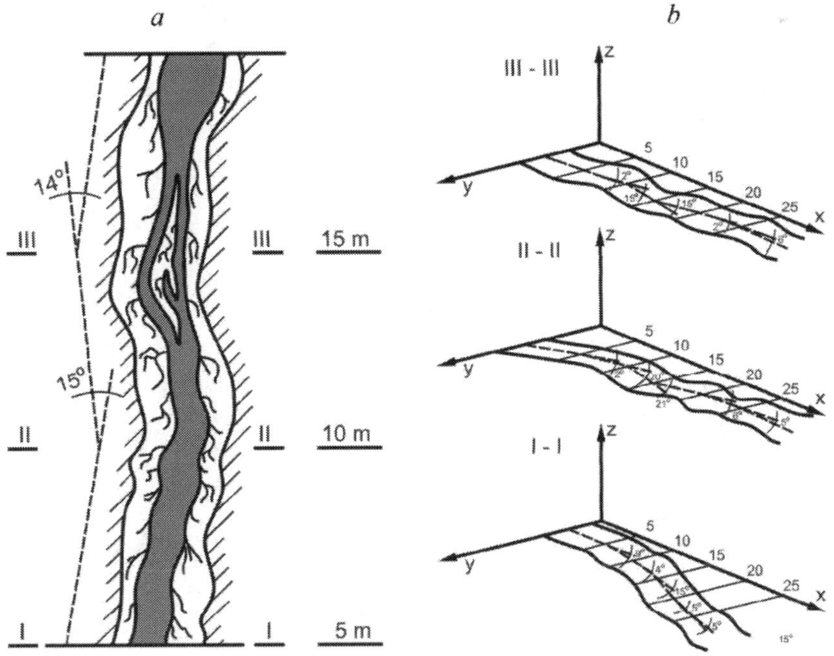

Fig. 1. Change in the dynamic axis of a stope in the process of shrinkage stoping of ore bodies: a – to the dip; b – across the pitch.

It should be noted that due attention was not paid to the analysis of axial section of a stope walls in the process of steeply inclined veins mining. It depends on the fact that enterprises of the industry initially mined approachable sites with the stable wall rocks and large content of commercial components.

Mining deepening complicates significantly occurrence of the steeply inclined ore bodies. Extraction of the reserves involve the deposit sites where wall rocks are of mean stability or less; moreover, commercial components need measures to upgrade rock mass being mined (i.e. decreased ore dilution, selective mining etc.).

For the first time ever, such measures involved data concerning the effect of drilling and blasting on the rock mass state (i.e. wall rock fissuring, stability, and roughness).

Hypsometric plans of wall surfaces are built while using methods of statistic processing of experimental data and results of comparative evaluations with the corresponding theoretical indices (Fig. 2). Density of their isolines matches the adopted drilling and blasting parameters in terms of ore shrinkage stoping. Analysis of the hypsometric plans obtained at Zaporizhzhia Iron-Ore Integrated Works has helped determine that in the context of traditional drilling and blasting operations, the amplitude between contour figures run at the level of peaks and valleys of wall surface of the ore bodies, being extracted, is 0.3 to 0.4 m; in some cases it reaches 0.5 to 0.55 m [3].

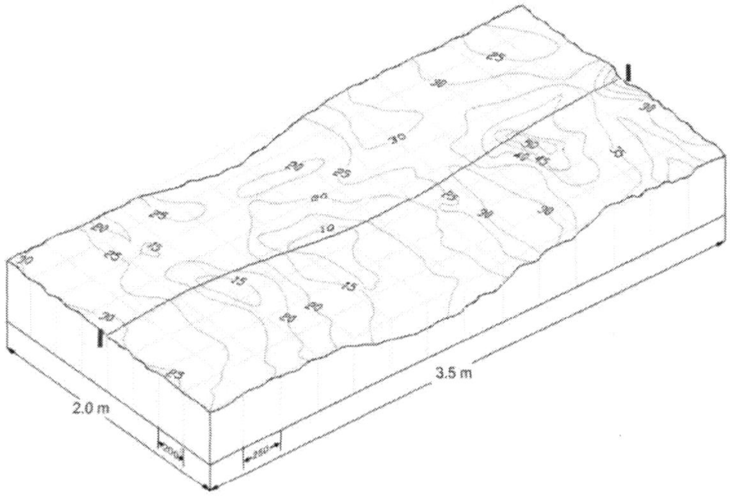

Fig. 2. Hypsometric surface of a stope walls after shrinkage stoping of ore.

It should be noted that shape of peaks and valleys depends considerably on the technological parameters of blasting and drilling as well as on physical and mechanical characteristics of rock mass being blasted.

Analysis of the developed hypsometric plans of wall surfaces of stopes of Zaporizhzhia Iron-Ore Integrated Works made it possible to state that changes in their axial sections are of stochastic nature. It has also been determined that both mining and geological as well as technological factors are involved in the formation of a stope wall contour. Physical and mechanical properties as well as their fracturing degree are among basic mining and geological factors. Following factors are the first ones among those reflecting specific character of shrinkage stoping of thin steeply inclined vein: diameter of a blast hole and its length; type of explosives used in the blast holes, their specific consumption, and concentration; arrangement of the blast holes within a stope and blasting technique etc.

It has also been determined that amplitude of peaks of valleys of a bottom wall and upper wall of the vein stipulates cooperation of load-haul-dumpers effecting significantly the losses of ore during its drawing from blocks.

From the viewpoint of interaction between wall rocks and mining equipment operating within narrow face area of a block with unsteady hypsometry, axial sections and vertical cross-sections of top wall and bottom wall of a stope simulating morphology of ore body occurrence across the pitch are of specific interest (Fig. 1b).

From the viewpoint of determination of ore losses and nature of its overhanging while drawing from the block, axial sections and cross sections of both top and bottom walls of a stope to the dip of ore body as well as amplitude of irregularities (i.e. peaks and valleys) of the vein top wall and bottom wall are of particular interest (Fig. 3).

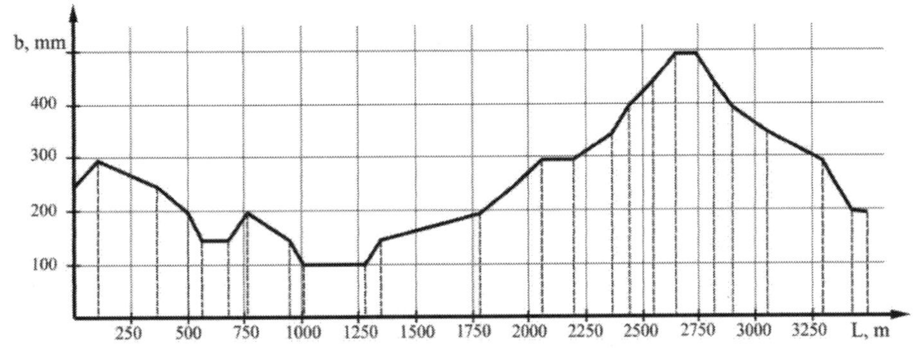

Fig. 3. Fragment of axial section of wall surfaces of the block along I-I line (Fig. 1).

Horizontal drilling-and-blasting technique, applied to the rise, is the specific feature to mine thin-vein steeply inclined deposits. The technique relies upon shrinkage stoping method making it possible to decrease ore dilution since the stope can follow the vein despite variability of its occurrence elements; it provides breaking of ore sufficient for its unrestricted drawing within narrow stoping zone; and it allows performing oriented discharge of the broken rock mass towards drawing-off level.

However, in addition to traditional techniques applied for ore breaking by means of rising blastholes, it is required to have rather stable wall rocks. Practices of thin-vein gold fields mining shows that significant ore dilution takes place when parameters of drilling and blasting operations mismatch the conditions of thin-vein deposit development.

Moreover, use of load-haul-dumpers in the context of thin-vein steeply inclined deposits mining is possible if only design parameters of drilling and blasting operations are met in full since the parameters provide optimum width of a stoping zone m_{opt} and minimum irregularities of the stope walls.

Studies [4] concerning the support of design parameters of dri may achieve ± 30% due to irregularities of wall surfaces and variable directions of ore body occurrence. That complicates seriously the process aimed at selection and substantiation of a type of load-haul-dumpers since minimal width and length are the basic design parameters determining their maneuverability within narrow stoping zone.

In this connection, Fig. 4 represents a scheme according to which it is proposed to determine a stoping zone width to substantiate minimal design parameters of load-haul-dumpers.

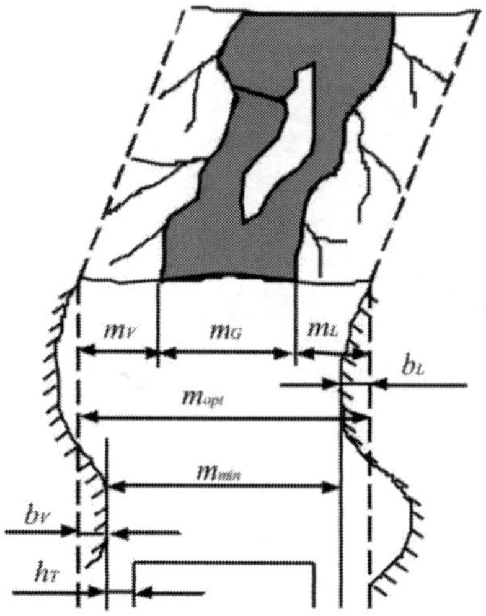

Fig. 4. A scheme to determine minimal width of a stoping zone: m_G is ore body thickness; m_L and m_V are thicknesses of the undercut rocks of top wall and bottom of the vein; m_{opt} is optimum width of the stoping zone; b_L and b_V are amplitudes between contour figures at the level of peaks and valleys of wall surfaces of the stope; m_{min} is minimum width of the stoping zone; and h_T is technological gap.

Optimum width m_{opt} involving a coefficient of thickness being mined k_M should be considered as a minimum width of the stoping zone:

$$m_{min} = m_{opt} \cdot k_M \; . \tag{1}$$

Optimum width of the stoping zone is determined taking into consideration recommendations from [5]; the width consists of the vein thickness and thickness of wall rocks being cut. Together with mineralization degree, the thickness of wall rocks being mined is considered as total thickness of rocks of top wall and bottom wall being cut:

$$m_{opt} = m_G + m_P \, ; \tag{2}$$

$$m_P = m_L + m_V \; . \tag{3}$$

In the context of Fig. 1, the coefficient of thickness k_M variation coefficient is determined as:

$$k_M = \frac{m_P}{\left(m_P + \dfrac{b_L + b_V}{2} \right) k_G} , \tag{4}$$

where m_P is total thickness of wall rocks being cut; b_L and b_V are heights of the ore vein top wall and bottom wall respectively; and k_G is hypsometric coefficient of the stope zone walls after ore breaking up by explosives.

A type of geometric (determinate) index of V.A. Bukrinski was taken as a description point to determine hypsometric coefficient [6]:

$$k_G = \frac{\int dl}{L} , \tag{5}$$

where $\int dl$ is integral over the contact curve; and L is the block width.

Indices of the stoping zone wall irregularities are associated with traditional technique of blasthole drilling from one drill setup when ore breaking up by explosives is evaluated in terms of significant release of rock mass lumps resulting in considerable dilution of the mined ore; besides, it complicates the ore drawing, increases expenditures for its crushing in the preparation process, and makes the rock mass unstable since fissures grow deep disturbing the rock mass integrity.

To decrease losses and degree of ore dilution, oriented blast-hole drilling is recommended for future use of the mining system with ore shrinkage. Processing description of drilling and blasting operations specifies bottom drilling from using three drill setups (Fig. 5). The procedure of hole drilling is as follows: first, central hole No. 1 is run in; then the drill is displaced right (left) at the distance of 0.2 – 0.3 m with 85° angle, periphery holes No. 2 and 3 are run in. According to blasting and drilling description, face area of a stope, formed in the process of ore breaking up by explosives, is characterized by minimum width and hypsometry of wall surfaces (amplitude between levels of peaks and valleys will be 0.2 to 0.25 m) simulating the vein walls [7].

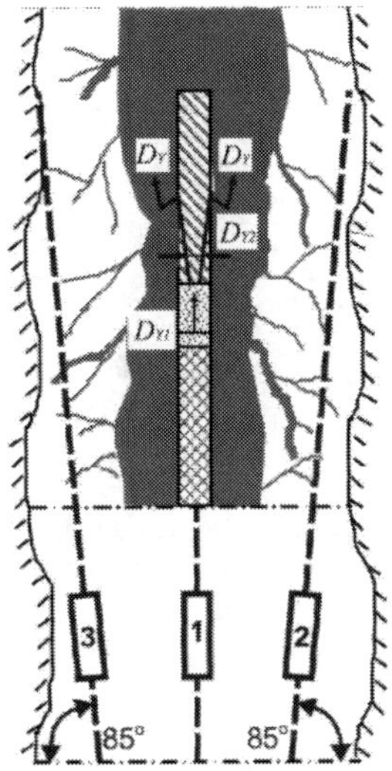

Fig. 5. Technique of ore shrinkage stoping in the context of oriented drilling of periphery holes ore breaking with schematic representation of propagation of shock waves and "compression" of lower part of tamping while blasting: D_Y is a shock-wave front propagating within the rock mass; D_{Y1} is a shock-wave front propagating within the low-density porous layer; D_{Y2} is a shock-wave front propagating within the tamping from the borehole hole wall.

When directions of periphery holes vary, the amount of ore losses increases along with the amount of inrushes. Moreover, the mined thickness of the ore bodies as well as topography of the stope also experiences changes. If drilling directions are met strictly then optimum parameters of stoping zone and crushing of wall rocks being cut are obtained. In addition, decrease in peripheral rock disturbance is achieved resulting in the decreased degree of ore dilution.

If the oriented drilling of peripheral holes is applied, then distance between horizontals run between extreme levels of peaks and valleys is 0.2 to 0.25 m when the site length is 2.0 m to the dip, and 3.5 m across the pitch. Parameters of the sites being studied are substantiated: if downdip, then by means of traditional hole length in the context of shrinkage stoping of thin steeply inclined veins; across the pitch, then by means of length of load-haul-dumper St 500HE Microscoop.

3 Results and discussion

The results of physical modeling of ore drawing process if shrinkage is applied have shown that ore loss amounts vary within 12 – 15%; they depend linearly on the ore wall hypsometry reflecting substantively actual amplitudes between levels of peaks and valleys being formed in the process of shrinkage stoping of ore bodies and on the vein bedding angle; on the moisture content of ore being drawn and fraction value of the material being produced (Fig. 6).

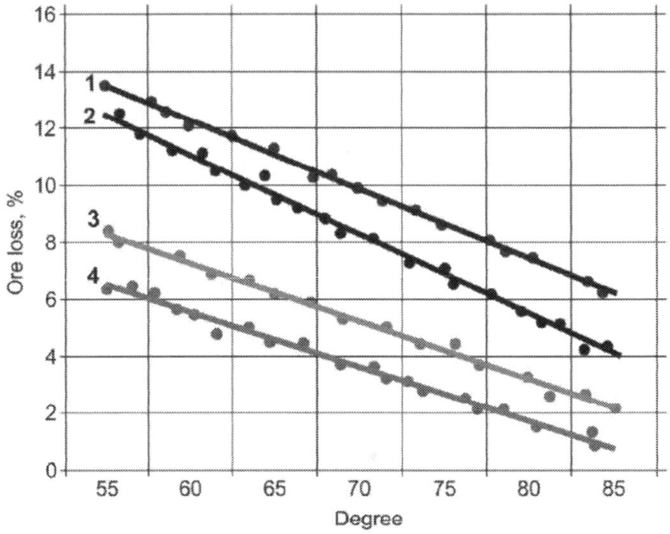

Fig. 6. Graphical dependences of changes in ore losses on the vein wall hypsometry formed as a result of shrinkage stoping of the ore drawing with 1.0 to 5.0 mm amplitude of peaks and valleys: 1 –300 – 400 mm if ore moisture is W=10%; 2 – 300 – 400 mm if ore moisture is W = 0%; 3 – 200 – 250 mm if ore moisture is W = 10%; 4 – 200 – 250 mm if ore moisture is W = 0%.

A new design of pin-point blasting charge with the improved tamping properties has been proposed for the oriented hole drilling and effectivization of charge blasting. In this context, low-density damping layer of polymer beads is placed between explosive column and tamping (Fig. 5). As experiments show, such a charge provides more qualitative blockage of detonation products within gaseous space as well as increase of total blasting pulse.

Experiments concerning the efficiency of blasting of charges with low-density layer have been carried out within the testing area of Industrial Production Enterprise "Kryvbasvzryvprom" [8]. In the context of clays, ammonite charges No. 6 ZhV with 200 g to 3 kg weight were blasted in 26 – 200 mm diameter holes and wells at the depth of 0.5 – 1.5 m. Foamed granulated polystyrol of 0.03 – 0.08 kg/m^3 was used as the damping layer between tamping and charge. In terms of each series of trial blasts, a diameter, charge height, and depth were constant; only polymeric layer varied (i.e. its height and acoustic stiffness). Explosive efficiency of such charges was estimated according to the dimensions of blasting cone. The experimental results have demonstrated that if parameters of polymeric layer are optimal then blasting efficiency may be increased up to 35%. In terms of height within a hole, low-density layer geometry depends upon physical and mechanical properties of the soil to be blasted, tamping, and of the properties of the layer material itself.

While analyzing processes taking place while blasting with the use of the above composite tamping, it was supposed that permutation of incompressible particles rather than their deformation happens within shock wave being formed within the damping space material, soil tamping part, and within the rock mass during blasting [9].

Determination of the shock wave parameters within the damping layer should involve velocity of the layer material as well as the shock wave [10]:

$$v_1 = \sqrt{\left(\frac{P_1}{\rho_{0P}}\right)\cdot\left(1-\frac{\rho_{0P}}{\rho_P}\right)} = \left(\frac{3}{8}\right)^{0.5} \cdot D \cdot \left[1-\left(\frac{P_1}{P_H}\right)^{0.33}\right]; \tag{6}$$

$$P_1 = \rho_{0P} \cdot v_1 \cdot D_{Y1},$$

where P_1 is pressure within the shock wave; D is detonation velocity; P_H is pressure within the detonation front; ρ_{0P} and ρ_P are the layer material density before detonation front and after it.

Owing to the advance of the extending detonation front over the rock mass in the line of a hole being shaped as for its front in terms of tamping (being provided by means of availability of damping polymeric layer within which the decrease in a shock wave parameters takes place), conditions for compression and consolidation of lower part of tamping material are developed. As a result, the shock wave, moving from the central part of the hole, will be expanded through nonhomogeneous continuum rather than a channel (a hole). That provides more qualitative blockage of explosion products within gaseous cavity since in this case when charge is blasting, tamping material travels simultaneously with upper part of the broken rock mass. Furthermore, advance tamping release is excluded as it is observed when charges with standard tamping are blasted. Moreover, the delay in the process of tamping release factors into pressure drop within the explosion cavity at the expense of extension of its walls decreasing the number of the explosion pulses obtained by the tamping while travelling.

At a first approximation, dependence of relative explosion efficiency V_{OTN} (ratio between volume released in terms of blasting charges with low-density interlayer and volume released in terms of interlayer-free charges) on the polymer height h (in terms of the charge diameters) is approached by means of the exponential function:

$$V_{OTN} = -A(h-0.5)^2 + 1.35, \tag{7}$$

where A is coefficient taking into consideration soil properties; it varies within 0.8 – 2.0 while passing from dry heavy-textured clay to loose sandy loam.

While selecting optimum height of damping layer, equation (7) is true if $0 < h < 0.8$. It has also been determined that when the density of soil being blasted increases, the efficiency of such a charge experiences certain decrease; if chambering ratio increases, it increases as well.

It is possible to determine rational height of damping layer in a hole according to the empiric formula:

$$h = k \cdot d \cdot \left(\frac{C_B}{C_H} - \frac{C_B}{C_0}\right)^{-1}, \tag{8}$$

where k is a coefficient involving soil properties; d is a well diameter, cm; C_B is sound velocity within soil tamping part, m/s; C_H is sound velocity within polymeric layer, m/s; C_0 is sound velocity within soil mass being blasted, m/s.

4 Conclusions

Basic advantage of a design of hole charges with low-density interlayer within tamping is as follows: they provide considerable efficiency explosion improvement expressed in the amount of soil outburst per a kilogram of an explosive. In this context, geometry of the damping interlayer turns out to be several times less to compare with, for instance, the dimensions of air gaps in the context of the available designs of release charges making it possible to use the amount of tamping material more productively.

Formation of a low-density material layer within a hole does not involve any processing difficulties being observed while forming charges with air gaps.

It is recommended to place low-density material intermediate tamping layer right after a hole was charged and live primers were inserted with the help of transportation and charging equipment where polymer is foamed within autonomous tanks while exhaust energy utilizing.

The work was carried out in the context of scientific projects "Scientific foundations to form unified system of energy saving and generating in the context of the objects of fuel and energy complex of Ukraine" (State registration No. 0117U001127) and "The development of background technology for environmentally friendly mineral extraction in the context of technogenically loaded mining regions of Ukraine" (State registration No. 0117U001134).

References

1. Khomenko, O., & Maltsev, D. (2013). Laboratory Research of Influence of Face Area Dimensions on the State of Uranium Ore Layers Being Broken. *Naukovyi Visnyk Natsionalnoho Hirnychoho Universytetu,* (2), 31-37.
2. Petlovanyi, M. (2016). Influence of Configuration Chambers on the Formation of Stress in Multi-Modulus Mass. *Mining of Mineral Deposits,* (10), 48-54. http://dx.doi.org/10.15407/mining10.02.048
3. Koroviaka, Ye.A. (2008). *Obgruntuvannia napriamiv udoskonalennia tekhnologii rozrobky tonkozhylnykh zolotorudnykh rodovychsh Ukrainy.* Dnipropetrovsk: Natsionalnyi hirnychyi universytet.
4. Shirin, L.N., Koroviaka, Ye.A., & Vilianski, V.N. (2004). Formation of the Minimum Width Clearing Area in Case of Small-Hole Breakage Thin-Stranded Steeply Dipping Deposits. *Naukovyi Visnyk Natsionalnoho Hirnychoho Universytetu,* (11), 11-13.
5. Agoshkov, M.I., Borisov, S.S., & Boiarski, V.A. (1983). *Razrabotka rudnykh i nerudnykh mestorozhdeniy.* Moskva: Nedra.
6. Bukrinski, V.A. (1985). *Geometriya nedr.* Moskva: Nedra.
7. Shirin, L.N., Koroviaka, Ye.A., & Vilianski, V.N. (2005). Substantiation of the Rational Parameters of the Drill Charge in the Case Breaking thin Steeply Dipping Strands. *Naukovyi Visnyk Natsionalnoho Hirnychoho Universytetu,* (1), 6-9.
8. Baranov, E.G., Vilianski, V.N., Oberemok, O.N., & Kurinnoi, V.P. (1990). Influence of Gap Parameters from Porous Low-density Materials on the Efficiency Downhole Charge Explosion. *Gornyy zhurnal,* (5), 72-76.
9. Liahov, G.M. (1982). *Volny v gruntakh i poristykh mnogokomponentnykh sredakh.* Moskva: Nedra.
10. Orlenko, L.P. (2004). *Fizika vzryva.* Moskva: Nauka (Fizmatlit).

Destabilization of the hard coal microstructure by a weak electric field

Oleksandr Molchanov[1], *Dmytro* Rudakov[2*], *Valerii* Soboliev[3], and *Oleksii* Kamchatnyi[1]

[1]Institute for Physics of Mining Processes National Academy of Science of Ukraine, 2a Simferopolska St., 49005 Dnipro
[2]National Mining University, Department of Hydrogeology and Engineering Geology, 19 Yavornytskoho Ave., 49005 Dnipro, Ukraine
[3]National Mining University, Department of Construction, Geotechnics, and Geomechanics, 19 Yavornytskoho Ave., 49005 Dnipro, Ukraine

Abstract. This study aims to analyse physical and chemical changes in hard coal samples under the influence of low-intensity electric fields in comparison to the fragments of ejected coal, as well as the coal samples selected from the zones of high and low outburst hazard. We used physical methods including X-raying, electron paramagnetic resonance, thermo-gravimetric analysis, differential scanning calorimetry, laser diffraction analysis of particle sizes, IR-spectrometry, nuclear magnetic resonance, and Raman spectroscopy. It has been shown that destruction of coal organic matter (COM) can be caused not only by mechanical impacts or thermal influences but also weak electric fields. Scientific novelty consists in the fact that for the first time we established the identity of the COM destruction mechanism of mechanical-chemical activation and weak electric fields influencing on the previously destabilized coal microstructure. The destruction mechanism is based on thermal field regularities in both cases. The results obtained are of practical significance for the technologies of coal conversion to other products. The research results can be useful in the development of methods for reducing outburst hazard in coal mines.

1 Introduction

Physical and chemical properties of coal formed naturally under the simultaneous influence of electric and magnetic fields, temperature, pressure, and mechanical impacts on a coal seam in different geological conditions [1 – 3]. Each of the aforementioned physical factors could play the leading role at different stages of coal formation with a certain probability in accordance with Le Chatelier-Brown's principle. The features of evolutional development of the coal seam including the degree of its coalification depend on changing these factors governing the thermodynamic and chemical equilibrium in the "coal – gas" system. The results of studying some physical properties of coal and the causes of appearance and retention of its unstable microstructure as an active state potentially capable of triggering

* Corresponding author: rudakovdv21@gmail.com

and developing physical and chemical processes cannot be unambiguously interpreted within the framework of classical thermodynamics under the complex influence of multiple factors.

The system consisting of nano-sized particles [4] was previously studied mainly using the methodology of analysing micro-sized objects [5, 6] without proper attention to the mechanisms of carbon nano-phase appearance. Such nano-sized phases as carbon, hydrocarbon chains, graphene [7] etc. that acquire new properties in case of physical-chemical transformations [8] and, therefore, may have a critical influence on forming physical properties of coal [9], as a rule, were not taken into account.

Chemical reactions that occur in coal result, particularly, in phase and structural transformations and, generally, in destruction independently on physical parameters responsible for keeping the system equilibrium. In some cases, such reactions eventually generate a coal of high outburst hazard with the microstructure that store a big amount of internal energy potentially capable to initiate COM transition to a gas [10]. The restored history of forming new properties of coal is in a good agreement with the evolutional character of this process. This is very important conclusion for studying gas-dynamic phenomena in coal mines, which refer to the most important scientific problems still without a physically justified solution [11].

In this aspect, studying the influences of weak electric and magnetic fields on phase transitions in mechanically pre-activated coal is of growing scientific interest because natural tectonic activation manifests itself as the complex of mechanical effects accompanied by the amplification of electric and magnetic fields, increasing local temperatures, i. e. physical factors capable of initiating active chemical processes.

The majority of previous experiments did not evaluate the inflows of weak intensity fields because of their seemingly obvious low-energy effect. However, some recent experimental studies refuted the ideas about the insignificant contribution of weak electric field energy to the initiation of any physical and chemical process. For example, the efficiency of weak electric [12] and magnetic [13] field influences was demonstrated on the examples of transformation in coal, initiation of phase transitions and structural transformations in minerals of the carbonate group (siderite, calcite) [14], and the fractional composition of diamond particles synthesized in high-pressure chambers [15] etc.

The relevance of these studies is contingent on the need for creation of a unified physical concept about the nature of coal formation and transition of coal solid components to the gas without high-temperature activation of chemical reactions. In addition, the issues of formation of various carbon phases from free atomic carbon remain topical [16]. This may be of practical interest in production of new materials with specific physical, chemical, and technical properties.

The objective of this study is to conduct an analysis of physical-chemical changes in hard coal by comparing the samples taken from different seams and samples exposed to weak-intensity electric fields. This will allow demonstrating that COM destruction may be caused not only by mechanical impacts but also weak electric fields.

Analysing the experiments we took into account the results of T.M. Khrenkova and her colleagues in the field of coal mechanical chemistry [17]. The results of a comparative analysis of coal physical-chemical characteristics were used to justify the hypothesis of forming new coal properties under microstructural and phase instability [12].

2 Materials and methods of research

We used in experiments two sorts of hard coal that are gassy coal (I) (carbon content 86.6%, hydrogen 5.7%, vitrinite reflection index $R_0 \geq 0.83\%$, $Y = 14$ mm, $V^{daf} = 36.2\%$) and fat coal (II) (carbon content 88.2%, hydrogen 3.9%, vitrinite reflection index $R_0 \geq 1.09\%$,

Y = 17 mm, V^{daf} = 31.3%). The samples were taken from the seams both of high and low hazard of coal and gas outburst. The samples of the average mass 1.28 g were prepared from coal crushed to the fractions of 100/200 μm; according to laser diffraction analysis the size of intact coal particles ranged from 111.7 to 214.5 μm. Before conducting the tests, the samples were dried at a temperature of 308 K for 48 hours, with the maximum temperature of heating during electrical-physical treatments not exceeding 320 K.

The experiments were conducted following the procedure described in [12]. The difference of the potential and the magnitude of the current were 1 – 300 V and 0.3 A respectively, each sample was processed about 4 hours. For X-ray phase analysis we used a DRON-3 unit. Numerous physical and chemical studies were carried out using the equipment for thermo-gravimetric analysis and differential scanning calorimetry TGA/DSC Mettler Toledo, the optical microscope LEICA DM ILM, the Shimadzu laser diffraction particle size analyser SALD-301V, and the calorimeter C-2000 IKA. Coal infrared spectra were recorded on a spectrometer ФСМ-1201 with a transmission in the spectral range of 500 – 5500 cm^{-1}. Nuclear magnetic resonance (NMR) studies were carried out following the method of the Institute for Physics of Mining Processes of the National Academy of Sciences of Ukraine (IPMP of NASU) [18, 19].

The NMR spectra of coal were recorded using a broad-line NMR spectrometer designed at the IPMP of NASU. The field strength of permanent magnet was 4600 E, the uniformity $2\ 10^{-6}$ E/cm, the resonant frequency 19.6 MHz.

3 Experiment results and discussion

The NMR ^{1}H spectra (Table 1) showed a significant increase in the characteristic time of methane desorption (Table 2) [19]. Gas saturation of the coal samples treated with an electric field decreased by only a few per cents compared to intact coal. Simultaneously, the rate of methane emission increased for gassy coal (I) by three times and for fat coal (II) by 1.7 times [20]. The decrease in gas saturation of treated coal samples is due to the additional gas release by COM treated with an electric field.

Table 1. Characteristics of coal samples treated by an electric field.

Material	I_1	ΔH_1	I_2	ΔH_2	S_1	S_2	S_1/S_2	I_2/I_1	θ
Gassy coal (I) treated	1.53	6.617	0.725	0.370	0.822	12.658	0.065	0.474	2.14
Gassy coal (I) intact	1.5877	6.347	0.731	0.472	1.054	12.635	0.083419	0.460618	2.27
Fat coal (II) treated	1.597	6.363	0.425	0.391	0.507	12.694	0.03994	0.266124	0.86
Fat coal (II) intact	1.532	6.11	0.318	0.668	0.641	11.377	0.056342	0.207572	1.09

Table 2. Change of methane desorption time T_{des} from the coal samples intact and treated by an electric field.

Coal	Gassy coal (I)		Fat coal (II)	
	Intact	Treated	Intact	Treated
T_{des}, min	25.9	79.5	39.1	68.0

In Table 1 we used the nomenclature applicable to the ^{1}H NMR spectrum and namely: I_1, I_2 are the amplitudes of the wide and narrow spectral lines, respectively; ΔH_1, ΔH_2 are their widths; S_1, S_2 are their areas; θ is the humidity before measurements, %.

The studies of gas-saturated samples losing the mass due to methane desorption were conducted at a temperature of 363 K using the gravimetric method. It has been shown that intact coal loses its mass at the rate higher than the samples treated with an electric field;

the gas saturation of treated samples decreases. This is due to the additional gas release by COM that is undergone to destructive processes as a result of electric field influence.

Similar results of studying the coal samples taken from the seams of high outburst hazard, i.e. coal with a disturbed microstructure are given in [18]. An increase in the coal temperature from 293 to 323 K with simultaneous treatment by the electric field stimulated COM destruction significantly. The methane desorption rate evaluated using the methodology [20] was practically the same for all samples taken from low and high outburst hazard zones.

Coal sampled from the zones of low outburst hazard increased its chemical activity after grinding. According to the EPR results, the concentration of paramagnetic centres (PMC) increased from $(0.5 - 0.7)\cdot 10^{19}$ PMC/g to $(0.9 - 3)\cdot 10^{19}$ PMC/g; additional treatment with an electric field increased the concentration to $(3.4 - 5)\cdot 10^{19}$ PMC/g. EPR line expansion for coal from high outburst hazard zones [12] is caused by free radicals and can be described by Lorentz's equation. EPR line expansion occurs in cases of mechanical, thermal and electrical destruction of coal, i.e., due mainly to the thermal mechanism of chemical bond rupture. The stability of bonds can be influenced by the high content of impurities in coal composition.

After mechanical-electric processing coal samples acquired the electret potential difference U_P evaluated by the compensation method. The electrochemical activity A_B for ejected fat coal (II) (the mine named after O.F. Zasiadko, Ukraine) was $4.3\cdot 10^{-5}$; in contrast $A_B = 0.7\cdot 10^{-2}$ for the non-hazardous zone.

The obtained correlation of changing the electret potential U_P in time (Fig. 1) indicates an ultra-slow relaxation of the electret state triggered by a weak field. The results of studying the untreated and treated coal samples taken from the zones of low and high outburst hazard as well as ejected coal (Table 3) showed that the relaxation time of the electret state increases with the concentration of structural defects in the crystallites and the polymer matrix of coal and electric charges, which is in good agreement with the results presented in [21]. The energy stored in defects of the coal nanostructure increases accordingly.

Table 3. Characteristics of the electret state of coal treated with a weak electric field.

Fat coal (II)	Initial electret potential U_P, mV	Electrochemical activity A_B	$n\cdot 10^{-18}$, PMC/g	Residual potential, $U_{P,res}$, mV
Low outburst hazard zones	2100	$7\cdot 10^{-3}$	60.0	0.88
High outburst hazard zones	2750	$9\cdot 10^{-3}$	83.0	0.78
Ejected	13	$4.3\cdot 10^{-5}$	6 – 14	0.63

The relaxation time of the electret state varies from $8\cdot 10^{4}$ to $2\cdot 10^{6}$ s for various samples, which exceeds the Maxwellian relaxation time (~30 s) several orders [22]. The concentration of defects increases as a result of mechanical grinding of coal and additional processing by an electric field. A comparative analysis of the electret state of coal from various zones and coal samples treated with an electric field reveals similar regularities of changing properties, especially for coal from the zones of high outburst hazard.

X-ray diffraction analysis allowed identifying two highly diffuse maxima on all diffractograms of studied coal samples, which corresponds to the angles 2θ of 24 and 43° [12]. In the vicinity of the first maximum, there are the lines with the values of 0.338, 0.371, 0.403 (weak line), 0.424 and 0.455 nm. In the vicinity of the second maximum in the diffractograms of all initial samples we identified the lines that can be attributed to the crystalline phase with an inter-planar distance $d = 0.198 - 0.200$ nm, which is close to the graphite line at $d = 0.202$ nm second by intensity.

Destructive processes in coal lead to an increase in the content of finer fractions, and correspondingly to an increase in the specific surface area. At the same time, the average particle size and content decrease by 3 – 8%, the share of the finest fraction increases, and the share of the coarsest fraction decreases accordingly (Fig. 2).

Diffractograms point at the increase of "amorphousness" of the coal microstructure, which is confirmed by features of particle size distributions and the results of EPR studies that indicate the increase in the PMC concentration. Coal from the outburst zone is known to have a much larger sorption surface than coal from the same seam but not ejected. An increase in the internal specific surface area of coal may be an after-effect of shear deformations, chemical reactions, and the effect of weak electric fields.

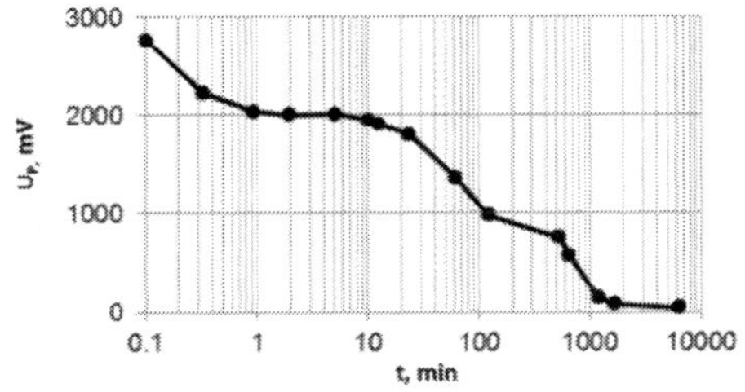

Fig. 1. Relaxation of the electret potential U_P in time in a sample of fat coal (II) taken from the zone of high outburst hazard ($\Delta U_P = 5\%$).

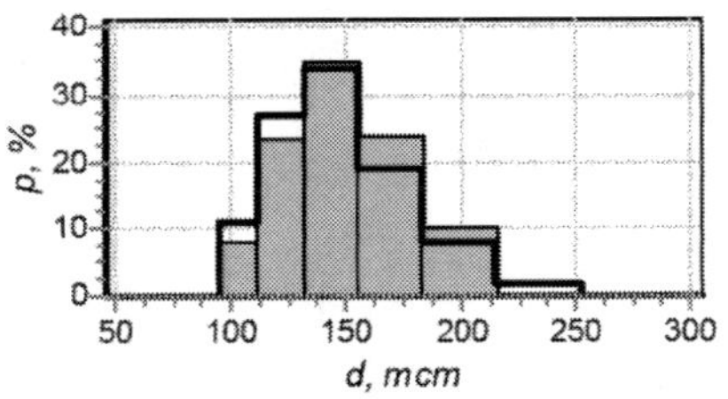

Fig. 2. Density of coal particle distribution p over particle size d: grey colour– untreated coal; bold line – coal treated with an electric field.

The IR-spectra of coal treated with an electric field are similar to the IR-spectra of coal taken from the seams of high outburst hazard. Destruction of bridge-type aliphatic chains correlates with a decrease in the optical density of bands 2860 – 2920 cm^{-1} (Fig. 3), which is in agreement with the valence and deformation vibrations of C-H bonds. These bonds relate to the structures that contain CH_2 and CH_3 groups and may be caused by reducing the band of 3000 – 3100 cm^{-1} in aromatic hydrocarbons.

A detailed analysis of IR-spectroscopy of fat and gassy coal presented in [12, 13] also indicates the destructive nature of processes induced by the treatment using weak electric fields.

The coal samples selected from the zones of high outburst hazard contain an excess of CH_3-groups with respect to the samples taken from low hazard zones. The studies [23, 24] established the correlation between the coal outburst hazard and the total content of CH_2 and CH_3 groups, as well as the ratio of carbon atoms with different hybridization of valent electrons (sp^2/sp^3). The spectra of coal of high outburst hazard show an increase in the carbon content for the state of sp^2-hybridization, which is possible with an increase of chain fragments containing CH-groups. Using the NMR methods, cross-polarization, and Raman spectroscopy we have established the physical effects of increasing the content of CH_3 groups in the zones of high outburst hazard and COM destruction that continues after the ejection. These effects are likely caused by the loss of stability of carbon and hydrocarbon chains with an increase in the electret potential and its further relaxation.

Numerical simulations showed that the influence of external electric charges leads to (1) bond breakage and (2) a random chain rupture after the decrease in the number of atoms in the chain to a certain critical value [25, 26]. These results indirectly confirm the hypothesis of mechanical-electric stimulation of COM transition to the gas.

Intensive methane formation is preconditioned by the achievement of a critical level of $-CH_3$ groups and a ratio of hydrogen atoms in the composition of the $=CH$ and $-CH_3$ groups of 4:1. As suggested by Ye. Ulianova this maintains an active process in the outburst zone.

 https://doi.org/10.1051/e3sconf/20186000023

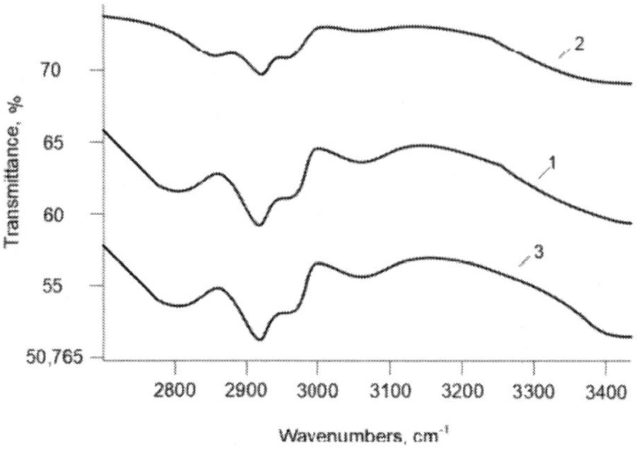

Fig. 3. A fragment of the IR spectrum of aliphatic saturated C_{al}-H groups: 1 – intact coal, 2 – coal treated with an electric field, 3 – coal treated with a magnetic field.

The additional bands with a frequency shift of 1190 cm^{-1} and 1430 cm^{-1} appear in the Raman spectra of coal samples from the zones of high outburst hazard (Fig. 4a) in comparison to the samples from low hazard zones in the D-band range. The band of the Raman spectrum of coal from high outburst hazard zones in the frequency shift range 1160 – 1190 cm^{-1} is associated with the vibrations of C-C bonds in =C-C= groups. A band with a frequency shift of 1436 cm^{-1} is caused by the appearance of polyin chains with ≡C-C≡ bonds [24, 27].

The Raman spectra of coal from the zones of high outburst hazard are decomposed into 6 components (Fig. 4a), the Raman spectra of coal from low outburst hazard zones in the coalification series up to the anthracite are maximally decomposed into 5 components, the Raman spectra of ejected coal contain two components only (Fig. 4b). Similar regularities are observed for thermally treated coal, which may be the evidence of the same mechanism of physical-chemical transformations in coal.

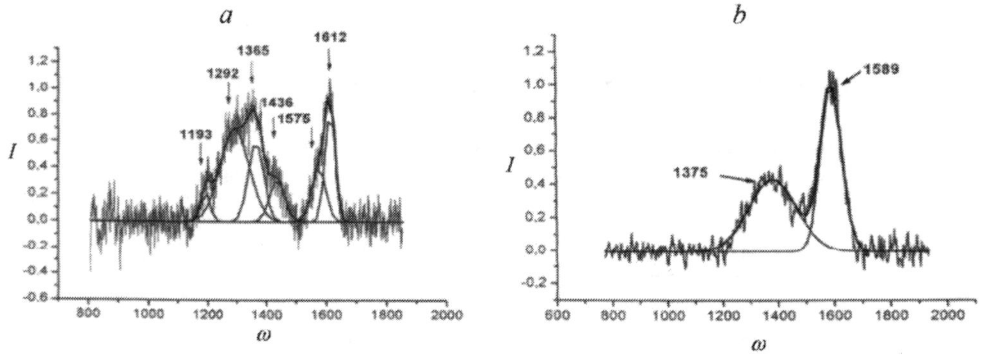

Fig. 4. Raman spectra of fat coal (II) taken from the seam h_6' of the mine named after O.O. Skochynskyi: a – in the zone of high outburst hazard; b – after the ejection. I is the intensity of Raman scattering, dimensionless, ω is the frequency shift, cm^{-1}.

The G-band maximum moves from the frequency shift of 1612 to 1589 cm^{-1}, which is associated with the destruction of the chain fragments containing CH-groups. The frequency shift of the G-band equals 1602 cm^{-1} for the Raman spectrum of a sample from a low outburst hazard zone. Changing features of the Raman spectra for coal from high outburst hazard zones and ejected coal confirms the conclusion made on the growth of the -

CH_3 group content in the zones of high outburst hazard and further COM destruction after the ejection due to reduction of the =CH-CH= and –CH_3 group contents.

Data generalization allows suggesting that methane forms during ejection mainly owing to the participation of hydrogen included in =CH groups (sp^2-hybridization) and –CH_3 groups (sp^3-hybridization).

4 Conclusions

As a result of studies of hard coal samples selected from the seams of high and low outburst hazard, and coal treated with a weak electric field, the following has been established.

Gas formation in coal treated with an electric field is a low-temperature process. Similar results can be achieved by heating coal to the temperatures exceeding 460 K.

Gas generation in coal stimulated by an electrical field can be caused by the following reasons:

– breakage of the weakest chemical bonds in carbon and hydrocarbon chains;

– complete destruction of chains if they contain less than a critical minimum number of atoms critical to chain stability;

– hydrocarbon moleculization with unpaired electrons with subsequent formation of stable gas molecules;

– desorption of chemisorbed molecules as a result of excitation of their connections to internal surfaces by passing electric current;

– breakage of graphene bonds by forming an unstable chain that breaks down to form atomic carbon followed by forming the bonds with oxygen and hydrogen.

The high rate of chemical reactions can be due to the involvement of surface active centres as catalysts.

The obtained results show that destructive processes in COM can be initiated both by mechanical and electrical influences. Moreover, destabilization and destruction of chemical bonds under electrical influences may have even greatest effect on methane formation. Such a factor as an electric field may have a strong influence on generation of an additional amount of methane, for example, in underground coal gasification [28], and the formation of volatile coal.

Authors are grateful to Ye. Ulianova and her colleagues for assistance in carrying out the research and active discussion of the results obtained.

References

1. Khomenko, O., Sudakov, A., Malanchuk, Z., & Malanchuk, Ye. (2017). Principles of rock pressure energy usage during underground mining of deposits. *Naukovyi Visnyk Natsionalnoho Hirnychoho Universytetu*, (2), 35-43.
2. Falshtynskyi, V.S., Dychkovskyi, R.O., Saik, P.B., Lozynskyi, V.H., & Cabana, E.C. (2017). Formation of thermal fields by the energy-chemical complex of coal gasification. *Naukovyi Visnyk Natsionalnoho Hirnychoho Universytetu*, (5), 36-42.
3. Caceres, E., & Alca, J.J. (2016). Rural Electrification Using Gasification Technology: Experiences and Perspectives. *IEEE Latin America Transactions*, *14*(7), 3322-3328. https://doi:10.1109/TLA.2016.7587637
4. Ginzburg, A.I., & Zhemchuzhnikov, Yu.A. (1960). *Osnovy petrologii ugley*. Moskva: Izdatel'stvo Akademii nauk SSSR.
5. Butuzova, L.F., Saranchuk, V.I., Gonchar, N.P., & Shurpach, V.I. (1992). Vozmozhnosti IK-spektroskopii pri issledovanii prirodnykh ugley i produktov ikh termodestruktsii. *Fiziko-khimicheskie svoystva uglya*, 20-32.

6. Obukhov, A.A., Frolkov, G.D., & Artem'yev, V.B. (2000). *Strukturno-khimicheskaya mekhanika ugley metamorficheskogo ryada plastov, opasnykh po vnezapnym vybrosam uglya i gaza.* Shakhty: YuRO AGN.
7. Ye, R., Xiang, Ch., Lin, J., Peng, Zh., Huang, K., Yan, Zh., Cook, N.P., Samuel, E.L.G., Hwang, Ch., Ruan, G., Ceriotti, G., Raji, A.-R.O., Marti, A.A., & Tour, J.M. (2013). Coal as an abundant source of graphene quantum dots. *Nature communications*, *4*(2943). https://doi.org/10.1038/ncomms3943
8. Jin, Ch., Lan, H., Peng, L., Suenaga, K., & Iijima, S. (2009). Deriving carbon atomic chains from grapheme. Phys. Rev. Lett., 102, 205501. https://doi:10.1103/PhysRevLett.102.205501
9. Yanovskiy, Yu.G., Nikitina, Ye.A., Karnet, Yu.N., & Nikitin, S.M. (2010). Modelirovanie deformatsii i razrusheniya grafena: razmernyy effekt, vliyanie defektov i modifikatsii poverkhnosti. *Fizicheskaya mezomekhanika*, *5*(13), 139-147.
10. Sobolev, V.V. (2003). K voprosu o prirode obrazovaniya vybrosoopasnykh ugley. *Sbornik nauchnykh trudov Natsionalnogo gornogo universiteta*, *1*(17), 505-511.
11. Antoshchenko, N.I., Radchenko, A.G., Ashikhmin, V.D., & Radchenko, A.A. (2015). Osobennosti proyavleniya vybrosoopasnosti ugley v ryadu metamorfizma. *Zbirnyk naukovykh prats Donetskoho natsinalnoho tekhnichnoho universytetu*, (1), 14-21.
12. Pivnyak, G.G., Sobolev, V.V., & Filippov, A.O. (2012). Phase transformations in bituminous coals under the influence of weak electric and magnetic fields. *Naukovyi Visnyk Natsionalnoho Hirnychoho Universytetu*, (5), 43-49.
13. Soboliev, V., Bilan, N., & Samovik, D. (2013). Magnetic stimulation of transformations in coal. *Annual Scientific-Technical Collection – Mining of Mineral Deposits*, 221-225. https://doi.org/10.1201/b16354-41
14. Sobolev, V.V., Bilan, N.V., & Khalimendik, A.V. (2017). On formation of electrically conductive phases under electro-thermal activation of ferruginous carbonates. *Naukovyi Visnyk Natsionalnoho Hirnychoho Universytetu*, 4(160), 27-35.
15. Sobolev, V.V., & Bondarenko, E.V. (1993). The change in granulometric composition of diamond crystals when treating synthesis products in electromagnetic field. *Sverkhtverdye materialy*, (4),57-58.
16. Sobolev, V.V. (2010). Synthesis of nanosized phases from an atomic carbon. *Advanced Materials Research*, 123-125; 791-794.
17. Khrenkova, T.M., & Chubarova, M.A. (1973). Mekhanokhimiya ugley. *Khimiya tverdogo topliva*, (1), 62-65.
18. Ul'yanova, Ye.V. (2012). *Strukturnye i kompozitsionnye perestroyki v iskopayemykh uglyakh.* Saarbrucken: Palmarium Academic Publishing.
19. Molchanov, A.N. (2011). Metodicheskie osobennosti issledovaniya protsessa desorbtsii metana iz iskopayemykh ugley. *Geotekhnichna mekhanika*, (94), 133-139.
20. Sobolev, V., Rudakov, D., Stefanovych, L., & Jach, K. (2017). Fizicheskoe i matematicheskoe modelirovanie usloviy vybrosa uglya i gaza. *Mining of Mineral Deposits*, *11*(3), 40-49 https://doi.org/10.15407/mining11.03.040
21. Panchenko, Ye.M. (2009). *Elektretnoe sostoyanie v oksidakh.* Moskva: Fizmatlit.
22. Electrets. (1980). Ed. G.M. Sessler. Berlin; Heidelberg; New York: Springer-Verlag.
23. Alekseyev, A.D., Molchanov, A.N., Ul'yanova, Ye.V., Zimina, S.V., & Pichka, T.V. (2012). Preobrazovanie lokal'noy struktury iskopayemykh ugley v rezul'tate vybrosa i pod deystviem vysokikh davleniy. *Fizika i tekhnika vysokikh davleniy*, *22*(1), 122-129.
24. Alekseyev, A.D., Ul'yanova,Ye.V., Trachevskiy, V.V., Ivashchuk, L.I., & Zimina, S.V. (2010). Primenenie metodov kombinatsionnogo rasseyaniya i yadernogo magnitnogo rezonansa dlya issledovaniya genezisa struktury uglerodnykh nanomaterialov prirodnogo proiskhozhdeniya. *Fizika i tekhnika vysokikh davleniy*, *20*(3), 126-140.
25. Sobolev, V.V. (2010). Zakonomernosti izmeneniya energii khimicheskoy svyazi v pole tochechnogo zaryada. *Dopovidi Natsionalnoi Akademii Nauk Ukrainy*, (4), 88-95.

26. Sobolev, V.V., Baskevich, A.S., Bilan, N.V., & Filippov, A.O. (2011). Ustoychivost' nanostruktury kamennogo uglya pri vozdeystvii elektricheskogo toka. *Naukovyi Visnyk Natsionalnoho Hirnychoho Universytetu*, (6), 80-84.
27. Alekseyev, A.D., Ul'yanova, Ye.V., Trachevskiy, V.V., Ivashchuk, L.I., Zimina, S.V., Borshch, T.V., & Shpak, A.P. (2010). Preobrazovanie struktury iskopayemykh ugley v geomasshtabnom tekhnogeneze. *Fiziko-tekhnicheskie problemy gornogo proizvodstva*, (13), 48-59.
28. Pivnyak, G., Dychkovskyi, R, Bobyliov, O., Cabana, C.E. & Smoliński, A. (2018). Mathematical and Geomechanical Model in Physical and Chemical Processes of Underground Coal Gasification. *Solid State Phenomena*, (277), 1-16. https://doi.org/10.4028/www.scientific.net/SSP.277.1

Results of the experimental research of the heat-transfer jet pressure to the rock surface during thermal reaming of the borehole

Oleksii Voloshyn[1*], *Iryna* Potapchuk[1], *Oleksandr* Zhevzhyk[2], and *Mykola* Zhovtonoha[3]

[1]Institute of Geotechnical Mechanics under the National Academy of Science of Ukraine, Department of Vibropneumatic Transport Systems and Complexes, 2a Simferopolska Str., 49005 Dnipro
[2]Dnipropetrovsk National University of Railway Transport named after Academician V. Lazarian, Department of Thermal Engineering, 2 Lazariana Str., 49010 Dnipro, Ukraine
[3]Dnipropetrovsk National University of Railway Transport named after Academician V. Lazarian, Department of Accounting, Auditing and Intellectual Property, 2 Lazariana Str., 49010 Dnipro, Ukraine

Abstract. The performed analysis of scientific sources confirms the existence of a small number of publications devoted to the experimental research of the gasdynamics and plasmodynamics of jets used as a heat-transfer medium in the thermal methods of mine rocks destruction. There are almost no experimental and theoretical publications related to the multiple-jet plasmotrons research. The expediency of own experimental researches performing has been substantiated concerning the lateral inflow of heat-transfer medium high-speed jets on the borehole surface. An experimental research has been made of the interaction between the heat-transfer medium high-speed jets and the surface of the borehole imitated by the through duct. The further prospects of this work are the following: to determine the gas velocity along the lateral surface of the through duct and the value of the heating capacity coefficient from the heat-transfer medium to the lateral surface of the through duct, which imitates the rock surface in the borehole. These parameters are required for creating a mathematical model of the brittle destruction of rocks.

1 Introduction

Taking into account productivity and expenditures for various methods of the borehole drilling and reaming realization, thermal methods of rock destruction are of the most interest.

The relevance of application of thermal methods for rock destruction is based on the wide range of facilities for realization of the heating or cooling processes of rock.

Forms of thermal influence on the rock have the unified physical basis, namely, change of power connections potential. Tensions of thermal expansion of rock minerals are proportional to the thermal expansion coefficient of minerals, Young's modulus and heating temperature. Since Young's modulus and thermal expansion coefficient of rock

* Corresponding author: OIVoloshyn1951@nas.gov.ua

minerals take different values, while heating inside the rock minerals besides tensions, predefined by the emergence of temperature gradient, the structural thermal tensions appear, that reach maximal values on the edges of mineral grains. Therefore most of thermal rock destruction products detach from the rock mass along the grain edges of minerals.

In the cases of application of thermal methods of rock destruction the destruction products detach from the rock mass under the influence of the shear and tensile thermal stresses.

It is known that the strength limit for shearing and stretching is approximately 7 – 10 times less than the strength limit for the compression. Therefore a thermal method of rock destruction is the most energy saving method of destruction.

While temperature increasing in a heating area along with the strength and aggregate hardness decreasing there is a reduction of rock fragility emerges that allows to use the thermal methods of rock destruction effectively not only in the processes of the borehole drilling but also in the processes of the boreholes reaming [1].

Most effective amid the processes of the borehole reaming are the thermal ones, in particular with application of the gas jet heating of rock with usage of arc electrical discharge.

Plasma burners have some advantages:

– widened adjustment range of thermal parameters and concentration of the jet power [2];

– diminished amount of hazardous gases emission;

– simplified system of the burners automation;

– fissure propagation at significant depth of the rocks in the process of their thermal destruction;

– higher values of the heat transfer coefficient and specific heat flux from the heat-transfer medium to the borehole surface.

Due to high velocity of the torch movement in a borehole an average velocity change of its temperature is provided for the range of 3000 – 5000 °C/s.

It should be noted that efficiency of thermal fragile rock destruction increases due to rock hardness augmentation and expenditures of this process realization tend to diminish. The highest efficiency of thermal method for rock destruction is reached in the borehole reaming process in a monolithic enough and well-drillable rock mass.

The analysis of scientific sources testifies that for the known devices with an arc electrical discharge for rock destruction the ranges of operating performance of thermal tools, effective modes of heating, mechanical loading and fragile rock destruction were determined, mostly, in an experimental manner. Prevailing majority of the known experimental investigations concern the determination of time and temperature values of rock destruction [3 – 9]. Few publications devoted to the question of the experimental investigation of gas dynamics and plasma dynamics of the jets, that are used as a heat transfer medium in the thermal methods of rock destruction [10, 11 – 13].

The reference [10] presents average values of the heat transfer coefficient obtained by means of experimental investigation from the heat transfer medium to the rock surface, however insufficient attention is paid to the gas dynamics and interaction of the heat transfer medium jet with the borehole surface.

The references [11, 12] provide an information about velocity of the heat transfer medium at the nozzle outlet and its value gains up to 120 m/s, that could be compared with velocity of plasma flow at the nozzle outlet of plasmatron, however as a heat transfer medium an incompressible liquid is used. It makes impossible to use the results of the experimental investigation for determination of the heat transfer coefficient from plasma jet to the rock surface.

In the reference [14] for fragile rock destruction the high-speed (Mach number within the range of 2 – 5) and high temperature (300 – 1100 °C) jets of the heat transfer medium were used. However the outflow of the heat transfer medium from a nozzle took place in the pulsatile mode, and efficiency of rock destruction depended on gauge pressure of the jet on the surface of the borehole, that was at the range of 5 – 20 MPa.

The device for creation of reservoir cavity in rocky ragstones is well-known [15]. The air (plasma) heated by plasmatron flows out through its nozzle along the axis of the borehole on the working face.

Among the flaws of device it should be noted that a plasma torch flowing along the axis of the borehole not enough effectively influences on the borehole surface during its reaming.

Well-known the device for electrical arc plasma drilling that contains flat nozzles located at interelectrode intervals that cross, and incorporated by the central channel of plasmatron is well-known [16].

Among the flaws of device it should be noted that flat nozzles are located on the end face of plasmatron that is why plasma torches spreading along the axis of the borehole not enough effectively influence on the borehole surface during its reaming.

Angular plasmatron that contains channel of plasma stream formation inclined with respect to the frame axis that allows to increase efficiency of thermal energy transmission inside the rock and mechanical influence of plasma stream on the borehole surface that ultimately increases the productivity of the borehole reaming in a zone of plasma stream channel forming is known [17].

Among the flaws of device it should be noted a necessity to revolve the plasmatron around its longitudinal axis in order to ream an inner surface of the borehole at a certain level.

A task to be decided in the present publication is the plasmatron creation that able to process to ream all inner surface of the borehole at a certain level without the plasmatron rotation around its longitudinal axis with the effective transmission of thermal energy inside a mountain the rock and mechanical influence of plasma stream on the borehole surface.

For such a plasmatron axes of plasma stream channels formation are inclined with respect to the frame axis in different directions at the same acute angles.

The features of the known experimental researches allow to determine velocity of the heat transfer medium on the surface of the borehole and values of the heat transfer coefficient only for the conditions indicated in those publications.

An efficiency of thermal methods of rock destruction is defined by specific heat flux that is transferred from the heat transfer medium to the rock surface.

Specific heat flux can be augmented due to increasing of the heat transfer medium temperature or heat transfer coefficient.

In case of rock destruction in the spallation mode, i.e. in case of thermal reaming of the boreholes, an increasing of the heat transfer medium temperature is inexpedient, as it can lead to the rock melting. In this case it is necessary to increase the heat transfer coefficient from the heat transfer medium to the rock surface due to the augmentation of the mass velocity of the heat transfer medium, i.e. increasing of the heat transfer medium rate and velocity. Thus, application of the high-speed jets of the heat transfer medium seems to be prospective in the processes of fragile rock destruction.

As the known results of experimental researches dissatisfy to the decision of tasks of the current work, fulfilment of own experimental research of the interaction of high-speed heat transfer medium jets with the borehole surface is necessary.

2 Methods

Experimental research of the interaction of high-speed jets of the heat transfer medium with the borehole surface is executed as a lateral surface of the cross duct imitated the rock surface in a borehole.

The scheme of the heat-transfer medium stream interaction with the side surface of the cross duct is shown in Fig. 1.

Compressed air from a compressor through the gate and regulative valves and prechamber flows through the air duct directly to the nozzle. A model that imitates the rock inside the borehole is executed as a cross duct channel (Fig. 2).

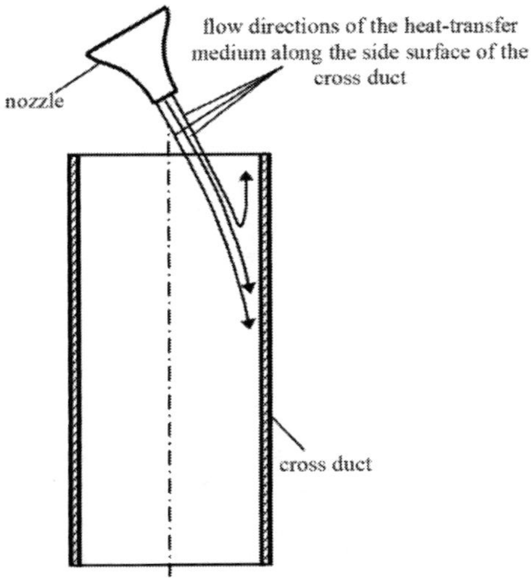

Fig. 1. The scheme of the heat-transfer medium stream interaction with the side surface of the cross duct.

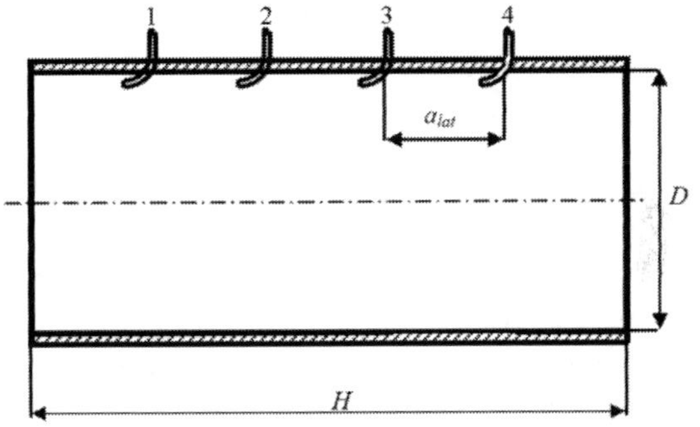

Fig. 2. The disposal of the drainage orifices in the cross duct; 1 – 4 – drainage orifices; a_{lat} – distance from the drainage orifices to the end of the cross duct; H – cross duct length; D – inner diameter of the cross duct.

For pressure determination in the air duct and prechamber the deformation standard manometers of 1227 model were used with a provisional scale of the MO type. The upper limit of measurement of the standard manometers is 1.6 MPa. The accuracy class of devices is 0.15 [18].

The essence of experimental investigation consisted of measurement of pressure on the lateral surface of the cross duct at the air jet impingement on the lateral surface. For this purpose the drainage orifices in the cross duct are made that connected by impulse tubes to the recording manometers [18].

The values of pressure on the lateral surface of the cross duct obtained during experimental investigation are necessary for determination of the gas velocity along the lateral surface of the cross duct what makes possible to determine the heat transfer coefficient from gas that imitates the heat transfer medium, to the lateral surface of the cross duct that imitates the rock surface in the borehole.

Drainage orifices 0.8 mm in diameter are executed in the normal line direction to the axis of the cross duct. The copper tubes of total pressure were set in the drainage orifices.

The edges of tubes are located on 0.3 mm above the inner surface of the duct and turned to the flow that impinges the duct, for measurement possibility of the total pressure of flow that enters the duct. For the velocities of flow less than $M < 2.3$ the error of total pressure measurement by a copper tube does not exceed 1% [19].

Rubber impulse tubes 2 mm in diameter from the drainage orifices of the cross duct brought to the recording manometers. The maximum error of manometers measurement equals 0.5% of the maximal pressure value that is measured [19]. Accuracy estimation of gauge pressure measurement as well as damping period of the wave processes in tubes were provided in accordance with well-known methodology [20].

Thermophysical and mechanical properties of mediums and materials that were used during experimental research were taken in accordance with [21].

An experiment for interaction of the gas jet with the lateral surface of the cross duct involved values as follows:

– nozzle outlet diameter (d_0) – 0.016 m;

– nozzle length (L_{noz}) – 0.016 m;

– inner diameter of the cross duct (D) – 0.08 m;

– cross duct length (H) – 0.15 m;

– distance between drainage orifices along the side surface of the cross duct (a_{lat}) – 0.030 m;

– relative distance from the outlet orifice plain to the inlet cross duct plain $\left(\bar{h} = h/d_0\right)$ – 12, 16, 20, 28;

– absolute air pressure at the nozzle inlet $P_{noz0} = 1.1$; 1.5; 1.7; 1.9; 2.1 MPa;

– an angle between the axis of nozzle and axis of the cross duct $\phi = 30^o$, 45°;

– mach number at the nozzle outlet (M_0) – 1.0.

The geometrical parameters of the cross duct and the nozzle are adopted in accordance with geometrical similarity to the technological and processing parameters of plasmatron and diameter of boreholes before the beginning of the thermal reaming process.

In more detail the methodology of experimental research and the scheme of the experimental set-up are described in [22].

An imitation of the plasma jet by means of air is absolutely acceptable, because an ionization level of the low temperature plasma used as the heat transfer medium in the processes of the rock thermal destruction does not exceed 1%.

3 Results and discussion

In Fig. 3 dependence of gauge pressure on the side surface of the cross duct, i.e. on the rock surface on absolute air pressure at the nozzle inlet and relative distance from the outlet orifice plain to the inlet cross duct plain for different values of an angle between the axis of nozzle and axis of the cross duct is shown.

Comparison of pressure values of the air stream on the inner surface of the cross duct a allows to make a conclusion that for other equal conditions pressure on the inner surface of the cross duct increases with the augmentation of pressure at the nozzle inlet that testifies an adequacy of measurement results.

Fig. 3 shows that increasing of pressure at the nozzle inlet causes an augmentation of pressure, and, accordingly, velocity of the heat-transfer medium on the rock surface.

In a point 1 air stream pressure on the surface of the cross duct reaches highest value since point 1 is the nearest one to the intersection of the stream axis with the surface of the cross duct.

Increasing of the distance from the outlet orifice plain to the inlet cross duct plain pressure on the inner surface of the cross duct diminishes and, accordingly, diminishes air velocity along the side surface of the cross duct.

Pressure distribution along the surface of the cross duct subjects to the nonlinear relationship.

Thus, the results of the experimental investigation allow to make a conclusion that the application of high-speed plasma jets and, actually, plasmatron as the thermal tool for rock destruction is a prospective direction of the scientific research.

https://doi.org/10.1051/e3sconf/20186000024

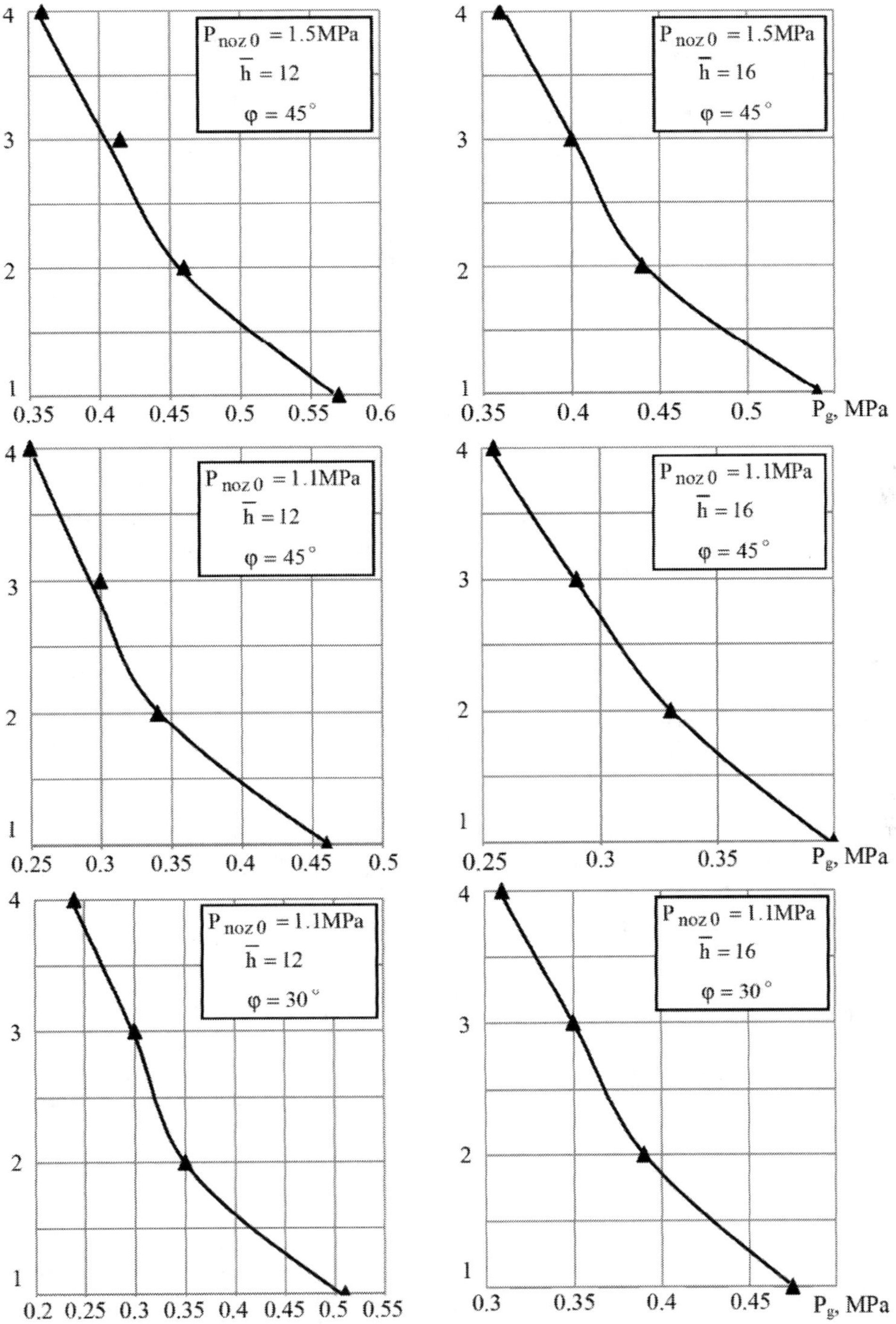

Fig. 3. Dependence of gauge pressure (P_g) of the heat-transfer medium on the side surface of the cross duct on absolute air pressure at the nozzle inlet and relative distance from the outlet orifice plain to the inlet cross duct plain.

4 Conclusions

The analysis of publications that proves the presence of small amount of publications devoted to the experimental research of gas dynamics and plasma dynamics of the jets that are in application as a heat transfer medium in the thermal methods of rock destruction.

The motivation of the fulfilment of own experimental researches of interaction of high-speed heat transfer medium jets with the inner surface of the borehole is made.

The experimental research of interaction of high-speed heat transfer medium jets with the inner surface of the borehole is made as a lateral inner surface of the cross duct imitated the rock surface in a borehole.

The further prospects of this work are next: the determination of gas velocity along the lateral surface of the cross duct and the value of the heat transfer coefficient from the heat transfer medium to the lateral surface of the cross duct that imitates the rock surface in the borehole.

Authors wish to gratitude to V.P. Teplyi, the Chief Project Designer in Department of Vibropneumatic Transport Systems and Complexes and to S.V. Shniakin, Principal Engineer in the Department of Vibropneumatic Transport Systems and Complexes of Institute of Geotechnical Mechanics named by M.S. Poliakov of National Academy of Sciences of Ukraine for their valuable support of the experimental investigation. Authors also wish to gratitude to L.T. Kholiavchenko, Ph.D, Senior Researcher in Department of Rock Breaking Problems Institute of Geotechnical Mechanics named by M.S. Poliakov for his valuable advices and suggestions relating to the experimental investigation.

References

1. Dmitriev, A.P., Goncharov, S.A., & Zilbershmidt, M.G. (2011). Contemporary problems of selective and energy saving rock destruction. *Gornyy informatsionno-analiticheskiy byulleten*, (1), 169-184.
2. Bulat A., Voloshyn, O., & Zhevzhik, O. (2013). Plasma reactor for thermochemical preparation of coal-air mixture before its burning in the furnaces. *Annual Scientific-Technical Collection – Mining of Mineral Deposits*, 39-44. https://doi.org/10.1201/b16354-9
3. Hoser, D., & von Rohr, P.R. (2018). Experimental heat transfer study of confined flame jet impinging on a flat surface. *Experimental Thermal and Fluid Science*, (91), 166-174. http://dx.doi.org/10.1016/j.expthermflusci.2017.10.014
4. Kant, M., Rossi, E., Madonna, C., Höser, D., & von Rohr, P.R. (2017). A theory on thermal spalling of rocks with a focus on thermal spallation drilling. *Journal of Geophysical Research: Solid Earth*, *122*(3), 1805-1815. http://dx.doi.org/10.1002/2016JB013800
5. Kant, M.A., Rossi, E., Duss, J., Amann, F., Saar, M.O., & von Rohr, P.R. (2018). Demonstration of thermal borehole enlargement to facilitate controlled reservoir engineering for deep geothermal, oil or gas systems. *Applied Energy*, (212), 1501-1509. http://dx.doi.org/10.1016/j.apenergy.2018.01.009
6. Rauenzahn, R., & Tester, J. (1989). Rock failure mechanisms of flame-jet thermal spallation drilling theory and experimental testing. *International Journal of Rock Mechanics and Mining Sciences & Geomechanics Abstracts*, *26*(5), 381-399. http://dx.doi.org/10.1016/0148-9062(89)90935-2
7. Rauenzahn, R.M., & Tester, J.W. (1991). Numerical simulation and field testing of flame-jet thermal spallation drilling – 2. Experimental verification. *International Journal of Heat and Mass Transfer*, *34*(3), 809-818. http://dx.doi.org/10.1016/0017-9310(91)90127-Z
8. Wilkinson, M.A., & Tester, J.W. (1993). Experimental measurement of surface temperatures during flame-jet induced thermal spallation. *Rock Mechanics and Rock Engineering*, *26*(1), 29-62. http://dx.doi.org/10.1007/BF01019868

9. Yan, C., Deng, J., Yu, B., Li, W., Chen, Z., Hu, L., & Li, Y. (2014). Borehole stability in high-temperature formations. *Rock Mechanics and Rock Engineering, 47*(6), 2199-2209. http://dx.doi.org/10.1007/s00603-013-0496-2
10. Yanchenko, G.A., Zharovkin, A.V., & Bulycheva, E.S. (2003). About the parameters of the thermal reaming of blastholes in the ferrous quartzites of Kostomuksha GOK at the use of emulsive explosives of "Sibirit-1200" type. *Gornyy informatsionno-analiticheskiy byulleten'*, (9), 1-5.
11. Babayan, E.V., & Chernenko, A.V. (2016). *Inzhenerne raschety pri burenii*. Moskva: Infra-Inzheneriya.
12. Osipov, P.F., & Skryabin, G.F. (2001). *Optimizatsiya rezhimov bureniya gidromonitornymi sharoshechnymi dolotami*. Yaroslavl': Medium-Press.
13. Gulin, V.V., & Ustimenko, T.A. (2014). Designing of generators of impulse jets on the basis of structural synthesis. *Vostochno-Yevropeyskiy zhurnal peredovykh tekhnologiy, 4*(7), 38-45.
14. Plugin, A.I. (2001). *Sposob obrazovaniya skvazhin i vyrabotok v geologicheskikh strukturakh*. Patent No. 2168598, Russian Federation.
15. Bulat, A.F., Makeiev, S.Yu., Osinnii, V.Ya., & Yemelianenko, V.I. (2014). *Ustanovka stvorennia rezervuarnoi porozhnyny v mitsnykh skalnykh porodakh*. Patent No. 94670, Ukraine.
16. Gizatullin, S.A., & Dautov, G.Yu. (1997). *Ustanovka elektrodugovogo plazmobura*. Patent No. 2100602, Russian Federation.
17. Lazhenitsyn, A.I., & Smirnov, A.A. (2012). *Plazmotron uglovoy*. Patent No. 2464746, Russian Federation.
18. Mulyev, Yu.V. (2003). *Manometry*. Moskva: MEI.
19. Andreev, V.V., Chekhlova, T.K., & Chuprov, D.V. (2008) *Izmereniya i pribory v fizicheskom eksperimente*. Moskva: RUDN.
20. Tarasova, V.N., Malynovskyi, A.S., & Rybak, M.F. (2006). *Metrolohiia, standartyzatsiia i sertyfikatsiia*. Kyiv: Tsentr navchalnoi literatury.
21. Bogoslovskiy, S.V. (2001). *Fizicheskie svoystva gazov i zhidkostey*. Sankt-Petersburg: SPbGUAP.
22. Voloshyn, O.I., Potapchuk, I.Yu., & Zhevzhyk, O.V. (2018). Influence of the heat-transfer stream pressure on the surface of the rock in a process of the thermal reaming of the borehole. *Naukovyi Visnyk Natsionalnoho Hirnychoho Universytetu, 164*(2), 53-59. http://dx.doi.org/10.29202/nvngu/2018-2/6

Specifics of hydro-loosening of coal seams with account of rocks displacement parameters

Vasyl Zberovskyi[1*], *Olena* Bubnova[1], and *Kateryna* Babii[1]

[1]Institute of Geotechnical Mechanics under the National Academy of Science of Ukraine, 2a Simferopolska St., 49005 Dnipro, Ukraine

Abstract. The analysis of efficiency decline of methods for hydraulic influence on outburst-prone coal seams in the conditions of large depths is represented. The models are considered for parameters calculation of high-pressure liquid injection, based on the hypotheses of the stress-strain state formation of the coal seam; the patterns of rock pressure manifestation as well as calculation discrepancy between the liquid injection pressure and the terms of large depths, are established. It is shown that in the parameters calculation of the high-pressure liquid injection, it is necessary to take the height of the column of rocks which lost their integrity and form the stratification zone in the underworked massif instead of depth *H* of the seam mining. It has been established that the efficiency of hydro-loosening depends not only on the injection pressure, but also on the location of borehole filtration part relative to the maximum of abutment zone.

1 Introduction

The most widespread and simple in technological performance, the method of gas dynamic phenomena (GDP) prevention is hydro-loosening of the coal seam selvedge in the mode of static liquid injection in the Donetsk Basin mines. The method is based on high-pressure injection and on the research results conducted in Makiivskyi naukovo-doslidnyi instytut (MakNDI) on the safety of works in the mining industry more than 60 years ago, when the mining operations were carried out at depths from 400 to 700 m.

With the deepening of mining operations, the technological schemes of the methods were revised, and the injection parameters remained unchanged [1 – 4]. In a normative technique for parameters calculation of hydro-loosening [3], the injection pressure of a liquid is determined by the formula:

$$P_f = (0.75 - 1.0)\gamma H, \tag{1}$$

where γ is the specific gravity of rocks; H is the depth of mining of the coal seam.

By the empirical coefficient (0.75 – 1.0), it has been determined the injection pressure range in the static mode, in which the unloading zone and the pack-off depth are determined experimentally. The parameter γH characterizes the rock pressure, which is created by the column of rocks, whose height is equal to the depth H of mining, according

* Corresponding author: avalansh@ua.fm

to hypothesis of O.M. Dynnyk.

Nowadays, there are quite enough theories and experimental data confirming the change in the molecular structure of coal under the influence of technogenic factors arising in the geological stratum. The shearing strains lead to structural coal changes with the formation of supermelonits and to generation of methane by the coal substance [5 – 7]. However, these processes are not considered in methods for calculation of a selvedge condition of an outburst-prone seam.

Therefore, in the conditions of large depths, despite the application of measures, the gas dynamic phenomena are still occurring with catastrophic consequences [2]. On the one hand, accidents arise because of a human factor, and on the other one, because of the insufficient study of the processes arising in a selvedge of the mined outburst-prone seam.

2 The results of calculating the pressure of the liquid injected into the coal seam

Let us consider the research results of mining-and-experimental operations of hydro-pulse impact on the example of hydro-loosening of i_3^1 seam of horizon 915 m in the conditions of JV "Sukhodolske-Skhidne", at depth of the seam mining $H \geq 1100$ m, $\gamma = 2.65$ t/cu. m.

According to (1), the range of liquid injection pressure should be equal to 22 – 29 MPa, and pack-off depth not less than the size of unloading zone. Whereas, the unloading zone and the safe depth of coal mining by the shearer depend on the type of support setting in mine working and the rate of its drivage. Thus, it has been established that at setting of the frame support and the rate of mine working drivage from 2 to 4 m per day, the unloading zone is equal from 5 to 3.5 m, respectively. When frame-and-roof bolting and roof-bolting, the size of the unloading zone is stabilized and is equal to 3.5 – 4.0 m. The given results of research show that the pack-off depth accepted based on experimental injections $l_p = 4.0$ m at a borehole length of $l = 6.0$ m, does not always correspond to the mining conditions.

The long-term MakNDI observations have established that in the case of hydro-loosening, the break of seam tightness or sloughing of one of its splits occurs at an injection pressure in the range of 20 – 30 MPa. The similar results have been obtained also in our example. Thus, during pack-off at a depth of 4 m and injection pressure of more than 25 MPa, the breakthrough of the liquid into mine working face happened in 4 – 5 min after injection.

The above mentioned allows to draw a conclusion that the normative technique for calculation of hydro-loosening parameters [3] does not correspond to conditions of large depths. It should also be noted that when calculating the injection pressure (1), an important pack-off depth parameter is not considered.

Therefore, the objective of the work is to analyze the research results of calculated and set in practice values of the static and pulse modes parameters of liquid injection with account of patterns of rocks displacement in the conditions of large depths.

It is well known that filtration properties of coal in selvedge of a seam are considerably influenced by the rock pressure value and stress state of the coal massif. Between water permeability of coal and rock pressure there is a distinct feedback – with increase in pressure the water permeability of a seam decreases and vice versa, as the rock pressure decreases, the water permeability increases.

The results analysis of practical application of GDP prevention methods based on high-pressure liquid injection in the Donbas mines makes it possible to note the following.

The decrease in coal permeability in the abutment zone is effectively used in the hydraulic sloughing of the selvedge of a seam. The technological parameters of the blast-hole [3] allow to locate its filtration part in the zone of abutment pressure maximum, and to regulate the additional loading force on the selvedge of a seam by liquid injecting under pressure of $P_f \geq 0.75\ \gamma H$. However, this process is connected with high probability of GDP initiation and is carried out remotely without the presence of people in mine working throughout all its length. If it does not cause difficulties in a stoping face, but in the preparatory mine working it leads to a decrease in the rate of its drivage by more than 50%.

The method of hydro-loosening also has its own features. So the location of the filtration part of the borehole, the length of which is 2.0 m relative to the abutment zone maximum, is of great technological importance. If, even its insignificant part is located in the unloaded zone, then in the process of injection there happens the breakthrough of liquid into the mine working face. If the filtration part of the borehole is located in the zone with an increased rock pressure or behind it, then because of low water permeability of the coal and, as a rule, the existence of carbonaceous inter-layers, natural pockets and cracks in the seam, the liquid breakthrough occurs in the unknown direction.

As we can see, the efficiency of static liquid injection depends on the location of the filtration part of the borehole relative to the boundary of the abutment zone. Nowadays, this problem is solved by control of unloading zone with a probable error of 0.5 m. It corresponds to the interval of drilled check blast-holes in determining the pack-off depth. The second task – to control the liquid filtration process in the coal seam at static injection – has not been solved yet due to the lack of control methods.

Thus, the application of methods with the mode of static liquid injection in conditions of large depths not only does not ensure the safety of works, but also becomes a constraint factor.

At the same time, the mining-and-experimental works on the hydro-pulse effect on the outburst-prone coal seams have shown that the application of pulse injection modes allows to solve this problem. However, the parameters of this method with account of the stress-strain state formation of the selvedge of a seam have not been sufficiently studied yet.

It is known that during mining operations in underworked massif there is a collapse of rocks with a constant main roof spacing. In the conditions of large depths, this leads to the formation of zones of compression, tension and stratifications (Fig. 1) [8 – 11].

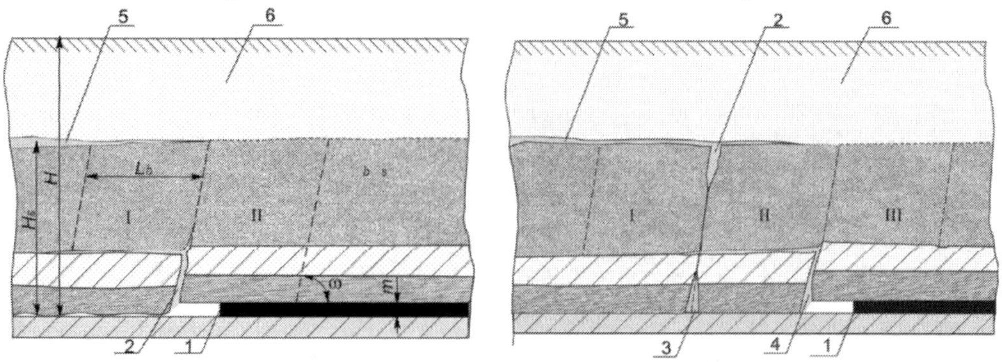

Fig. 1. The scheme of tension, compression and stratifications zones formation during rocks breakage at large depths: a – during collapse of the main roof block I; b – formation of stratification and compression zones [9]: 1 – stoping face; 2 – tension zone; 3 – compression zone; 4 – tension zones between blocks; 5 – stratifications zone; 6 – deformation zone of rocks without their discontinuity.

In the geological strata of the coal seams suite, the displacement of rocks initiates the technogenic transformations that exert an impact on the whole massif of the mine field. The structural transformations of coal with phase transitions of coal-gas occur in the outburst-prone seams. In tension zones, there is a possibility of free methane filtration that leads to degassing of the massif and formation of the unloaded safe zone. The compression zones on the contrary become the high rock pressure zones (HRPZ) in which the transformed outburst-prone coal seams acquire the stress-strain state with a high probability of self-destruction.

The generally accepted hypothesis of the stress-strain state formation of a coal seam is the model of Dynnyk O.M.:

$$\sigma_y = \gamma H. \tag{2}$$

It is accepted in the model that the pressure is created by the rocks massif, whose column height is equal to the depth H of mining. The research of authors [8, 9] substantiated that in the expression (2) the conditions of large depths are not considered when the stratification zone is formed in the underworked massif. At a certain point of time, namely in the period of formation of tension and compression zones in the underworked massif, rock pressure has a value equal to:

$$\sigma_m = \gamma H_x, \tag{3}$$

where H_x is the height of the column of rocks, which at the moment have lost the integrity with the massif, m.

The height of the column and the rock pressure value can be determined by calculating the rate of deformations development in the massif according to the results of mine-surveying observations of displacement process development. In the work [9] it is shown that in the conditions of large depths, when calculating the rock pressure value, only that part of rocks should be considered which has been separated from the underworked massif because of displacement. Based on the long-term research and experience of surveying observation, the height of this column at depths more than 1000 m, depending on the rocks lithification degree, ranges from 600 to 800 m.

With account of the defined features of rock pressure manifestation, we will consider how the parameters of liquid injection into coal seam will change. For this purpose, we will use the model for calculating the liquid injection pressure under static and pulse injection modes (Fig. 2) [12, 13].

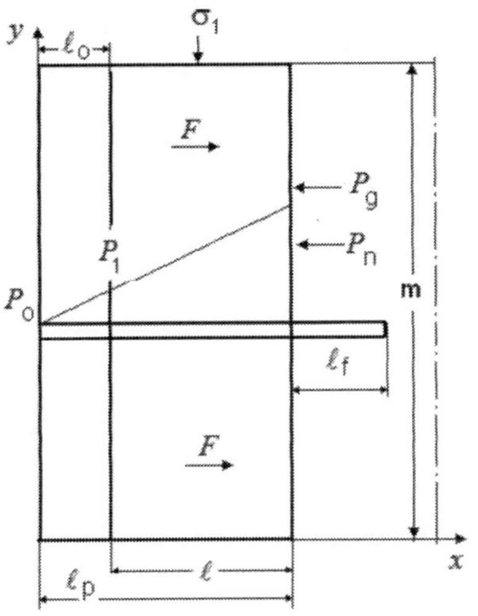

σ_1 – rock pressure forces, MPa;
$\sigma_1 = \gamma H$; $\gamma \approx 2.65$ t/cu.m;
P_n – injection pressure, MPa;
P_g – gas pressure in seam, MPa;
$P_g = 1.0 - 2.0$ MPa;
F – friction forces, MPa; $F = f\sigma_1 = f\gamma H$;
f – friction coefficient of rocks, in massif $f = 0.5$;
P_0 – liquid pressure on the surface of mine working face, $P_0 = 0$;
P_1 – liquid pressure on a seam edge, MPa;
m – seam thickness, m;
l_p – pack-off depth, m;
l_0 – destroyed seam edge, m $l_0 = 0.1$ m;
l_f – depth of filtration chamber, m;
l – depth of seam resistance zone, m.

Fig. 2. The scheme for calculating the pressure of liquid injection into coal seam with account of hydraulic sloughing and pack-off depth criteria.

In our example, the calculation parameters of hydro-loosening are as follows: depth of occurrence $H = 1100$ m; specific gravity of rocks $\gamma = 2.65$ t/cu.m; seam thickness $m = 1.75$ m; depth of the borehole drilling – 6.0 – 8.0 m; pack-off depth l_g = 2.5 – 5.5 m; friction coefficient of rocks in the massif $f = 0.5$; height of the column of rocks, which have lost their integrity with the massif $H_x = 600$ m.

Consider the results of calculating the hydro-loosening parameters of coal seams according to the normative technique [3] in the static mode of liquid injection. As it was noted earlier, the calculated range of liquid injection pressure by the formula (1) is equal to 22 – 29 MPa. At the same time, the change of technological parameters is considered by the empirical coefficient in the range of 0.75 – 1.0.

It has been established that when hydro-loosening through boreholes with 6.0 m length and 4.0 m pack-off depth, with an injection pressure of 23 MPa and more, a breakthrough of liquid and a pressure drop occurred in 10 – 15 minutes. The hydro-loosening without a breakthrough of liquid into mine working occurred at a pressure of no more than 22 MPa, which is less than the range established by the formula (1).

These results have confirmed once again that the problem of hydraulic fracturing and hydraulic sloughing during hydro-loosening, has been neither theoretically, nor practically solved.

At the same time, based on the analysis of the numerous experimental data obtained in the process of hydro-loosening in the conditions of large depths, the calculated pressure of the hydraulic sloughing $P_n \geq 0.75\ \gamma H$ [3] always corresponds to boundary value of the liquid injection pressure. Moreover, the efficiency of hydro-loosening depends on the location of the rock pressure maximum ahead of the mine working face and on the pack-off depth. Then, based on the condition of forces equilibrium on the contacting surfaces "seam – enclosing rocks" (Fig. 2), the phenomenon of hydraulic sloughing at the pack-off depth will occur at $P_1 = 0$ [12, 13]:

$$(P_n - P_1)m = 2f\gamma H\left(l_p - l_0\right). \quad (4)$$

It follows from the condition (4) that if in the process of liquid injection into the coal seam the stress-strain state of the massif and the energy of the gas accumulated in the seam exceed the contact friction forces, in the coal seam there will necessarily occur the hydraulic sloughing and hydraulic fracturing of interlayers composing the coal seam.

Then, according to the model (Fig. 2), the condition of hydraulic sloughing at the accepted depth of pack-off will be written as follows:

$$P_n \geq \frac{2 \cdot f \cdot \gamma \cdot H}{m} \cdot l_p = \frac{2 \cdot 0.5 \cdot 2.65 \cdot 1100}{1.75} \cdot l_p = 16.65 \cdot l_p, \quad (5)$$

and with account of condition of rocks stratification, at height H_X (3):

$$P_n \geq \frac{2 \cdot f \cdot \gamma \cdot H_X}{m} \cdot l_p = \frac{2 \cdot 0.5 \cdot 2.65 \cdot 600}{1.75 \cdot l_p} = 9.08 \cdot l_p. \quad (6)$$

By these formulas, depending on the pack-off depth, the calculation of injection pressure (Table 1) is made.

Consider the received calculation results with respect to the results established in practice.

When hydraulic sloughing the selvedge of a seam, the pack-off depth of the blast-hole should be from 2.0 to 5.5 m, and its filtration part 0.3 m [3]. In practice, the blast-holes from 2.5 to 3.5 m length are commonly used. With these parameters, the self-destruction of the seam is most effectively realized in the abutment zone in terms of the technology of hydraulic sloughing.

From the table, when calculating by the formula (5) the injection pressure from 41.6 to 58.3 MPa, and by the formula (6) from 22.7 to 32 MPa, corresponds to these parameters. According to the long-term MakNDI observations [1, 2], at depths more than 1000 m the hydraulic sloughing occurs at an injection pressure in the range of 20 – 30 MPa. These values coincide with the results calculated by the formula (6).

Table 1. Calculated values of hydraulic sloughing pressure of the selvedge of a seam depending on the pack-off depth and stratification zone of rocks.

According to the formula (5)		According to the formula (6)	
L_g, m	P_n, MPa	L_g, m	P_n, MPa
2.5	41.6	2.5	22.7
3.0	49.9	3.0	27.2
3.5	58.3	3.5	31.8
4.0	66.6	4.0	39.2
4.5	66.4	4.5	40.9
5.0	83.2	5.0	45.4
5.5	91.6	5.5	53.9

Now let us compare the calculated data with the results of hydro-loosening of i_3^1 seam of horizon 915 m in JV "Sukhodolske-Skhidne" of PJSC "Krasnodonvuhillia" [14]. With a pack-off depth of the technological borehole of 4.0 m, according to the unloading zone size from 3.5 to 4 m, its filtration part of 2.0 m was located behind the abutment zone maximum. The breakthroughs of liquid into the mine workings face were fixed at an injection pressure of more than 23 MPa. At an injection pressure of more than 25 MPa, the hydraulic fracturing of the seam and drop in injection pressure occurred in 4 – 5 minutes, and the signs of hydraulic sloughing in the mine working face were not observed.

In accordance with the set values at these parameters of technological boreholes (Table 1), according to the normative technique (5) of calculation, the hydraulic fracturing (hydraulic sloughing) can occur at an injection pressure of more than 66 MPa, and when calculating by formula (6) at a pressure of 39 MPa. As we can see, the results of calculation of liquid injection pressure by the formula (5) almost three times exceed the values set in practice. The results obtained by the formula (6) most closely correspond to them.

The importance of conducted research is that by comparing the calculation results and experimental data, the discrepancy of calculation model of liquid injection pressure is determined, based on the hypothesis of stress-strain state formation of coal seam under the pressure, which is created by the height of rocks column equal to the depth H of mining.

3 Conclusions

Comparison of the calculation results of the pressure of liquid injected into the coal seams with the results established during the work has shown that in the conditions of large depths the generally accepted hypothesis of the stress-strain state formation of a coal seam under pressure which is created by the height of a column equal to the depth H of mining, does not correspond to practical results.

It has been shown that when calculating the parameters of high-pressure liquid injection, it is necessary to take the height of the column of rocks which lost their integrity and form the stratification zone in the underworked massif instead of depth H of the seam mining.

For the first time in the practice of hydro-loosening of coal seams, the calculations have been made for determination of liquid injection pressure depending on pack-off depth with account of rocks displacement features at large depths.

The authors of the article express their gratitude to the engineers of PJSC "Krasnodonvuhillia" mines for the assistance in conducting mining and experimental research.

References

1. Timofeev, E.I., Mhatvari, T.Ya., Churadze, M.V., & Ryzhkov, M.F. (2003). Metodika vybora optimal'nykh parametrov gidrorykhleniya vybrosoopasnykh ugol'nykh plastov. *Sbornik nauchnykh trudov Makeyevskogo Nauchno-Issledovatel'skogo Instituta*, (1), 81-90.
2. Boyko, Ya.N., Nikiforov, A.V., & Rubinskiy, A.A. (2009). Povyshenie effektivnosti gidrorykhleniya vybrosoopasnykh ugol'nykh plastov v podgotovitelnykh vyrabotkakh i nishakh. *Sbornik nauchnykh trudov Makeevskogo Nauchno-Issledovatel'skogo Instituta*, *2*(24), 52-57.
3. SOU 10.1.001740088.011:2005. (2005). *Pravyla vedennia hirnychykh robit na plastakh, skhylnykh do hazodynamichnykh yavyshch*. Kyiv: Ministerstvo vuhilnoi promyslovosti Ukrainy.
4. Zberovskiy, V.V., & Androshchuk, K.V. (2002). O degazatsii i sdvizhenii geologicheskoy tolshchi pri osvoenii mestorozhdeniy na bol'shikh glubinakh. *Gornyy informatsionno-analiticheskiy biulleten' Moskovskogo Gosudarstvennogo Gornogo Universiteta*, (6), 70-72.
5. Zberovskiy, V.V. (2004). K voprosu transformatsii tverdykh uglevodorodov i sistemy "ugol' – gaz". *Gornyy informatsionno-analiticheskiy biulleten' Moskovskogo Gosudarstvennogo Gornogo Universiteta*, (8), 188-192.
6. Sobolev, V.V., Polyashov, A.S., Zberovskiy, V.V., Angelovskiy, A.A., & Chugunkov, I.F. (2013). *Sistema "ugo'l – gaz" v uglevodorodakh ugol'nogo genezisa*. Dnepropetrovsk: Art-Press.
7. Chetverik, M.S., & Androshchuk, Ye.V. (2004). *Theoriya sdvizheniya massiva gornykh porod i upravleniya deformatsionnymi protsessami pri podzemnoy vyemke uglya*. Dnepropetrovsk: RIA "Dnepr-VAL".
8. Chetveryk, M.S., Bubnova, O.A., Babiy, K.V., & Batur, M.O. (2016) Shvydkist rozvytku deformatsii u pidroblenomu hirskomu masyvi pry vyimanni korysnykh kopalyn na osnovi marksheiderskykh sposterezhen. *Heotekhnichna Mekhanika*, (130), 3-12.
9. Zberovskiy, V.V., Pazinich, A.V., Polyakov, Yu.E., Potapenko, A.A., & Angelovskiy, A.A. (2011) Model' predel'nogo sostoyaniya ugol'nogo plasta pri nagnetanii zhidkosti, *Zbirnyk Naukovykh Prats Natsionalnoho Hirnychoho Universytetu*, (36(1)), 194-199.
10. Zberovskiy, V.V. & Kostandov, Yu.A. (2011). Predel'noe sostoyanie vybrosoopasnykh ugol'nykh plastov pri ikh gidrorykhlenii s uchetom soprotivlyaemosti uglya sdvigu. *Zbirnyk Naukovykh Prats Natsionalnoho Hirnychoho Universytetu*, (36 (2)), 36-43.
11. Zberovskiy, V.V., Pazinich, A.B. & Polyakov, Yu.E. (2010). Otsenka effektivnosti gidroimpul'snogo vozdeystviya po gazovomu faktoru v zaboyakh podgotovitelnykh vyrabotok. *Heotekhnichna Mekhanika*, (89), 126-133.
12. Khomenko, O.Ye., Sudakov, A.K., Malanchuk, Z.R., & Malanchuk, Ye.Z. (2017). Principles of rock pressure energy usage during underground mining of deposits. *Naukovyi Visnyk Natsionalnoho Hirnychoho Universytetu*, *2*(158): 34-43.
13. Kovalevska, I., Barabash, M., & Snihur, V. (2018). Development of a research methodology and analysis of the stress state of a parting under the joint and downward mining of coal seams. *Mining of Mineral Deposits, 12*(1), 76-84. https://doi.org/10.15407/mining12.01.076
14. Kovalevs'ka, I., Vivcharenko, V., & Snigur, V. (2013). Specifics of percarbonic rock mass displacement in longwalls end areas and extraction workings. *Annual Scientific-Technical Collection – Mining of Mineral Deposits*, 29-33. https://doi.org/10.1201/b16354-7

Ash as an alternative source of raw materials

Olena Svietkina [1*], *Hanna* Tarasova[1], *Olha* Netiaha[1], and *Svitlana* Lysytska[1]

[1]National Mining University, Department of Chemistry, 19 Yavornytskoho Ave., 49005 Dnipro, Ukraine

Abstract. The objective of the work is to study the aluminosilicate fractionation from fly ash, physical and mechanical properties of fly ash derived from the Thermal Power Plants (TPP) wastes. Ash, carbon concentrate (unburned carbon), ash concentrate and products of their treatment with reagents were tested by optical methods. The particle morphology of the objects of research was studied with the scanning electron microscope REM-100. The composition of the ash phases was investigated using the X-ray diffractometer DRON-2. A dispersed analysis of the TPP fly ash suggests a conclusion that it is advisable to separate particles of a narrow grain-size class within the range from 40 to 150 μm with an ash content of about 33%. The first product may be enriched by flotation method. Such a coal product may be used as a reducing medium in metallurgical processes, agglomeration, etc. The calorific capacitance of the concentrate is about 6000 kcal/kg (25000 kJ/kg).

1 Introduction

The ash of TPP is formed as a result of coal combustion and, on the one hand, is a source of environmental pollution; but on the other hand, it is a source of raw materials needed for industry.

In Ukraine, thousands of tons of ash, containing a number of useful components, have accumulated in the TPP dump sites. Moreover, more than 10 million tons of fly ash is added annually to the TPP dump sites. Despite the relatively high content of phase constituents with high consumer properties in a number of ashes, it is only possible to attribute them to technogenic minerals by convention, since now there are no available cost effective methods of extracting the useful components. Therefore, ash from thermal power plants is used to a limited extent.

Products of TPP ash separation may be used in various industries: metallurgy, power engineering, chemical industry, in the production of building materials. An important is also the allocation of the ash storage priority for the integrated use of their resources in various sectors of the economy based on a comprehensive comparative analysis of their technological and economic indicators by the integrated assessment method [1].

In metallurgy, attempts are being made to replace the breeze coke with the high-temperature char and anthracite in the process of iron ore concentrate agglomeration. Obviously, a high-carbon flotation concentrate can be used in order to replace partially the

* Corresponding author: svetkina.yelena@gmail.com

breeze coke.

In the production of metallized pellets, the fuel additive (breeze coke) makes up 14% (grain-size <100 microns – 70%, ash content – 18.35%, sulfur content – 0.65%). Thus, without lowering the quality of metallized pellets, the high-temperature char may be replaced by a flotation concentrate of ash [2].

As a component of pulverized fuel oil compositions for blast furnace processes, it is possible to use low-ash flotation concentrate. In addition, a method was developed for metal siphoning into a mold, which uses a concentrate with carbon content of 35 – 60% as a heat-insulating coating. It is important that the use of this material allows abandoning graphite for the same purposes [3].

In the energy sector, the flotation concentrate can be used as a fuel additive to boiler fuel. The use of concentrate can significantly reduce the heat loss from mechanically unburned carbon when burning the coal of rank AS; in addition, the operating parameters of the boiler plant will not deteriorate under the condition of continuous dosing of the concentrate [4].

The method has been developed for sorption of heavy metal cations, which allows capturing them from industrial effluents. In this regard, a concentrate froth of the cleaner flotation is the most effective due to its high carbon content. The cleaner flotation waste with ash content of 40 – 50% may be used as a sorbent of organic acids and various petroleum products. An ash-based sorbent, obtained through changes in the boiler furnace temperatures and coarsening the grind of raw materials, is sufficiently effective for the purification of surfactant-containing sewage [5].

Waste of TPP ash flotation is more widely used in the production of building materials than the initial ash. This can be explained by the adjustment of carbon content and granulometric compositions of the ash mineral fraction.

Flotation waste is successfully used in the manufacture of various concretes and concrete mixtures in order to save cement without deterioration in the strength characteristics of the concrete. Moreover, they can be used to obtain special concretes, for example, foam concrete, light-weight and autoclaved cellular concretes. With the addition of ash mineral fraction, the consumption of cement is significantly reduced [6]. In the process of producing bitumen-concrete mixtures, it is recommended to use the product of the TPP ash flotation processing as a mineral powder; the mixture thus obtained is cheaper than that with a limestone mineral powder. It is not inferior in its technological and strength parameters. In the manufacture of wall products, a modified ash of low-reactive coal added to clay may serve as a raw material. The obtained burned bricks have higher strength characteristics than the material with no modified ash admixtures [7].

The promising direction in the use of ash wastes is their processing into various filling agents for building materials. The low density fillers are made by granulating the ash with the subsequent firing of the obtained pellets. Ash, together with sandy loam and a waste admixture from the lead-dressing and processing enterprise, is part of the mixture for fly-ash aggregate. A method for obtaining artificial aggregate has been proposed, which provides the separation of ash into two classes with subsequent processing of the class containing about 4.5% of unburned coal residuals.

The TPP ash with high degree of unburned carbon must be subjected to flotation concentration in order to obtain a carbonaceous fraction (concentrate) and a mineral fraction (waste). The unburned particles of coal can be separated by direct flotation using nonpolar and heteropolar surface-activated substances (surfactants) at a low pulp pH (pH 6 – 8 and pH 3 – 6.5 at the first and the second stage, respectively). Several recleanings are used to obtain products of the specified quality. A method has been developed for ash flotation with the use of MK reagent as a collector in the presence of a depressor. The method ensures obtaining conditioned products without recleanings, with 95 – 98% carbon yield to the concentrate. The preliminary hydraulic classification of ash into 40 μm class enhances the efficiency of its separation.

The effect of the raw material grinding on the process of ash flotation was studied. Due to the increased contrast range of the phases being separated, a concentrate with a high content of the combustible fraction has been obtained, with a slight change in the ash content of the waste. The inverse flotation of the TPP ash allows obtaining the products with physical-chemical properties that differ from the properties of the concentrates obtained by direct ash flotation. Hence, the carbon-bearing ash fraction has a high specific surface area, which allows it to be used as an effective sorbent, in contrast to a direct flotation concentrate which has a slightly less effective specific surface area of the particles due to collector adsorption on their surface. The inverse flotation waste manifests clearly expressed hydrophobic properties due to the collector adsorption on the surface of the particles, which allows using it as a hydrophobic filler of polymer composite materials with no additional surface modification [8].

To obtain the raw material for manufacturing heat-insulating materials, fly ash was subjected to the following operations: a preliminary flotation was performed to separate the inorganic and organic parts, followed by activation of the inorganic part by two methods, i.e. chemical and mechanochemical activation. The consumer properties of the raw materials were assessed by studying their phase composition, physical-chemical and physical-mechanical properties, and their compliance with the technical specifications for the further redistribution processes of the products obtained. The objective of the work is aluminosilicate fractionation from fly ash, study of physical and mechanical properties of fly ash derived from the Thermal Power Plants (TPP) wastes.

2 Methods

Ash, carbon concentrate (unburned carbon), ash concentrate and products of their treatment with reagents were tested by optical methods. The particle morphology of the objects of research was studied with the scanning electron microscope REM-100. The composition of the ash phases was investigated using the X-ray diffractometer DRON-2.

Mechanochemical activation of the materials was performed in a vertical vibration mill which main advantage is in a shock-vibrating impact upon the material to be destructed [9]. As a rule, mechanical activation happens at the moment of chemical processing in the point of contact between solid substance particles, where chemical processes occur due to the local heat emission, and simultaneous formation of rather high pressures and shear stresses within that area of a crystal.

Shock-vibrating activation of the material makes it possible not only to perform surface shear but also to effect the particles with the penetration of deformation zone into its whole volume. That ensures not only surface activation of the material but also the damage of internal structures of the particles and deeper penetration of active zone; it opens up new opportunities of solid state synthesis – development of the materials having maximum selective conductivity, mechanical strength, chemical and thermal stability

3 Results and discussion

Coal ash is formed from all mineral impurities in the coal bands, rock seams of the roof and bottom, which are entrapped in the marketable coal during its mining [10]. The major impurities are clay minerals and quartz, while carbonates, sulfides, iron oxides and other minerals occur in minor amounts. The phase composition of the ash is mainly determined by clay minerals, which are transformed into other aluminosilicates (glass phases, possibly mullite, cristobalite) during high-temperature processing in boiler furnaces. It should be taken into account that this is a large group of minerals with compositions varying in a wide range [11, 12].

The chemical composition of fly ash from DTEK Prydniprovska TES is presented in Table 1. Table 2 shows the chemical composition of ash processing products.

Table 1. Chemical composition of fly ash.

Product	Content, % of dry matter										
	SiO_2	Fe_2O_3	CaO	MgO	SO_3	Al_2O_3	TiO_2	P_2O_5	K_2O	Na_2O	C
Fly ash	49.66	12.32	2.18	1.27	0.44	12.87	0.53	0.90	3.10	0.67	12.00

Table 2. Chemical composition of ash enrichment products.

Component	SiO_2	CaO	MgO	Al_2O_3	Fe_2O_3	ZnO	MnO	TiO_2	Na_2O	K_2O	V	S_{total}	C
Carbon concentrate	16.7	0.91	0.53	6.63	3.20	0.01	0.04	0.29	0.23	1.25	0.01	0.40	69.8
Ash cake	55.8	2.7	1.8	22.1	10.7	0.03	0.08	0.96	0.75	4.17	0.02	0.06	0.83

Electron microscopic studies and X-ray diffraction analysis of the morphology of the ash phases and products of its enrichment have revealed the features of their surface microstructures [13].

Fig. 1 and 2 shows the general view of ash particles and ash concentrate of the X-product sample. The material consists of isolated aluminosilicate spheres, carbon particles and their intergrowths.

Fig. 1. General view of TPP ash particles (initial), magnification 517x: 1 – spheres; 2 – coal particles.

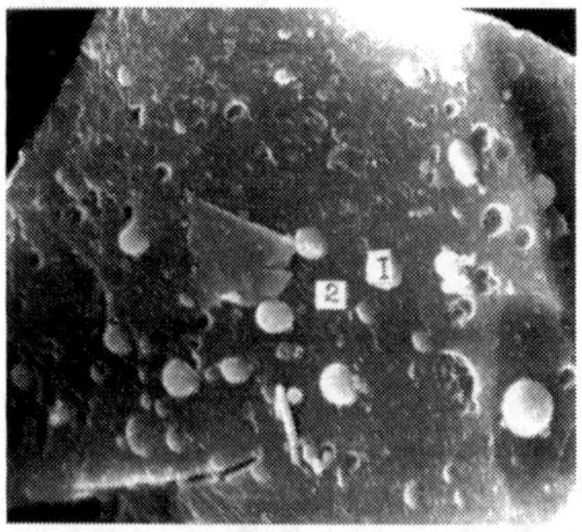

Fig. 2. General view of TPP ash concentrate, magnification 929x: 1 – spheres; 2 – coal particles.

Colonies of spheres located more deeply (in microcracks and microcavities) are not separated by this chemical activation, their separation is possible through mechanical action. After the treatment of the TPP ash concentrate with a 40% solution of H_2SO_4, the sample material predominantly consists of coal particles (Fig. 3). Treating the samples with acids results in a partial separation of the coal particles and shallowly embedded spheres from the concentrate surface, so that its surface acquires a cellular microstructure. After the treatment of the product with a 40% H_2SO_4, the coal liberation occurred and the surface of the carbon particles has been cleared well of aluminosilicates (Fig. 3 and 4).

Fig. 3. General view of TPP ash particles (after treatment with 40% H_2SO_4), magnification 1305x: 2 – coal particles.

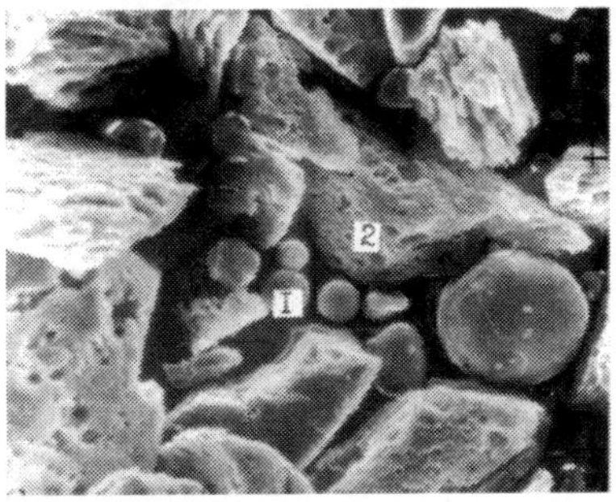

Fig. 4. Ash concentrate after treatment with 40% H_2SO_4, magnification 842x: 1 – spheres; 2 – coal particles.

The treatment of the X-product with a mixture of acids (HCl + H) (1:3) almost completely cleaned up the material from the aluminosilicate component. There are single spheres. The surface of the coal particles is microporous, microcellular (Fig. 5 and 6).

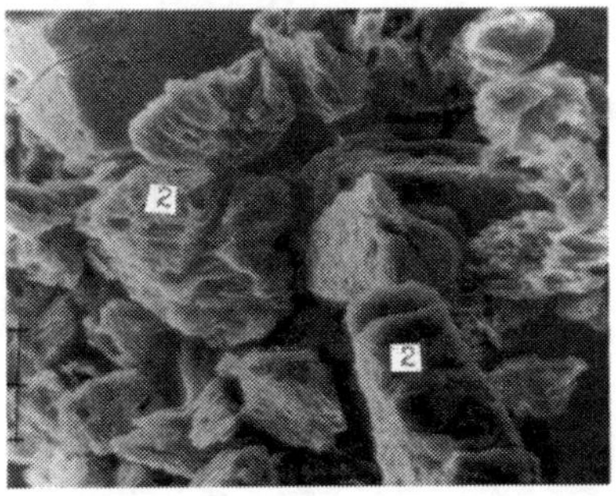

Fig. 5. General view of ash particles of TPP (after treatment with HCl + H), magnification 1305x: 2 – coal particles.

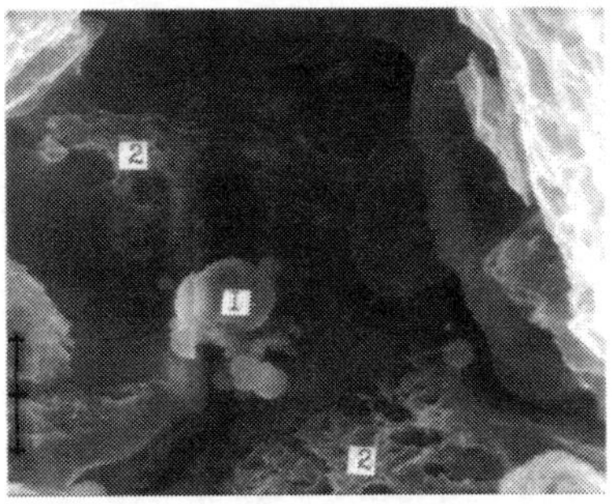

Fig. 6. General view of the carbon concentrate of TPP ashes, magnification 20x: 1 – spheres; 2 – coal particles.

The chemical composition of the mineral matter of the ash enrichment products practically coincides with the chemical composition of the waste, with the exception of the carbon content. An ash processing technique and a plant operating in conjugation with TPP will allow obtaining products of specified quality irrespective of the conditions of coal combustion at the TPP. This, on the one hand, excludes the building of new ash dumps at the TPP, and, on the other hand, allows the production of sorbents of oil products and heavy metals, heat-insulating coatings for steel casting, low-sulfur reductants, fuels, and raw materials for building materials [14].

The carbonaceous concentrate is also an effective sorbent of nonpolar and surface-active substances (surfactants), including petroleum products.

The resulting ash cake (mineral faction of ash) may have an ash content of 90 – 95%. The decarburized aluminosilicates can find a variety of applications in the production of building materials, composites, in metallurgy and other industries.

The studies have confirmed the effectiveness of the use of enriched ash in the preparation of clay-ash claydite and brick, high-quality ash gravel, contact-condensate mixtures, as well as in the construction of road bases and coverings.

As a result, the coal concentrate may contain less than 20% of ash with a sulfur content of about 0.4%. The sulfur content exceeds the European requirements. Such a coal product may be used as a reducing medium in metallurgical processes, agglomeration, etc. The calorific capacitance of the concentrate is about 6000 kcal/kg (25000 kJ/kg).

4 Conclusions

It has been discovered that in the case of high-temperature coal processing (within the range of 800 – 1300 °C), the sorbents of multipurpose use can be obtained.

It has been established that the most effective low-cost sorbents may be obtained under the TPP conditions with the controlled combustion of coal of certain grades.

Concrete obtained on the basis of alumo-silicate spheres separated from fly ash can be used for the production of light-weight concrete, external wall structures, small building blocks, and also in monolithic housing construction.

The use of preliminary shock-vibrating activation allows increasing the reactivity of aluminosilicates due to the formation of new effective bonds, that improves the physical-mechanical properties of binding materials based on them.

Authors express their gratitude to the private enterprise "Ukrpromtsert" for the help in providing experimental equipment during the work performance.

References

1. Petlovanyi, M.V., & Medianyk, V.Y. (2018). Assessment of coal mine waste dumps development priority. *Naukovyi Visnyk Natsionalnoho Hirnychoho Universytetu*, (4), 28-35. https://doi.org/10.29202/nvngu/2018-4/3
2. Stavitskaya, S., Strelko, V., Nikipelova, Ye., Vikarchuk, V., Tsyba, N., Petrenko, T., & Alekseenko N. (2009). Sintez i issledovanie novykh kombinirovannykh materialov na osnove modifitsirovannogo uglya i prirodnykh oksidov. *Ukrainskiy himicheskiy zhurnal*, *75*(11), 29-34.
3. Dubinin, M. (1981). Adsorbtsiya parov vody i mikroporistaya struktura uglerodnykh adsorbentov. *Izvestiya Akademii Nauk SSSR*, (1), 9-21.
4. Belyaeva, O., Krasnova, T., & Golubeva, N. (2017). Vliyanie usloviy predvaritel'nogo progreva aktivnykh ugley na adsorbtsiyu organicheskikh soedineniy iz vodnykh rastvorov. *Sorbtsionnye i khromatograficheskie protsessy*, *17*(6), 947-953.
5. Svetkina, O., Netiaga, O., & Tarasova, H. (2015). Sorbents of purify mine waters. *Theoretical and Practical Solutions of Mineral Resources Mining*, 111-117. http://dx.doi.org/10.1201/b19901-77/
6. Adeeva, L., & Borbat, V. (2009). Zola TETs – perspektivnoe syr'ye dlya promyishlennosti. *Vestnik Omskogo Universiteta*, (2), 141-151.
7. Ovchinnikov, R., & Avakyan, A. (2013). Vliyanie razlichnykh metodov aktivatsii na svoystva kislykh zol TES kak aktivnoy dobavki v betony. *Izvestiya vuzov*, (6), 100-103.
8. Yeroshkina, N., Korovkin, M., & Korovchenko, I. (2015). Ispolzovanie zoly TES v tekhnologii geopolimernykh stroitelnykh materialov. *Molodoy uchenyy*, (7), 117-120. https://moluch.ru/archive/87/17049/
9. Franchuk, V. (1975). Vibratsionnaya tekhnika v malykh proizvodstvakh. *Heotekhnichna Mekhanika*, (85), 290-296.
10. Petlovanyi, M.V., Lozynskyi, V.H., Saik, P.B., & Sai, K.S. (2018). Modern experience of low-coal seams underground mining in Ukraine. *International Journal of Mining Science and*

Technology. Article in press. https://doi.org/10.1016/j.ijmst.2018.05.014

11. Svetkina, E. (2013) Intensifikatsiya protsessa obogascheniya pri vibroudarnoy aktivatsii mineralov. *Naukovyi Visnyk Natsionalnoho Hirnychoho Universytetu*, (2), 38-43.
12. Ishkov, V., & Koziy, E. (2017). Distribution of toxic and potentially toxic elements in the coal of the layer c7h of the "Pavlogradskaya" mine of Pavlogradsko-Petropavlovskiy geological and industrial district. *Visnyk of Taras Shevchenko National University of Kyiv*, (79), 59-66. https://doi.org/10.17721/1728-2713.79.09
13. Svetkina, O., Netiaga, O., & Tarasova, H. (2016). Physical and mechanical characteristics of binders based on TPP waste. *Visnyk Natsionalnoho Tekhnichnoho Universytetu "Kharivskyi politekhnichnyi instytut"*, *22*(1194), 177-182.
14. Rimkevich, V., Pushkin, A., & Churushova, O. (2015). Complex processing of coal ash of TEC. *Gornyy informatsionno-analiticheskiy byulleten'*, (6), 250-259.

The study of melting process of the new plugging material at thermomechanical isolation technology of permeable horizons of mine opening

Andrii Sudakov[1*], *Andrii* Dreus[2], *Diana* Sudakova[1], and *Oleksandr* Khamininch[2]

[1]National Mining University, Department of Technic Prospecting of Deposits, 19 Yavornytskoho Ave., 49005 Dnipro, Ukraine
[2]Oles Honchar Dnipro National University, Department of Fluid Mechanics and Energy and Mass Transfer, 72 Haharina Ave., 49010 Dnipro, Ukraine

Abstract. The article presents the results of experimental and theoretical studies, the purpose of which was to substantiate the technology of drilling wells isolation using new thermoplastic composite material. The basis of the proposed material is gravel, and secondary polyethylene terephthalate acts as a binding material. The use of the proposed insulation material avoids a number of disadvantages specific for traditional grouting mortars. The technology of material application provides its melting in a well by thermomechanical drilling. The article deals with the issues, related to the substantiation of the optimal formulation of a thermoplastic composite material based on secondary polyethylene terephthalate, and the determination of rational operating parameters of thermomechanical drilling, which allow to melt effectively the material at the bottom of a well. The possibility of material application for the insulation of absorbing horizons in borehole conditions has been proved. Based on the analysis of the heat balance at the bottom of a well, the calculation procedure has been proposed and the dependences of the velocity and time of thermomechanical melting of the grouting thermoplastic composite material on the operating parameters of drilling, thermal properties and geometric characteristics of the drill bit, have been determined.

1 Introduction

Drilling of exploring and production wells in order to explore deposits and extract mineral resources in the area of iron ore and coal basins is conducted in a high degree of development and metamorphism, in strong and fractured rocks [1 – 10]. The rocks of the developed horizons are in a highly stressed state [11 – 18], what makes the technology of mine workings constructions more complicated [19 – 22]. Absorption of drilling fluids is one of the most common complications in drilling wells [23]. Elimination of this complication, leads to significant material and time costs. The main method of controlling

* Corresponding author: sudakovy@ukr.net

the absorption of drilling fluids is the isolation of fractured walls of wells with the help of oil wells. Most of these materials are produced on a water basis with the introduction of mineral-binding or synthetic substances, mainly cement mortars. The effectiveness of such solutions is low due to the fact that they have a high sensitivity to dilution with water [24]. In practice, when the absorption of washing liquid is eliminated with cement mortars, tons, tens of tons of cement are used [25].

The low efficiency of conventional well pumping technology forces researchers to look for both new materials and technologies to prevent drilling fluid losses [26 – 30]. In particular, the promising direction is the use of grouting mortars based on thermoplastic materials with a low melting point, the melt of which can easily penetrate into the absorption channels of the washing liquid and harden there.

These materials are sufficiently durable waterproofing material, which have high corrosion resistance in aggressive environments. One of this kind of technologies ensuring an increase in the reliability of insulation works, improvement of working conditions and a significant reduction in material costs for the tamponing of complication zones was developed at the National Technical University [31].

The idea of the work is to use briquetted plugging thermoplastic composite materials (TPCM) delivered to the absorption zone followed by thermomechanical melting. The resulting melt of the TPCM penetrate into the absorption channels of the washing liquid solidifies there and forms an impermeable insulation shell around the borehole of the borehole.

Therefore, for the practical implementation of this technology, it is necessary to determine the rational parameters of the thermomechanical melting process. In this article are presented the results of experimental bench and theoretical studies of the thermomechanical melting of TPCM serving as a justification for the thermomechanical technology of isolating the absorbing horizons of boreholes.

2 Methods of the research

Bench testings. Technological modes of the process of thermomechanical melting of the oil-filled material during the plugging of the permeable horizon with the use of TPCM, which determined by physical modeling methods in bench conditions [32 – 39].

With this purpose there are developed and invented experimental bench with the count of the similar criteries (Fig. 1), including the ZIF-650M drill rig with a continuously variable drive 4, the absorbing horizon model 3 and the measuring and computing complex 1, the model of the laboratory thermomechanical drill bit (Fig. 2), which allowed for thermomechanical melting at the bottom of the well to create power from 1.7 to 5.8 kW.

Bench examinations were conducted in two stages.

At the *first stage*, the optimal combination of the regime parameters of the board process for the various TPCM formulations was worked out. For this purpose, TPCM blocks with different component ratios were made.

At the *second stage*, during thermomechanical melting of TPCM under the conditions of the absorbing horizon model, it was determined that:

– melting temperature of the melt of TPCM;

– penetrating ability and spreading of the melt along the channels of absorption of the washing liquid, depending on the crack opening.

TPCM is a cylindrical briquette (Fig. 3), consisting of a solidified mixture of gravel and secondary crystalline polyethylene terephthalate (PET).

Theoretical studies of the process of thermomechanical melting of TPCM were carried out using mathematical modeling methods.

3 Results of the research

In a result of bench studies, optimal combinations of the mode parameters of the thermomechanical melting technology of TPCM have been determined, that ensure the contact heating and melting of the oil well with minimal time and high drilling rates.

Fig. 1. General view of the stand:1 – measuring and computing complex; 2 – the model of absorbing horizon with spacers; 3 – a string of drill pipes; 4 – base of the drilling rig.

Fig. 2. Thermomechanical drill bit: 1 – friction plate; 2 – matrix; 3 – drill bit body.

Fig. 3. General view of TPCM blocks.

As a conclusion of the research, it follows that:

– the rate of thermomechanical melting, regardless of the binder to filler ratio, as well as the TPCM formula for the regime parameters, is of the same order of magnitude (Table 1):

– at a ratio of PET to gravel of 1:1, the rate of thermomechanical drilling (melting) varies from 0.4 to 0.9 m/h;

– at a ratio of PET to gravel of 1:2, the rate of thermomechanical drilling (melting) varies from 0.4 to 0.7 m/h;

– at a ratio of PET to gravel of 1:3, the thermomechanical drilling speed (melting) varies from 0.8 to 1.7 m/h;

– at a ratio of PET to gravel of 1:4, the rate of thermomechanical drilling (melting) varies from 0.3 to 0.7 m/h;

– when TPCM is melting, the temperature of its heating did not exceed the critical value (3000C) at which the astringent is destructed;

– the time of heating of the TPCM to its melting point depends on the TPCM formulation;

– the temperature rise in the sample of TPCM to the melting temperature takes place in the contact zone of the tool working part.

Table 1. Results of determination of technological parameters of thermomechanical drilling TPCM.

Composition	n, min^{-1}	P, daN	The beginning of melting, s	Drilling depth, mm	Drilling time, s	Average time of drilling, s	Thermomechanical drilling speed, m/h	Composition	n, min^{-1}	P, daN	The beginning of melting, s	Drilling depth, mm	Drilling time, s	Average time of drilling, s	Thermomechanical drilling speed, m/h
PET + Gravel: Concentration 1: 1; d <0.5mm	300	500	90	50	467	461.0	0.4	PET + Gravel: Concentration 1: 3; d <0.5mm	300	500	8	50	357	355.7	0.5
			91		445						8		355		
			88		471						8		355		
		700	50	50	361	358.7	0.5			700	8	50	142	143.7	1.3
			50		352						8		140		
			52		363						8		149		
	500	500	40	50	326	323.0	0.6		500	500	6	50	155	154.3	1.2
			40		320						5		152		
			41		323						6		156		
		700	25	50	269	269.7	0.7			700	5	50	119	117.7	1.5
			23		270						6		121		
			25		270						6		113		
	700	500	27	50	320	327.0	0.6		700	500	5	50	142	142,3	1.3
			27		336						5		144		
			27		325						5		141		
		700	11	50	204	203.3	0.9			700	5	50	101	104.0	1.7
			10		200						5		106		
			11		206						5		105		
PET + Gravel: Concentration 1: 2; d <0.5mm	300	500	15	50	417	418.7	0.4	PET + Gravel: Concentration 1: 4; d <0.5mm	300	500	10	50	555	553.7	0.3
			14		420						10		550		
			15		419						9		556		
		700	11	50	310	307.7	0.6			700	8	50	492	490.7	0.4
			10		305						8		490		
			10		308						8		490		
	500	500	18	50	357	355.0	0.5		500	500	7	50	408	410,7	0.4
			18		350						6		403		
			17		358						6		421		
		700	12	50	309	311.3	0.6			700	5	50	314	317.3	0.6
			13		315						5		320		
			12		310						5		318		
	700	500	10	50	260	258.7	0.7		700	500	5	50	250	259.7	0.7
			9		261						5		266		
			9		255						5		263		
		700	5	50	248	246.3	0.7			700	5	50	251	250.7	0.7
			5		250						5		250		
			5		241						5		251		

The results of bench modeling have shown the possibility of obtaining a monolithic plugging stone, the appearance of which is shown in Fig. 4.

Taking into account the results of bench studies of the thermomechanical drilling speed (melting) of TPCM, TPCM with a component ratio of 1:1 is recommended for use in downhole conditions.

Theoretical research. For the theoretical determination of the thermomechanical melting speed of TPCM, a mathematical model was developed on the basis of the heat balance equation on the contact surface (Fig. 5).

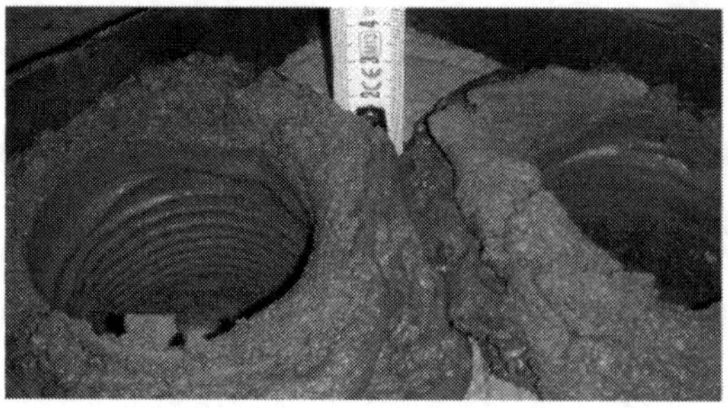

Fig. 4. The view of the melt (plaster stone) obtained at a rotation frequency of 500 min^{-1} and an axial load equal to 700 daN, for TPCM with a component ratio of 1:3.

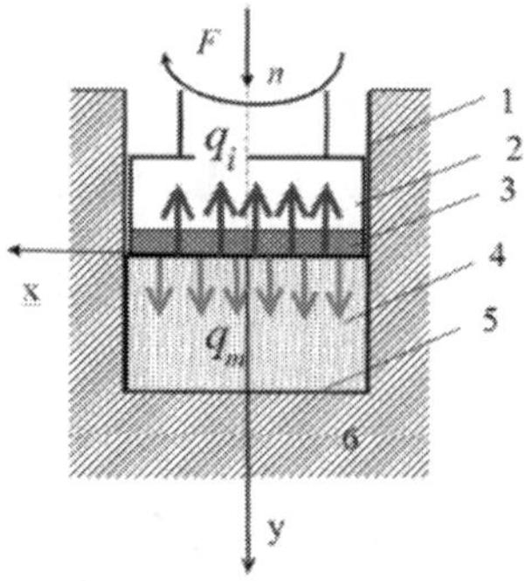

Fig. 5. Scheme to calculation of thermomechanical melting by friction: 1 – well walls; 2 – thermomechanical drill bit; 3 – friction plate; 4 – block TPCM; 5 – hole bottom; 6 – rock.

The heat flow is generated at the contact line of the friction plate 3 and the surface of the TPCM briquette during the operation of the thermomechanical drill bit 2. The generating heat flow is distributed to the body of the TPCM briquette due to thermal conductivity. Under the influence of thermal energy – TPCM heats up. After the temperature of the contact surface has reached the value of the phase transition (melting point) – the material begins to melt. The molten part of the material is crushed into the porous walls of the well 1, due to the pressure created by the drill bit.

To determine the rate of thermomechanical drilling (melting), the heat balance equation is used in the form:

$$q_m = q_{melt} + q_\lambda, \tag{1}$$

where q heat going to the melting of the surface layer of the material with a thickness of d_ξ for a time d_τ, q_λ is the heat flow, going to the heating of the inner layers of the material, determined by the Fourier law. The heat flow q_{melt} is given by:

$$q_{melt} = \rho L \frac{d_\xi}{d_\tau}, \tag{2}$$

where L is the latent heat of fusion of TPCM. The speed of thermomechanical drilling is defined as $V = d_\xi / d_\tau$.

The heat flow on the working surface is determined from:

$$q_m = \frac{\mu k_m \pi F D n}{S}, \tag{3}$$

where μ is the coefficient of friction; k_m is the coefficient taking into account the fraction of the heat of friction going to the heating and melting of the TPCM; F is the axial load; D is the diameter of the friction surface; S is the friction surface area; n is the drill bit rotation speed.

The solution of the non-stationary problem for determining the surface temperature of an isotropic semi-infinite body when heated by a heat flux is presented in [40]. Using this solution and (3) it is easy to obtain an expression that establishes a relationship between the process parameters of the process and the parameters of the thermophysical processes during tool operation:

$$Fn = \frac{\lambda D}{2\mu k_m \sqrt{\pi a \tau_{ff}}}\left(t_f - t_0\right), \tag{4}$$

where τ_f is the surface heating time to the melting point; t_f the melting temperature of the TPCM; a is the coefficient of thermal diffusivity of the TPCM.

Since the temperature on the surface after the beginning and throughout the melting process is constant and equal to t_f. In this case, the temperature field in a semi-infinite body is described by the expression [41 – 43]:

$$t = t_f + \left(t_{0f} - t_f\right)\operatorname{erf}\left(\frac{y}{2\sqrt{a\tau}}\right), \tag{5}$$

where t_{0f} is the temperature field, which was formed in the workpiece at the beginning of the melting process. We note that the entrainment of the mass of the melt from the contact surface leads to a restructuring of the temperature field at each instant of time, and expression (5) can not be used for a rigorous investigation of thermal processes in thermomechanical melting. Therefore, further reasoning will be approximate. Using (5) we can determine:

$$q_\lambda = -\lambda \left.\frac{\partial t}{\partial y}\right|_{y=0} = -\lambda\left(t_{0f} - t_f\right)\frac{e^{-\frac{y^2}{4a\tau}}}{\sqrt{\pi a \tau}} = -\frac{\lambda\left(t_{0f} - t_f\right)}{\sqrt{\pi a \tau}}. \tag{6}$$

After all, from equation (1), taking into account (4), (6) we obtain:

$$q_{melt} = \frac{2\lambda\left(t_f - t_0\right)}{\sqrt{\pi a \tau_f}} - \frac{\lambda\left(t_{0f} - t_f\right)}{\sqrt{\pi a \tau}} = \frac{\lambda}{\sqrt{\pi a}}\left(\frac{2\left(t_f - t_0\right)}{\sqrt{\tau_f}} + \frac{t_f - t_{f0}}{\sqrt{\tau}}\right). \tag{7}$$

Therefore the drilling speed is expressed by:

$$V = \frac{\lambda}{\rho L \sqrt{\pi a}}\left(\frac{2\left(t_f - t_0\right)}{\sqrt{\tau_f}} + \frac{t_f - t_{0_f}}{\sqrt{\tau}}\right). \tag{8}$$

For the practical use (8), it is necessary to know the temperature t_{0_f}, which is the function of the coordinate. Suppose that the temperature in the bulk of the thermoplastic material at the beginning of the melting process is slightly different from the initial temperature $t_{0_f} \approx t_0$. This assumption is based on the fact that the heating time τ_f, as a rule, is one order of magnitude shorter than the time of thermomechanical drilling, the thermal conductivity coefficient of TPCM is a small value. In this case, formula (8) can be written in the form:

$$V = \frac{\lambda(t_f - t_0)}{\rho L\sqrt{\pi a}}\left(\frac{2}{\sqrt{\tau_f}} + \frac{1}{\sqrt{\tau}}\right). \quad (9)$$

Determining τ_f through the regime parameters of drilling using (4), we obtain:

$$V = \frac{1}{\rho L}\left(\frac{4\mu k_m Fn}{D} + \frac{\lambda(t_f - t_0)}{\sqrt{\pi a \tau}}\right). \quad (10)$$

The average speed of thermomechanical drilling during the time T is defined as:

$$V_{av} = \frac{1}{\mathrm{T}}\int_0^{\mathrm{T}} V(\tau)\mathrm{d}\tau, \quad (11)$$

where $V(\tau)$ it is defined by the expression (10).

After integrating (10) with the range in time from 0 to T, taking into account the condition $\xi = 0$ for $\tau = 0$, we obtain the expression for drilling depth for the period T:

$$h = \frac{1}{\rho L}\left(\frac{4\mu k_m Fn}{D}\mathrm{T} + \frac{2\lambda(t_f - t_0)}{\sqrt{\pi a}}\sqrt{\mathrm{T}}\right). \quad (12)$$

The results of calculating the depth of drilling are shown in Fig. 6 and the rate of thermomechanical melting of TPCM in the form of the nomogram in Fig. 7, for a thermomechanical drill bit with a diameter of 59 mm.

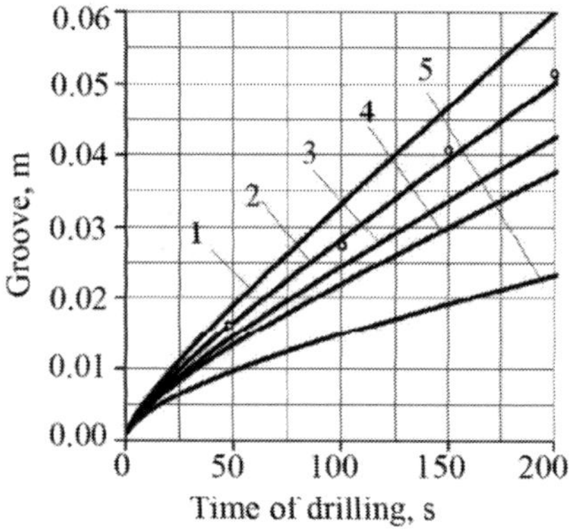

Fig. 6. Calculation dependence of the change in the depth of the thermomechanical melting of the borehole in time on tool dimensions, with an axial load of 700 daN and a rotation speed of 700 min^{-1}: 1 – 46 mm; 2 – 59 mm; 3 – 76 mm; 4 – 93 mm; 5 – 269 mm; o – experimental results.

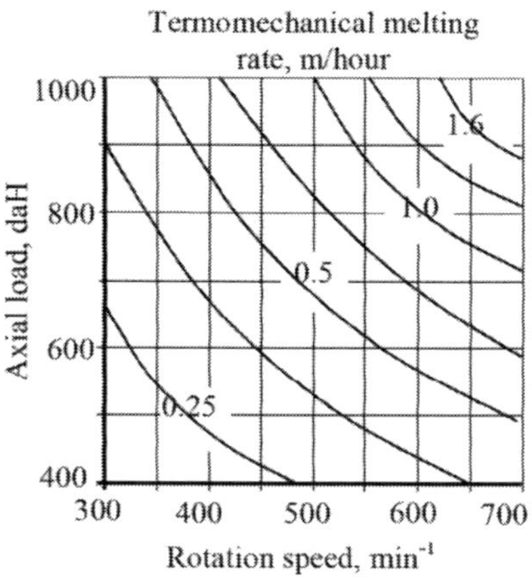

Fig. 7. The calculated dependence of the stationary thermomechanical melting rate of TPCM on the value of the regime parameters.

With an axial load of 700 daN and a speed of 700 min^{-1}, the design speed of thermomechanical melting with a drill bit with a diameter of 46 mm is – 1.2 m/h; 59 mm – 1.0 m/h; 76 mm – 0.84 m/h; 93 mm – 0.74 m/h; 269 mm – 0.46 m/h. The discrepancy between the results of analytical and bench studies did not exceed 10%.

Taking into account the thermophysical characteristics of the TPCM and the drill bit, the rate of thermomechanical melting of TPCM and the regularity of the temperature propagation in the TPCM are determined, as well as the regime parameters of the thermomechanical technology for eliminating the absorption of the washing liquid.

4 Conclusions

The possibility of using TPCM as a plugging material for drilling wells has been proved.

As a result of experimental studies and numerical calculations, it is shown, that the rate of thermomechanical melting of TPCM depends on the ratio of the constituent components.

For application in undergroud conditions, TPCM with a component ratio of 1:1 is recommended, in combination with an axial load of at least 700 daN with a drill bit rotation speed of 700 min^{-1}.

Approximate analytical relationships are obtained for calculating the rate of thermomechanical melting of the TPCM at the bottom, depending on the operating parameters of the drilling, the geometry of the drill bit, and the thermophysical properties of the material.

Authors express gratitude for the help and consultations during work performing for director of LTD Industrially-geological group "Dniprohidrobud" of Oleh Mostinets.

References

1. Bondarenko, V., Kovalevs'ka, I., & Fomychov, V. (2012). Features of carrying out experiment using finite-element methodat multivariate calculation of mine massif – combined support system. *Geomechanical Processes During Underground Mining*, 7-13. https://doi.org/10.1201/b13157-3
2. Bondarenko, V.I., Kharin, Ye.N., Antoshchenko, N.I., & Gasyuk, R.L. (2013). Basic scientific positions of forecast of the dynamics of methane release when mining the gas bearing coal seams. *Naukovyi Visnyk Natsionalnoho Hirnychoho Universytetu*, (5), 24-30.
3. Dychkovskyi, R.O., Lozynskyi, V.H., Saik, P.B., Petlovanyi, M.V., Malanchuk, Ye.Z., & Malanchuk, Z.R. (2018). Modeling of the disjunctive geological fault influence on the exploitation wells stability during underground coal gasification. *Archives of Civil and Mechanical Engineering*, *18*(4), 1183-1197. https://doi.org/10.1016/j.acme.2018.01.012
4. Bondarenko, V., Maksymova, E., & Koval, O. (2013). Genetic classification of gas hydrates deposits types by geologic-structural criteria. *Annual Scientific-Technical Collection – Mining of Mineral Deposits*, 115-119. https://doi.org/10.1201/b16354-21
5. Kovalevska, I., Symanovych, H., Barabash, M., & Snihur, V. (2017). Research intro rock pressure manifestations in interstratal rocks during descending and simultaneous mining of C_9 and C_{10}^{top} coal seams. *Mining of Mineral Deposits*, *11*(1): 50-56. https://doi.org/10.15407/mining11.01.050
6. Khomenko, O., Kononenko, M., Myronova, I., & Sudakov, A. (2018). Increasing ecological safety during underground mining of iron-ore deposits deposits. *Naukovyi Visnyk Natsionalnoho Hirnychoho Universytetu*, (2), 29-38.
7. Bondarenko, V., Svietkina, O., & Sai, K. (2018). Effect of mechanoactivated chemical additives on the process of gas hydrate formation. *Eastern-European Journal of Enterprise Technologies*, *1*(6(91)), 17-26. https://doi.org/10.15587/1729-4061.2018.123885
8. Kovalevs'ka, I., Symanovych, G., & Fomychov, V. (2013). Research of stress-strain state of cracked coal-containing massif near-the-working area using finite elements technique. *Annual Scientific-Technical Collection – Mining of Mineral Deposits*, 159-163. https://doi.org/10.1201/b16354-28
9. Lozynskyi, V.H., Dychkovskyi, R.O., Falshtynskyi, V.S., & Saik, P.B. (2015). Revisiting possibility to cross disjunctive geological faults by underground gasifier. *Naukovyi Visnyk Natsionalnoho Hirnychoho Universytetu*, (4), 22-28.

10. Sdvizhkova, Ye.A., Babets, D.V., Smirnov, A.V. (2014). Support loading of assembly chamber in terms of western donbas plough longwall. *Naukovyi Visnyk Natsionalnoho Hirnychoho Universytetu*, (5), 26-32.
11. Falshtyns'kyy, V., Dychkovs'kyy, R., Lozyns'kyy, V., & Saik, P. (2013). Justification of the gasification channel length in underground gas generator. *Annual Scientific-Technical Collection – Mining of Mineral Deposits*, 125-132. https://doi.org/10.1201/b16354-23
12. Kovalevs'ka, I., Illiashov, M., Fomychov, V., & Chervatuk, V. (2012). The formation of the finite-element model of the system "undermined massif – support of stope". *Geomechanical Processes during Underground Mining*, 73-79. https://doi.org/10.1201/b13157-13
13. Kyrychenko, Y., Samusia, V., Kyrychenko, V., & Goman, O. (2012). Experimental investigation of aeroelastic and hydroelastic instability parameters of a marine pipeline. *Geomechanical Processes During Underground Mining: School of Underground Mining 2012*, 163-167.
14. Falshtynskyy, V., Dychkovskyy, R., Lozynskyy, V., & Saik, P. (2012). New method for justification the technological parameters of coal gasification in the test setting. *Geomechanical Processes During Underground Mining*, 201-208. https://doi.org/10.1201/b13157-35
15. Saik, P.B., Dychkovskyi, R.O., Lozynskyi, V.H., Malanchuk, Z.R., & Malanchuk, Ye.Z. (2016). Revisiting the underground gasification of coal reserves from contiguous seams, *Naukovyi Visnyk Natsionalnoho Hirnychoho Universytetu*, (6), 60-66.
16. Pivnyak, G., Dychkovskyi, R., Smirnov, A., & Cherednichenko, Y. (2013). Some aspects on the software simulation implementation in thin coal seams mining. *Energy Efficiency Improvement of Geotechnical Systems*, 1-10. https://doi.org/10.1201/b16355-2
17. Kovalevska, I., Barabash, M., & Snihur, V. (2018). Development of a research methodology and analysis of the stress state of a parting under the joint and downward mining of coal seams. *Mining of Mineral Deposits, 12*(1), 76-84. https://doi.org/10.15407/mining12.01.076
18. Khomenko, O. (2012). Implementation of energy method in study of zonal disintegration of rocks. *Naukovyi Visnyk Natsionalnoho Hirnychoho Universytetu*, (4), 44-54.
19. Zhanchiv, B., Rudakov, D., Khomenko, O., & Tsendzhav, L. (2013). Substantiation of mining parameters of Mongolia uranium deposits. *Naukovyi Visnyk Natsionalnoho Hirnychoho Universytetu*, (4), 10-18.
20. Khomenko, O., Kononenko, M., & Myronova, I. (2013). Blasting works technology to decrease an emission of harmful matters into the mine atmosphere. *Annual Scientific-Technical Collection – Mining of Mineral Deposits*, 231-235. https://doi.org/10.1201/b16354-43
21. Vladyko, O., Kononenko, M., & Khomenko, O. (2012). Imitating modeling stability of mine workings. *Geomechanical Processes During Underground Mining*, 147-150. https://doi.org/10.1201/b13157-26
22. Khomenko, O., Tsendjav, L., Kononenko, M., & Janchiv, B. (2017). Nuclear-and-fuel power industry of Ukraine: production, science, education. *Mining of Mineral Deposits, 11*(4), 86-95. https://doi.org/10.15407/mining11.04.086
23. Cook, J., Growcock, F., Guo, Q., Hodder, M., & van Oort, E. (2011). Stabilizing the wellbore to prevent lost circulation. *Oilfield Review, 23*(4), 26-35.
24. Ashok Kumar Santra, B.R. Reddy, & Mfon Antia. (2007). Designing Cement Slurries for Preventing Formation Fluid Influx After Placement. *International Symposium on Oilfield Chemistry*, 28 February-2 March, Houston, Texas, U.S.A. SPE-106006-MS. https://doi.org/10.2118/106006-MS
25. Fedorov, B., Ratov, B., & Sharauova, A. (2017). Development of the model of petroleum well boreability with PDC bore bits for Uzen oil field (the Republic of Kazakhstan) *Eastern-European Journal of Enterprise Technologies, 3*(1(87), 16-22. https://doi.org/10.15587/1729-4061.2017.99032
26. Brazhenenko A.M., Goshovskiy S.V., & Sudakov, A.K. (2007). *Tamponazh gornykh porod pri burenii geologorazvedochnykh skvazhin legko-plavkimi materialami*. Kyiv: Ukrainskyi derzhavnyi heoloho-rozviduvalnyi instytut.

27. Yu, H., Li, L., Zheng, J., Ji, W., Qin, X., Fu, X., & Gao, W. (2016). New Method of Steam Channeling Plugging in Horizontal Wells of Heavy Oil Steam Stimulation. In *SPE Latin America and Caribbean Heavy and Extra Heavy Oil Conference*. SPE-181173-MS.
28. Liu, Y., Song, T., & Xu, Y. (2016). A new evaluation method for micro-fracture plugging in high-temperature deep wells and its application: A case study of the Xushen Gas Field, Songliao Basin. *Natural Gas Industry* B, 3(2), 158-164.
29. Sudakov, A.K., Dreus, A.Yu., Khomenko O.Ye., & Sudakova. D.A. (2017). Analitic study of heat transfer in absorbing horizon of boreholes in the formation of protection cryogenic plugging material. *Naukovyi Visnyk Natsionalnoho Hirnychoho Universytetu*, *3*(159), 32-46.
30. Sudakov, A., Dreus, A., Ratov, B., & Delikesheva, D. (2018). Theoretical bases of isolation technology for swallowing horizons using thermoplastic materials. *News of the national academy of sciences of the republic of Kazakhstan*, *2*(428), 72-80.
31. Kuzin, J., Isakova, M., Sudakova, D., & Mostinets, O. (2017). Isolation technology for swallowing zones by thermoplastic materials on the basis of polyethyleneterephthalate. *Naukovyi Visnyk Natsionalnoho Hirnychoho Universytetu*, (1), 34-39.
32. Kozhevnikov, A.A., Dreus, A.J., Lysenko, K.Ye., & Sudakov, A.K. (2013). Research of heat transfer in cryogenic gravel filter at his transporting on barrel of drillhole. *Naukovyi Visnyk Natsionalnoho Hirnychoho Universytetu*, (6), 49-54.
33. Dreus, A., Sudakov, A.K., & Kozhevnikov, A.A. (2016). Investigation of heating of the drilling bits and definition of the energy efficient drilling modes. *Eastern-European Journal of Enterprise Technologies. Technologies*, *3*(7(81)), 41-46.
34. Ziborov, K.A., Protsiv, V.V., Blokhin, S.Ye., & Fedoriachenko, S.O. (2014) Applicability of computer simulation while designing mechanical systems of mining rolling stock. *Naukovyi Visnyk Natsionalnoho Hirnychoho Universytetu*, (6), P. 55-59.
35. Sudakov, A.K. Khomenko O.Ye., Isakova M. L., & Sudakova, D.A. (2016). Concept of numerical experiment of isolation of absorptive horizons by thermoplastic materials. *Naukovyi Visnyk Natsionalnoho Hirnychoho Universytetu*, *5*(155), 12-16.
36. Protsiv, V., Ziborov, K., & Fedoriachenko, S. (2013) On formation of kinematical and dynamical parameters of output elements of the mine vehicles in transient motion. *Naukovyi Visnyk Natsionalnoho Hirnychoho Universytetu*, (4), 65-70.
37. Dreus, A.J., Sudakov, A.K., Kozhevnikov, A.A., & Vahalin, J.M. (2016). Study on thermal strength reduction of rock formation in the diamond core drilling process using pulse flushing mode. *Naukovyi Visnyk Natsionalnoho Hirnychoho Universytetu*, *3*(153), 5-9.
38. Khomenko, O.Ye., Sudakov, A.K., Malanchuk, Z.R., & Malanchuk, Ye.Z. (2017). Principles of rock pressure energy usage during underground mining of deposits. *Naukovyi Visnyk Natsionalnoho Hirnychoho Universytetu*, *2*(158), 34-43.
39. Kozhevnykov, A.O., Dreus, A.Yu., Baochang, Liu, & Sudakov, A.K. (2018). Drilling fluid circulation rate infuence on the contact temperature during borehole drilling. *Naukovyi Visnyk Natsionalnoho Hirnychoho Universytetu*, *1*(163), 35-43.
40. Franchuk, V.P., Ziborov, K.A., Krivda, V.V., Fedoriachenko, S.O. (2017). On wheel rolling along the rail regime with longitudinal load. *Naukovyi Visnyk Natsionalnoho Hirnychoho Universytetu*, (3), 62-67.
41. Polezhaev, Yu.V., & Yurevich, F.B. (1976). *Thermal protection*. Moskva: *Energiya*.
42. Sundén, B. (2012). *Introduction to heat transfer*. Southampton: WIT Press.
43. Dreus, A.Yu., & Lysenko, K.Ye. (2016). Computer simulation of fluid mechanics and heat transfer processes at the working face of borehole. *Naukovyi Visnyk Natsionalnoho Hirnychoho Universytetu*, (5), 29-35.

Aboveground storage of hydrocarbon gas hydrates

Mykola Zotsenko[1*], *Larysa* Pedchenko[1], and *Andrii* Manhura[1]

[1]Poltava National Technical Yuri Kondratyuk University, Department of Oil and Gas Exploitation and Geotechnics, 24 Pershotravnevyi Ave., 36011 Poltava, Ukraine

Abstract. The work objective is to substantiate the technical bases for the production of gas hydrate blocks directly on-site for further transportation and storage in above ground storages, as an alternative to transporting and storing gas in underground gas storages. The theoretical bases have been considered and processes of gas hydrate blocks producing, physical-mechanical characteristics of artificial hydrates with the purpose of determining the conditions of their storage and prolonged storage were tested in laboratory conditions. The construction of above ground gas hydrate storage in the form of a building, partially deepened in a ground, which is separated from the environment by a wall of gravel cement elements made by the mixing method were substantiated. It is proved that such a constructive solution for a long-term storage of gas hydrate blocks is the most economical in terms of energy consumption to maintain internal negative temperature.

1 Introduction

Gas hydrates – solid combination of a cage type of variable class – clathrates. The clathrate compound is described by the formula $M \cdot n \cdot H_2O$, where *M* is the molecule of the gas involved in hydrate formation; *n* – the number of water molecules belonging to one included molecule of gas and depends on the type of gas, pressure and temperature [1 – 3]. Gas hydrates are formed by the inclusion of gas molecules in the cavity of the frame (crystalline lattice) and temperature, built up by H_2O molecules under certain, for each component of gas, thermodynamic conditions [4]. With their violation gas molecules that were held in the lattice water skeleton by the weak forces of Van der Waals, leave it and hydrate is decomposed into gas and fresh water with a significant absorption of heat. The water molecules skeleton of methanhydrate is a cubic structure similar to ice and many properties similar to it. Its density varies within 0.908 – 0.917 g/cm^3 depending on the type of structure of the gas hydrate and the filling of its cavities CH_4 [5]. Specific gas content CH_4, at full filling of cavities of gas hydrates of the cubic structure of the I-th type with density 0.916 g/cm^3, when decomposition is 170.6 m^3 [6].

The thermodynamic stability of gas hydrates is a basic condition for the medium of formation and preservation of hydrates and, accordingly, an important criterion for detecting zones of possible hydration formation (ZPHF). Violation of the thermodynamic

* Corresponding author: zotcenco@hotmail.com

conditions in the system leads to the decomposition of hydrates.

The main parameters of the reservoir, which should be characterized in this case for the allocation of ZPHF, – temperature, pressure, composition of gas and mineralization of reservoir water, as well as lytofascial composition (permeability) of the reservoir [7]. High mineralization of reservoir water not only inhibits, but sometimes also eliminates the possibility of hydrotreatment processes [7], since the cavities of the lattice water skeleton are occupied by salt molecules. It is generally known that $CaCl_2$ in concentration 25 – 35% (weight) is an active inhibitor of man-made gas hydrate formation and is widely used in oil-field practice to prevent it.

In the work of M. Pedchenko [8 – 10] the method of hydrate production of the accompanying petroleum gas for the purpose of its transportation and storage is offered (Fig. 1).

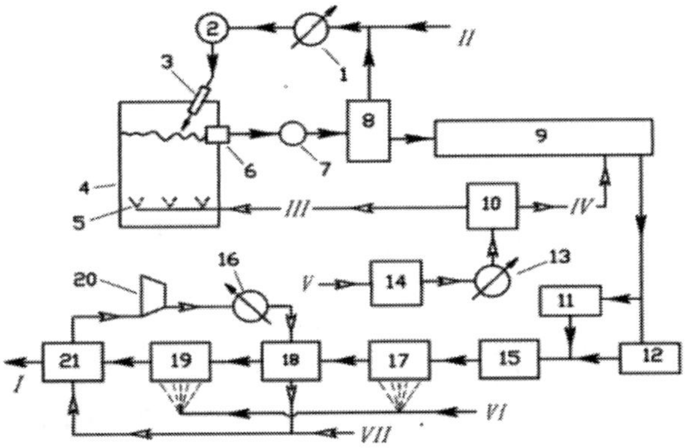

Fig. 1. Method of production of gas hydrate blocks: 1 – cooling of water; 2 – water pumping; 3 – feed water stream into the reactor; 4 – formation of gas hydrates; 5 – bubbling gas; 6 – selection of gas hydrates; 7 – vibration treatment; 8 – separation; 9 – concentration and cooling; 10 – crushing of hydrate; 11 – pelleting of hydrate; 12 – separation of gas; 13 – cooling of gas; 14 – preparation of gas; 15 – formation of blocks; 16 – cooling of air; 17 – the first water supply to the blocks; 19 – the second water supply on the blocks; 18 – cementing surface; 21 – freezing of the ice layer; 20 – air circulation; stream: I – blocks; II, VI – water; III – methane; IV – condensate; V – gas; VII – air.

The method involves the production of gas hydrates in the form of gas hydrate blocks and includes: preparation of gas and water; formation of gas hydrate; separation of the maximum amount of gas and water from the water – gas hydrate mixture, which was not included in the composition of the gas hydrate (separation); dehydration of a part of a film water (in order to concentrate the gas hydrate in the target product) by its binding in a gas hydrate at a contact with a part of a raw gas; freezing of gas hydrate; formation of a part of granules hydrate mass and crushing of another; formation of a gas hydrate block (GB) from a mixture of crushed and granulated gas hydrate; preserving of GB with ice crust (by double coating the surface of the water and freezing the flow of cooled gas at atmospheric pressure). As a way to increase the stability of the gas hydrate, it was chosen to preserve it with a layer of more stable gas hydrate. To do this, a layer of gas hydrate with the highest content of ethane, propane and *i*-butane is formed on the surface of hydrated mass particles. Gas enriched with ethane, propane and *i*-butane, has been previously obtained by fractionation of raw gas during its separation after compulsory cooling. Due to the temperature and pressure conditions of hydrate formation process in the reactor and in the process of drying gas components that were not included in the composition of the gas hydrate, but these conditions are captured in the form of droplets in the intercrystal space. Taking into account the insignificant specific content of such components in the gas entering the process, the presence of these inclusions does not affect the stability of the formed gas hydrate blocks. Experiments [8] show that these components are safely preserved in the volume of the gas hydrate mass, even after the pressure has been dropped to atmospheric.

In order to create optimal conditions for intensive hydration formation during the drainage process, namely prevention of water freezing, it is assumed in this work that the temperature of the water – gas hydrate mixture should be maintained at 275 K. This is done by taking into account the balance of the enthalpy of gas streams of dehumidification and cooling of the gas hydrate mass entering the drainage, and the enthalpy of hydration formation. The optimum drainage parameters are maintained taking into account the humidity of the gas hydrate mass, the composition of the drying gas and the rate of hydration so that the maximum amount of water is bound to the gas hydrate even before the temperature drop is lower than 273 K. Heat utilization of hydrate formation and cooling of the gas hydrate mass is carried out mainly by forced circulation of a portion of the drying gas through an external heat exchanger, which is not indicated in the diagram. The rate of hydration formation in contact with the gas with ice is insignificant in comparison with the process of interaction between gas and water [11]. Therefore, the gas flow, after cooling it to a temperature below 273 K, initially cooled the gas hydrate mass, which was drained and preserved more stable gas hydrate. After heating to 273 K, the gas during the dehumidification is actively included in the hydrate. Moreover, gas and gas hydrate in these successive processes move toward each other. In the device for carrying out the drainage process 9 between the zone of hydration and the cooling zone there is a zone of quality control of this process. Since the temperature of the gas hydrate mass is below 273 K it requires the maximum amount of water to be connected to the gas hydrate, then, accordingly, the humidity of the gas hydrate mass in the intermediate zone is an indicator of the quality of the drying process. In the above mentioned device for the formation of hydrate, gas is fed after separating it from the homologues of methane in the form of condensate (for use in the process of drying). Due to the change in the composition of the gas (the loss of parts of the heavier components), the thermobaric conditions in the reactor for the implementation of the process are supported more rigid than for the initial gas composition. To improve the quality of the formed mixture and the completion of the hydration process, it was suggested that the water – gas hydrate mixture withdrawn from the reactor should be subjected to vibration treatment for the destruction of the hydrated crust bubbles. The process of hydration heat removing from the reactor zone is carried out from the water – gas hydrate mixture outside the reactor. Moreover, heat removal, removal from the reactor of the formed hydrate and supply to the reactor of the required amount of water was combined into one process. The purpose of the operation for the formation of hydrate structures is the maximum reduction of porosity to increase the specific content of hydrate and mechanical strength, to prevent the heterogeneity of their structure, reduce energy consumption and the duration of the production cycle. The production of hydrate monolithic blocks includes the following steps: the separation of dried and frozen gas hydrate into two streams with a bulk ratio of the first and second 48 and 52% respectively; grinding of the first stream hydrate; pressing of granules from hydrate of the second stream in the size of 20 – 30 mm; uniform mixing of granular and crushed hydrate; pressing of formed mixture in the form of cylinders or hexagonal prisms of appropriate sizes; formation of ice blocks produced on the surface. Consequently, the gas hydrate structures [8] formed by the method are suitable for long-term storage and transportation under atmospheric pressure and a slight negative temperature.

2 Principles of artificial gas hydrate blocks formation for their transportation and storage

Having analyzed the work [8 – 13], from the theoretical substantiation of the parameters of the process of intensive hydration formation, one can conclude that the process of gas hydrated blocks production, as well as granulated gas hydrate, consists of several technological operations. Among them, the most responsible is the formation of blocks. The paper [8] substantiates the necessity of conducting experiments for setting parameters, which describes the process of the gas hydrate mass compressing during the formation of gas hydrate blocks at a temperature not exceeding 263 K. The aim of the work was to study the rheological properties of the gas hydrate mass and to check the adequacy of the proposed method and technology for the production of large-sized gas hydrate blocks with the aim of their transportation and long-term storage. The process of a gas hydrate mass formation is carried out by its compression. When applying force to the body on one axis, in the absence of lateral walls, there are transverse forces of stretching F_{str}. Between the faces of the cylindrical sample diameter d and height h, and press plates they are balanced by friction force F_{fric}. By the length of the sample there is a place where the forces of stretching F_{str} reach the highest value, and frictional forces F_{fric} – the smallest ($h/2$). With an increase in height h the forces of F_{str} increase and F_{fric} forces decrease. Hence, the compression strength of samples with proportions $d \approx h$ is considerably higher than the variant with parameters $d < h$.

In the case of the presence of vertical walls during compression of a gas hydrate mass of cylindrical shape, the transverse forces of stretching F_{str} that occur in this case will be counterbalanced by the forces of the resistance of the lateral walls. As a result, the surface layer of the hydrate mass is better compressed than internal one (Fig. 2). That is, the formed gas hydrate block will have an inhomogeneous porosity.

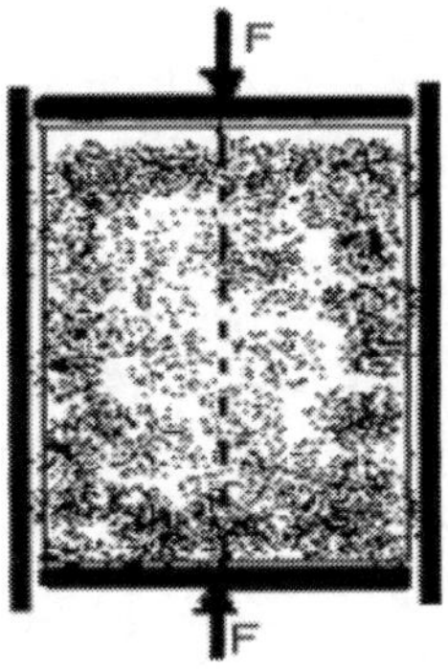

Fig. 2. Scheme of sample compression (heterogeneous porosity of gas hydrate mass after pressing).

Taking into account this and the need to create energy-efficient technology, it was proposed to form blocks of large sizes in the form of cubes from a mixture of crushed and granular gas hydrate of minimum porosity, taken in the appropriate proportion [8, 9]. Such a design of gas hydrate structures, as the basis of the proposed gas hydrate technology, allows simultaneously to solve problems of their uniform porosity (homogeneity) and maximum possible dense stacking during transportation and storage. It has been proved that during the formation of the gas-hydrate mass deformation goes beyond the limits of elasticity, the volume of the sample decreases. External forces perform work aimed at breaking the bonds between the crystals, reducing the volume of pore space, and counteract the forces of cohesion (regelation) between crystals of gas hydrate and ice at negative temperatures. In cube, as resistance to external forces, internal stress arises. Its value depends not only on the size of the deformation of the body, but also on the speed with which the external force acts on the subject. When forming by stroke, the internal stress will not be sufficiently large in comparison with the stress that occurs during slow compression. The internal stress arising during compression is manifested in the form of internal friction and will be directed against the action of external forces – in the direction of reducing the impact on the sample. Part of the work at sample deformation is spent on its heating.

However, according to the proposed [8, 9] technology of production of gas hydrate blocks, at the exit of the installation, their temperature should not be higher than 258 K. Taking into account the physical properties of the gas hydrate, it was concluded that the duration of the cooling process of the manufactured blocks and, as a consequence, the time of their location in a suitable installation can not meet the conditions of industrial production. Therefore, the gas hydrate mass is proposed to cool down to the time of formation in blocks. Moreover, the level of supercooling should take into account the thermal balance of the operation, the optimum temperature of storage of hydrate blocks under atmospheric pressure and possible losses of energy during transportation and storage.

As a result of gas hydrate sample deformation, its partial dissociation will occur, even under thermobaric conditions that are consistent with its stability. This process occurs at low and high temperatures. Dissociation is a violation of the integrity of the crystal lattice and microcracks. At a temperature higher than 273 K, there is the formation of water, the growth of nuclei of hydrate formation. When the sample is compressed, water migrates to intercrystalline space and causes its rapid deformation [8].

Comparison of mechanical strength of methane hydrate and ice of similar porosity showed that the relative rate of deformation of ice is 4.5 times higher [15]. For the formation of methane hydrate at a temperature of 253 K, it was noted that it is necessary to add 10 times more effort than the formation of ice granules [8, 16, 17]. At temperatures below 273 K, methane hydrate is 20 times stronger than ice. In addition, the difference increases with a temperature decrease [8].

In the process of compression of the gas hydrate mass, in addition to dissociation, there were several processes. Under the influence of pressure due to the melting of crystals at the point of contact and subsequent freezing, the process of regelation is manifested. In the case of breaking of the structure of crystals, they are freezing or "sticking". Due to these processes, the cracks are "treated" over time and the mass that is shrinked becomes stronger. After a certain time (5 – 10 hours) strength and hardness of it increase nine fold [18].

It can be concluded that the process of compressing the sample of gas hydrate is accompanied by the following processes: the destruction of the crystalline lattice of gas hydrate and ice; the conversion of part of mechanical energy into heat; partial dissociation of gas hydrate with the evolution of gas and free water (including supercooled) and the absorption of energy, necessary for its dissociation; the formation of gas hydrate from water and gas with the release of hydrate formation energy; heat exchange between porous space and crystals. Moreover, these processes take place simultaneously and mainly in the places of structural units contact. The predominance of one or another process depends on the initial temperature of the sample and gas pressure in the pores. Therefore, at a temperature higher than the point of crystallization of water, and atmospheric pressure, the processes of dissociation and destruction of crystals will prevail. Accordingly, at low temperatures, the processes corresponding to recrystallization predominate: re-formation (but in another form and elsewhere) of the gas hydrate from the products of dissociation and the crystallization of supercooled water.

Also, the compression of gas hydrate during its formation is accompanied by a sharply expressed creep process, which, based on the nature of the gas hydrate, consists of a series of successive phase transitions on the verge of structural units. These transitions are subjected to certain laws and can not pass immediately. Therefore, taking into account the features of this type of phase transitions will enable to further optimize the parameters of the technological process of formation of gas hydrate blocks.

In order to substantiate the parameters of gas hydrate blocks formation in [8], it was proposed to establish the features of the process of the gas hydrate mass compressing, which are based on the rheological model, characterizing its stress-deformed state. One of its components is the work of the internal forces of elasticity, which arise as a reaction to the attached external forces. It is determined by the selected or experimentally established rheological model, which characterizes the stress-deformed state of the gas hydrate mass as a shrinking object.

The gas hydrate mass compressing experiment, which consisted of a mixture of powdered and granular gas hydrate, was conducted in laboratory conditions in order to establish its rheological parameters. The object of study in [8] was also the parameters of deformation (shrinking) of the sample with a gradual increase in load and its dependence on porosity and initial temperature.

Also, in the considered work, the experimental procedure was developed. It was based on the variation of the parameter h (deformation values) with step and constant values of the internal stress σ^* as a one-factor experiment. For this purpose, after each application of the force to the sample for one minute, the magnitude of its shrinkage was measured.

According to the results of the laboratory experiment, the regression equation (power function) was determined for the mathematical description of the dependence of the gas hydrate sample shrinkage on the magnitude of the created step load:

$$P = 0.103 \cdot h^{1.3}, \tag{1}$$

where P – pressure on the sample, MPa; $a = 0.103$ and $b = 1.3$ – empirical coefficients; h – shrinkage of the gas hydrate block (investigated factor), mm. Consequently, the shrinkage of the gas hydrate block increases with increasing compressing force (load) (correlation coefficients $r = 0.99$ and variation $\upsilon = 0.035$).

As a result of uniaxial compression, the length of gas hydrate samples decreased by 21.3 – 21.6 mm, while the average porosity of the block was within 0.06 – 0.08. Obtained as a result of forming cylindrical forms of gas hydrate ($d = 81$ mm, $h = 102.6$ mm) had a uniform distribution of sample granules in a volume without the manifestations of their destruction or deformation. On the basis of the obtained results of measurements of samples shrinkage h for each case of applying constant load σ_i^* ($\sigma_i^* = \text{const}$) a rheological diagram was depicted in Fig. 3.

These experimental research shows that for every such case when ($\sigma_0 = \text{const}$) relative deformation varies slightly (within $0.0108 \le \varepsilon_0 \le 0.0121$). By way of regression analysis of the OA section data (Fig. 3) it is established that the stress dependence σ_0 from the instantaneous deformation ε_0 is linear and described in the same way as Hooke's law:

$$\sigma_0 = E_0 \varepsilon_0, \tag{2}$$

where σ_0 – internal stress, Pa; E_0 – instantaneous deformation module, Pa; ε_0 – instante deformation.

Experimentally obtained value of the instantaneous deformation module of the gas hydrate mass E_0 is given in Table 1.

Table 1. Rheological parameters of the gas hydrate mass, which are determined experimentally.

Instantaneous deformation	Deformation of creep	
Instantaneous deformation module E_0, MPa	Module of long deformation E_l, MPa	Coefficient of viscosity η, MPa·s
26 – 369	134 – 674	1781 – 8905

Analysis of creep deformation change with time for $\sigma_i^* = \text{const}$ is carried out on the basis of AB senousction data (Fig. 3). At constant stress of the sample there was a dampening with time – creep.

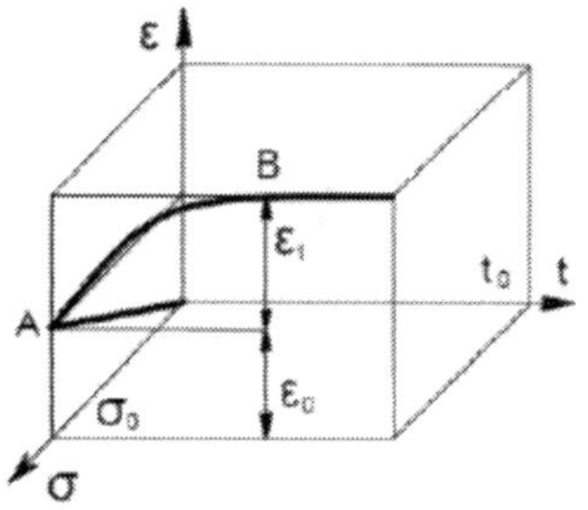

Fig. 3. Rheological model of gas hydrate compression at constant loading ($\sigma_i^* = \text{const}$).

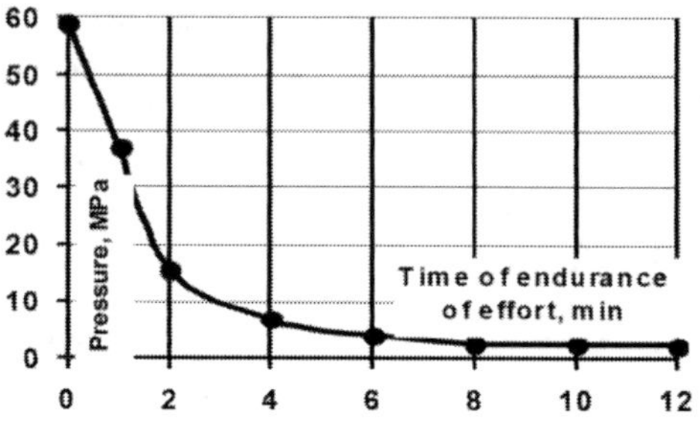

Fig. 4. Dependence of the pressure required to form a gas hydrate sample on the holding time.

The approximated exponential dependence (with coefficients of correlation $r = 0.973$ and variation $v = 0.045$), the change in the relative creep deformation of gas hydrate sample with time, according to experimental data, has the form:

$$\varepsilon_1 = 0.0089(1 - e^{-0,076\tau}), \tag{3}$$

where $A' = 0.0089$ and $B' = -0.076$ – empirical coefficients; ε_1 – relative deformation of creep; τ – time, s.

On the basis of experimental data processing it was established that the deformation change with time is described by the model of the elasto-viscous Kelvin-Voigt body (parallel connection of elastic and viscous elements) [19]:

$$\sigma^* = \varepsilon_1 E_1 + \eta \frac{d\varepsilon_1}{d\tau}, \tag{4}$$

where E_1 – long-term deformation module, Pa; η – coefficient of viscosity, Pa·s.

Using a rheological compressing model for a gas hydrate mass was calculated the following: a long-term deformation module E_1 and coefficient of viscosity η (Table 1).

With the help of the established deformation model of the elastic-viscous body, the pressure required to form the hydrate blocks of the required porosity was calculated and the time for this process to be performed was determined. On the basis of the obtained data, a graph of the pressure dependence created per unit area was constructed from the time of the application of the force (Fig. 4).

As can be seen from the graph (Fig. 4), the pressure required to compress the sample of the gas hydrate to the installed porosity (0.08 – 0.1), with the extension of time of the force bearing is sharply reduced. The duration of forming within the range of 7 – 9 min is optimal. Comparing with the short-term process (stroke), holding for 8 min reduces the pressure of the formation of the gas hydrate. Consequently, the process of creep is in a series of successive phase transitions on the verge of structural units. The process is subjected to certain laws and can not pass instantly. Therefore, taking into account the features of this type of phase transitions will enable to further optimize the parameters of the technological process of gas hydrate blocks formation.

3 Methodical peculiarities of designing and construction of aboveground storage of hydrocarbon gas hydrates

The next stage of the implementation of the new method of methane gases transporting and storing in the gas hydrate state is the production of an industrial design for the manufacture of gas hydrate blocks based on technology, the basis of which was set forth above. In our view, the production model of such a device should be placed on a mobile chassis in order to carry out work on the manufacture of gas hydrate blocks directly at facilities for the complex preparation of gas exploitation wells of oil and gas fields.

For convenient storage, gas hydrate blocks have to be manufactured in the form of cubes, with a side of 0.6 m. As we know from the properties of gas hydrates in 1 m^3 of methane hydrate at a pressure of 2.6 MPa and a temperature of 273 K, will contain 164 m^3 of gas. The volume of gas that will be emitted due to the dissociation of the cubic gas hydrate block, with a side of 0.6 m and a volume of 0.216 m^3, will be 35.5 m^3. According to statistics, in 2017 160 mln m^3 of gas was spent for the needs of the industry in Poltava, which is 3.1 mln for the week. It is proposed to create a aboveground storage facility for gas volume of $3.5 \cdot 10^6$ m^3, capable of meeting the necessary industrial needs. The number of gas hydrate blocks, after which the dissociation will receive a weekly gas volume, will amount that is 100 000 units. For reasons of the need to keep a certain temperature in the storage, it must be deepen in the ground as shown in Fig. 5a.

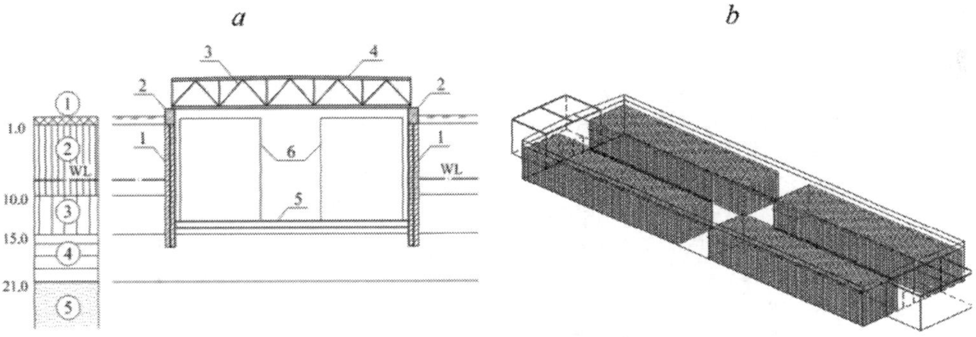

Fig. 5. Section (a) and general view in isometry (b) ground gas storage in hydrate state: 1 – walls from gravel cement elements; 2 – reinforced concrete belt; 3 – steel truss; 4 – thermally insulated coating; 5 – concrete floor; 6 – shelving of hydrate block.

Since the entrance to the cooling room externally involves a wind-poch, therefore, in addition to the storage of gas hydrate blocks, there are two transfer chambers: loading and unloading. Transfer chambers are transitional premises in which the loading of gas hydrate blocks is carried out for further transportation to the storage facility and, accordingly, their unloading, for subsequent delivery to the dissociation site. For greater convenience, loading and unloading works are carried out using gantry cranes located above the transfer chambers. The loading and unloading of gas hydrate blocks takes place through the sliding roof of the entrance and exit chambers. Sliding roof consists of two sliding elements, which for different tasks open independently of each other. The contour of the storage room consists of gravel cement elements, which are made from the surface of the Earth by mixing method to the excavation of the foundation pit. After the soil cement hardened, the inner volume of the soil is removed. Facing material for interior walls of the storage facility is plaster, with the addition of hydrophobic substances, with a thickness of 0.02 m.

The walls between the storage room of gas hydrate blocks and gateways are made of bricks. Brick walls are designed from clay ordinary solid bricks, plastic molding of the brand is not lower than M100. The brick for the walls of the refrigerator buildings should have a mark of frost not lower than M25. Facing material for brick walls is also used plaster 0.02 m thick.

The cover of the storage is a construction made of metal trusses with a span of 30 m, which is needed for the complete coverage of the storage. This construction is mounted on a reinforced concrete plinth. The base is arranged around the perimeter of the storage, 1 m high, and 1.2 m wide.

In determining the size of the ground storage of methane gas in the hydrated state, the volume of gas hydrate blocks that will be stored in it are taken into account. To this volume space for loading and unloading is added. And, finally, the volume of transfer chambers for receiving and sending hydrated blocks is added too. That is, the total volume of the storage will be 43907 m^3, in this volume it is possible to place four assemblages of blocks of $45 \times 10.12 \times 12 = 5508$ m^3. After specifying the size of the storage, hydrate blocks in it occupy a volume of 22032 m^3, which corresponds to 102 000 blocks, or 3.6 mln m^3 of hydrocarbon gas. The general view of the storage with hydrate blocks in isometry is shown in Fig. 5b. For loading and unloading works in the storage, the telescopic loader of the British company – JCB model 535 – 125 – was chosen as the most rational one. It is characterized by a low center of gravity and a large outflow of the loader arm at its reduced cross-section, it is designed to work with massive loads at a height. Additional benefits of the JCB 535 – 125 include the steering, which has three modes of operation to control all four wheels, reduced by 250 mm radius of rotation, high maneuverability and geometric patency.

To prevent the spreading of structured water on the floor, it is assumed to be installed along the perimeter of the gutter system for drainage of water. The water through the trenches will flow to the drainage basins located in the corners of the storage and in each of the gateways. From the collectors, structured water is pumped off to the surface.

In the refrigerated facility, alarm systems should be provided: safety, fire, security and alarm systems for inadmissible level of gas contamination of premises. Emergency ventilation should have triggers both in and out of the ventilated areas, as well as automatically when the concentration of gas in the air above the maximum permissible level is increased. In all cases, the signal is provided in a room with 24-hour staffing.

Since the storage of gas hydrates is a structure in which there is a methane hydrate, it is necessary to install a supply and exhaust ventilation system. The sypply system supplies fresh air to the storage, and the exhaust system removes gaseous air from the room. By way of inducing air, the ventilation system refers to a system with mechanical induction (with the use of ventilating fans, ejectors, etc.). The connecting link between the storage of gas hydrate blocks and the entrance and exit transfer chambers is the sealed gates, which are additionally covered with thermal-insulating material.

The storage facility for gas hydrate blocks is a building with negative temperatures in the premises, which is not erected in permafrost soils, therefore, it is necessary to design a storage facility, taking into account the prevention of freezing of soils, which are the basis for foundations and floors. For this purpose, it is proposed to use artificial heating systems such as non-freezing liquids and electric heating. Soil protection systems for freezing should be provided under all rooms with negative temperatures, such as: storage and transfer chambers.

Cooling of the storage room to a temperature of 253 – 268 K should be carried out using bivalent absorption heat pumps with a low-temperature source of energy – soil heat, as it provides a relatively uniform heat source temperature and high efficiency of the heat pump. Installation of heat exchangers of selected heat pumps is envisaged at the stage of creation of soil-cement elements of the second row of the enclosing structure of the storage facility.

According to preliminary data, the estimated cost of the proposed construction of a land storage of hydrocarbon gases in a hydrated state of 3.5 million m^3 is about 65 mln UAH. Taking into account that the lifetime of the storage building is 100 years, according to preliminary calculations, the estimated cost of manufacturing gas hydrates, storage and subsequent dissociation of natural gas in a hydrate state during the month is 223 UAH per 1000 m^3 of free gas.

4 Conclusions

As a result of theoretical and experimental researches, the bases of creation of gas hydrate blocks from hydrocarbon gases by the transformation of gas into a hydrated mass under certain thermobaric conditions, densification of hydrate mass by a press to the given values of density and its shape, with subsequent preservation of a block by a layer of ice were developed. Thus, the conditions for the transportation and storage of hydrocarbon gas at atmospheric pressure and the negative temperature 253 – 26 K are created.

As an alternative to underground gas storage facilities, a 3.5 mln m^3 of aboveground storage facility is proposed. The dimensions of the storage are 123×31×16, the internal space is 43907 m^3. The storage is deepened in the soil by placing its exterior walls without a preliminary excavation by means of their erosion from the surface of the Earth by the mixing method from the gravel cement elements. After hardening of the soil cement, the storage volume is released from the soil, insulating and finishing works are performed, the premises are covered and the necessary equipment is installed. Heat pumps are used in the storage to create a negative temperature. Estimated cost of construction is 65 mln UAH. According to preliminary calculations, the estimated cost of manufacturing gas hydrates, storage and subsequent dissociation of natural gas in a hydrate state during the month is 223 UAH per 1000 m^3 of free gas.

This work has become possible due to financial and organizational support within the frames of the state budget research project under the auspices of the Ministry of Education and Science of Ukraine “Application of gas hydrate technology in the development of traditional and gas hydrate gas deposits” (State registration No. 0113U00857), “Research of influence of thermodynamic parameters of phase transitions in systems with gas hydrates on the efficiency of gas hydrate technology” (State registration No. 0115U002420). The authors express their gratitude to the supervisor of these projects – Doctor of Sciences, Professor, Head of the Underground Mining Department at the National Mining University (Dnipro) Volodymyr Bondarenko for his support in conducting the research.

References

1. Carrol John (2009). *Natural Gas Hydrates*. A Guide for Engineers, 2nd Ed. Oxford: Gulf Professional Publishing.
2. Makogon, Yu.F. (2001). Prirodne gidraty: otkrytie i perspektivy. *Gazovaya promyshlennost'*, (5), 10-16.
3. Makogon, Yu.F. (2010). Gazogidraty. Istoriya izycheniya i perspektivy osvoeniya. *Geologiya i poleznye iskopaemye mirovogo okeana*, (2), 5-20.
4. Melnikov, V., & Gennadinik, V. (2018). Cryodiversity: the world of cold on the Earth and in the

solar system. *Philosophy and Cosmology*, (20), 43-54. https://doi.org/10.29202/phil-cosm/20/4

5. Levik, O.I. (2000). Thermophysical and Compositional Properties of Natural Gas Hydrate. Trondheim: Norwegian University of Science and Technology, Department of Petroleum Engineering and Applied Geophysics.
6. Dawe, R., Thomas, M., & Kromah, M. (2003). Hydrate Techology for Transporting Natural Gas. *Engineering Journal of the University of Qatar*, (16), 11-18.
7. Musayev, R.M. (1977). Izuchenie izmeneniya entalpii sistemy "gaz – voda" pri obrazovanii odnoy tverdoy fazy gidratov. *Trudy Vsesoyuznogo nauchno-issledovatel'skogo proektnogo instituta po podgotovke k transportirovke i pererabotke prirodnogo gaza*, (4), 42-48.
8. Pedchenko, L.A., & Pedchenko, M.M. (2013). *Research of reology properties of gas hydration mass in the process it of forming*. In *Conference reports materials*: *Problems of energy saving and nature use 2013* (pp. 127-134). Budapest.
9. Pedchenko, L., & Pedchenko, M. (2012). Substantiation of Method of Formation of Ice Hydrate Blocks with the Purpose of Transporting and Storage of Hydrate Gas. *Naukovyi Visnyk Natsionalnoho Hirnychoho Universytetu*, (1), 28-34.
10. Pedchenko, L.O., & Pedchenko, M.M. (2013). *The method production of hydrates of associated petroleum gas the purpose of their transportation and storage*. Patent No. 101882, Ukraine.
11. Bondarenko, V., Kovalevska, I., Astafiev, D., & Malova, O. (2018). Examination of Phase Transition of Mine Methane to Gas Hydrates and their Sudden Failure – Percy Bridgman's Effect. *Solid State Phenomena*, (277), 137-146. https://doi.org/10.4028/www.scientific.net/SSP.277.137
12. Lyubchak, O., Kolodiy, I., & Khokha, Y. (2017). Thermobaric conditions of oil and gas contents and predicting of hydrocarbon phase state (on the example of the bilche-volytsa oil-and gas-bearing area. *Visnyk of V.N Karazin Kharkiv national university-series geology geography ecology*, (46), 15-22. https://doi.org/10.26565/2410-7360-2017-46-02
13. Bondarenko, V., Svietkina, O., & Sai, K. (2017). Study of the formation mechanism of gas hydrates of methane in the presence of surface-active substances. *Eastern-European Journal of Enterprise Technologies*, *5*(6 (89)), 48-55. https://doi.org/10.15587/1729-4061.2017.112313
14. Yakushev, V., Gerasimov, Yu., Kvon, V., & Istomin, V. (2008). *The Current State of the Gas Hydrate Technology*. Rewiew. Inf. OOO "Gazprom RPI".
15. Laura A. Stern, Stephen H. Kirby, & William B. Durham. (1996). Peculiarities of Methane Clathrate Hydrate Formation and Solid-State Deformation, Including Possible Superheating of *Water Ice. Science*, *273*(5283), 1843-1848. https://doi.org/10.1126/science.273.5283.1843
16. Durham, W.B., Kirby, S.H., Stern, L.A., & Zhang, W. (2003). The strength and rheology of methane clathrate hydrate. *Journal of Geophysical Research*, *108*(B4), 2-11. https://doi.org/10.1029/2002JB001872
17. Bondarenko, V., Lozynskyi, V., Sai, K., & Anikushyna, K. (2015). An overview and prospectives of practical application of the biomass gasification technology in Ukraine. *New Developments in Mining Engineering 2015: Theoretical and Practical Solutions of Mineral Resources Mining*, 27-32. https://doi.org/10.1201/b19901-6
18. Tushinskiy, G.K. (1971). *Inzhenernaya glatsiologiya*. Moskva: Moskovskiy gosudarsnvennyy universitet.
19. Zotsenko, M.L., Kovalenko, V.I., Yakovlev, A.V., Petrakov, O.O., Shvec, V.B., Shcola, O.V., Bida, S.V., & Vinnikov, Yu.L. (2004). *Inzhenerna heolohiia: Mekhanika hruntiv, osnovy ta fundamenty*. Poltava: Poltavskyi natsionalnyi technichnyi universytet.

Simulation of deposit parameters in underground development mining

Vladimir Golik[1], *Yuriy* Razorenov[2], *Volodymyr* Morkun[3*], and *Nataliia* Morkun[3]

[1]North Caucasus Mining Metallurgical Institute, 85 Pobedy St., 308015 Vladikavkaz, Russia
[2]South Russian State Polytechnic University, 132 Prosveshcheniya St., 346428 Novocherkassk, Russia
[3]Kryvyi Rih National University, 11 Vitaliia Matusevycha St., 50027 Kryvyi Rih, Ukraine

Abstract. The article is aimed at improving development mining to prepare an ore body for stoping by access ramps to provide comfortable conditions and high technical and economic indices in underground mining. Efficient parameters of underground mining are chosen in the course of simulating data on the mining theory and practice considering ore losses and dilution on the basis of critical analysis of uranium mining enterprises' activities. The research provides data on geological and engineering zoning of an ore deposit and physical-mechanical properties of ore bearing rocks. The advanced experience is systemized and there is provided system analysis of modern development mining schemes with access ramps (ring, spiral, one-way inclined, central inclined and across the strike). The research recommends schemes of development mining and substantiates their advantages. There are quantitative indices of physical simulation of development variants as to drawn ore quality according to criteria of soil location in ore draw points. The scientific novelty implies developing the criterion of optimality and ranking variants of development mining according to technical-economic and geomechanical indices considering some technological factors as well as the number of stopes operating simultaneously on the level. The study consists in increasing authenticity of development projects through applying complex schemes of access ramps according to the complex criterion of increasing mining depths, equipment application, ventilation and underground mine capacity.

1 Introduction

Conditions of deposit localization and mining often affect efficiency indices. They can be taken into account while designing deposit mining [1 – 4].

High mining efficiency and relatively small amounts of development and face-entry operations accompanied by complex mechanization of operations cause high ore losses and dilution of some mining variants.

Along with undisputable advantages of certain development variants, first of all, due to reduced amounts of excavation, they are characterized by high ore losses and dilution [5, 6].

Repeated fragmentation is observed in the drawn ore that has an impact on ore losses and dilution in mining. Computational methods used to determine development mining

* Corresponding author: morkunv@gmail.com

parameters allow some discrepancies in setting quality indices. Ore losses and dilution can be determined directly only during the operation stage.

With increasing the stoping depth at most ore deposits, mining and geological conditions such as the ore body thickness, the dip angle and forms of rock pressure manifestation change.

To maintain the designed capacity of a deposit and achieved labour efficiency there arises a necessity to improve parameters of development mining before stoping.

For the most widely used development schemes through access ramps and sublevel drifts under changing mining-geological conditions, the amount of development and face-entry operations increases greatly reaching 80% of the total.

Improvement of development schemes implies increased rates of increasing stoping depths, application of mobile equipment, provision of stope ventilation and increased mine capacity [7 – 9].

2 Materials and methods

The research aim is achieved through applying methods of analyzing the ore mining theory and practice and simulating enterprises' mining indices.

Through application of a complex method, the authors develop the optimality criterion serving the basis for ranking development variants considering technological factors allowing them to increase authenticity of development mining projects with access ramps.

To develop recommendations on choosing efficient development schemes, geological and engineering ore zoning is performed.

Any operating mining enterprise and coal mine in particular, represent the difficult dynamic system that functioning in continuously changing conditions of the internal and external environment.

3 Results

Horizontal rock movements at the distance of 30 m from the deposit are conditioned by rock subsidence of the hanging wall over the stoping zone. Vertical rock movements are influenced upon by stoping operations. Table 2 presents physical and mechanical properties of rocks.

Table 1. Basic properties of ore-bearing rocks.

Rocks	Density, 1×10^{-3}, g/m^3	Compression strength, MPa	Internal friction angle, degree	Longitudinal wave velocity, m/sec	Transverse wave velocity, m/sec	Acoustic stiffness, 1×10^6, kg/m^{2}sec	Poisson ratio
Quartz porphyry	2.76	152.6	59	4440	2630	12.1	0.22
Conglomerate	2.85	98.3	53	4210	2490	12.0	0.22
Interstratified	2.86	143.8	58	4690	2700	13.3	0.25
Rocks	2.86	102.0	56	4550	2580	13,0	0.26

In the ring-type scheme of mining development, level drifts are located either within the ore body or in the hanging and footwall rocks. The disadvantages of the variants include low ore stability and increased amounts of development mining.

With LHD machines, the haulage length is 50 – 60 m.

The distance between stope ore passes is:

$$l_{ore} = l_{del} - l_{flt} = 50 - 20 = 30 \text{ m}, \quad (1)$$

where l_{ore} is the distance between ore passes, m; l_{del} is the haulage distance, m; l_{flt} is remoteness of ore passes from the fault zone, m.

In case of development mining with ramps in the deposit centre, sublevel drifts are replaced with ore drifts. Ore is mined from flanks to the centre towards the ramp with two stopes per a level simultaneously (Fig. 1).

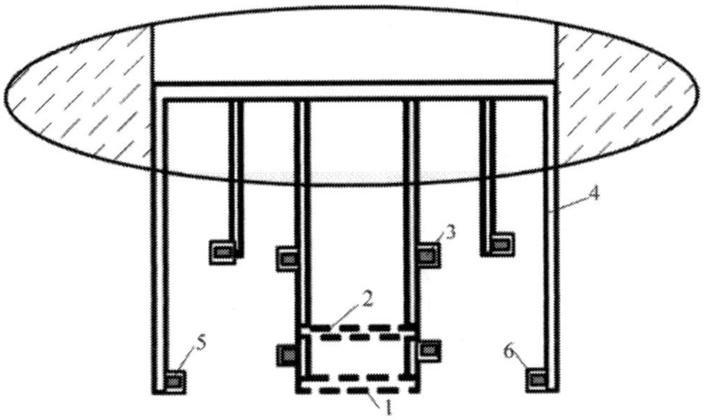

Fig. 1. Development mining of a level with an access ramp in the centre: 1 – inclined mine working of a ramp; 2 – horizontal mine working; 3 – haulage sublevel ort; 4 – ventilation-haulage ort; 5 – ore pass; 6 – raise.

As to mine working localization, an access ramp can be spiral and one-way. The spiral ramp is characterized by a smaller drifting capacity as compared to the one-way one.

The spiral ramp length is:

$$L_{inc} = (\frac{h_{ss}}{\sin\alpha} + l_r)n \, , \tag{2}$$

where L_{inc} is the ramp working length; h_{ss} is a sublevel height; α is a slope angle; l_r is the curve length between road turns; n is the number of sublevels.

The one-way ramp length is:

$$L_{inc} = (\frac{h_{ss}}{\sin\alpha} + l_r)n + 3L_{hor} \, , \tag{3}$$

where L_{hor} is the length of the ramp drift on a sublevel, m.

$$l_{hor} = \frac{h_{ss}}{tg\alpha} = \frac{12.5}{tg\alpha} = 71. \tag{4}$$

The disadvantage of the spiral ramp location parallel to the strike consists in the fact that the final mining point on sublevels increases making operations less flexible.

The one-way ramp does not have this disadvantage, yet, it is characterized by increased amounts of drifting. The spiral ramp with extra drifting on a sublevel is characterized by minimum production costs. Two sublevels are driven towards the ramp, and one sublevel is driven to the mine working connecting the deposit centre and the ramp.

The length of the spiral ramp with a horizontal drift on the sublevel is:

$$L_{inc} = (\frac{h_{ss}}{\sin\alpha} + l_r)n + 3L_{hor} \, . \tag{5}$$

Development mining with a ramp in the centre increases the development duration and makes control of rock pressure more complicated (Fig. 2).

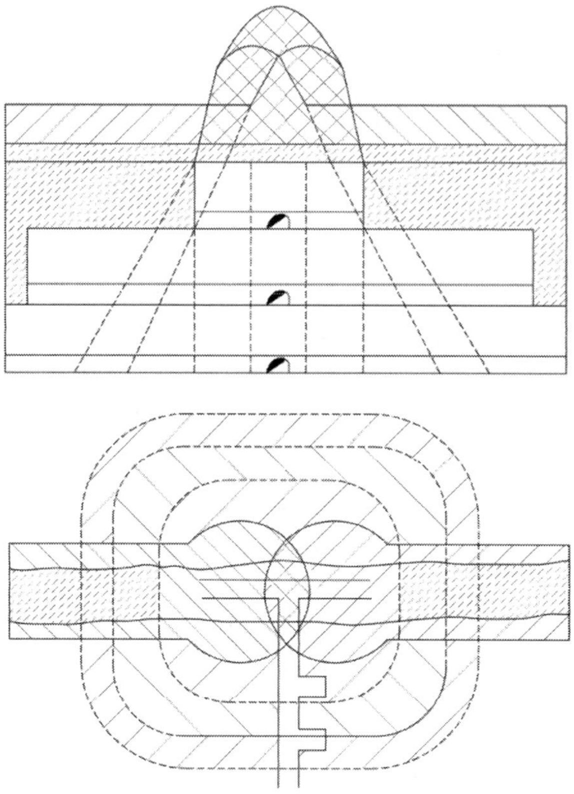

Fig. 2. Distribution diagrams of increased rock pressure in the vicinity of an access ramp.

As to rock pressure control, it is more efficient to mine an ore deposit from the centre to flanks, although this increases the number of development workings because of an extra access ramp.

There is no need to increase the length of a haulage drift of the main level if an access ramp is across the strike.

When a level is developed with two access ramps, extra haulage drifts cannot be used. For unidirectional inclined workings on the flanks, drifts connecting ramps are along the ore, while those connecting orts are used for ventilation and ore haulage.

Development mining variants with access ramps increase the development duration and limit the number of stopes functioning simultaneously on a level.

To avoid these drawbacks access ramps are located at the distance of a quarter of an ore body length from the body flanks (Fig. 3).

Per each ramp in 16 stopes, increased intensity of level driving reduces the time for drift presence in the zone of rock pressure development.

In level driving with three access ramps in the centre and on the flanks at the distance of 1/6 of the body length, drifts are driven to each ramp on both flanks simultaneously with up 24 faces. The duration of level drifting and working maintenance costs decrease (Table 3).

Counter-stope mining onto a ramp is characterized by the fact that stress values rise to up to dangerous levels at the final stage of mining operations.

Centre-to-flanks or flank-to-flank stoping variants are reasonable if haulage units are focused at the ore body flanks.

Considering stope ventilation, decreased distances of ore haulage and collapse prevention, ore passes should be located at the hanging wall 5 – 10 m off the ore body.

 https://doi.org/10.1051/e3sconf/20186000029

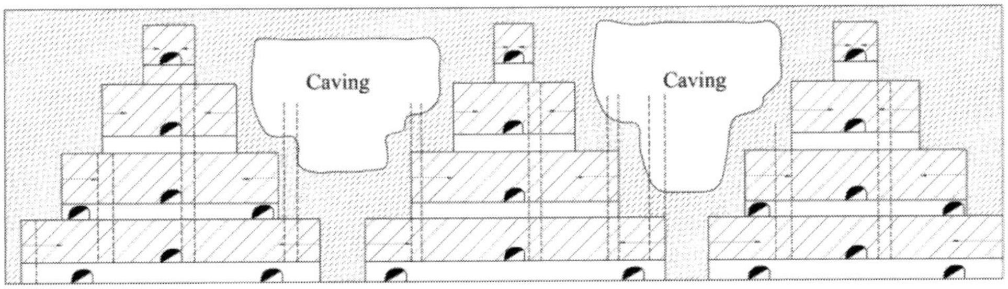

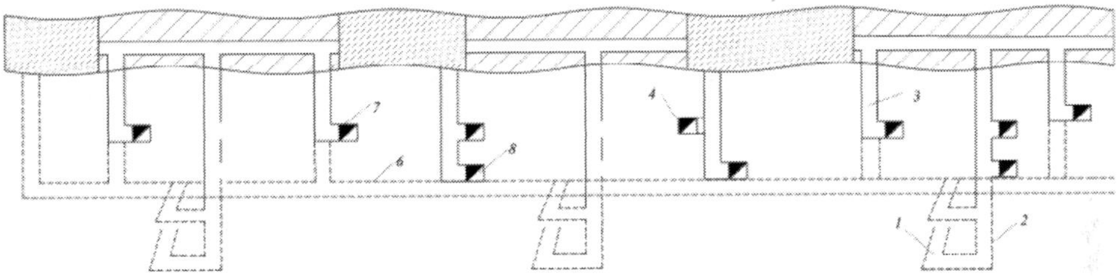

Fig. 3. Level development with three ramps: 1 – inclined drift; 2 – horizontal drift; 3 – haulage sublevel ort; 4 – ventilation-haulage ort; 5 – haulage crosscut of the main level; 6 – haulage drift of the main level; 7 – ore pass; 8 – raise.

Table 3. Drifting amount in access ramp development, m^3.

Mine workings	Access ramp position as to the ore body			
	centre	flanks	centre and flanks	optimal
Inclined	2450	4900	7340	2200
Horizontal	340	680	1020	
Breakthrough	160 – 800	1600	2400	1300
Haulage orts	2280 – 3360	2240 – 3500	2240 – 2280	2280
Ventilation-haulage orts	43004800	3800 – 4300	2160	–
Haulage crosscuts	3600	2160 – 3360	2640	3020
Haulage drifts	3880	3880 – 5600	3800	4840
Ore passes	3040	234 – 3740	2100	2340
Raises	2720 – 3180	2730 – 3640	2730	2130
Sublevel drift breakthrough	–	1920	–	1750
Sublevel drifts	–	–	–	15800
Total workings, m^3	25140 – 26800	25100 – 32400	27200	39800
Ore, m^3/1000 m^3 rocks	111 – 125	14 – 117	127	185

All the variants do not use development operations at the footwall, a loop-haulage system on main levels and the increased sublevel height (12.5 m). The variants differ in stoping directions and the number of access ramps.

The following development schemes are recommended:

– one access ramp between ore bodies at the flank and a hoister on the other flank to develop centre-to-flank stoping;

– one access ramp between ore bodies to develop flank-to-flank stoping;

– two hoisters at both flanks.

In case of inclined sublevel caving and abut ore drawing, sublevel inclined drifts are both haulage and face-entry.

In sublevel mining, concentration of operations increases. If the slope angle is 10 in inclined drifting, the number of workings increases (9 against 6).

With the equal number of sublevels, the developed height of the ore body reduces by 1.5 times, ore drawing conditions improve due to the reduced influence of the working abut and natural drainage occurs.

The efficiency of caving is determined by the quantity of ore losses and dilution [10 – 12].

The dependency of ore losses and dilution on the slope angle of the mine working has been investigated.

To simulate the process, crushed ore of 0 + 6 mm was used, the particle size taken geometrically similar to the actual one at a scale of 1:50. Overlying rocks are crushed 5 – 10 mm rocks. The block model of 160 × 160 × 300 mm enabled us to change the slope angle of a mine working. 0°, 5°, 10° working models were studied. Each series was performed three times.

To investigate the ore drawing quality with changing slope angles of a stope as to the 10° mine working, there were simulated 90°, 100° and 110° variants. In the model, the ore was covered by first two layers 80 cm thick and the third one 40 cm thick. On the surface of each layer, there were marks to form an ore drawing figure. The ore was drawn in portions of 200 g. Each portion was bolted to determine dilution values. The ore drawing was over when dilution reached 75% (Table 4).

Table 4. Ore drawing simulation results.

Caving step	Horizontal drift			5° draw point		10° draw point	
	lot number	rock weight, g	ore weight, g	rock weight, g	ore weight, g	rock weight, g	ore weight, g
1st	1	0	200	0	200	0	200
	2	53	147	65	135	47	153
	3	88	112	78	122	88	112
	4	127	73	105	95	112	88
	5	128	72	118	82	128	72
	6	148	52	145	55	128	72
	7	150	50	165	45	145	55
Total	–	694	706	666	734	648	752
2nd	1	0	200	0	200	0	200
	2	63	137	30	170	38	162
	3	100	100	92	108	70	130
	4	135	65	132	68	105	95
	5	148	52	148	52	143	57
	6	150	50	155	45	146	54
Total	–	596	604	557	634	502	698
3rd	1	0	200	0	200	0	200
	2	28	172	28	172	23	177
	3	60	120	67	133	72	128
	4	108	92	98	102	100	100
	5	143	57	132	68	135	65
	6	150	50	152	48	147	53
Total	–	308	492	309	491	308	492
4th	1	0	200	0	200	0	200
	2	50	150	55	145	53	147
	3	110	90	107	93	105	95
	–	148	52	147	53	150	50
Total	–	308	492	309	491491	308	492
Total, g	–	2107	2493	2009	2591	1935	2665
%%	–	45.8	54.2	43.7	56.3	42.1	57.9

The rock mass drawing from the model is specified by the following:

– at the 1st caving step abut the working, there was no side contact with the rock, conditions for ore sliding outside the caving zone were perfect. 75% ore dilution in the last doze increased with 1400 g drawing;

– on the contrary, at the 4th caving step, there was a contact with perfect ore sliding at the abut plane of the working. That is why, the ore came into motion at the beginning of the drawing and ore dilution increased up to 75% with 800 g drawing. There were no conditions to create a durable ellipsoid of ore drawing.

Table 5 presents research results of the dependency of ore losses on localization of the soil of ore draw points.

Table 5. Ore losses in abut ore drawing, %.

Caving step	Slope angle of the draw point, degree		
	0	5	10
1st	50.4	52.4	53.7
2nd	50.3	53.6	57.6
3rd	57.6	60.2	60.3
4th	61.5	61.4	61.5

With the 10° draw point, ore losses were by 5% less as compared to the horizontal surface of the working.

The research results are matched with those by foreign and national mining scholars [13 – 16]. The data on development mining before stoping can be used while improving underground mining technologies [17 – 20].

4 Conclusions

Improvement of development mining before stoping by means of access ramps is an efficient method of providing the designed mine capacity and labour efficiency.

The most widely spread schemes include the following: an access ramp between ore bodies on the one flank and a hoister on the other; an access ramp between ore bodies with stoping moving from the one flank to the other and two hoisters on both flanks.

Efficient parameters of development mining are chosen through simulating according to the criterion of ore losses and dilution chiefly.

The authors are grateful to Professor Kondrat'ev Yu.I., "The North Caucasian Mining and Metallurgical Institute (State Technological University)", for valuable comments in writing this article and discussing the obtained results.

References

1. Golik, V., Komashchenko, V., & Morkun, V. (2015). Feasibility of using the mill tailings for preparation of self-hardening mixtures. *Metallurgical and Mining Industry*, (3), 38-41.
2. Golik, V., Komashchenko, V., & Morkun, V. (2015). Innovative technologies of metal extraction from the ore processing mill tailings and their integrated use. *Metallurgical and Mining Industry*, (3), 49-52.
3. Kaplunov, D., Rylnikova, M., & Radchenko, D. (2015). Problema ispolzovaniya vozobnovlyaemykh istochnikov energii v khode razrabotki mestorozhdeniy tverdykh poleznykh iskopayemykh. *Fiziko-tekhnicheskie problemy razrabotki poleznykh iskopayemykh*, (1), 88-96.
4. Neverov, A. (2014). Geomekhanicheskaya otsenka kombinirovannoy geotekhnologii pri otrabotke moshchnoy pologoy rudnoy zalezhi. *Fiziko-tekhnicheskie problemy razrabotki*

poleznykh iskopayemykh, (1), 119-132.

5. Golik, V., Komashchenko, V., & Kachurin, N. (2015). Kontseptsiya kombinirovaniya tekhnolohiy razrabotki rudnykh mestorozhdeniy. *Izvestiya Tul'skogo gosudarstvennogo universiteta*, (4), 76-88.
6. Golik, V., Komashchenko, V., Morkun, V., & Zaalishvili, V. (2015). Enhancement of lost ore production efficiency by usage of canopies. *Metallurgical and Mining Industry,* (4), 325-329.
7. Morkun, V., Morkun, N., & Pikilnyak, A. (2015). Adaptive control system of ore beneficiation process based on Kaczmarz projection algorithm. *Metallurgical and Mining Industry*, (2). 35-38.
8. Golik, V.I., Polukhin, O.N., Petin, A.N., Komashchenko, V.I. (2013). Ekolohicheskiye problemy razrabotki rudnykh mestorozhdeniy KMA. *Gornyy zhurnal*, (4), 91-94.
9. Morkun, V., Morkun, N., & Pikilnyak, A. (2014). The gas bubble size distribution control formation in the flotation process. *Metallurgical and Mining Industry*, (4), 42-45.
10. Golik, V., Komaschenko, V., Morkun, V., & Khasheva, Z. (2015). The effectiveness of combining the stages of ore fields development. *Metallurgical and Mining Industry*, (5), 401-405.
11. Golik, V., Komashchenko, V., & Morkun, V. (2015). Geomechanical terms of use of the mill tailings for preparation. *Metallurgical and Mining Industry,* (4), 321-324.
12. Golik, V., Komaschenko, V., Morkun, V., & Burdzieva, O. (2015). Metal deposits combined development experience. *Metallurgical and Mining Industry*, (6), 591-594.
13. Morkun, V., Morkun, N., & Pikilnyak, A. (2014). Iron ore flotation process control and optimization using high-energy ultrasound. *Metallurgical and Mining Industry*, (2), 36-42.
14. Golik, V., Komashchenko, V., Morkun, V., & Burdzieva, O. (2016). Simulation of rock massif tension at ore underground mining. *Metallurgical and Mining Industry,* (7), 76-79.
15. Haifeng, W., Yaqun, H., Chenlong, D., Yuemin, Z., Youjun, T. & Cuiling, Y. (2012). Development of Mineral Processing Engineering Education in China University of Mining and Technology. *Advances in Computer Science and Engineering*, 77-83.
16. Bakhtavar, E., & Shahriar, K. (2007). Optimal ultimate pit depth considering an underground alternative. *Proceedings of Fourth Aachen International Mining Symposia-High Performance Mine Production*, 213-221.
17. Golik, V.I., Razorenov, Y.I., Ignatov, V.N., Khasheva, Z.M., & Shulgaty, L.P. (2016). The History of Russian Caucasus Ore Deposit Development. *The Social Sciences, 11*(15), 3742-3746.
18. Dmitrak, Y., & Kamnev, Y. (2016). Vedushchiy proektno-izyskatelskiy i nauchno-issledovatelskiy institut promyshlennoy tekhnologii. Put' dlinoy v 65 let. *Gornyy zhurnal*, (3), 6-12.
19. Kachurin, N., Stas, G., Kalaev, S., & Korchagina, T. (2016). Geoecological assessment of the effectiveness of environmental and nature protection measures in under the earth. Mining coal. *The Bulletin of Tula State University. Earth Sciences*, (3), 63-81.
20. Kachurin, N., Stas, G., Levin, D., & Fisher, W. (2016). Aerodynamics of waste dumps of coal mines. *The Bulletin of Tula State University. Earth Sciences*, (1), 23-33.

Analysis of patterns of the open-pit mine water influx formation in the conditions of the Inhulets iron ore deposit using a three-dimensional geofiltration model

Vasyl Tymoshchuk[1*], *Yevheniia* Sherstiuk[1], and *Tamara* Morozova[2]

[1]National Mining University, Department of Hydrogeology and Engineering Geology, 19 Yavornytskoho Ave., 49005 Dnipro, Ukraine
[2]National Mining University, Department of Physics, 19 Yavornytskoho Ave., 49005 Dnipro, Ukraine

Abstract. The purpose is to study the patterns of groundwater influxe formation within the north-eastern side of the Private Joint-Stock Company "Inhuletskyi hirnycho-zbahachuvalnyi kombinat" (PJSC INHZK) open-pit mine in the conditions of disturbed hydrodynamic regime of the eastern adjacent side of the open-pit mine by the developed three-dimensional geofiltration model. The patterns have been determined of the open-pit mine water influx formation on the basis of a complex analysis of the actual data related to geological and hydrogeological structure of the iron ore deposit, the intensity and character of groundwater influx within the eastern side of the open-pit mine, and the results of numerical simulation of geofiltrational processes. By the results of inverse problems solutions, the geofiltration model reflects adequately the geological, hydrogeological, and geotechnical conditions of the studied territory where the technogenically disturbed hydrodynamic regime have been formed; the calculated values of the open-pit mine water influx within the eastern side of the open-pit mine and its north-eastern site, have been determined. It has been found that the values of groundwater influx and their distribution within the eastern side of the open-pit mine and its north-eastern site depend on the manner of the water-permeable alluvial strata occurrence on the eroded surface of seat clays, and the existence of a technogenic aquifer near the rock-disposal dump base at the adjacent site of the open-pit mine. The results of this research form the basis for a substantiation of technical solutions for dewatering the north-eastern side of the PJSC INHZK open-pit mine in the conditions of further development of mining operations, and for determining the hydrodynamic parameters of dewatering facilities.

1 Introduction

The open-pit mining of iron ore deposits in the conditions of technogenically disturbed hydrodynamic regime is often accompanied by the formation of large volumes of open-pit mine water influxes. This fact complicates mining works and leads to the development of

* Corresponding author: vasyl.tymoshchuk@gmail.com

negative engineering-geological processes and phenomena in the sides of the open-pit mines [1 – 4]. The investigation of peculiarities of the open-pit mine water influx formation has been performed for the geological and hydrogeological conditions of the Inhulets iron ore deposit, which is developed by PJSC INHZK. The total volume of groundwater influx to the PJSC INHZK open-pit mine is 510 – 670 m^3/h, including water influx from the alluvial aquifer within the eastern side of the open-pit mine, about 90 – 120 m^3/h.

The development of landslide deformations and collapse phenomena in the area of occurrence of sandy-clay alluvial deposits and underlying sedimentary formations of the Neogene-Paleogene age should be distinguished among the number of consequences caused by flooding the eastern side of the open-pit mine, in particular of its north-eastern site [5, 6].

One of the factors influencing the nature of water seepage in the north-eastern side of the open-pit mine is the location within this area of the No. 3 rock-disposal dump. It creates an additional hydro-geomechanical loading onto the rock massif near the adjacent side of the iron ore open-pit mine.

The presence of significant open-pit mine water influx in the conditions of complex geological and hydrogeological structure of the iron ore deposit, as well as the existing spatial and temporal uncertainty of groundwater influx distribution within the north-eastern side of the open-pit mine complicates the development of effective technical solutions aimed at ensuring its hydra-geomechanical stability.

In this regard, the objective of the research is to study the patterns of formation of the open-pit mine water influx and its distribution within the eastern side of the PJSC INHZK open-pit mine and its north-eastern site for substantiation of effective technical solutions in their dewatering.

The object of research is geofiltration processes within the laminal rock massif in the conditions of the existing interconnection between the system components of the 'north-eastern side of the PJSC INHZK open-pit mine – No. 3 rock-disposal dump – floodplain of the Inhulets River' (Fig. 1).

In general, the mining and geological conditions of the north-eastern side of the PJSC INHZK open pit mine are determined by the complex hydrodynamic and geomechanical interaction of the natural and technical elements of the system 'iron ore open-pit mine – No. 3 rock-disposal dump – alluvial deposits within the floodplain of the Inhulets River'. In addition, this interaction takes place in the conditions of geological environment disturbed in whole by mining works.

During the entire period of the Inhulets iron ore deposit mining, the groundwater influx into the open-pit mine was formed by drainage of the groundwater in the crystalline massif fractured zone, the terrigenous-carbonate deposits of the Paleogene-Neogene age and the quarternary alluvial aquifer of the Inhulets River floodplain.

Within the eastern side of the operating open-pit mine, the predominant share in the total volume of open-pit mine water influx belongs to the alluvial aquifer, which serves as the groundwater transit from the Inhulets River to the contour of their drainage in the interval of alluvial deposits developed by the side of the open-pit mine.

To protect the eastern side of the open-pit mine from the groundwater influx formed from the alluvial aquifer, the Inhulets River was directed and withdrawn to a distance of up to 400 m, while along the eastern side of the open-pit mine the seepage cutoff wall has been constructed. It is a polyvinyl chloride film enclosed in a trench and filled with a bentonite clay solution with a designed permeability of 0.001 m/day.

The construction of the seepage cutoff wall has provided a sufficiently effective protection of the eastern side of the open-pit mine from water seeping from the Inhulets River, as evidenced by the difference in groundwater levels recorded in the observation wells on both sides of the seepage cutoff wall. However, the effectiveness of this wall over time has been significantly reduced due to numerous its discontinuity.

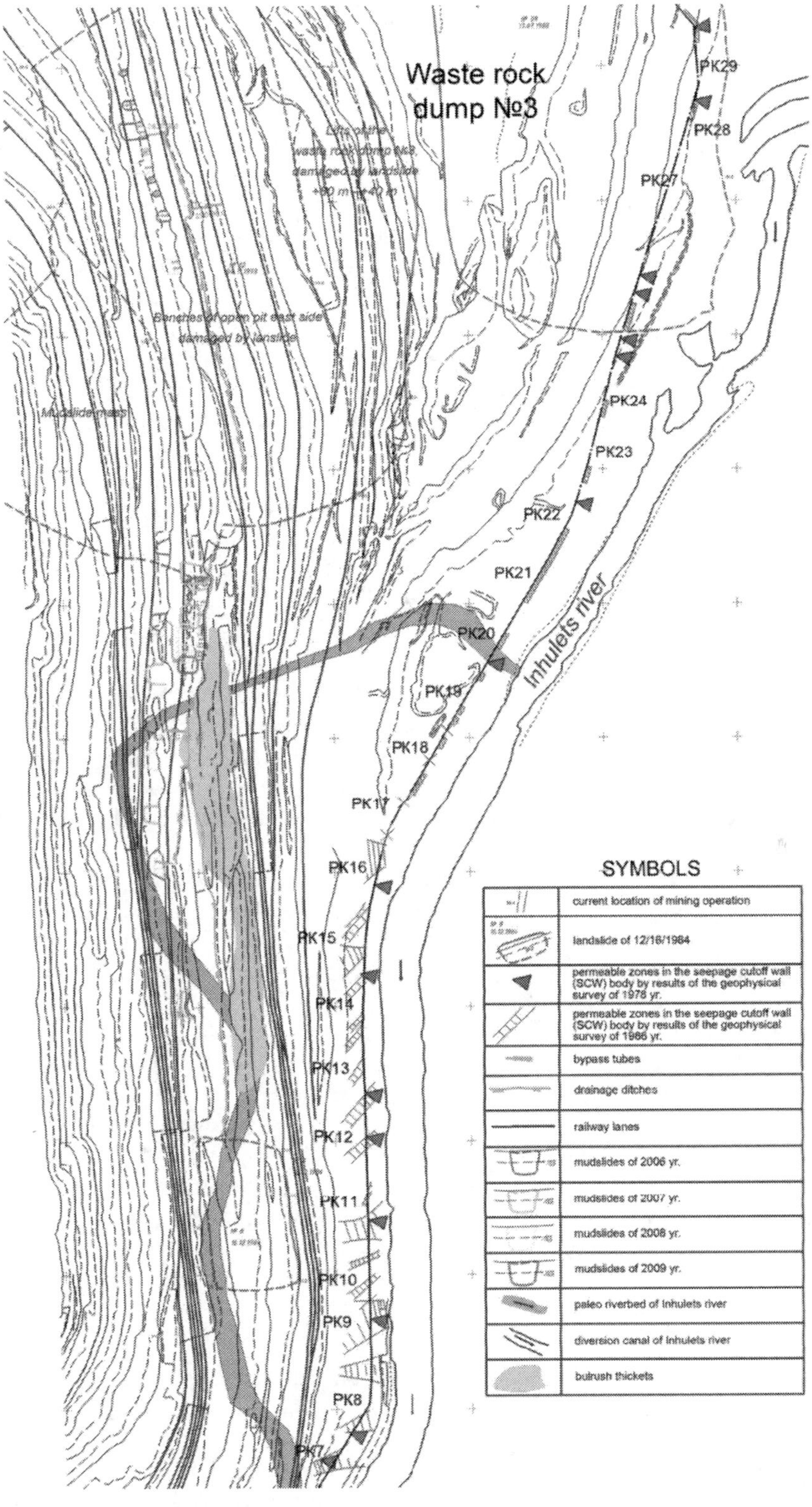

Fig. 1. The character of groundwater influx manifestation and related to it deformation processes within the eastern side of the open-pit mine (according to PJSC INHZK).

The analysis of the engineering survey materials, regime observations data and the open-pit mine water influx dynamics observations have shown that to date the main sources of formation of the open-pit mine water influx within its north-eastern side are the seepage water from the Inhulets River, through which the dynamic resources formation of the alluvial aquifer, as well as infiltration recharge of the quarternary aquifer within the No. 3 rock-disposal dump location in the Inhulets River floodplain.

The decisive influence of a seepage from the Inhulets River on the open-pit mine water influx formation is evidenced by the influx dependence on the mutual position of the riverbed and the contour of the open-pit mine within its north-eastern side (see Fig. 1). The participation of seepage water in the alluvial aquifer recharge is also confirmed by the dynamical increase in flooding the eastern side of the open-pit mine and the chemical composition of groundwater, involved in the formation of open-pit mine water influx.

Analysis of the dependence of groundwater influx volume on climatic factors indicates the existence of a sufficiently stable relation between the open-pit mine water influx and the amount of atmospheric precipitation, as well as their distribution during a year in the conditions of the existing regime of the Inhulets River influence on mining operations. The determined dependence also allows to consider the formation of accumulative water within the No. 3 rock-disposal dump as one of the factors contributing to the flooding of the north-eastern side of the open-pit mine at the site of its junction with the No. 3 rock-disposal dump located within the Inhulets River floodplain.

Indirect evidence of the accumulative water influence on the alluvial aquifer hydrodynamic regime at the site of the No. 3 rock-disposal dump is the landslide processes development within the north-eastern side of the open-pit mine and in the dump itself, which occurred during the whole time of the deposit mining. Thus, in the early 1980's, within the eastern side of the open-pit mine, there were several local landslides with a total volume of about 1 million cubic meters, the cause of which was the water-logging of weak clay rocks, which formed the benches of the open-pit mine sides with absolute elevations: +24 m, +12 m, ± 0.0 m, –15 m.

The role of alluvial deposits in the floodplain of the Inhulets River is largely determined by its structure: the soil layer consists of consertal sandstone with pebble and gravel of quartz, and other solid crystalline rocks. They have a high hydraulic permeability, which ensures the transit of seepage water from the Inhulets River to the drainage boundary, that is to the eastern side of the PJSC INHZK open-pit mine.

A characteristic feature of the groundwater influx distribution within the eastern side of the open-pit mine and its northeastern site is their belonging to lower elevation area of Paleogene clay layer top up to absolute elevations less than +20.0 m, as a rule, from +10.0 to + 12.0 m. Thus, according to PJSC INHZK data, intensive groundwater influx (up to 50 m^3/h as of June 21, 1989) within the north-eastern side of the open-pit mine and related to it large-scale landslide deformations at the junction with the No. 3 rock-disposal dump have coincided with the period of entering the open-pit mine side into the zone of Paleogene clay layer top (bottom of alluvial consertal sandstone) at elevations of –6.0 to + 7.0 m.

2 Modeling

In conditions of complex hydrodynamic interaction of the geotechnical systems elements, the application of mathematical model technique has become widespread in the study of geofiltration processes [3, 7 – 9].

To study the conditions of the open-pit mine water influx formation and to determine the patterns of its distribution within the north-eastern side of the open-pit mine, we have created a geofiltration model of the studied area by means of the software complex MODFLOW 2009.1. When developing this model, the actual data were taken into account:

the observations of groundwater influx within the eastern side of the open-pit mine; peculiarities of the alluvial strata occurrence at the junction of the open-pit mine with the No. 3 rock-disposal dump; patterns of permeability changes of water-bearing strata within the studied area; the character of hydraulic relation between the Inhulets River and water-bearing strata, as well as the predictive infiltration recharge elevation of the technogenic aquifer in the rock-disposal dump area.

The numerical model approximates the spatial groundwater flow with constant density in a porous medium. This model is described by the partial differential equation:

$$\frac{\partial}{\partial x}\left(k_{xx}\frac{\partial h}{\partial x}\right)+\frac{\partial}{\partial y}\left(k_{yy}\frac{\partial h}{\partial y}\right)+\frac{\partial}{\partial z}\left(k_{zz}\frac{\partial h}{\partial z}\right)+W=S_s\frac{\partial h}{\partial t}, \quad (1)$$

where k_{xx}, k_{yy} and k_{zz} – hydraulic conductivities in directions of coordinate axes X, Y and Z; (L/T); h – potentiometric head (L); W – volumetric flux per unit, representing sources and sinks of water (T^{-1}): for inflow $W > 0$, for outflow – $W < 0$; S_s – specific storage of porous environment (L^{-1}); t – time (T).

An equation (1), together with the boundary and initial conditions, describes a three-dimensional transient groundwater flow in a heterogeneous and anisotropic medium, provided that the main directions of the hydraulic conductivities coincide with the directions of the coordinate axes [10 – 12].

The calculation mapping of the studied area has been done in accordance with the peculiarities of the geological structure, the character of occurrence of water-bearing rocks and waterproof rocks, as well as their hydraulic properties, the existence of a hydraulic relation between the surface water and groundwater, specifics of aquifer recharge and discharge along their area and along their contours [8, 9].

The dimensions of the simulated area have been determined within the territory of 4600×2400 m, respectively, in the meridional and latitudinal directions with a total area of a model of 11.0 km^2. The simulated area is represented by a grid of blocks with dimensions of 25×25 m in order to approximate thoroughly the current configuration of the Inhulets River, the contour of the north-eastern side of the open-pit mine and the seepage cutoff wall, as well as of the hypsometry of rock layers and groundwater levels in the alluvial aquifer.

The model structure has an eight-layer thickness (Fig. 2), according to the rocks stored into the No.3 rock-disposal dump, the geo-lithological structure of water-bearing deposits in alluvial aquifer, that include the Paleogene clays (later Eocene) and coal and clay strata of the middle Eocene, the lower layer of which according to the geological structure of the considered area is a fractured zone of ore-bearing crystalline massif.

As the hydrodynamic boundaries on the external contours of the model, the remote boundaries with constant recharge have been assigned, corresponding to the boundaries of the third kind (throughflow) $Q = f(H)$, according to the general hydrodynamic scheme of the studied area.

The values of hydraulic heads on the model contours are determined from the data of exploratory drilling in accordance with groundwater elevations within the studied sites of the simulated area, and this relates to a decisive influence of geomorphological structure on its hydrodynamic regime. The water conductivity values on the external contours of the calculated layers have been determined from the averaged values of the lithological varieties permeability within separate sites of the model area and range from $3.6 \cdot 10^{-5}$ to 102.46 m^2/day.

On the contour sites of the simulated area, where the inflow/outflow is absent in accordance with the groundwater flow direction, the contours with the boundary condition of the second kind ($Q = 0$) have been assigned.

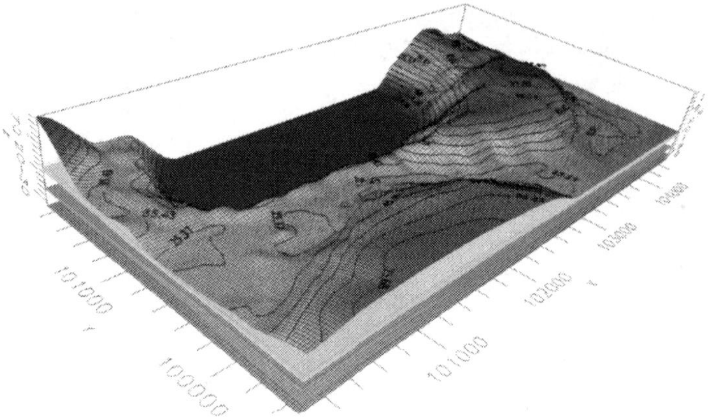

Fig. 2. Structure of the simulated area within the eastern side of the PJSC INHZK open-pit mine.

The inner hydrodynamic boundary of the model is the Inhulets River. Within the studied area the Inhulets River has its absolute elevations of 22.05 – 22.50 m, its width varies from 50.0 to 75.0 m, averaging 60.0 m, the depth varies from 1.6 to 2.0 m, reaching 2.5 m at separate sites.

In terms of hydrodynamics, the Inhulets River in the numerical model is represented by the boundary conditions of the third kind, which consider the interconnection of surface water and groundwater determined by the riverbed conductivity. The parameters of water conductivity of riverbed deposits within the calculated blocks of the model have been assigned in accordance with the alluvial deposits permeability with values from 0.000625 to 4312.5 m^2/day.

The drainage contour of alluvial deposits within the north-eastern side of the open-pit mine is determined by the position of the bottom hypsometry of consertal sandstone layer underlaid by the Paleogene clay, and opened by a side of the open-pit mine.

The seepage cutoff wall (SCW) operation was simulated by setting a low-permeable 'wall' at a distance of 25 – 50 m from the western bank of the Inhulets River, deepened along its entire length to the siltstone clay layer top (calculated layer 6). In accordance with the design parameters of the seepage cutoff wall, its width was assigned equal to 0.7 m with its material permeability of 0.05 m/day.

The calculated values of hydraulic conductivity, accepted in accordance with the structure of the numerical model and specified by the results of inverse problems solution, are given in Table 1.

Table 1. Calculated parameters of permeability and porosity of a numerical model

Calculated layer	Permeability and porosity parameters		
	permeability coefficient, K ($K_X = K_Y = K_Z$), m/day	specific yield μ, %.	specific storage μ*, m^{-1}
Layer 1 – rock-disposal dump	0.5 – 0.7	10.0	0.0001
Layer 2 – loam and sandy loam	0.001 – 0.7	10.0	0.0001
Layer 3 – fine sand	0.001 – 0.84	15.0	0.00001
Layer 4 – sandy clay	0.001	2.0	0.00001
Layer 5 – consertal sandstone	6.9 – 13.0	20.0	0.00001
Layer 6 – siltstone clay	0.00001	1.0	0.00001
Layer 7 – coal clay	0.0005	1.0	0.00001
Layer 8 – weathered shale	0.00005	1.0	0.00001

The infiltration value within the studied area ranges from 13 to 42 mm/year [13]. At the location of the No. 3 rock-disposal dump, composed of oxidized and semi-oxidized jaspilite and overburden rocks, the infiltration recharge value was assumed equal to 50 mm/year at the stage of the inverse problem solution – taking into account the extra recharge in accordance with the recommendations for industrial areas of the mining and metallurgical complex, with the possibility of its further adjustment.

According to the calibration data, the deviation of the calculated groundwater elevations in the alluvial aquifer comparatively to the measured ones does not exceed 1.364 m (Fig. 3). The standard error of the estimate defined by the results of performed calculations is 0.126 m. The significance of statistical estimates is characterized by a value of $R^2 = 0.737$ m and its standardized value of 8.126%.

In terms of considered conditions, the above estimates indicate a satisfactory convergence of the modeling results, taking into account the existing heterogeneity of the alluvial aquifer permeability and the inconstant character of the spatial occurrence of water-bearing layers and waterproof layers, separating them.

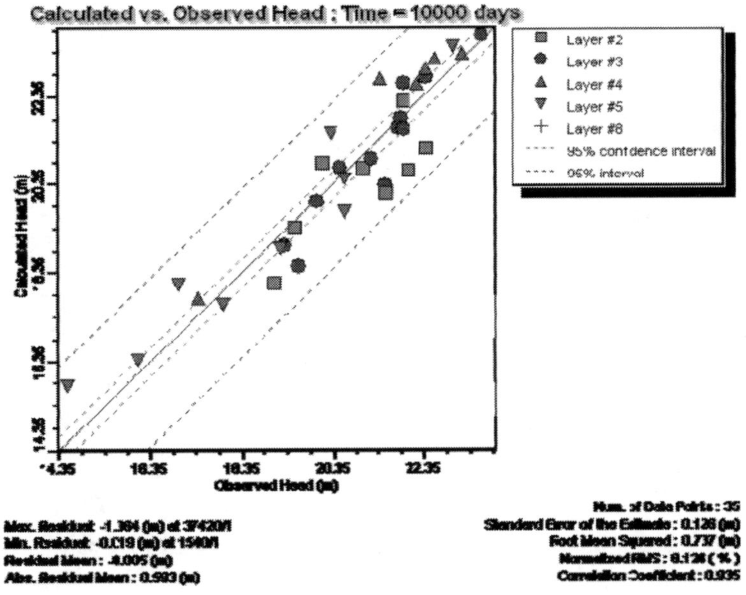

Fig. 3. Calibration plot of measured hydraulic heads plotted against MODFLOW-simulated heads as of April-May 2015 – inverse problem solution, transient flow regime.

3 Results

MODFLOW-simulated groundwater heads at the baseline of the rock-disposal dump and the alluvial aquifer within the studied territory, which were obtained from the inverse problems solution in accordance to the accepted hydrodynamic scheme for the period of engineering surveys are presented in Fig. 4 – 6.

According to modeling results, the formation of the hydrodynamic regime of the alluvial aquifer in the studied area occurs due to infiltration of atmospheric precipitation, lateral inflow from the watershed sites (eastern boundary of the simulated area), and inflow of seepage water from the Inhulets River; and discharge occurs by leaving the permeable rocks on their drainage contour within the eastern side of the open-pit mine.

At the same time, the seepage from the Inhulets River has the significant value for the alluvial aquifer recharge, which is located directly in the drainage zone of the open-pit mine.

MODFLOW-simulated groundwater influx on the discharge contour of the alluvial aquifer is well correlated with the measured values of the groundwater influx within the eastern side of the iron ore open-pit mine.

According to PJSC INHZK data, the total groundwater influx towards the eastern side of the open-pit mine as of 2015 reached 90 – 120 m^3/h or 2160 – 2880 m^3/day. Based on the data of MODFLOW-simulated inverse problem solution, the total value of drainage water intake for this contour is 2374.9 m^3/day or 98.9 m^3/h, which indicates a satisfactory convergence of the calculated groundwater influx values with the measured data within the eastern side of the open-pit mine. Taking into account the losses due to the bottom of alluvial aquifer at the southern site of the side of the open-pit mine at the level of 551.6 m^3/day (22.9 m^3/h), the total estimated influx towards the eastern side is about 121.8 m^3/h.

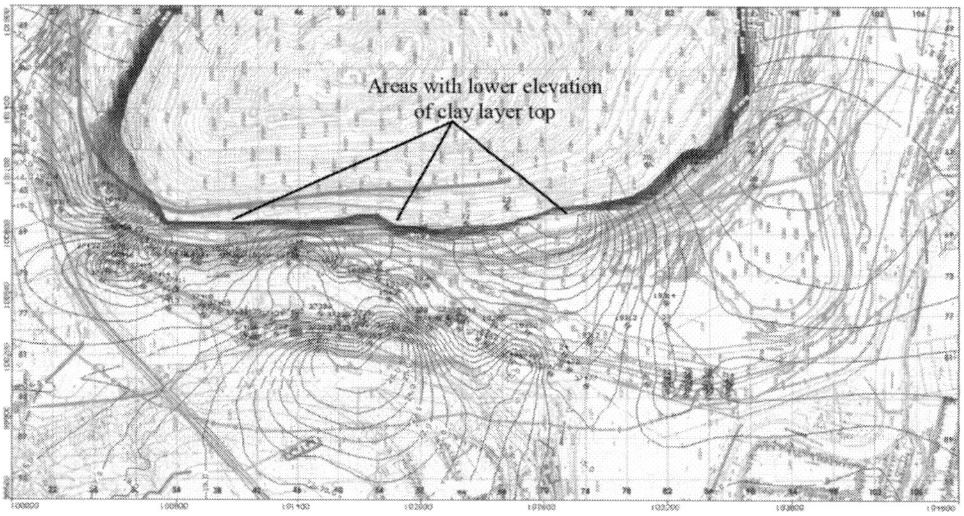

Fig. 4. Elevation of Paleogene clay layer top within the eastern side of the PJSC INHZK open-pit mine, m.

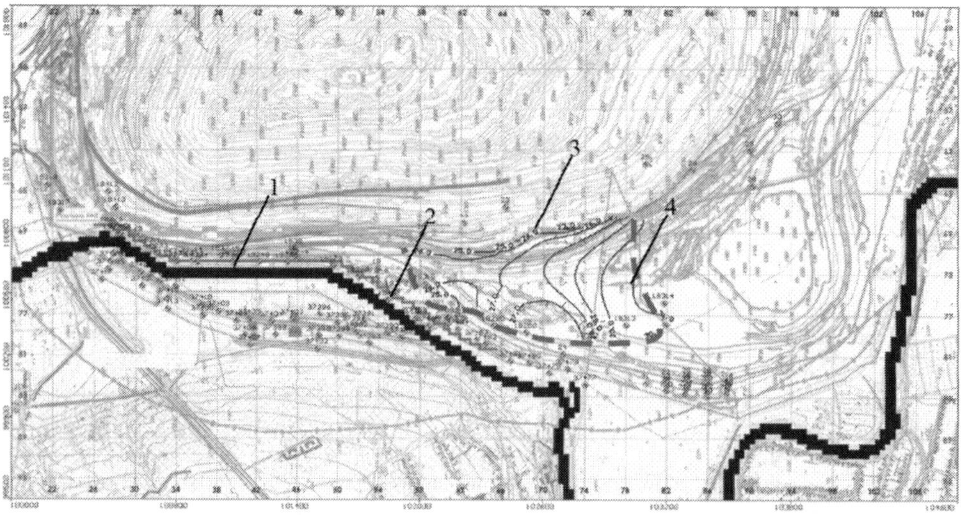

Fig. 5. MODFLOW-simulated groundwater heads at the baseline of №3 rock-disposal dump as of April-May 2015 – inverse problem solution, transient flow regime, m: 1 – current position of the Inhulets River; 2 – seepage cutoff wall; 3 – groundwater flow direction; 4 – technogenic aquifer.

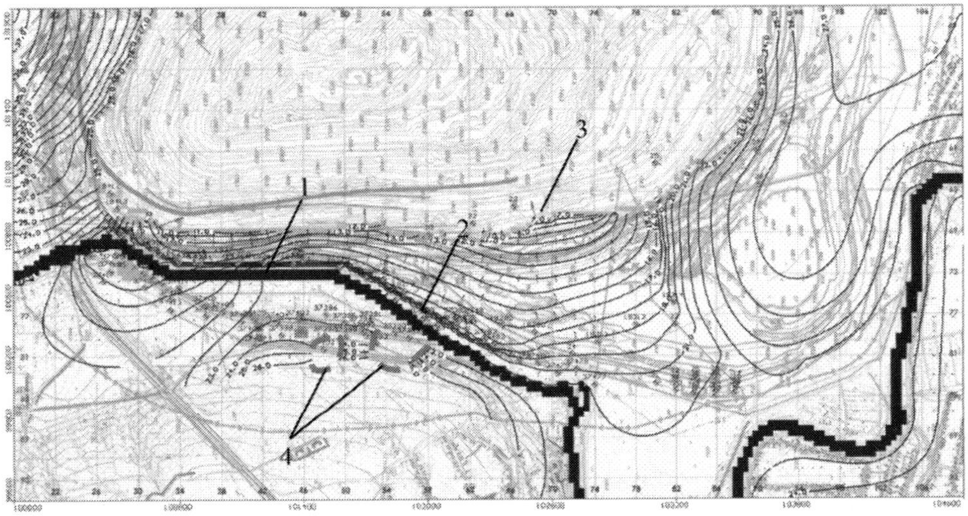

Fig. 6. MODFLOW-simulated groundwater heads of alluvial aquifer as of April-May 2015 – inverse problem solution, transient flow regime, m: 1 – current position of the Inhulets River; 2 – seepage cutoff wall; 3 – groundwater flow direction; 4 – “dry” zones.

The distribution of calculated groundwater influx values along the open-pit mine contour is characterized by relative concentration of groundwater influx within the sites 30 – 42 (831.29 m^3/day), 50 – 58 (436.87 m^3/day) and 62 – 74 (587.28 m^3/day) of mine surveying axes (s.a.). The position of these sites coincides with the location of the relatively reduced elevation of the Paleogene clay layer top (Fig. 7). The values of the MODFLOW-simulated groundwater influx within the allocated areas reach the values of 13.2 m^3/h (s.a. of 38 – 42) and 9.5 – 10.4 m^3/h (s.a. of 50 – 54, 66 – 70).

Significant groundwater influx in the area of s.a. 30 – 42 is related to the proximity of the southern site of the open-pit mine eastern side to the existing bed of the Inhulets river. According to modeling results, the total groundwater influx value at this site is 34.6 m^3/h, which is consistent with the groundwater influx in situ observations – the visually estimated groundwater influx value within the mentioned axes is in the range of 33.0 – 56.0 m^3/h.

As defined by the modeling results, the presence of a local zone for descending discharge of the alluvial aquifer into the fractured zone of the underlying parent rock, is characteristic for this site. This fact is evidenced by the surface-level deformations of the alluvial aquifer, which are confined to the areas with lower elevations of the clay layer top.

Relatively low calculated values of groundwater influx within 58 – 90 mine surveying axes – 785.16 m^3/day (32.71 m^3/h) are explained by the relative removal of the Inhulets River from the northern site of the open-pit mine eastern side. The current situation is confirmed by the actual observation data: at the end of April 2015, the groundwater influx into the open-pit mine along the mine surveying axes 50 – 70 did not exceed 21.30 m^3/h. The water influx, determined for this period, has been recorded after elimination of water manifestations at the horizon +12.0 (measurements on 04/02/2015) and a decrease in groundwater influx at the horizon –30.0 (surveying axes 65 – 70) from 68.18 to 15.36 m^3/h according to the measurements, respectively on 25/03/2015 and 04/02/2015.

We have carried out the comparative assessment of the groundwater influx to the north-eastern site of the open-pit mine during different periods of iron ore deposit mining and defined the direct correlation of their values on the hypsometric of the layer bottom of heterogeneous gravel sands within the contour of their development by the eastern side of the open-pit mine.

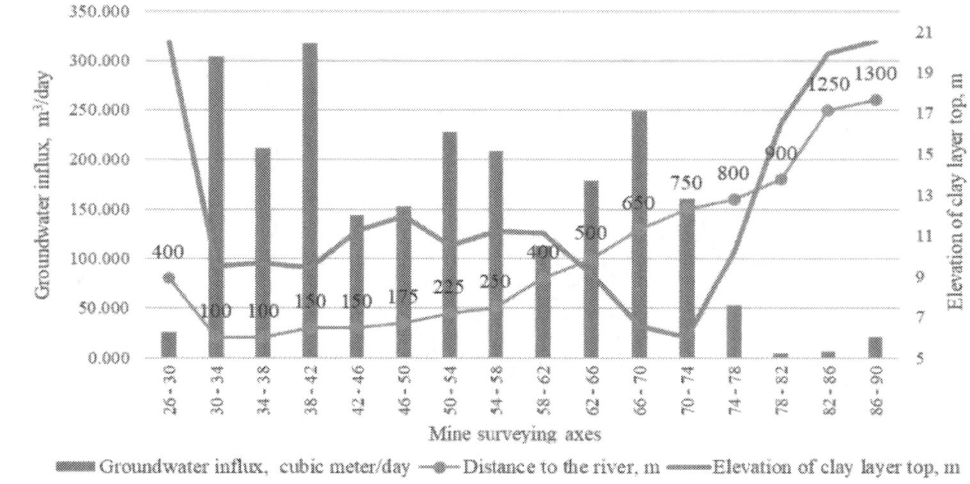

Fig. 7. Comparative diagram of the layer top elevation of Paleogene clay, the values of groundwater influx into the open-pit mine and the distance from the open-pit mine contour to the Inhulets River along the eastern side of the open-pit mine.

The above is confirmed by the results of hydrodynamic modeling, according to which the concentration of groundwater influx in ranges of 50 – 58 and 62 – 74 surveying axes coincides with the lowered absolute elevations up to 10.4 – 12.0 m and 5.5 – 10.4 m of the layer bottom of gravel sand lying on the Paleogene clay within the north-eastern site of the open-pit mine side.

The groundwater influx distribution within the open-pit eastern side indicates that influx values depend on the bottom hypsometry of the heterogeneous gravel sands, lying on the eroded surface of Paleogene clay, the presence and distribution of waterproof alluvial clay layer in the roof of the gravel sand, the hydraulic permeability of them at particular sites of the Inhulets River floodplain and the infiltration value of the alluvial aquifer recharge both within the floodplain and at the site of the №3 rock-disposal dump location.

According to the modeling data, it has been found that the presence of strata with reduced hydraulic permeability at the base of the No.3 rock-disposal dump in conditions of increased infiltration recharge leads to the water-saturated zone formation in the lower range of the rock dump with the general direction of its surface slope towards the south-west and south-east and with the absolute elevations of 24.1 – 31.4 m.

With the value of infiltration recharge within the rock dump of 50 mm/year, the groundwater influx values from the technogenic horizon on the contour of the open-pit mine in the range of s.a. 54 – 82 reach 28.27 m^3/day (1.2 m^3/year).

4 Conclusions

When carrying out the research, a spatial geofiltration model has been developed of the eastern side of the PJSC INHZK open-pit mine and its north-eastern site, which, in accordance with the general hydrodynamic scheme, represents the main patterns in the recharge and discharge of the alluvial aquifer formed within the studied area. The main patterns that we have taken into account are the constant recharge of the alluvial aquifer by the seepage water of the Inhulets River and the additional infiltration recharge from the No. 3 rock-disposal dump located within the floodplain, in the zone where the alluvial aquifer is drained by the eastern side of the open-pit mine.

The values of groundwater influx and the particularity of its distribution, determined by the results of modeling, are consistent with the measured water influx values and its distribution both within the entire eastern side of the open-pit mine and within its north-eastern site. With a total volume of groundwater influx towards the eastern side of the open-pit mine at a level of 2374.9 m^3/day or 98.9 m^3/h, its maximum is 316.8 m^3/day (13.2 m^3/year) in the mine surveying axes interval 38 – 42 and gradually decreases in the northern direction to 159.8 – 249.1 m^3/day (6.7 – 10.4 m^3/h). This is determined by the relatively remote location of the side of the open-pit mine, being a drainage of the alluvial aquifer, from the Inhulets River.

A characteristic feature in the distribution of groundwater influx values within the eastern side of the open-pit mine is their dependence on the manner of heterogeneous gravel sands occurrence on the eroded surface of Paleogene clay. The natural growth of the calculated groundwater influx within the eastern side of the open-pit mine corresponds to areas of lowered (up to 5.5 – 12.0 m) elevations of layer bottom of the gravel sand aquifer. With the existing stage of mining, there are two such sites in the eastern area of the open-pit mine – within the s.a. 50 – 58 and 62 – 70, where the values of MODFLOW-simulated groundwater influx reach 249.09 m^3/day (10.38 m^3/h).

According to the modeling results, it has been established that the hydrodynamic impact of the No.3 rock-disposal dump within the north-eastern side of the open-pit mine is determined by both the infiltration recharge value at the site of the rock-disposal dump location and the possible occurrence of a layer with low-permeable pressed rocks in its base, which leads to the water-saturated zone formation in the lower range of the rock dump with the general direction of its surface slope towards the south-west and south-east and with the absolute elevations of 24.1 – 31.4 m.

The research was carried out with the cooperation and technical support of the specialists from the Private Joint-Stock Company "Inhuletskyi hirnycho-zbahachuvalnyi kombinat" Head Geological Department.

References

1. Ahkozov, Yu.L., Chupriy, S.V., & Mechnikov, Yu.P. (2006). Vliyanie tektonicheskogo faktora na inzhenerno-geologicheskie svoystva porod osadochnogo chekhla Inguletskogo mestorozhdeniya (Krivorozhskiy basseyn). *Heoloho-mineralohichnyi visnyk, 2*(16), 100-103.
2. Bahrii, I., Blinov, P., & Bielokopytova, N. (2002). *Heoekolohichni problemy Kryvorizkoho baseinu v umovakh restrukturyzatsii hirnychodobuvnoi haluzi.* Kyiv: Feniks.
3. Rudakov, D., Timoschuk, V., Perkova, T., & Sherstyuk, E. (2011). Identifikatsiya trekhmernoy geofiltratsionnoy modeli tekhnogenno nagruzhennoy territorii tsentral'noy chasti Krivorozhskogo basseyna. *Naukovyi Visnyk Natsionalnoho Hirnychoho Universytetu*, (5), 21-25.
4. Timoshuk, V., Demchenko, J., & Sherstuk, Y. (2010). The role of natural and technogenic components in failure of geomechanical stability of the territories which are in the influence zone of mining objects. *New Techniques and Technologies in Mining,* 189-192. https://doi.org/10.1201/b11329-31
5. Stetsenko, A. (2018). Lithology and geochemistry of sedimentary rocks in Pavlivsk field in the western part of the Kryviy Rih structure. *Visnyk of V.N Karazin Kharkiv national university-series geology geography ecology*, (48), 92-103. https://doi.org/10.26565/2410-7360-2018-48-07
6. Tadyeyev, O. (2017). To the problem of elimination of invariance loss effects in the evaluation of deformation fields of the earth using GNSS-data. *Geodynamics, 1*(23), 34-45. https://doi.org/10.23939/jgd2017.02.034
7. Sadovenko, I., Zahrytsenko, A., Podvihina, O., & Dereviahina, N. (2017). Water balance control within rock mass using the capacity of water-bearing formations. *Naukovyi Visnyk Natsionalnoho Hirnychoho Universytetu, 4*(160), 19-27.

8. Timoshchuk, V.I., & Sherstyuk, Ye.A. (2012). Geofiltration regularities in the areas loaded by gravitation in tailings and waste rock dumps. *Naukovyi Visnyk Natsionalnoho Hirnychoho Universytetu*, (4), 30-36.
9. Tymoshchuk, V., Tishkov, V., & Sherstyuk, E.A. (2013). Hydrodynamic substantiation of water control measures at the site of sludge collector in Yasinovaya valley, Dnipropetrovsk oblast. *Naukovyi Visnyk Natsionalnoho Hirnychoho Universytetu*, (3), 5-10.
10. Kinzelbach W. (1986). *Groundwater modeling*. Amsterdam: Elsevier.
11. McDonald, M.G., & Harbaugh, A.W. (1984). A modular three-dimensional finite-difference ground-water flow model. *U.S. Geological Survey Open-File Report* 83-875.
12. Schlumberger. (2009). Visual MODFLOW 2009.1 User's Manual. 1st ed. Waterloo, Ontario, Canada: Schlumberger Water Services.
13. Grinevskiy, S., & Novoselova, M. (2010). Zakonomernosti formirovaniya infil'tratsionnogo pitaniya podzemnykh vod. *Vodnye resursy, 37*(6), 1-12.

Scientific-practical results of monitoring anthropogenic influence upon environment of mining territories

Nikolay Kachurin[1], *Vitaliy* Komaschenko[2], *Vladimir* Golik[3], and *Volodymyr* Morkun[4*]

[1]Tula State University Department of Geotechnologies and Underground Constructions, 92 Lenina Ave., 300012 Tula, Russia
[2]Belgorod State National Research University, Trade-union St., 117485 Belgorod, Russia
[3]North Caucasus Mining Metallurgical Institute, 85 Pobedy St., 308015 Vladikavkaz, Russia
[4]Kryvyi Rih National University, Computer Science, Automation and Control Systems Department, 11 Vitalii Matusevych St., 50027 Kryvyi Rih, Ukraine

Abstract. The objective of this paper is to improve the ecological monitoring system and to take effective technological measures by way of wastes neutralization. The scientific-practical results of monitoring the anthropogenic influence upon the environment of mining-industrial territories were discussed. The global environmental problems of modernity are connected only with the anthropogenic pollution of the atmosphere. Pollution by hazardous substances at various enterprises harms health and disrupts the dynamic balance of natural systems. It is especially typical for mining-industrial region territories. The solution of the problem of environmentally optimal distributing resources for industrial enterprises of the mining-industrial region will allow making maximal profit from each enterprise by assumption of reducing anthropogenic influence upon environment up to the maximum permissible value, which is specified by normative documents. The patterns of formation of mining enterprises influence consequences on the atmosphere have been improved with the use of experimental and theoretical studies and in order to optimize the resources distribution for preventive measures, which can ensure the atmospheric safety of the mining-industrial region. The proposed methodological principles for a comprehensive assessment of the atmosphere of the mining-industrial region allow to implement an integrated approach to forecasting the intensity of atmospheric pollution, the economic efficiency of production and atmospheric air condition control, using the basic requirements of the environmental imperative for the territory under consideration.

1 Introduction

Increasing scale of environmental influencing imparts special topicality for problem of creating adequate environmental-economical mathematical models. Increasing scale of anthropogenic influence upon natural environment and its negative consequences and else

*Corresponding author: pikilnyak@gmail.com

https://doi.org/10.1051/e3sconf/20186000031

possibilities of optimizing this influencing require in-depth study. The search of scientific substantiated forms and scale of industrial human activity providing rational using natural resources and getting necessity useful productions without pernicious influence upon natural environment acquires specific significance [1 – 10].

Provisional evaluations show that long-range planning without taking into account environmental effects is impossible. On the other hand, existing environmental models have specific character and are impossible to reflect complex interconnections but can show consequences of anthropogenic influences. For example, Kuznetsk Basin (Russia) coal industry influences upon environment essentially. Retrospective analyzing and statistical evaluating show that intensity of anthropogenic influencing Company "Prokop'evskugol'" enterprises upon atmosphere is equal to 6 – 7 kg of dust-gas emission per 1 ton of produced coal [11 – 15].

2 Materials and methods

There are three stages of the process getting and using knowledge. These are observation, mathematical modelling and evaluating adequacy. Scheme of the process getting and using knowledge about the environmental condition of the concerned territory is universal [16 – 19].

The conceptual formula has the following form: $R_{ENV} = P_{ANT} D$, where R_{ENV}, P_{ANT} are risk and probability negative anthropogenic influence to environment accordingly; D is the detriment of negative anthropogenic influence to the environment. This formula can be written in the following form:

$$R_{ENV} = P\{I_{INF} > M_{PEL}\} P_{GL} D, \tag{1}$$

where I_{INF} is the intensity of negative anthropogenic influence to the environment; $MPEL$ is a maximal level of influence to the environment; $P\{I_{INF} > M_{PEL}\}$ is the probability of appearance negative anthropogenic influence to the environment, which more than maximal level; P_{GL} is the probability of failure at the environmental protection system.

The detriment of negative anthropogenic influence to the environment has the following form:

$$D = \int_{\Omega} p_{Pop}(x, y, z) P_{Inh}(x, y, z)\, d\Omega, \tag{2}$$

where $p_{Pop}(x, y, z)$ and $P_{Inh}(x, y, z)$ are accordingly distribution of population density of at the concerned territory and probability of negative influence to inhabitants in all zone of negative anthropogenic influence to the environment; Ω is a zone of negative anthropogenic influence to the environment.

The basic theoretical principle is the realization of arrangements, which allow providing follow environmental conditions. The theoretical function of distribution identifies the probability that life-activity without environmental conditions infringements is less than the specified time. It's time of environmentally favourable life-activity. Consequently, the basic theoretical principal can be formulated follow form. The theoretical function of distribution identifies the probability that life-activity without environmental conditions infringements is less than specified time t (time of environmentally favourable life-activity) that is why:

$$P\{I_{INF} > M_{PEL}\} \times P_{GL} = 1 - Q\{T < t\}, \tag{3}$$

where Q and T are probability and period of the life-activity for concerned territory without environmental infringements distribution law of probability for life-activity in concerned territory without environmental infringements has the following form:

$$Q(t) = \exp\left[-\int_0^t \lambda(\tau)d\tau\right], \qquad (4)$$

where Q and λ are the probability of the life-activity for concerned territory without environmental infringements and intensity of the times on negative anthropogenic influence to environment accordingly.

Period of time when the life-activity for concerned territory without environmental infringements can be written in the following view:

$$T = \int_0^\infty \left\{\exp\left[-\int_0^t \lambda(\tau)d\tau\right]\right\} dt\,. \qquad (4)$$

The problem of managing the environmental safety of life-activity at the concerned territory is the realization of arrangements, which allows providing follow conditions:

$$R \to \min;\; Q \to \max;\; T \to \max;\; \int_0^t \lambda(\tau)d\tau \;\to \min. \qquad (5)$$

One of the arrangements is a reference-information system. Regional reference-information system on normative-legal environmental protection control of Tula region was created. The geoecological monitoring system of Prokop'evsk-Kisilevsk district was created in Kuznetsk Basin. Functional possibilities of automatic monitoring system make it possible to solve following problems:

– researching content and structure of existent database on influence to the environment in the districts of locating enterprises by coking coal mining;

– improvement of natural laws by forming powder-gas emission, liquid and solid wastes and disturbing soil during coking coal mining;

– improving method of evaluating influence to the environment during coking coal mining;

– developing complex software tools making it possible to automatize of environmental influence quality estimation during coking coal mining;

– prognosis estimations of environmental consequences for coal mining development in the Prokop'evsk-Kisilevsk district and in Kuznetsk Basin as a whole.

Analysis information on Prokop'evsk district mines makes it possible to get follow empirical dependence:

$$J_i = \alpha_{1i} + \alpha_{2i}A + \alpha_{3i}E + \alpha_{4i}t\,, \qquad (6)$$

where J_i is the intensity of influence to environmental components; α_{1i}, α_{2i}, α_{3i}, α_{4i} are empirical coefficients; A is production capacity of the mine, thousand t/year; E is power consumption at the concerned mine during the accounting period, thousand kW/year; t is time, years; i is an index of influence to environment kind.

Forecasting indexes of anthropogenic influence upon the environment by Coal Company "Prokop'evskugol'" and by enterprises of Kuznetsk Basin Coal industry until the 2025 year are shown in Tables 1 and 2.

The second region for using proposed monitoring system was Tula Region. Basic mines of Moscow Coal Basin located in Tula Region. Functional possibilities of automatized monitoring system made it possible to substantiate follow theoretical foundations:

– one-parametric dependences of powder-gas emission into the atmosphere from consuming electrical energy dynamics in mining industry regions have got linear form;

– powder-gas emission into atmosphere dependence from consuming electrical energy dynamics is described by the differential equation of logistic kind;

– functioning "consuming electrical energy – powder-gas emission into atmosphere" system at a priori indeterminacy conditions is simulated with using local mathematical models;

– anthropogenic influence upon the atmosphere of mining industrial region is characterized by consuming electrical energy, which makes it possible to evaluate part of every industrial branch at total powder-gas emission.

Table 1. Forecasting indexes of anthropogenic influence upon the environment by Coal Company "Prokop'evskugol'".

Name of the index	Forecast index of anthropogenic influence for existent environmental protection activity			
	2010	2015	2020	2025
Coal mining, mln ton	4.114	19.775	23.0	27.0
Emitting pollutants into the atmosphere, thousand ton	26.741	128.537	149.500	175.50
Emission liquid wastes, mln m^3	14.152	68.026	79.12	92.88
Area of derangement soil, hectare	23.73	114.102	132.71	155.79
Mass of solid industrial wastes, mln ton	22.97	106.088	127.65	149.85

Table 2. Forecasting indexes of anthropogenic influence upon the environment by enterprises of Kuznetsk Basin Coal industry until the 2025 year.

Name of the index	Forecast index of anthropogenic influence for existent environmental protection activity			
	2010	2015	2020	2025
Coal mining, mln ton	190	218	240	270
Emission pollutants into the atmosphere, thousand ton	560.5	643.1	708	796.5
Emission liquid wastes, mln m^3	260.3	298.7	328.8	369.9
area of derangement soil, hectare	1265.4	1451.8	1598.4	1798.2
mass of solid industrial wastes, mln ton	1001.3	1148.8	1264.8	1422.9

Calculation scheme of anthropogenic influence upon environment ensuing analysis material-energetic flows is shown in Fig. 1.

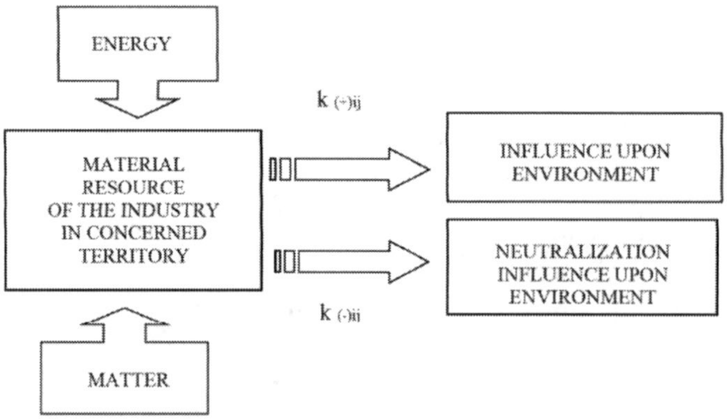

Fig. 1. Scheme of influence upon the environment.

It is possible two scripts. A first the territory has got the unlimited material resource. A second there is a competition of material resource in this territory. Let intensity of influence upon the environment is I_{ij}, where i is the index of the environmental element; j is the index of intensity of environmental influence. Mathematical models of influence upon industry development region environment are shown in Table 3.

Table 3. Mathematical models of influencing industry upon the regional environment.

Characteristics of the territory	The differential equation of influence upon the environment	Initial condition
1. Unlimited material resource	$\frac{dI_{ij}}{dE} = \xi_{ij}\left(I_{ij}^{(\infty)} - I_{ij}\right)$	$I_{ij}\big\vert_{E=0} = 0$
2. Competitive activity for material resource	$\frac{dI_{ij}}{dE} = \varepsilon_{ij}\left(I_{ij}^{(\infty)} - I_{ij}\right)I_{ij}$	$I_{ij}\big\vert_{E=0} = I_{ij}^{(0)}$

3 Results

Solving these differential equations are theoretical foundations of evaluating environmental situation [20 – 26].

Solving differential equations of influence upon environment has got the following form:

– with unlimited material resource:

$$I_{ij}(E) = I_{ij}^{(\infty)}\left[1-\exp\left(-\xi_{ij}E\right)\right], \tag{6}$$

– with competitive activity for material resource:

$$I_{ij}(E) = \frac{I_{ij}^{(\infty)}}{1-\left(1-\varphi_{ij}\right)\exp\left(-\varepsilon_{ij}I_{ij}^{(\infty)}E\right)}, \tag{7}$$

where E is consumption of electrical energy; $\xi_{ij} = k_{ij}^{(+)} + k_{ij}^{(-)}$; $\varepsilon_{ij} = K_{ij}^{(+)} + K_{ij}^{(-)}$; $\varphi_{ij} = I_{ij}^{(\infty)} / I_{ij}^{(0)}$; $k_{ij}^{(+)}$, $k_{ij}^{(-)}$, $K_{ij}^{(+)}$ and $K_{ij}^{(-)}$ are constants of energy intensity for influence upon environment and neutralization of it accordingly; $I_{ij}^{(\infty)}$ is equilibrium intensity of influence upon the environment; $I_{ij}^{(0)}$ is the initial intensity of influence upon the environment.

Let's use following designations: $RI_1(e_1) = RI_2(e_2) = I_{ij}(E)/I_{ij}^{(\infty)}$; $e_1 = \xi_{ij}E$; $e_2 = \varepsilon_{ij}I_{ij}^{(\infty)}E$.

Diagrams of dependences for relative intensity of influence upon environment from the energetic parameter for two scripts are shown in Fig. 2 and 3.

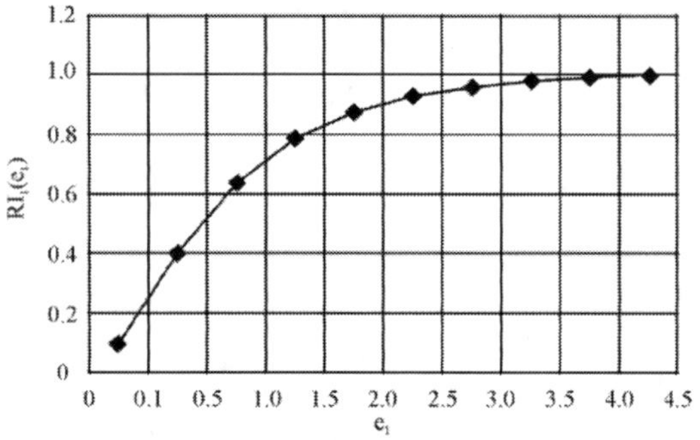

Fig. 2. Diagram of dependence for relative intensity of influence upon environment RI_1 from parameter e_1 (with the unlimited resource).

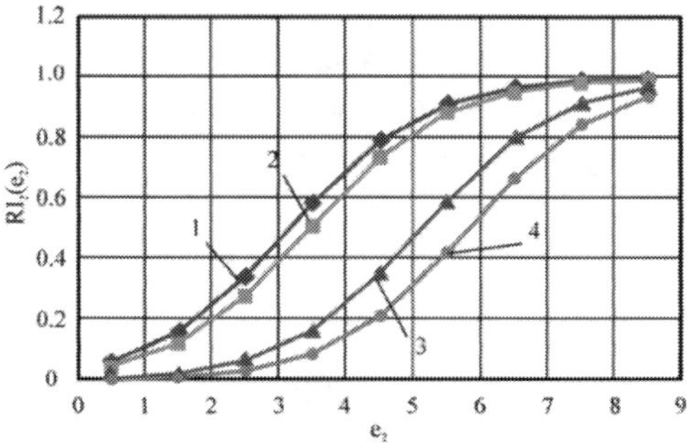

Fig. 3. Diagrams of dependence for relative intensity of influence upon environment RI_2 from parameter e_2 (with competitive activity for the material resource) for φ_{ij} is equal to follow: 1 – 15; 2 – 20; 3 – 100; 4 – 200.

Forecast estimates show that by increasing coal mining this influencing upon atmosphere index on Company "Prokop'evskugol'" enterprises will rise almost in five times in the 2025 year. Forecast indexes of anthropogenic influencing will rise very much in comparison with crisis time highly considerably. Polluting substance mass emitting into the atmospheric air will rise on 55%.

4 Conclusions

In conclusion, we want to say follow. People are a major population of the Earth. Our planet is Human ecological niche. Conditions of Human Environment define the quality of their life. Improving environmental monitoring system and creating effective technological actions by neutralization wastes are based on adequate mathematical models of an environmental condition for a territory. These models have to reflect the connection between indexes of public health, influence upon the environment and basic demographic characteristics. Consequently, it is environmental rationally and economical reasonably considering atmospheric air as a natural resource of the mining-industrial region.

Therefore, optimizing technical-economical indexes by atmospheric factor must be based at principals of regulating by institutes of agreement and consolidating mining enterprises with enterprises of others industrial branches. Proposed methodical principals of the complex evaluating atmosphere of mining-industrial region allow realizing integrated approach to forecasting intensity of polluting atmosphere, the economic efficiency of production and atmospheric air condition control with using basic demands of the environmental imperative for considered territory.

Authors express gratitude for the help and consultations during work performing for rector of Kryvyi Rih National University Professor Mykola Stupnik.

References

1. Golik, V., Komashchenko, V., & Morkun, V. (2015). Feasibility of using the mill tailings for preparation of self-hardening mixtures. *Metallurgical and Mining Industry*, (3), 38-41.
2. Golik, V., Komashchenko, V., & Morkun, V. (2015). Innovative technologies of metal extraction from the ore processing mill tailings and their integrated use. *Metallurgical and Mining Industry*, (3), 49-52.

3. O'Sullivan, D., & Newman, A. (2014). Extraction and Backfill Scheduling in a Complex Underground Mine. *Interfaces, 44*(2), 204-221.
4. Wang, Weijing, Huang, Shaofeng, Wu, Xiaobo, & Ma, Qingfei. (2011). Calculation and Management for Mining Loss and Dilution under 3D Visualization Technical Condition. *Journal of Software Engineering and Applications*, (4), 329-334.
5. Rakishev, B.R. (2013). Complex usage of ore in the enterprises of non-ferrous metallurgy of Kazakhstan. *Mining Journal*, (7), 67-69.
6. Sekisov, G.V., & Rasskazov, I.Y. (2014). Creation of a research and production mining and processing complexes for innovative supporting of mining industry. *GIAB*, (9), 113-121.
7. Kantemirov, V.D. (2014). Technologic features of the development of new raw material bases. *GIAB*, (6), 369-373.
8. Sokolov, I.V., & Smirnov, A.A. (2012). About the parameters for computation of ore reserves. *Mining Journal*, (4), 56-61.
9. Vorobev, V.V. (1995). Efficiency of gaseous additives for explosive materials. In *Proceedings of the 1st Ukrainian scientific conference "Industrial explosives and means of their initiation"* (pp. 13-15). State Research Institute of Chemical Industry.
10. Komir, V.M., Vorobiev, V.V., & Napadaylo, V.I. (1991). *The effect of gas-forming additive substances in the explosive material on the effectiveness of rock destruction. Blasting work, 90/47.* Moscow: Nedra.
11. Haifeng, W., Yaqun, H., Chenlong, D., Yuemin, Z., Youjun, T., & Cuiling, Y. (2012). Development of Mineral Processing Engineering Education in China University of Mining and Technology. *Advances in Computer Science and Engineering. AISC 141.* Springer-Verlag, Berlin Heidelberg, 77-83.
12. Bakhtavar, E., & Shahriar, K. (2007). Optimal ultimate pit depth considering an underground alternative. In *Proceedings of Fourth Aachen International Mining Symposia-High Performance Mine Production* (pp. 213-221). Aachen, Germany.
13. Capilla, A., & Delgado, A. (2015). *The destiny of the Earth's mineral resources.* London: World Scientific Publishing Co. Pte. Ltd.
14. Benardos, A., Athanasiadis, I., & Katsoulakos, N. (2014). Modern earth sheltered constructions: A paradigm of green engineering. *Tunnelling and Underground Space Technology*, (41), 46-52.
15. Bowman, S. (2016). Interferometric Synthetic Aperture Radar (InSAR). Back-ground and Application: guidelines for in-vestigating geologic hazards and preparing engineering geology reports, with a suggested approach to geologic-hazard ordinances in Utah. *The University of Utah*, 198-203.
16. Morkun, V.,Tron, V., & Goncharov, S. (2015). Automation of the ore varieties recognition process in the technological process streams based on the dynamic effects of high-energy ultrasound. *Metallurgical and Mining Industry*, (2), 31-34.
17. Morkun, V., Morkun, N., & Pikilnyak, A. (2015). The study of volume ultrasonic waves propagation in the gas-containing iron ore pulp. *Ultrasonics*, (56), 340-343.
18. Morkun, V., & Tron, V. (2014) Ore preparation energy-efficient automated control multi-criteria formation with considering of ecological and economic factors. *Metallurgical and Mining Industry*, (5), 8-10.
19. Morkun, V., & Tcvirkun, S. (2014). Investigation of methods of fuzzy clustering for determining ore types. *Metallurgical and Mining Industry*, (5), 11-14.
20. Morkun, V., Morkun, N., & Pikilnyak, A. (2014). The gas bubble size distribution control formation in the flotation process. *Metallurgical and Mining Industry*, (4), 42-45.
21. Morkun, V., Morkun, N., Pikilnyak, A. (2014). Ultrasonic facilities for the ground materials characteristics control. *Metallurgical and Mining Industry*, (2), 31-35.
22. Morkun, V., Morkun, N., Pikilnyak, A. (2014). Iron ore flotation process control and optimization using high-energy ultrasound. *Metallurgical and Mining Industry*, (2), 36-42.

23. Morkun, V., Morkun, N., & Pikilnyak, A. (2014). Modeling of ultrasonic waves propagation in inhomogeneous medium using fibered spaces method (k-space). *Metallurgical and Mining Industry*, (2), 43-48.
24. Morkun, V., Morkun, N., & Pikilnyak, A. (2014). Simulation of the Lamb waves propagation on the plate which contacts with gas containing iron ore pulp in Waveform Revealer toolbox. *Metallurgical and Mining Industry*, (5), 16-19.
25. Morkun, V., Morkun, N., & Tron, V. (2015). Formalization and frequency analysis of robust control of ore beneficiation technological processes under parametric uncertainty. *Metallurgical and Mining Industry*, (5), 7-11.
26. Morkun, V, Semerikov S., & Hryshchenko, S. (2014). Environmental competency of future mining engineers. *Metallurgical and Mining Industry*, (4), 4-7.

Parameters of selective martite ores disintegration in structured ore bodies deposits by borehole hydraulic monitors

Mykola Stupnik[1*], *Viktor* Tarasiutin[1], and *Pavlo* Fedorenko[1]

[1]Kryvyi Rih National University, Department of Underground Mining of Mineral Deposit, 11 Matusevycha Ave., 50027 Kryvyi Rih, Ukraine

Abstract. The work is aimed at determining parameters of advance borehole stoping by hydraulic monitors leading to rational use of reserves structured according to ore types at deposits by means of advance stoping with selective disintegration of high-grade martite ores. The research methods include experimental investigations of parameters of breaking a high-grade martite ore massif by high-pressure water jets of borehole monitors; laboratory studies of technological properties of hydrodisintegrated products and their concentratibility; theoretical generalization of experimental data. The research has resulted in developing experimental borehole hydraulic monitors and determining basic technical parameters of high-pressure water jets; determining regularities of hydraulic monitor disintegration of high-grade martite ores considering structural and mineralogical characteristics of the ores and hydrodisintegration modes; proving the fact that the process of high-grade martite ore disintegration by monitors is also the process of ore disintegration to the level of ore mineral grain release providing the higher quality disintegration product after subsequent dewatering than that of the initial one. The scientific novelty of the research consists in determining a criterion of hydrodisintegration of martites, conditions of forming a required fractional composition of monitor disintegration products through controlling stability of a pulse of the water jet action on the stope face. The practical relevance of the monitor breaking process in conditions of mining units at deep levels of Kryvbas underground mines consists in providing advanced stoping of rich martite ores through raises and obtaining a new kind of product – martite superconcentrate.

1 Introduction

Nearly a quarter of reserves of high-grade martite ores with iron content of 65 – 69% occurs relatively compactly and in accordance with natural mineralogical zonation at deep levels of Kryvbas mines within ore bodies structured by ore types [1 – 4]. Chemical and mineral characteristics of such ores allow treating them as a prospective raw material base for iron ore superconcentrate production with stable chemical and technological parameters and wide suitability/consumer properties [5, 6]. Structural and morphological features of

* Corresponding author: mstupnik2012@gmail.com

high-grade martite massifs, their instability and low strength do not allow traditional selective mining without significant qualitative and quantitative losses when applying common sublevel caving options based on blasting and areal ore drawing.

Advance stoping of high-grade instable ores applying the borehole hydraulic technology is one of the resource saving directions of mining structured ore bodies. The technology provides for disintegration of rock massifs by monitors through boreholes [7]. Implementation of the new principle in underground iron ore disintegration enables: provision of sufficient stability of formed stopes without disturbing continuity of the massif of leaner and stronger ores with their further mining as compensatory areas; enhancement of the applied mining system structures; increase of mining operations concentration, safety and efficiency. The new method of breaking in rich iron ore deposits of Kryvbas will enable production of a new iron ore product and increase competitiveness of Kryvbas underground mines on metallurgical markets.

Water jet energy is known to be used to disintegrate millions of tons of various rocks, ores and coals: when washing out weak overburden rocks by surface direct-flow monitors in open-pit mining; when mining coals by underground direct-flow monitors in collieries; at remote disintegration of construction materials, chemical raw materials, non-ferrous and rare earth ores by borehole monitors from the daylight surface or in underground mines.

Analysis of the theory and practice of hydraulic monitor disintegration of rocks show [8 – 11] that organization principles and methods of calculating process parameters are mainly based on the general concept of the disintegration mechanism and empirical regularities obtained in particular mining and geological conditions of coal deposits. Technical and technological parameters of rock disintegration by monitors high-pressure water jets and its efficiency are determined depending on mining and technological properties of rocks, hydrodynamic characteristics and methods of water jet impacts on rocks.

The process of disintegrating rich iron ores by monitor waters jets through boreholes still remains understudied. This deters experimental and industrial testing and practical application of the method under challenging engineering-geological conditions of Kryvbas iron ore underground mines.

The work is aimed at determining parameters of borehole stoping by hydraulic monitors leading to rational use of reserves structured according to ore types in ore bodies by means of advance stoping with selective disintegration of high-grade martite ores.

Investigation of the process of disintegrating martite ores by high-pressure water jets involved solution of the interrelated problems:

– development of experimental underground borehole monitors for breaking weak iron ores and determination of hydrodynamic and geometric characteristics of free high-pressure water jets;

– experimental investigation of the mechanism characteristics and determination of basic parameters and regularities of disintegrating high-grade martite ores by monitor water jets through boreholes;

– determination of chemical and mineralogical characteristics and technological properties of monitor disintegrated products.

According to [2, 4], high-grade martite ores at deep levels of Krivbass underground mines are structurally fine-grained, highly porous, mainly two component systems containing 90 – 95% of martite and 5 – 10% of quartz in volume. Their characteristics are as follows: iron content – 66.6 – 68.8% (average 67.9%); silica content – 1.0 – 2.0% (average 1.45%); density – 3.50 – 3.99 (average 3.68) 10^3 kg/m^3; porosity – 20 – 38% (average 30%); uniaxial compressive resistance – 0.8 – 10 MPa (average 4.6 MPa).

The stable relationship between uniaxial compressive resistance and porosity is described as follows:

$$\left[\sigma_{cr}\right]=\left[\sigma_{q}\right]\cdot(1-\eta)^{2}\cdot e^{-12\eta}, \tag{1}$$

where $[\sigma_q]$ is uniaxial compressive resistance of oxidized quartzite (jaspilites or hornfels) similar to porous martite ores by genesis, composition and structure, MPa; η – general porosity of martite ores, fr.unit.

Rotating water jets form a worked out out areas in martite massifs through cutting down the raise until a quasi-cylindrical room of the designed sizes is formed (Fig. 1).

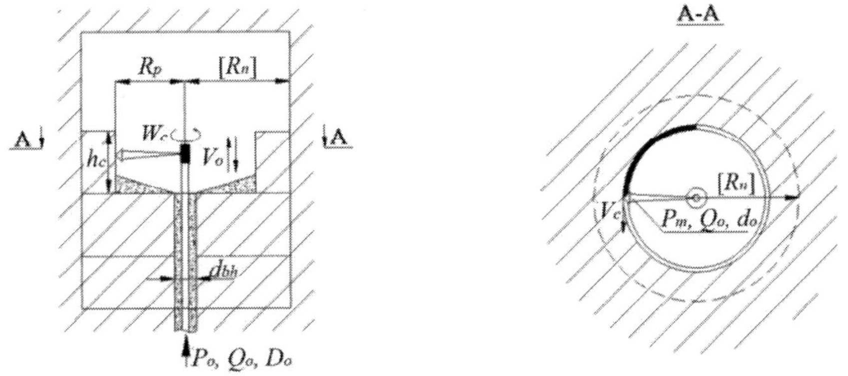

Fig. 1. Calculated diagram of monitor disintegration of martite ores by rotating jets: P_o, Q_o, D_o are pressure, consumption and diameter of the technological water flow respectively; d_{bh} is a borehole diameter; R_c, $[R_m]$ are current and maximum radii of disintegration; h_c is height of the disintegrated ore layer; W_v, V_v are angular and linear velocity of a jet travel in a face respectively; V_o is a velocity of jet reciprocal movements; P_m is a axial dynamic pressure of a jet; d_o is a nozzle diameter.

Breaking of iron ores by monitors is a nonstationary process. Its parameters depend on hydrodynamic characteristics of a water jet varying with the distance to ore exposures, velocity of jet travels in a face and structural and mineralogical features of the ore massif.

2 Methods

The process of monitor disintegration of high-grade loose iron ores has undergone experimental investigations in mining geological and technical conditions of mining units of 950 – 1240 m levels at "Rodina" and "Yuvileina" iron ore underground mines, Kryvyi Rih iron ore basin.

General methods of investigation involved: choosing representative areas of high-grade martite ores in structured iron ore bodies; developing experimental underground plants; choosing methods of controlling and measuring technical and technological parameters of breaking; conducting experimental and methodological work on disintegrating ores by high-pressure water jets through boreholes from underground workings; sampling and determining fractional and chemical compositions of hydraulic breaking products; determining characteristics of their concentration; analyzing and generalizing experimental results and determining technical and technological parameters of martite ores hydraulic breaking.

3 Results and discussion

To conduct field investigations a prototype and experimental plants were designed and built. The parameters of the prototype are as follows: water pressure in the nozzle P_o varies from

1 to 8 MPa at the water consumption $Q_o = 2.9$ m³/h through the initial jet diameters d_o varying from 3.0 to 4.5 mm. Parameters of the experimental monitor (Fig. 2) are as follows: water pressure in the nozzle P_o varies from 0.5 to 2.5 MPa at the water consumption $Q_o = 38$ m³/h the initial jet diameters d_o varying from 10 to 15 mm.

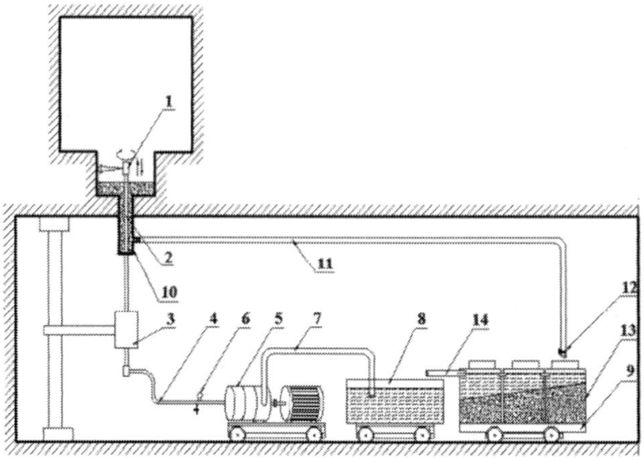

Fig. 2. Diagram of the experimental monitor plant: 1 – monitor head, 2 – monitor barrel, 3 – monitor unit, 4 – high-pressure hose, 5 – pump, 6 – manometer, 7 – water pipe, 8 – water tank, 9 – container platform, 10 – production hole preventer, 11 – pulp line, 12 – discharge device, 13 – hydrodisintegrated products, 14 – water overflow.

Monitors were tested in various modes of initial parameters. The experimental data demonstrated that monitor efficiency is characterized by the integral parameter of jet compactness – its initial part length L_i where the axial dynamic pressure P_m equals the pressure of the fluid going out of the jet forming nozzle P_o. For borehole monitors the value L_i is determined from the expression:

$$L_i = (95 - 50Re)d_o, \tag{2}$$

where d_o is the jet forming nozzle outlet diameter, m; $Re = V_o d_o / \nu$ is the Reynolds number; $V_o = 42\sqrt{P_o}$ is the initial velocity of the outgoing jet, m/s;ν is kinematic viscosity of the fluid (for water at $t = 20$ °C, $\nu = 1.01 \cdot 10^{-6}$).

Changes of the jet parameter P_m (Fig. 3) beyond L_i are normally described by the following dependency:

$$P_m = P_o\left(L_i / L_e\right), \tag{3}$$

where L_e is the length of the effective part of the jet, m.

For experimental borehole monitors, the length of the effective part of the jet L_e should be assumed equal to (8 – 10)L_i. Here, the average working diameter D_w of the disintegrating part of the jet throughout its effective part L_e will equal (6 – 8) d_o.

The primary objective of experiments in underground conditions is to determine principal features of the monitor disintegration mechanism and elicit dominant influencing factors through visual and instrumental observations.

It is determined that the axial dynamic pressure in the monitor jet which makes:

$$\left[P_m\right] = 0.25\left[\sigma_{cr}\right], \tag{4}$$

or considering (1)

$$\left[P_m\right]=0.25\left[\sigma_q\right]\cdot(1-\eta)^2\cdot e^{-12\eta}, \quad (5)$$

can serve as a criterion of disintegration of highly porous martite ores with mass content of iron of over 65%.

Main results of the experiments with the prototype are given in Tables 1 and 2.

Table 1. Sieve composition of high-grade martite ore fragmented by various processes.

Size, mm	Blasting destruction of ore by hole charges			Ore milling			Monitor disintegration of ore		
	yield, %	*Fe* mass fraction, %	release, %	yield, %	*Fe* mass fraction, %	release, %	yield, %	*Fe* mass fraction, %	release, %
+2.0	45	67.4	0.0	3.0	63.8	0.0	2.8	62.7	0.0
–2.0+1.0	10.2	68.1	0.0	9.6	65.5	0.0	11.3	68.3	2.0
–1.0+0.5	10.9	68.3	0.0	13.7	68.0	3.0	16.8	68.8	5.0
–0.5+0.25	7.2	68.1	0.0	13.0	66.7	8.0	14.8	67.8	10.0
–0.25+0.1	9.0	68.0	0.0	3.3	67.3	50.0	26.5	68.9	96.0
–0.1+0.074	4.8	66.1	4.0	14.8	68.6	60.0	12.7	67.2	100.0
–0.074+0.044	2.8	66.1	10.0	6.3	67.7	85.0	7.0	67.9	100.0
–0.044	1.5	66.0	8.0	36.3	67.1	90.0	8.1	68.7	100.0
Total	100	67.7	6.0	100	67.2	20.5	100	68.3	45.0

As is seen from Table 1, milling produces by 25% fewer free martite grains than disintegration by monitors. The –0.044 mm size yield is larger by 28%. This can be explained by the fact that in milling ore minerals are overground due to the spongy nature of martites. Full ore particle release is achieved at grinding to the size of less than 0.25 mm which is comparable with martite grains.

Thus, borehole monitor disintegration is a combined process of ore preparation that includes milling and sizing of disintegrated products.

Table 2. Fractional composition of integrated ore samples.

Size, mm	Initial water jet pressure, MPa			
	1.5	2.5	3.5	4.0
	Yield, %			
+1.0	18.2	28.9	30.0	30.3
–1.0 + 0.5	11.8	14.2	24.3	26.0
–0.5 + 0.25	25.2	19.7	13.7	16.2
–0.25 + 0.1	28.3	24.4	15.8	13.0
–0.1	22.5	12.8	8.2	4.5
Average particle size, mm	0.514	0.663	0.739	0.763

Table 2 demonstrates that at the same velocity of water jet movements along the face increase of the jet dynamic pressure results in the increased yield of the average size of hydrodisintegrated product particles due to the yield of smaller and released martite fractions.

One of the basic parameters that characterizes the process of porous martite ores disintegration by monitors is depth of technological disturbances of ore massifs at exposures. The experiments prove that at the jet dynamic pressure of 6 – 8 MPa, technogenic disturbance does not exceed 0.1 – 0.2 m, i.e. is 20 – 30 times less than at the ore massif disintegration by borehole explosive charges.

The primary objective of the field experiments was to determine parameters of the disintegration process and their dependencies on initial hydrodynamic characteristics of a water jet and its mode of impacting the ore massif.

It is determined that in martite ores (porosity of 20 – 35%) monitor jets with travel velocity of 0.5 – 3.0 m/s and formed by a 10 mm nozzle under water pressure of 2.5 to 3.0 MPa provide the following parameters: pulp production – 40 to 60 t/h, solid fraction production – 12 to 25 t/h, specific water consumption – 0.7 to 3.0 m^3/t; pulp density – 1.3 to 1.5 t/m^3, radii of the jet disintegration impact on the ore massif – from 2.0 to 3.0 m.

The kinetic process of disintegration in the same ore massif is known to run in a number of ways depending on the type, character and value of the actual load. The results of monitor breaking demonstrate the following. The jet interacts with the disintegrated massif and in the subsurface part of the limited volume of the latter there occur and develop fields of secondary stresses. The fields are caused by normal and tangential components of the jet impact rate and spreading and by continuous change of the contact surface geometry. This conditions the disintegration process kinetics.

It is quite obvious that a high-pressure hydraulic impact on a certain exposure surface area/ exposed area in a limited ore volume results in development of all systems of pores and fractures, as well as new input microfractures in places of intergrain contact exposures – an interphase boundary or defects of intergrain surfaces. The fluid flow impacted by pressure heads for fractures and reaches their tops provoking fracture development. At first, as the velocity is rather great, the fluid reaches the fracture top in a short time provoking its growth. However, the fluid flow in the narrow fracture is characterized by appreciable fluid losses. Fluid pressure in longer fractures decreases quickly and becomes insufficient for their disruption but still acts as a lever or a wedge increasing concentration of tension stresses in the fracture top. This effect is of considerable significance for fracture elongation at the next action of the jet.

Thus, in a certain volume of rocks emerging hydrostresses cause ordered disintegration in the form of ore jointings, grain aggregates and mineral grains proper. Considering the above mentioned, it can be concluded that fractional analysis of jointings resulted from disintegration by monitors can serve as a source of information on character and features of the monitor disintegration process and for developing recommendations on its streamlining. Apparently, grains and aggregates will be formed in a thin, commensurate with sizes of grains and grain aggregates, layer near the surface of jet contact exposures as a result of disintegration by rupture – washout of intergrain contacts. As a result of ruptures, the lump part will be formed some distance inward from the exposure. It follows from the above mentioned that disintegration mechanism is a volume-differential one and depends on jet parameters and the ore massif state. At that, one part of water selectively disintegrates the ore until grains and their aggregates, another part disintegrates the ore massif to lump fractions and takes them to the pulp-preparation place, another part of water penetrates into technogenic fractures to the depth of up to 0.1 – 0.15 m and stays there as a hydraulic wedge till the next action cycle.

The given disintegration mechanism scheme demonstrates that monitor disintegration in highly porous martite ores by a travelling fluid jet will mainly occur by development and integration of fracture systems caused by tension stresses and wedging actions of the fluid. The breaking stress depth will be proportional to the fluid pressure and their lifetime will depend on the fluid action time. Tentatively, the disintegration mechanism is characterized by one of the strength parameters of the environment, e.g. compressive resistance $[\sigma_{cr}]$ that correlates with martite grain sizes and intergrain bond strength.

Thus, the technology of rock disintegration by monitors through boreholes satisfies the conditions of selective disintegration the physical significance of which consists in providing disintegration along boundaries of mineral grains and jointing fractures as a result of shear and tension stresses. Monitor disintegration can replace the whole technological line of disintegrating ore materials at the stage of preparation for concentration and result in energy savings and economic benefits.

Considering the obtained results of the maximum use of the effective part of a jet, the following condition should be fulfilled:

$$\left[P_o\right]=2.5\left[\sigma_{cr}\right]. \tag{6}$$

The maximum efficient jet part:

$$\left[L_e\right]=\frac{P_o \cdot L_o}{0.125\left[\sigma_{cr}\right]}, \tag{7}$$

for the type of mined ores is the basic structural parameter when choosing hydraulic mining locations.

The experiments show that, all things being equal, the disintegration degree (fractional composition of disintegrated products) crucially depends on the velocity V_v of a jet travel velocity in the face, i.e. on the jet-massif contact time t_c.

The travelling jet-massif contact time is determined as follows:

$$t_c = D_p / V_v. \tag{8}$$

For the jet travelling round the monitor in the borehole at the distance L_e:

$$V_v = 2\pi L_e n, \tag{9}$$

where n is the angular velocity of a jet rotating round the axis of the borehole monitor barrel, rps.

The analysis show that the process of disintegrating the massif by a travelling water jet is characterized by the specific impulse of the power dynamic pressure of a jet on the ore massif equal to

$$I_m = P_m \cdot t_c, \tag{10}$$

or

$$I_m = \frac{K}{L_e^2 n}, \tag{11}$$

where $K = 2P_o L_i d_o$ is a monitor dimension-type factor.

It is determined that for the required fractional composition of disintegrated products there exists a threshold value of the specific impulse of the axial dynamic pressure, I_m which integrally reflects structural-textural and mineralogical features of ores, geomechanics of the ore massif, technology and parameters of the jet impact on the massif. The value $[I_m]$ for concrete engineering and geological conditions and the required degree of disintegration is determined experimentally.

As is seen from (11), the jet rotation velocity n is a controlled parameter ensuring the required mode of ore massif disintegration by monitors at varying and controlled L_e.

Considering the fact that for every monitor dimension-type, parameters P_o, L_n, d_o possess concrete values and that the required fractional composition is controlled by the constant value $[I_m]$, expression (11) is transformed into:

$$L_e^2 \cdot n = \frac{K}{\left[I_m\right]} = \left[W\right] \approx const, \tag{12}$$

where $\left[W\right] = 2P_o \cdot L_H \cdot d_o / \left[I_m\right]$ is the value constant for concrete conditions and integrally reflecting engineering and mineralogical features of the massif, the ore disintegration degree, monitor parameters and the disintegration technology.

Thus, for the given geological and technological characteristics of the ore massif the procedure of calculating basic parameters of disintegration by monitors include:

– choosing a monitor type and parameters and calculating basic structural and hydrodynamic characteristics of borehole monitor jets;

– calculating technological parameters of production holes and dead areas;

– choosing the required mode of disintegration by monitors to provide either breaking or disintegration or selective disintegration;

– calculating technical and economic parameters of ore massif disintegration for the required product type.

Iron ore concentrate obtained by gravity concentration of hydrodisintegrated products on concentrating tables (Table 3) was characterized by mass content of iron of up to 70% and that of silicon dioxide of up to 0.1%. The characteristics satisfy requirements of iron powder, ferrite and battery paste production. Martite superconcentrate tested by the *Ferroceram* company satisfies all requirements for production of high class barium ferrite.

Table 3. Results of gravity concentration of high-grade martite ore at different ore preparation methods.

Ore preparation method	Concentrate				Product			
	yield, %	mass fraction, %		recovery, %	yield, %	mass fraction, %		recovery, %
		Fe	SiO_2			*Fe*	SiO_2	
disintegration by monitors	84.1	69.7	0.2	85.8	15.9	60.9	–	14.2
milling	73.5	69.0	0.5	74.5	26.5	62.2	–	24.5

5 Conclusions

The investigation resulted in:

– determining parameters, regularities and dependencies disintegration of high-quality martite ores by monitors through boreholes considering mining and geological characteristics of the ore massif, hydrodynamic parameters and methods of monitor jet application, and allowing rational control of ore granulometry, form and cavity sizes;

– proving the possibility of selective high-grade martite ore disintegration in Kryvbas underground mines which creates exceptionally favourable opportunities for organizing production of matrite superconcentrate with extremely high technological and economic parameters of a new level product for iron powder and high-class ferrite metallurgy.

Borehole disintegration of martite ores by monitors is promising and expedient when developing new technologies of selective mining of weak instable ores from structured iron ore deposits. It can also be applied for unmanned mining in vertical separated workings and creating rooms for various purposes.

This work was conducted within the projects "Determination of regularities of the stress-strain state of rocks disturbed by workings with the purpose of developing resource-saving ore mining technologies" (State registration No. 0115U003179).

References

1. Stupnik, N.I., Kalinichenko, V.A., Kolosov, V.A., Pismenniy, S.V., & Fedko, M.B. (2014). Testing complex-structural magnetite quartzite deposits chamber system design theme. *Metallurgical and mining industry*, (2), 89-93.

2. Stupnik, N., Kalinichenko, V., & Pismennyi, S. (2013). Pillars sizing at magnetite quartzites room-work. *Mining of Mineral Deposits*, 11-15. http://dx.doi.org/10.1201/b16354-4.
3. Tarasyutin, V.M. (2015). Geotehnology features of high quality martite ore from deep mines of Kryvyi Rih basin. *Naukovyi Visnyk Natsionalnoho Hirnychoho Universytetu*, (1), 54-60.
4. Yevtekhov, V.D., Paranko, I.S., & Yevtekhov, Ye.V. (2001). *Heolohiia i korysni kopalyny Kryvorizkoho zalizorudnoho baseinu*. Kryvyi Rih: Kryvorizkyi tekhnichnyi universytet.
5. Balashov, A.G., & Anisimov, V.N. (2004). Bogataya ruda KMA v metallurgicheskom proizvodstve. *Gornyy zhurnal*, (1), 69-70.
6. Kravtsov, Ye.N., & Kravtsov, N.K. (2004). Razrabotka tekhnologii polucheniya zhelezorudnogo kontsentrata iz rud podzemnoy dobychi s massovoy doley kremnezoma do 0.03%. *Razrabotka rudnykh mestorozhdeniy*, 98-100.
7. Arens, V.Zh. Babichev, N.I., & Bashkatov A.D. (2007). *Skvazhinnaya gidrodobycha poleznykh iskopayemykh*. Moskva: Izdatelstvo "Gornaya kniga".
8. Britan I.V. (2007). Ispytanie skvazhinnogo gidrodobychnogo oborudovaniya dlya otboyki rudy v zatoplennykh kamerakh. *Gornyy informatsionno-analiticheskiy byulleten'*, (7), 286-291.
9. Malanchuk, Z.Ye., & Boblyakh, S.N. (2007). Doslidzhennia protsesu rozmyvu zernystykh fosforytiv hidromonitornym strumenem. *Visnyk Kryvorizkoho Tekhnichnoho Universytetu*, (16), 33-38.
10. Petrichenko, V.P. Kolesnikov, V.I., & Pinchuk, A.V. (2005). Fizicheskie sposoby i sredstva razuplotneniya i erliftirovaniya rykhlykh rud KMA. *Gornyy informatsionno-analiticheskiy byulleten'*, (3), 283-287.
11. Stets, S.Ye. (2006). Eksperymentalne doslidzhennia protsesu hidrozmyvu tseolitsmektytovykh tufiv. *Visnik Natsionalnoho Universytetu Vodnoho Hospodarstva*, *4*(36), 365-370.

Hereditary model of loose mined rock layer deformation in disintegrators

Volodymyr Nadutyi[1*], *Oleksandr* Tytov[1], and *Ivan* Cheberiachko[2]

[1]National Mining University, Department of Mining Machines and Engineering, 19 Yavornytskoho Ave., 49005 Dnipro, Ukraine
[2]National Mining University, Department of Mining Mechanics, 19 Yavornytskoho Ave., 49005 Dnipro, Ukraine

Abstract. The objective of this work is to substantiate the hereditary character of the stresses and deformations accumulation in loose mined rock during the disintegrators working surfaces convergence. It is shown, in particular, that the layer porosity after compression and the side thrust coefficient depend not only on the applied pressure, but also on the initial porosity. It is substantiated the mechanism of interrelation between the stresses and deformations of a layer. The experimental researches on fine–grained granite compression in a rigid matrix have been fulfilled. The approximation of obtained results by polynomial functions has been carried out. The coefficients have been determined by the least-square method. The close agreement of the analytical and the experimental values has been recorded. The initial porosity parameter was put into the equation of dependences of the current porosity and the side thrust coefficient on pressure. The mechanism of the forces distribution on separate particles contacts has been explained in terms of the material compacting hystory. The practical value of the research is in calculating the disintegrators operating parts parameters with account of the mined rock layer condition prior to the act of crushing.

1 Introduction

It is possible to enhance the efficiency and to reduce the energy consumption of coal, limestone, iron ore, granite and other mining rocks disintegration by correct choice of the disintegrators operating parts parameters. This requires clear understanding of the disintegration process and interrelations of the mined rock condition with the loading parameters.

Today, the questions of a separate rock piece deformation, clamped by two crushing plates, have been explicitly studied. With that, the stressed condition of the rigid body with a given shape was studied [1]. The calculation of this task does not cause problems for modern computer programs [2, 3].

But, there is a sufficient number of disintegrator designs, which use the principle of simultaneous crushing of several material layers, when a separate rock piece interacts

* Corresponding author: nadutyvp@yandex.ua

mostly with other similar pieces [4]. This method leads to selective mined rock crushing, better disclosure of minerals and less energy consumption.

The number of works devoted to the “in the layer” crushing is small. They are characterized by taking into account the internal friction of loose material, manifested in arising the side thrust coefficient. The side thrust coefficient is expressed in accordance with Mohr-Coulomb model by known formula [5]:

$$\xi = \frac{\sigma_3}{\sigma_1} = f(c, \beta), \tag{1}$$

where σ_1, σ_3 are the primary stresses; c is the cohesion coefficient; β is the internal friction coefficient.

The deformation zone of material between two parallel plates is studied in the contextually similar work [6], with an assumption on constant value of the side thrust coefficient within the whole zone area. But, the obtained theoretical formulas of stresses distribution on crushing plates have satisfactory accuracy only under conditions of low pressures. Their errors increase together with the pressure growth.

In the article [7] based on the analysis of different structure mining rocks reaction on volumetric strain, it is shown, that the stresses distribution tends to hydrostatic one, and the side thrust coefficient tends to unity if the pressure rises unlimitedly.

One can draw the same conclusions by the analysis of very widespread empirical Hoek-Brown failure criterion of mining rocks [8]:

$$\sigma_1 = \sigma_3 + R_c \cdot \left(m_b \cdot \frac{\sigma_3}{R_c} + s \right)^a , \tag{2}$$

where R_c is the tensile strength in uniaxial compression; m_b, s and a are the empirical coefficients, which consider the structural peculiarities of rock massif together with Mohr’s enveloping curve.

Also, it is concluded in the work [8], that this Hoek-Brown failure criterion is applicable in conditions, when the rock massif is mainly in cohesive condition. Therefore, GSI (Geological Strength Index), which influences m_b and a coefficients values, must be not less than 20.

On the one hand, Hoek-Brown failure criterion could be used to describe the behavior of piece material state under high pressure, because the value GSI = 20 corresponds to the disintegrated rock massif consisted of separate rock pieces.

But, on the other hand, there will be some difficulties:

– the porosity of loose material is sufficiently changed during its compression, and this, in turn, influences on its deformation characteristics;

– the porosity of material may be simply determined, but it is difficult to correlate this value with GSI parameter;

– there is a problem to record the “destruction” point of a sample having zero cohesion coefficient, and therefore the destruction of which is not connected with pieces’ disintegration;

– the increase in velocity of loading is also of great importance during disintegration [9];

– the availability of moisture has great impact on the friction forces of material particles, especially in case of high porosity [10, 11].

According to N. Gersevanov “hydraulic capacity” principle [5], the current porosity of loose material layer is unambiguously determined by the sum of primary stresses on three coordinates and, thus, it does not depend on the loading history. This principle is valid only for pressure range up to 0.1 – 1.0 MPa, which are not sufficient for rock massif disintegration.

It has been shown in work [12], that for the case of flat zone of deformation and pressure range up to 4 MPa the initial porosity factor influences the current porosity value not less than the applied pressure.

Real disintegrator constructions are designed to provide average pressures along working surfaces up to several MPa, and locally up to several tens or even hundreds MPa.

Hence, it is necessary to set the influence of applied pressure (in a wide range) and of initial porosity on the current porosity and the side thrust coefficient of mined rock layer in order to describe the disintegrators working process. This is an actual scientific problem.

2 Analytical substantiation of the hereditary deformation model of mined rock layer

Every act of mined rock disintegration is followed by different modes of particles interaction. Firstly, mutual slippage of particles occurs, being typical for compacting of very loose material. As the process progresses, sharp edges begin to be cut. Then, the body of particles is destructed with the outflow of splinters into existing pores. At the next stage of compression, the destruction stops and changes to elastic deformation. As a rule, disintegrators are designed to reach the stage of particles bodies destruction [13].

The centers of particle masses do not move with essential absolute displacement values during the destruction act, and the average number of contacts of each particle with neighboring particles rises constantly with reduce in the layer porosity.

Suppose that the average coordination number for the layer is determined as follows:

$$N = k_1 - k_2 \cdot e, \tag{3}$$

where e is the fractional porosity; k_1 and k_2 are the empirical coefficients.

We will take into account the degree of forces development at particles contacts connected with the porosity reduction as the hereditary accumulation of stresses.

Let us consider the process of quasistatic deformation of an elementary volume of material. Suppose, that, initially, the fractional porosity of layer was equal to e_0 and had an identical value within the whole layer. According to (3), every particle had in average definite number of contacts N_0.

We assume, that gradient of force increment for the contacts existed at initial time moment, is expressed by dependence:

$$\frac{dP_0}{dx} = s_0 + s_1 \cdot (x - x_0) + s_2 \cdot (x - x_0)^2 + ... + s_n \cdot (x - x_0)^n, \tag{4}$$

where x is the current contact deformation coordinate; x_0 is the coordinate of the beginning of the contact deformation; n is the maximum approximation exponent.

As further convergence of particle mass centers proceeds, which is corresponding to initial contact deformation x_1, an additional number of contacts with corresponding increment of coordination number ΔN_1 is formed. Then, we will have ΔN_2 increment for x_2 deformation, and so on. The gradient of force increment, corresponding to γ – sequential queue of contacts, will be as follows (scheme in Fig. 1):

$$\frac{dP_\gamma}{dx} = s_0 + s_1 \cdot (x - x_\gamma) + s_2 \cdot (x - x_\gamma)^2 + ... + s_n \cdot (x - x_\gamma)^n. \tag{5}$$

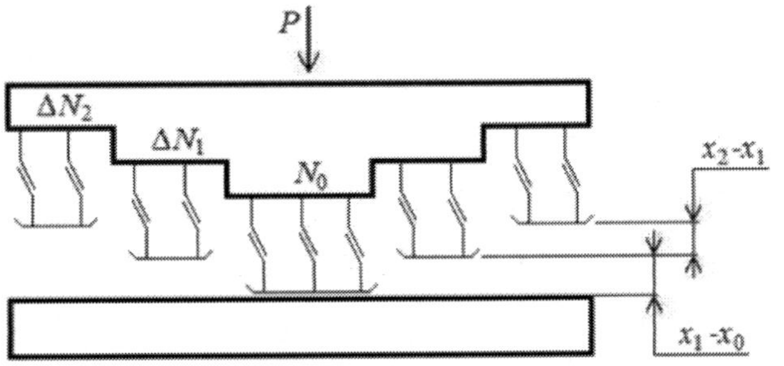

Fig. 1. The scheme of contacts development in loose material during compacting.

The total gradient of force increment on the particle contacts is determined by expression:

$$\frac{dP}{dx} = N_0 \cdot \frac{dP_0}{dx} + \Delta N_1 \cdot \frac{dP_1}{dx} + \Delta N_2 \cdot \frac{dP_2}{dx} + \ldots + \Delta N_\gamma \cdot \frac{dP_\gamma}{dx} + \ldots \quad (6)$$

It is very difficult to calculate the real number of contacts in a layer of material and to determine the contact forces. Hence, we move on from x and P values to e value and force parameter λ, using expressions

$$de = -w \cdot dx\,; \quad (7)$$

$$d\lambda = q \cdot dP\,, \quad (8)$$

where w and q are some constants.

By analogy to equations (5) and (6), the gradient of increment of the maximum primary compressive stress in a layer will be written as follows:

$$\frac{d\lambda}{de} = -(k_1 - k_2 \cdot e_0) \cdot \left[a_0 + a_1 \cdot (e_0 - e) + a_2 \cdot (e_0 - e)^2 + \ldots + a_n \cdot (e_0 - e)^n\right] + \\ - k_2 \cdot \left[a_0 \cdot (e_0 - e) + a_1 \cdot \frac{(e_0 - e)^2}{2} + a_2 \cdot \frac{(e_0 - e)^3}{3} + \ldots + a_n \cdot \frac{(e_0 - e)^{n+1}}{n+1}\right] \cdot \quad (9)$$

The solutions of the last equation are the following hereditary functions:

$$\lambda = f(e, e_0)\,; \quad (10)$$

$$e = \varphi(\lambda, e_0)\,. \quad (11)$$

Thus, the interrelation between the fractional porosity coefficient and the force parameter has been set analytically, taking into consideration the initial condition of loose mined rock layer.

3 Experimental research technique

The fine-grained granite of size less than 0.5 mm has been put to the research by uniaxial compression in a rigid cylindrical matrix. Slow increase in pressure, in steps, was made for various mass of material, to a value of approximately 40 MPa. The normal pressure values on a punch and the values of punch settlement were measured. The results were easily converted to the current fractional porosity dependences on the pressure, which are presented in Fig. 2.

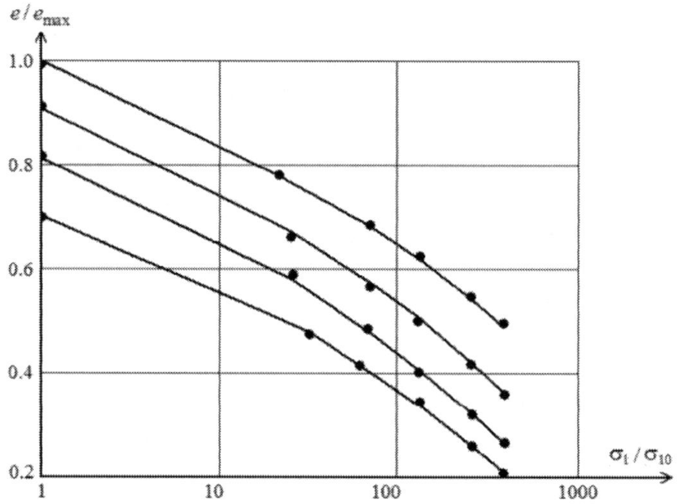

Fig. 2. Odometric dependences for fine–grained granite: *lines* – theoretical dependences; *points* – experimental values.

The diagrams show almost linear dependence of the fractional porosity on the axial pressure logarithm. Besides, it can be seen the essential influence of the initial fractional porosity on its current value, because the curves shifting value is almost constant within the whole pressure range.

Initially, the matrix was filled with material without preliminary compaction. Different initial values of the fractional porosity may be explained by influence of friction on the side surfaces of the matrix, because the material layer height was in the range from 0.5 to 4.1 diameters. The friction factor would be expected to influence on the side surfaces of the matrix and also on the stresses distribution in the layer at higher pressures.

The linear model of loose material layer deformation in a rigid matrix is shown in Fig. 3.

The balance equation of material elementary layer, having z coordinate value, is as follows:

$$d\sigma_1 \cdot \frac{\pi D^2}{4} = -f \cdot \xi \cdot \pi D \cdot dz \,, \qquad (12)$$

where f is the coefficient of material friction on the side surfaces of a matrix.

Taking into consideration the above reasons, the side thrust coefficient is supposed to be determined by equation:

$$\xi = 1 - (1 - \xi_0) \cdot \exp\left(b_1 \cdot \lambda + b_2 \cdot \lambda^2 + \ldots + b_m \cdot \lambda^m\right); \qquad (13)$$

$$\lambda = \ln\left(\frac{\sigma_1}{\sigma_{10}}\right), \qquad (14)$$

where σ_{10} is the basic value, m is the maximum approximation exponent.

If $\sigma_1 \geq \sigma_{10}$ and all polynomial coefficients are negative, we will obtain $\xi = \xi_0$ in case $\sigma_1 = \sigma_{10}$, and also $\xi \to 1$ in case $\sigma_1 \to \infty$, corresponding to the hydrostatic pressure distribution.

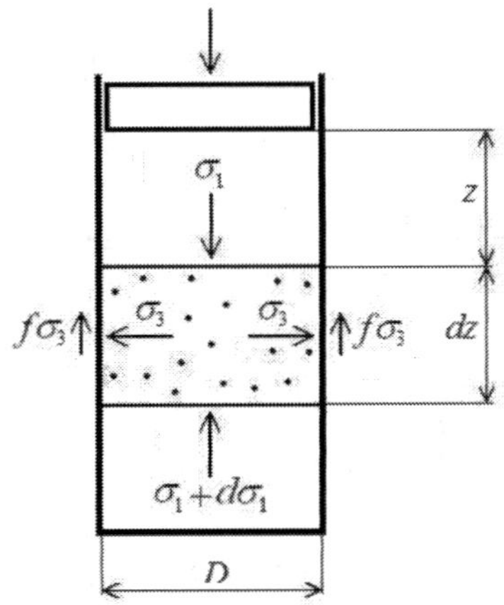

Fig. 3. The linear model of the layer deformation in a rigid matrix.

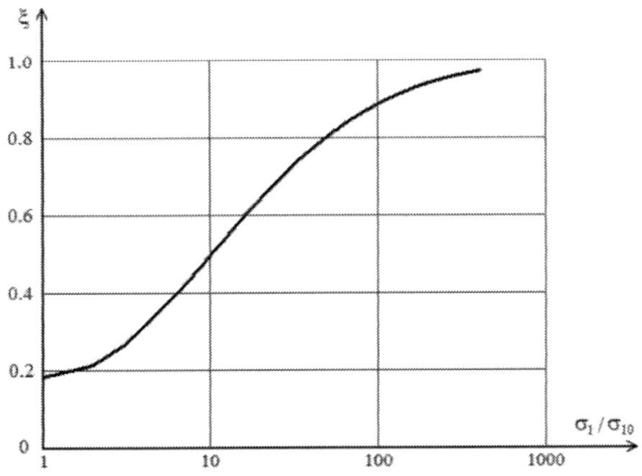

Fig. 4. Dependence of the side thrust coefficient on the maximum primary stress level.

Expression (11) is presented according to Pascal triangle of u exponent as follows:

$$e = \sum_{i=0}^{u} \sum_{j=0}^{u-i} c_{i,j} \cdot e_0{}^{i} \cdot \lambda^{j}. \qquad (15)$$

The processing of the experimental research data was reduced to the solution of equation (12) with the determination of the coefficients of b and c families by the least-square method. Therein, the values of k_1, k_2 and f are given as known values, and also

$$S = \sum_{v} \left(\frac{e_{vt} - e_{ve}}{e_{ve}} \right)^2 \to \min, \qquad (16)$$

where e_{vt} is the average value within the layer according to (15), e_{ve} is the similar experimental value.

The values calculated by formula (15) are plotted in Fig. 2 (see above) in the form of points.

4 Results and discussion

It can be noticed the close agreement of the analytical and experimental values of the fractional porosity in previous research (Fig. 2, see above). This proves the adequacy of the hereditary model of loose material layer deformation.

The analysis of expression (13) character, which is a part of the described model and demonstrates the ratio of the reactive side pressure and the active vertical pressure on the deformed loose layer, is of particular interest.

The results in Fig. 4 show the rise of the side thrust coefficient in case of the pressure increase, and its asymptotic approximating to unity, that corresponds to the given above other authors' data.

In such a way, the behavior of loose material deviates progressively from Mohr-Coulomb model and approximates to a hydrostatic model with an increase in active pressure.

5 Conclusions

The impossibility to use existing dependences of the side thrust coefficient of loose material for the description of destruction of several layers of particles in disintegrators has been substantiated.

The hereditary model has been developed for accumulating stresses and deformations, during compression of a loose mined rock layer, with account of the initial porosity of a layer.

Experimental research on compression in a rigid matrix of fine-grained granite has been carried out, which confirmed the adequacy of the developed model.

The dependences of the side thrust coefficient on the maximum primary stress and the initial porosity have been determined, based on a linear model of the material deformation in the matrix. They confirm, that the distribution of stresses in the material is close to hydrostatic one at high compression pressure values.

The obtained results allow to specify sufficiently the existing techniques for calculating the disintegrators operating parts parameters as regards 'in the layer' disintegration, in order to enhance their efficiency and to reduce energy consumption.

The authors express their gratitude to professor, Head of Higher Mathematics Department of National Mining University, Olena Sdvyzhkova for the help and consultations during the work performance.

References

1. Andreev, S.E. (1980). *Droblenie, izmelchenie i grokhochenie poleznykh iskopaemykh.* Moskva: Nedra.
2. Kolosov, D., Dolgov, O., & Kolosov, A. (2013). The stress-strain state of the belt on a drum under compression by flat plates. *Annual Scientific-Technical Collection – Mining of Mineral Deposits*, 351-357. https://doi.org/10.1201/b16354-65
3. Belmas, I.V., Kolosov, D.L., Kolosov, A.L., & Onyshchenko, S.V. (2018). Stress-strain state of rubber–cable tractive element of tubular shape. *Naukovyi Visnyk Natsionalnoho Hirnychoho Universytetu*, (2), 60-69.
4. Revnivtsev, V.I., Denisov, G.A., Za1rogatskii, L.P., & Turkin, V.Y. (1992). *Vibratsionnaya desintegratsiya tverdykh materialov.* Moskva: Nedra.
5. Dalmatov, B.I. (1988). *Mekhanika gruntov, osnovaniya i fundamenty (vkliuchaya spetsialnyy kurs inzhenernoy geologii).* Leningrad: Stroyizdat.
6. Nadutyy, V.P., & Titov, A.A. (2016). Analiz napriazhennogo sostoyaniya ploskoy zony deformirovaniya vibratsionnykh desintegratorov s uchetom vybrosa materiala. *Vibratsii v tekhnitsi ta tekhnolohiiakh*, *3*(83), 54-58.
7. Babets, D.V., Sdvyzhkova, O.O., Larionov, M.H., & Tereshchuk, R.M. (2017). Estimation of rock mass stability based on probability approach and rating systems. *Naukovyi Visnyk Natsionalnoho Hirnychoho Universytetu*, (2), 58-64.
8. Sdvyzhkova, O.O., Babets, D.V., Kravchenko, K.V., & Smirnov, A.V. (2016). Determining the displacements of rock mass nearby the dismantling chamber under effect of plow longwall. *Naukovyi Visnyk Natsionalnoho Hirnychoho Universytetu*, (2), 34-42.
9. Sdvyzhkova, O., & Patynska, R. (2016). Effect of increasing mining on longwall coal mining – Western Donbass case study. *Studia geotechnika et mechanika, 38*(1), 91-99. https://doi.org/10.1515/sgem-2016-0010
10. Bondarenko, A.A. (2012). Mathematical modelling of soil dredger absorption process in the underwater bottomwhole. *Metallurgical and Mining Industry*, (3), 79-81. orcid.org/0000–0002–7666-6752

11. Bondarenko, A.A. (2012). Laws of determination of fine materials suction limits in submarine suction dredge face. *Naukovyi Visnyk Natsionalnoho Hirnychoho Universytetu*, (4), 59-64. orcid.org/0000–0002–7666–6752
12. Nadutyy, V.P., & Titov, A.A. (2017). Analiz vliyaniya uplotniayemosti sypuchey gornoy massy na protsess eye deformirovaniya mezhdu ploskimi poverkhnostiami dezintegratorov. *Vibratsii v tekhnitsi ta tekhnolohiiakh, 1*(84), 35-40.
13. Revnivtsev, V.I., Barzukov, O.P., Ivanov, N.A. and others (1984). Osnovnye zakonomernosti izmeneniya sostoyaniya sloya sypuchego materiala pri szhatii. *Obogashchenie poleznykh iskopaemykh*, (4), 3-6.

Scientific and methodological foundations to develop numerical hydrodynamical models of mine fields in Donbas

Alina Zahrytsenko[1*], *Olena* Podvigina[1], and *Nataliia* Dereviahina[1]

[1]National Mining University, Department of Hydrogeology and Engeenering Geology, 19 Yavornytskoho Ave., 49005 Dnipro, Ukraine

Abstract. The paper considers specific character of hydrogeological problems arising during the whole cycle of a mining enterprise operation; the most reliable their solutions are implemented in the context of numerical hydrodynamical models of mine fields. Basic approaches to schematize laminated carbonic formation through simplification of natural rock mass permeability have been substantiated; prediction algorithm of its changes during temporal transformation has been developed. Numerous identification problems solving in the process of modeling mine fields has helped understand that water inflow amount cannot be correlated with the extension of mining area. Within the areas with roof caving, permeability of carbonic formation is a time variable which value increases 10 – 15 times in the context of rock displacement, halves after 5 – 10 years and nears natural values after 15 – 20 years. In the context of Western Donbas, the modularized schemes have been tested while solving problems concerning filtration control within mine fields and their surface influence areas with high confidence of hydrogeological forecasts.

1 Introduction

Formation of technogenic mode of ground water within areas of underground mining influence is one of the most complicated processes among the whole system of anthropogenic effects on the underground hydrosphere as for predictability and controllability. In this connection, it is practical to use numerical geofiltrational modeling to evaluate hydrodynamical situation within mine fields. The evaluation is supported by the development of powerful computer facilities and corresponding software (Modflow, FEFLOW, HydroGeoSphere, MODHMS and others). Recent research by foreign authors [1, 2] are intended to study the processes of contaminant migration within geotechnical systems, modeling of saturated and unsaturated flows, and transportation of different-density solutions and heat.

However, modeling nature of such a complicated and multifactor geotechnical object as a mine originates a number of debating points at each stage of its functioning. The points are connected with the methodology of adequate simulation with the help of a model of

* Corresponding author: alinanik@bigmir.net

geomechanically disturbed rock mass under different standard conditions of coal mines.

Mironenko V.A. , Shestakov V.M. , Zhernov I.E. , Gavich I.K., Lomakin E.A. and others contributed much to the development of both theory and methodology of numerical modeling for mining hydrogeology. Recent studies of scientific team under the guidance of Norvatov Yu.A. concern geofiltrational studies in the process of operation and liquidation of mining enterprises. Paper [3] has developed typification of naturally technogenic structures on the basis of a character of hydraulic subsurface system-technogenic system connection; it involves rock mass disturbed by water-supply fissures; disintegrated rock within worked-out area; and unfilled stopes.

Several papers [4, 5] schematize naturally technogenic complex having a system of mine workings with up to 400 m thickness as a one-layer formation with indefinite temporal changes in the rock mass permeability parameters. Such simplification prevents from making a point of dynamics of inflow formation within certain seams and determining their factors generation. The hydrodynamic situation is described by means of parameters of pressure distribution in the context of final period of a mine operation; total cavitation volume of technogenic system is evaluated analytically on the ratio between a volume of all worked-out seams and a volume of technogenic complexes.

The authors agree as for the prevailing effect of compressibility of carbonic formation on the velocity of underground water level recovery in the context of flooding of mines; the fact is supported by research carried out for Central Donbas [6, 7] as well as for other regions of Ukraine [8, 9]. Saturation velocity of disturbed rock mass is important while controlling a mine field flooding process; parameters of temporal permeability variations required to solve problems of water control in the process of a field development and to use the georesource during postdevelopment periods remain uncertain.

Thus, the *studies* are intended to substantiate scientific and methodological approaches to schematize natural hydrodynamical conditions of mine fields used to simulate technological process of coal seams mining and hydromechanical transformation of rock mass making it possible to determine factors of inflow formation and regularities concerning changes in temporal permeability of worked-out area.

The process involves a number of numerical experiments for the conditions of Western Donbas to develop a solution algorithm for identification problems to make decisions concerning filtration control within mine fields and areas of their surface effect.

2 Results and discussion

The model compliance with the study object and, hence, reliability of the obtained hydrogeological prognoses, depend upon the performance of schematization of hydrogeological conditions determined by formulation of practical tasks. The whole range of problems concerning water control in the context of coal-mining regions going with periods of mining, flooding, and postdevelopment, can be divided into two groups: mining and technological, and hydroecological (Fig. 1).

The above-mentioned peculiarities cannot cover the whole range of hydrogeological problems arising within mine fields differing in their hydrogeological conditions but being typical for the majority and making it possible to formulate general principles concerning the development of geofiltration schemes of mine fields. Simplification of environmental conditions as well as operative factors is performed in accordance with particular order; moreover, it consists of several stages (Fig. 2).

Many schematization techniques are well-known and have been published in the papers [10, 11]. Turn our attention to the moments being important for the process of inverse problem solving and for the improvement of total reliability of modeling results. From the methodological viewpoint, substantiation of optimum number of simulation layers and modeling of laminated system of rocks are the most difficult problems.

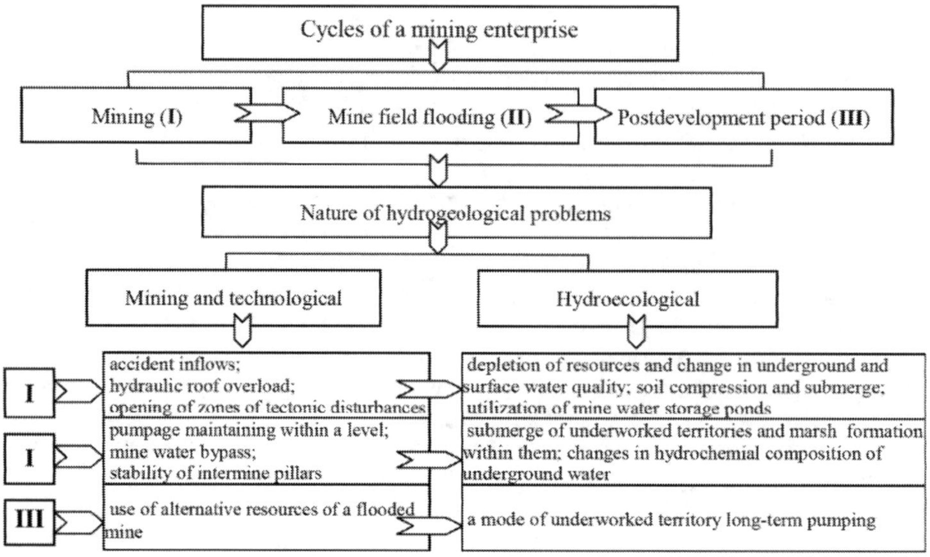

Fig. 1. Specific character of hydrogeological problems for a coal mine.

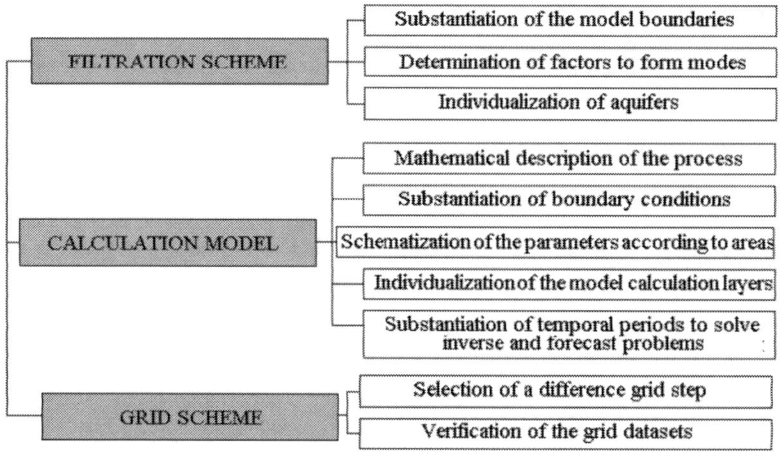

Fig. 2. Stages to develop geofiltrational scheme of a mine field model.

Layered water-bearing structure within the model is shown by means of weight-average filtrational parameters in calculation blocks as well as dependency parameters in the context of calculation layers [12]. The number of calculation layers for each model is determined individually taking into consideration complexity of water-bearing structure, angle of rocks and fault planes of tectonic disturbances, and distance between levels of coal seams mining. To simulate actual hydrodynamic situation within mine fields of Western Donbas, carbonic formation with interbedding of argillites, aleurites, sandstones, limestones, and coal is schematized in the form of seams being extracted which thickness is determined by means of height of a zone of aquiferous fissures.

Most of all, parameters of sedimentary formation or a surface of water-bearing level, neighbouring a zone of intensive fissuring, are schematized with the help of hydraulic formulation basing upon a supposition by Miatiiev-Girinski (in the context of permeable layers, underground water moves horizontally; in the context of slightly permeable ones, the movement is vertical). Schematization of parameters of water-bearing formation of coal measures is implemented hydrodynamically. In this context, nonhomogeneous filtration area is divided into solid components by means of a system of mutually orthogonal planes in *X*, *Y*, and *Z* coordinates.

A system of interbedding of well-permeable layers and slightly permeable ones is averaged out by means of weighted parameters along *X*, *Y*, and *Z* axes. A level (pressure) within each simulation layer belongs to a centre of a solid component.

If *rocks occur vertically* (bedding angle is $\alpha = 90^o$, Fig. 3) and the layer thickness is h_i, then the weighted values of filtration coefficient k_x and k_y are calculated according to the dependences corresponding to the filtration:

– in parallel with the bedding

$$K_{y_i} = K_{z_i} = \frac{\sum k_i h_i}{\sum h_i} = \frac{\sum k_i h_i}{\Delta x}, K_{y_{i+1}} = K_{z_{i+1}} = \frac{\sum k_{i+1} h_{i+1}}{\sum h_i} = \frac{\sum k_{i+1} h_{i+1}}{\Delta x}; \quad (1)$$

– transversely to the bedding

$$K_{x_i} = \frac{\sum h_i}{\sum \frac{h_i}{k_i}} = \frac{\Delta x}{\sum \frac{h_i}{k_i}}, \quad K_{x_{i+1}} = \frac{\sum h_{i+1}}{\sum \frac{h_{i+1}}{k_{i+1}}} = \frac{\Delta x}{\sum \frac{h_{i+1}}{k_{i+1}}}. \quad (2)$$

Filtration coefficient $K_{z_i, z_{i+1}}$ along Z axis between blocks of neighbouring calculated layers is computed on the dependence according to filtration being transverse to K_{z_i} and K_{z_i+1} layers:

$$K_{z_i, z_{i+1}} = \frac{\Delta z_i + \Delta z_{i+1}}{\frac{\Delta z_i}{K_{z_i}} + \frac{\Delta z_{i+1}}{K_{z_{i+1}}}}. \quad (3)$$

Relation parameter of neighbouring calculated layers (TR) along Z axis is computed according to the formula:

$$TR = \frac{\Delta z_i + \Delta z_{i+1}}{\frac{\Delta z_i}{K_{z_i}} + \frac{\Delta z_{i+1}}{K_{z_{i+1}}}} \times \frac{1}{0.5\left(\Delta z_i + \Delta z_{i+1}\right)}. \quad (4)$$

Judging by the dependences, weighed values of filtration coefficients within blocks along *X* and *Y* axes will differ owing to the direction of underground water movement.

If *layers occur horizontally* (Fig. 4), then weighed values of filtration coefficients k_x and k_y within each layer (block) of the model are calculated on the dependence in parallel with bedding:

$$K_{x_i} = K_{y_i} = \frac{\sum k_i h_i}{\sum h_i} = \frac{\sum k_i h_i}{\Delta z}, K_{x_{i+1}} = K_{y_{i+1}} = \frac{\sum k_{i+1} h_{i+1}}{\sum h_{i+1}} = \frac{\sum k_{i+1} h_{i+1}}{\Delta z_{i+1}}. \quad (5)$$

Hence, along *Z* axis, i.e. transversely to the bedding, k_z is:

$$K_{z_i, z_{i+1}} = \frac{\Delta z_i + \Delta z_{i+1}}{\left(\sum \frac{h_i}{k_i} + \sum \frac{h_{i+1}}{k_{i+1}}\right)}. \quad (6)$$

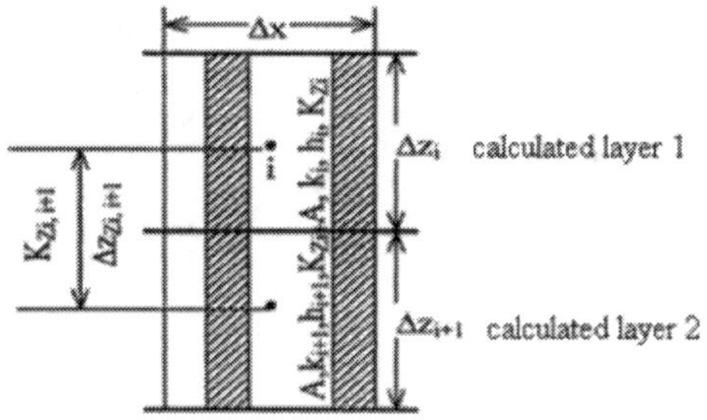

Fig. 3. Scheme of vertical arrangement of layers within the calculated blocks.

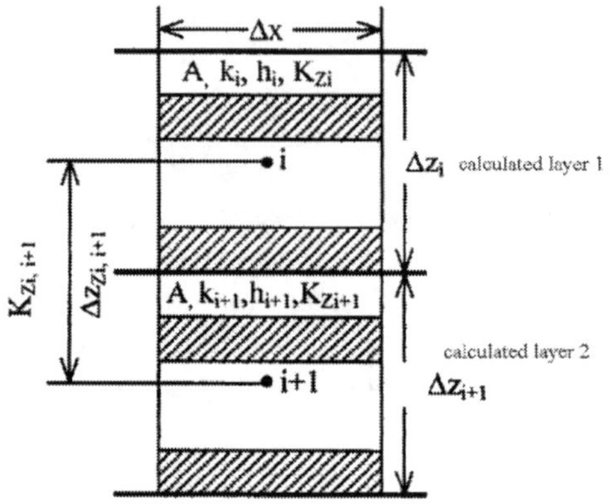

Fig. 4. Scheme of horizontal layer bedding within the calculated blocks.

Relation parameter of neighbouring calculated layers along Z axis (TR) is computed according to dependence (4).

If *layers occur aslope* (Fig. 5), then average values of filtration coefficients k_x and k_y along X and Y axes for each block of the model are equal to:

$$K'_{x_i} = K_{x_i} \cdot \cos^2 \alpha + K_{z_i} \cdot \sin^2 \alpha;$$
$$K'_{y_i} = K_{x_i}; \tag{7}$$
$$K'_{z_i} = K_{x_i} \cdot \sin^2 \alpha + K_{z_i} \cdot \cos^2 \alpha.$$

Filtration coefficient $K'_{z_{i,i+1}}$ along Z axis between blocks of the calculated layers is computed transversely to layers with perpendicular with K'_{z_i} and $K'_{z_{i+1}}$ filtration coefficients:

$$K'_{z_{i,i+1}} = \frac{\Delta z_i + \Delta z_{i+1}}{\left(\dfrac{z_i}{K'_{z_i}} + \dfrac{z_{i+1}}{K'_{z_{i+1}}} \right)}. \tag{8}$$

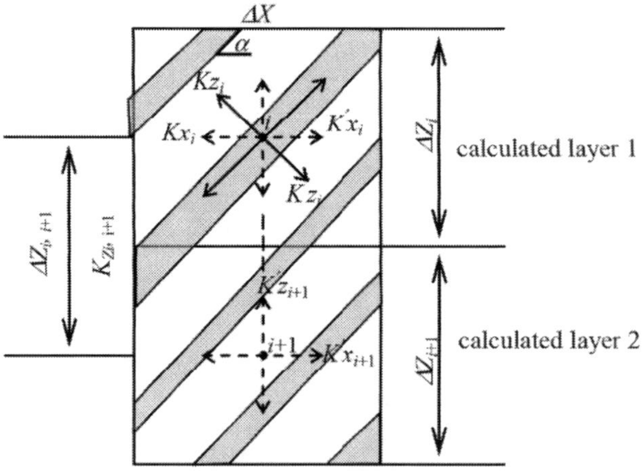

Fig. 5. Scheme of inclined bedding of layers within the calculated blocks.

Array of *anisotropy coefficient* is introduced within the model to consider differences in filtration coefficients along X and Y axes. In this context, the anisotropy coefficient is a product of coefficients of natural anisotropy and that one obtained as a result of schematization of rock filtration properties.

It should be noted that natural carbon permeability increases in the process of a stope roof caving; it decreases when rocks compact. Thus, to reflect real processes taking place during mining, temporal changes have to be set within calculated blocks of the model: filtration coefficients of the calculated layers; underground water-mine workings relation parameters; neighbouring calculated layers – dewatering coefficients ratio. Meanwhile, in the process of technogenic transformation, their quantitative values are determined while *solving inverse identification problems.*

Simulation of long-term periodical chronology of coal seams mining [13] with typical water inflow dynamics is the feature of a mine field verification model. That makes it possible to determine water inflow formation factors in terms of each seam and evaluate quantitatively changes in permeability parameters of any worked-out rock mass in time.

Numerous identification problems solving in the process of modeling mine fields in the context of such mining enterprises as M.I. Stashkova mine, "Samarska", "Zakhidno-Donbaska", "Dniprovska" mines [14] has helped understand that water inflow amount cannot be correlated with the extension of mining area. Within the areas with roof caving, permeability of carbonic formation is a time variable which value increases 10 – 15 times in the context of rock displacement, halves after 5 – 10 years and nears natural values after 15 – 20 years (Fig. 6).

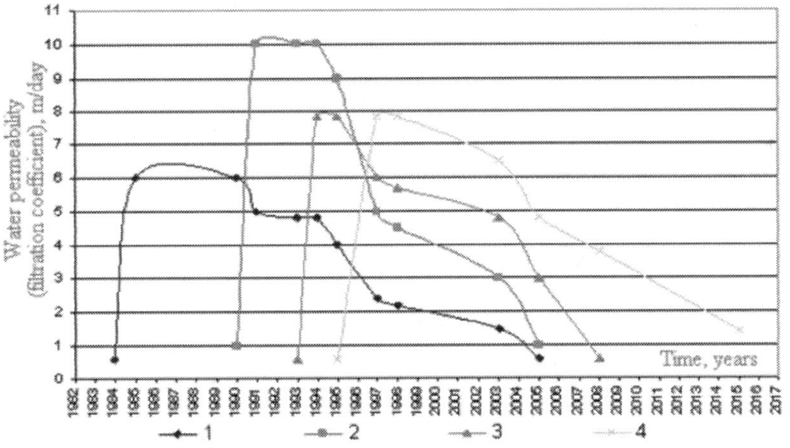

Fig. 6. Changes in water permeability in terms of C_5 seam within (1 – 4) areas worked out during different time periods.

Identification of the models also has determined that permeability of high-amplitude tectonic faults (Bohdanovskyi and Prodolnyi) corresponds to the parameters of filtration shields (Table 1) where filtration coefficient is $10^{-4} – 10^{-7}$ m/day.

Table 1. Hydrodynamical parameters of a zone of tectonic faults.

Filtration and capacitive parameters	A zone of tectonic faults	Behind fault area
Filtration coefficient, m/day (C_6 / C_1 seam)	10^{-4} / 10^{-6}	10^{-2} / 10^{-2}
Coefficient of elastic filtration characteristic (C_6 / C_1 seam)	10^{-6} / 10^{-7}	10^{-5} / 10^{-5}

3 Conclusions

Factors to form water inflows as well as regularities concerning timely changes of the disturbed rock mass permeability have been determined at the stage of model identification on the dynamic similarity while simulating technological process to mine coal seams and to perform hydromechanical transformation of rock mass. That made it possible to develop adequate models of vine fields with prognostic decisions of high integrity (more than 90%) as for the control of hydrodynamic situation both at the surface and underground, at the stage of intensive mining operations and at the final stage of liquidation and rock mass flooding.

The research has been carried out with financial support from "DTEK Pavlohradvuhillia" PJSC, Shahtoupravlinnia "Ternivske" and "Dniprovske. The authors are grateful for the Companies management and assistance in surveying and hydrogeomechanical monitoring in underground mine workings.

References

1. Barnett, B., Townley, L., Post, V., Evans, R., Hunt, R., Peeters, L., Richardson, S., Werner, A., Knapton, A., & Boronkay, A. (2012). Australian groundwater modeling guidelines. *Waterlines Report Series, June,* (82), 204.
2. Shuchita Garg, S.K. (2016). Singh Modeling of arsenic transport in groundwater using MODFLOW: A case study. *Inational journal of geomatics and geosciences*, *6*(4).
3. Norvatov, Yu.A., Petrova, I.B., & Stepanova, A.S. (1999). Tipizatsiya usloviy obrazovaniya vodopritokov v gornykh vyrabotkakh pri nablyudenii za zatopleniem ugol'nykh shakht. *Gornaya geomekhanika i marksheyderskoe delo*, 117-121.
4. Norvatova, O.I. (2005). *Metodika prognoza gidrogeologicheskikh usloviy dlya stroitel'stva ugol'nykh shakht na baze chislennogo modelirovaniya geofil'tratsionnykh protsessov (na primere Vostochnogo Donbassa)*. Phd. Vsesoyuznyy nauchno-issledovatel'skiy marksheyderskiy institut.
5. Savel'ev, D.I. (2014) *Gidrogeologicheskoe obespechenie meropriyatiy po predotvrashcheniyu negativnykh posledstviy zatopleniya ugol'nykh shakht* Ph.D. Natsionalnyy mineral'no-syr'yevoy universitet "Gornyy".
6. Sadovenko, I.A., & Zagritsenko, A.N. (2003). Obosnovanie parametrov i sposobov upravleniya geofil'tratsionnym rezhimom ugleporodnogo massiva pri zatoplenii shakht. *Sbornik nauchnykh trudov Natsionalnogo Gornogo Universiteta*, *17*(1), 48-51.
7. Sadovenko, I., Rudakov, D., & Podvigina, O. (2010). Analysis of hydrogeodynamics in a mining region during exploitation till closure of coal mines. *New Techniques and Technologies in Mining: School of Underground Mining*, 61-69. https://doi.org/10.1201/b11329-12
8. Liashchova, M. (2016). Microfacies types and paleontological characteristics of the visean carbonate rocks (the southern zone of Dnieper-Donets basin). *Visnyk of Taras Shevchenko National University of Kyiv. Geology*, *2*(73), 38-45. https://doi.org/10.17721/1728-2713.73.01
9. Pribilov, V. M. (2017). Assessment of technological pollution sources influence on underground water on the territory of Kharkiv region. *Visnyk of V.N Karazin Kharkiv national university-series geology geography ecology*, (47), 198-205. https://doi.org/10.26565/2410-7360-2017-47-26
10. Timoshchuk, V., & Sherstyuk, Y. (2012). Geofiltration regularities in the areas loaded by gravitation in tailings and waste rock dumps. *Naukovyi Visnyk Natsionalnoho Hirnychoho Universytetu*, (4), 30-35.
11. Sadovenko, I., Zahrytsenko, A., Podvihina, O., & Dereviahina, N. (2017). Water balance control within rock mass using the capacity of water-bearing formations. *Naukovyi Visnyk Natsionalnoho Hirnychoho Universytetu*, *4*(160), 19-27.
12. Malashkevych, D., Sotskov V., Medyanyk V. & Prykhodchenko D. (2018). Integrated Evaluation of the Worked-Out Area Partial Backfill Effect of Stress-Strain State of Coal-Bearing Rock Mass. *Solid State Phenomena*, (277), 213-220. https://doi.org/10.4028/www.scientific.net/ssp.277.213

13. Fomichov, V., Sotskov, V., Pochepov, V. & Mamaikin, O. (2018). Formation of a calculation model determining optimal rate of stping face movement with a large deformation of a rock massif. *ARPN Journal of Engineering and Applied Sciences*, *13*(7), 2381-2389.
14. Sadovenko, I., Zahrytsenko, A., Podvigina, O., Dereviahina, N., & Brzeźniak, S. (2018). Methodical and Applied Aspects of Hydrodynamic Modeling of Options of Mining Operation Curtailment. *Solid State Phenomena*, (277), 36-43. https://doi.org/10.4028/www.scientific.net/SSP.277.36

Integrated analysis of geofiltrational parameters in the context of underground coal gasification relying upon calculations and modeling

Oleksandr Inkin[1*], *Volodymyr* Tishkov[1], *Nataliia* Dereviahina[1], and *Vadym* Sotskov[2]

[1]National Mining University, Department of Hydrogeology and Engineering Geology, 19 Yavornytskoho Ave., 49005 Dnipro, Ukraine
[2]National Mining University, Department of Underground Mining, 19 Yavornytskoho Ave., 49005 Dnipro, Ukraine

Abstract. Objective of the paper is to substantiate and implement the integrated approach aimed at the studies of filtration and mass-transfer processes within roof rocks of coal seams in the context of their underground gasification. Integrity of the studies is substantiated by the use of analytical calculations as well as physical and numerical modeling. Zones with different permeability have been determined and values of specific water inflow have been identified basing upon the performed numerical modeling and involving multicomponent transformations within roof formation.The research data actuality is in the studies of spatial and temporal dynamics of rock technogenic rock permeability above gasification channel depending upon their geological structure which will favour the substantiation of efficient engineering solutions to control a process of underground coal gasification under difficult hydrogeological conditions.

1 Introduction

Underground coal gasification (UCG) is the method owing to which both coal mining and its processing takes one stage. The technology is a process when coal seams, considered as unprofitable for open-pit mining and underground mining, can be transformed into combustible gas to be used in the form of fuel, to generate electrical energy and to be used as chemical raw material with the help of wells. The UCG technology application relies upon a wide range of engineering and interfacing disciplines intended to provide stable process of coal seams gasification.

The paper represents integrated studies of filtration and migration processes within roof rocks of a coal seam in the context of its gasification taking into consideration the effect of temperature capabilities of the coal under gasification and rocks, enclosing them, as well as geomechanical and hydrogeological characteristics; moreover, technological conditions of gas generator operation are also involved. Integrity and multifactor nature of the processes under consideration determine a sequence of the approach to the research structure [1 – 4].

[*] Corresponding author: inkin@ua.fm

2 Analytical approach to study temperature field distribution within rock mass during underground coal gasification (UCG)

It is known that modes of conductive and convective rock mass heating may arise within underground gas generator in terms of different underground water pressure-gas pressure ratios. Consider a situation when pressure within gas generator is less than water pressure; thus, gas effusion doesn't originate and heat transfer is of conductive nature, mass transfer with surrounding rock mass is minimal being carried out at the expense of diffusion.

In terms of static position of two rock mass phases (i.e. rock + water), thermal flow from gas generator may be reduced to axisymmetrical consideration of temperature field described by an equation of the type [5]:

$$T(r,t)-T_0=\frac{q}{\lambda}R_2\left\{\frac{R_2^2}{R_2^2-R_1^2}\left[2\frac{at}{R_2^2}-\frac{1}{4}\left(1-2\frac{r^2}{R_2^2}\right)-\frac{R_1^2}{R_2^2}\left(\ln\frac{r}{R_1}+\frac{R_2^2}{R_2^2-R_1^2}\ln\frac{R_1}{R_2}+\frac{3}{4}\right)\right]+\right.$$

$$\left.+\sum_{n=1}^{\infty}\frac{\pi}{\mu_n}\frac{I_1\left(\mu_n\frac{R_1}{R_2}\right)I_1(\mu_n)}{I_1\left(\mu_n\frac{R_1}{R_2}\right)I_1(\mu_n)}\cdot\left[I_0\left(\mu_n\frac{r}{R_2}\right)Y_1\left(\mu_n\frac{R_1}{R_2}\right)-Y_0\left(\mu_n\frac{r}{R_2}\right)I_1\left(\mu_n\frac{R_1}{R_2}\right)\right]\cdot e^{-\mu_n^2\frac{at}{R_2^2}}\right\},\quad(1)$$

where q is specific heat flow capacity; R_1 and R_2 are radius of a gas generator radius and its depth relative to earth's surface respectively; r is distance from the gas generator axis to a reference point; t is time baseline period; T_0 is background rock mass temperature; μ_n are characteristic first-order numbers of Bessel function for boundary values R_1 and R_2; λ is heat conductivity factor of water-saturated rock mass.

Heat-transfer problem considering relative displacement of one of the rock mass phases in the context of analytical version is extremely difficult.

One-dimensional solution in terms of finite differences for static case and taking into consideration liquid phase transfer is represented by means of the equations:

$$T_i^t=\frac{1}{2+\frac{\Delta x^2}{a\Delta t}}T_{i-1}^t+\frac{1}{2+\frac{\Delta x^2}{a\Delta t}}T_{i+1}^t+\frac{\Delta x^2}{2a\Delta t}T_i^{t-\partial t};\quad(2)$$

$$T_i^t=\frac{\frac{\lambda}{\Delta x^2}-\frac{C_wV_w}{\Delta x}}{\frac{C_r}{\Delta t}-\frac{C_wV_w}{\Delta x}+\frac{2\lambda}{\Delta x^2}}T_{i+1}^t+\frac{\frac{\lambda}{\Delta x^2}}{\frac{C_r}{\Delta t}-\frac{C_wV_w}{\Delta x}+\frac{2\lambda}{\Delta x^2}}T_{i-1}^t+\frac{\frac{C_r}{\Delta t}}{\frac{C_r}{\Delta t}-\frac{C_wV_w}{\Delta x}+\frac{2\lambda}{\Delta x^2}}T_i^{t-\Delta t},\quad(3)$$

where a is temperature conductivity coefficient; Cr and Cw are heating capacities of rock and water respectively; and Vw is actual velocity of water flow; the mentioned specifications have already been involved.

To use the solutions correctly, certain features of the set problem should be mentioned.

Nonlinear temperature within axisymmetrical thermal flow is: temperature gradients decay at the distance of 3 – 4 radii of a thermal source reduced to a cylindrical form. In the context of the layer, practical evaluations can not involve difference between one-dimensional and axisymmetrical flows.

Formula (1) is true for boundary second-type conditions when a function of a heat flow is known. In terms of sensible temperature difference between thermal source (T_s) and absorptive medium (T_0), the function becomes constant:

$$q = \sigma \cdot C(T_s - T_0), \quad (4)$$

where σ is Stefan-Boltzmann constant; C is a coefficient depending upon capability of the medium to absorb thermal energy.

Underground gas generator emits thermal flow which temperature achieves 1000 °C; as a rule, it means that condition (4) has been applied. It is obvious that within rock masses, being typical for coal deposits, a value of thermal flow according to condition (4) can be obtained from equation (1) on the experimental temperature measurements. To do that, use actual data from [6].

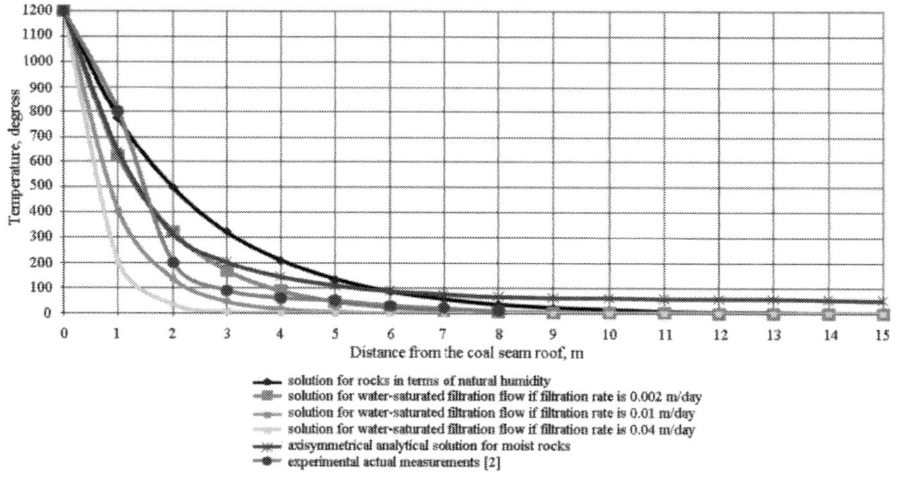

Fig. 1. Demonstrates results of variants calculations showing dynamics of temperature field within roof rocks of underground gas generator.

Results of variant calculations of dynamics of a temperature field in roof rocks of an underground gas-generator, confirmed by convergence with the actual data [6] create a possibility to evaluate the influence of hydrogeological conditions on it more differentially. The highest temperature distribution indexes in a five-meter contour are typical for a condition of rocks at natural humidity, i.e. when a massif is pre-drained. The smallest range of distribution of a temperature field is typical for conditions of water-saturated filtration flow at a filtration rate of 0.4 m/day, which is typical for location of sandstone rocks in a roof of a gas-generator. The study of distribution of a temperature field allows applying accurate corrections to change in physical and mechanical properties of rocks when creating numerical mathematical models.

3 Experimental procedure and results of the experiments in the context of physical modeling

Experimental module to study a process of distribution of temperature field of overburden rocks and their permeability (Fig. 2) was developed and manufactured taking into consideration the calculated similarity coefficients. Modeling of coal seam degassing process was performed within clear thermal-resistant tube with 1 m length and 0.04 m diameter. Compressor was used to supply air into combustion chamber. Air consumption was similar during the whole experiment. Pressure difference was recorded with the help of a manometer; temperature was recorded with the help of microthermocouple elements.

Temperature within a medium being modeled (i.e. fine sand) was controlled from a combustion zone to combustion products extracted to atmosphere through a hydroseal. Gas consumption was recorded with the help of flow rate meter.

Source excess pressure providing air supply while igniting, was $1.013 \cdot 10^5$ Pa. The ignition was provided by means of thermal heat of coal. After the coal started firing, the ignition device removed and combustion rate was supported with the help of air supply. After filtration, the gas was extracted through a hydroseal to the atmosphere [7].

The experiment involved visual observation of changes in the state of the coal seam and overburden rocks. Its period was limited by time of coal combustion (i.e. 45 min); then the medium, which has already been saturated by gasification products, is ejected from the thermal-resistant tube for the analysis of filtration properties.

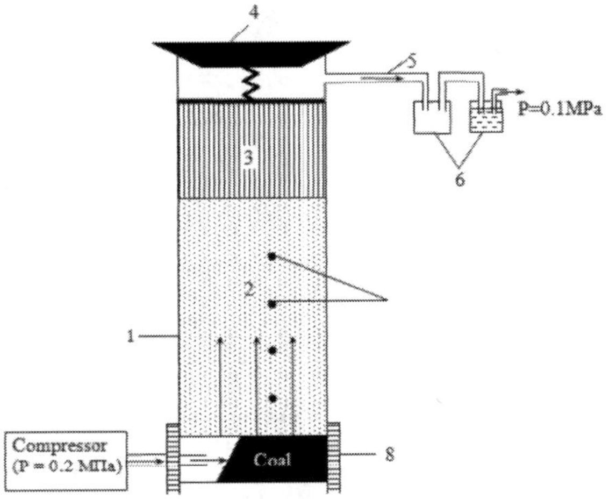

Fig. 2. Scheme of experimental facilities: 1 – clear thermal-resistant tube; 2 – roof rock of a coal seam (sand); 3 – laboratory cotton; 4 – a stopper; 5 – interconnector; 6 – moisture collector and hydroseal; 7 – thermocouple elements; and 8 – ignition device.

To evaluate filtration parameters of the medium, its particle composition was studied with the help of grain-size analysis. Since effective diameter is 0.1 to 3 mm and heterogeneity coefficient is less than 5, nomogram of N.N. Bindeman has been applied to determine a filtration coefficient of the medium being modeled [8]. A value of the filtration coefficient was $k_f = 3$ m/day. Permeability coefficient k_p has been identified according to the dependence [9]:

$$k_p = \frac{k_f \cdot \mu}{\rho \cdot g}, \tag{5}$$

where k_f is filtration coefficient; μ and ρ are dynamic viscosity and water density. The calculated value of sand permeability within the experimental facilities was determined as $4.46 \cdot 10^{-11}$ m^2.

Permeability coefficient was also evaluated according to data obtained during the experiment relying upon formula (6) [10]:

$$K_p = \frac{2\mu P_{at} QL}{F\left(P_1^2 - P_2^2\right)}, \tag{6}$$

where P_{at} is atmospheric pressure ($1.013 \cdot 10^5$ Pa); Q is consumption of air being pumped to a combustion chamber ($4 \cdot 10^{-5}$ m^3/s; L and F are a seam thickness and sectional area of overburden rocks (0.5 m and $1.25 \cdot 10^{-3}$ m^2); μ is average dynamic viscosity of filtering gas ($1.481 \cdot 10^{-5}$ Pa·s); and P_1, and P_2 are intake pressure and output pressure on the exit from a layer of gas-permeable rocks respectively ($2.026 \cdot 10^5$ Pa and $1.013 \cdot 10^5$ Pa).

The k_p value, determined on (6), was $3 \cdot 10^{-12}$ m^2. Difference between the obtained values of permeability coefficient is allowable taking into consideration empiric data of granular sand composition. The obtained experimental value should be used in the context of subsequent calculations.

In the course of the experiment, temperature was recorded in certain points of the model at the distances of 5, 10, 12, 15 and 20 cm from the combustion source. The temperature was recorded continuously with the help of identical thermal couples connected to multichannel recorder H-307. The thermal couples have been calibrated relative to hyperthermal temperature gauge TG-31A No. 433 with ±2 °C error. Fig. 3 and 4 demonstrate over-time temperature variations within the filtering medium.

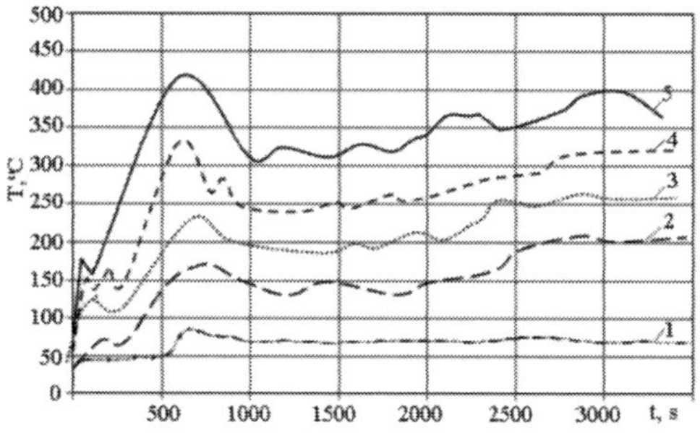

Fig. 3. Temperature variations within the filtering medium: 1, 2, 3, 4 and 5 are temperature variations in a roof at the distance of 20, 15, 12, 10 and 5 cm from the top edge of the combustion zone respectively.

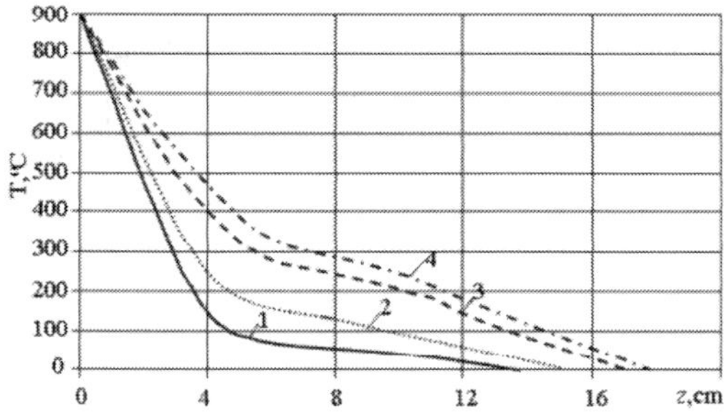

Fig. 4. Heating intensity of overburden rocks: 1, 2, 3, and 4 are temperature distribution within a roof after 200, 400, 600 and 2500 s from the experiment beginning respectively.

The coal seam state as well as overburden rocks state was observed visually synchronous with temperature recordation; as a result, the following was determined:

– a zone of coal combustion is convex towards air motion; its geometry depends upon a supply rate;

– combustion zone-overburden rocks contact is unstable; overlying sand material penetrates into the combustion zone;

– in due course, the filtering medium (i.e. sand material) becomes grey with point distribution of black microinclusions;

– through the microscope, films of resinous substances, thickening within irregularities, are seen at the surface of the sand grains;

– an arch is formed right above the zone of coal combustion; its stability depends upon the number of carboniferous inclusions and dimensions of the zone is determined with the help of the coal-combustion process duration.

The data, obtained during the experiments, have helped conclude that rocks of immediate roof, occurring closer to the combustion zone, become of higher temperature and start varying first. Temperature increase within overburden rocks is followed by changes in their physical and mechanical characteristics. After rock heating, evaporation of natural and bound moisture starts. The evaporation originates at the temperature of 100 °C; when the temperature achieves 200 °C and exceeds it, chemical moisture is liberated from roof rocks. Temperature increase up to 600 °C results in agglomeration and further decomposition of certain elements of overburden rocks (Fig. 5).

 https://doi.org/10.1051/e3sconf/20186000035

Fig. 5. Samples of sandy deposits and argillaceous deposits of Dniprobas after 400 and 650 °C temperature field effect on rock mass.

Analysis of curves (Fig. 2 and 3) means that roof rock heating process is intensive during the first 600 seconds from coal ignition within a combustion zone. Further heat transfer is a very slow process. The phenomenon can be explained by the fact of the decreased thermal conductivity of overburden rocks at the expense of changes taking place in their physical state (i.e. agglomeration, expansion, pore mudding) under the effect of high temperature and chemical effect of escaping gas. Insignificant temperature variations during the experiment can be explained by the unstable burning process and, consequently, temperature fluctuations within inlet boundary.

In the context of the medium, being modeled, vertical temperature distribution is nonlinear process. Distribution of temperature within certain areas (see Fig. 3) corresponds to the processes of rock variations, condensation of filtering gas, and circulation of gas flows with constant temperature.

Both physical and filtration properties of the medium, saturated during the experiment, were calculated and determined from its different areas by means of a technique of cutting cylinders [11]; Table 1 contains the results.

Table 1. Structure and properties of overburden rocks saturated by gasification products (according to the data of physical modeling).

Zone	Distance from a coal seam roof, m	Prevailing processes	Permeability, m^2	Poriness, %	Density, kg/m^3
3	> 2.5	Filtration gas flow (undisturbed rocks)	$3\cdot10^{-12}$	33	1510
2	0.5 – 2.5	Filtering gas condensation (pores are filled with liquid carbohydrates)	$1.6\cdot10^{-12}$	27	1630
1	< 0.5	Rock melting (slagged and agglomerated rock with the inclusion of unburnt combustibles)	$9.2\cdot10^{-13}$	21	1740

The analysis shows that migration of gasification products (GP) within filtering medium results in nonuniform filling of pore space with the formation of several zones. First of all, GP distribution depends upon temperature distribution. Poriness of overburden rocks and their permeability vary owing to mechanical blocking of pores by means of unburnt combustibles as well as physical and chemical GP-rock interaction. A zone of thermally altered rocks with solid carboniferous inclusions is the closest to the degassed area. In consideration of geometrical similarity coefficient ($C_l = 25$), thickness of the zone will not be more than 0.5 m under full-scale conditions. Above the zone (at the distance of 0.5 – 2.5 m) a condensation zone is located which porous space is filled with resinous products. Within the overlying zone of undisturbed rocks, constant-temperature gas flows circulate.

The determined regularities coincide qualitatively with the data obtained at Shahtinsk station "Pidzemhaz" [12, 13]. Thus, after gasification of a coal seam "Rozovyi" with 0.4 m thickness, upper share of overburden rocks was melted layer containing unburnt combustibles with up to 20 cm thickness. Above the layer (at the distance of 0.2 – 1.5 m) rocks transferred gradually from melted (i.e. red colour) to undisturbed (deep blue ones). Temperature above the layer being characterized was 700 – 800 °C; it dropped gradually down to 500 °C. Then, the temperature became 150 – 200 °C within a zone where the rocks transferred from heat-dried state to water-saturated one.

Zones of immediate roof annealing of "Rozovyi" coal seam were located closer to a rock boundary of the underground gas generator to compare with the zones selected for the experiment. Their eliminating can be explained by slower decrease in the parameters of gas being filtered within the medium under modeling (sand material) and, consequently, by its greater permeability and heat transfer to compare with "Rozovyi" seam roof (shale).

4 Numerical modeling of filtration parameters in the context of two-layer rock formation within a roof of a gas generator

If a coal seam roof contains water-proof argillaceous rocks (even with carbonaceous component) and up to 0.01 m/day values of filtration coefficient being typical for them (Fig. 5), the values of specific water inflow in the context of shear deformations of the rock formation [14, 15] are within 0.11 – 0.23 m^2/day when gas generator operates.

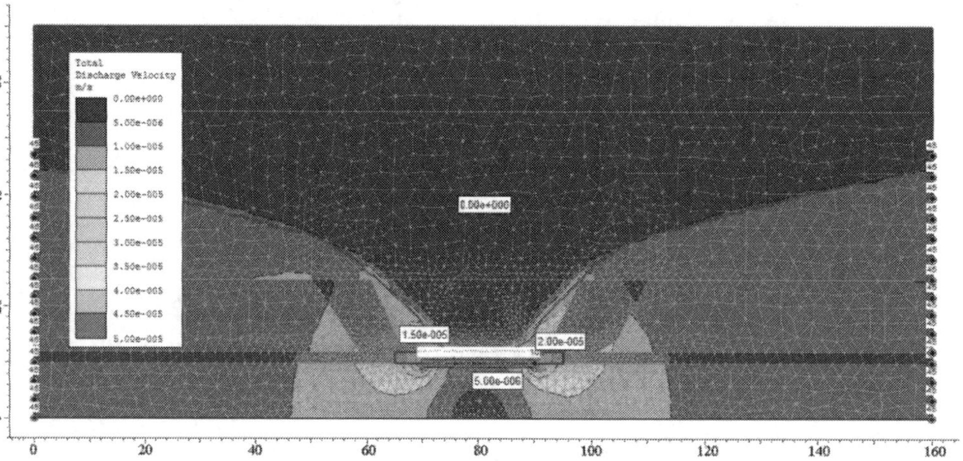

Fig. 5. Distribution of filtration rates within foliated rock mass disturbed by operation of the gas generator if waterproof layer is available within a roof of a coal seam, m/s.

The evaluation of changes in hydrodynamic mode of the complicated rock formation, containing gas generator, has shown that formation of filtration parameters depends directly on the changes in rock lithology under the effect of temperature field, geomechanical processes, and residual hydraulic pressure.

On the basis of numerical simulation, under conditions of a complex lithological structure of roof rocks [16, 17], it is possible to determine the values of specific water inflows into the channel of a gas-generator at various stages of its formation with sufficient accuracy. The solution of this problem is possible only with a complex approach based on previously obtained data on the transformation of physical and mechanical properties of rocks under the influence of temperatures, and a change of geofiltration parameters in a geomechanically disturbed massif.

5 Conclusions

It is established that the zone of intensive thermal transformation of rocks reaches 2.5 m, which is confirmed by the results of physical simulation. The decrease of the temperature field to 100 °C occurs in an interval from 1.5 to 4.5 m, and is non-linearly distributed vertically. These patterns coincide qualitatively with the data obtained at Shakhtynska station of "Pidzemhaz".

The temperature increase in covering rocks is accompanied by a change in their physical and mechanical characteristics. Evaporation of natural and bound water begins after the warming of rocks. At a temperature above 200 °C, the release of chemical moisture from the roof rocks begins, and a temperature increase to 600 °C leads to sintering and further disintegration of individual components of covering rocks and a change in their permeability.

The physical modeling results have helped determine permeability coefficient for different roof zones above gasification channel. Difference between the obtained values of the permeability coefficient is allowable on the basis of granular sand composition.

Zones with different permeability have been determined and values of specific water inflow have been identified basing upon the performed numerical modeling and involving multicomponent transformations within roof formation.

Publications are based on the research provided by the grant support of the State Fund For Fundamental Research (project No. 54918).

References

1. Nourozieh, H., Kariznovi, M., Chen, Z., & Abedi, J. (2010). Simulation Study of Underground Coal Gasification in Alberta Reservoirs: Geological Structure and Process Modeling. *Energy Fuels*, *24*(6), 3540-3550. https://doi:10.1021/ef9013828
2. Otto, C., & Kempka, T. (2017). Prediction of Steam Jacket Dynamics and Water Balances in Underground Coal. *Gasification Energies*, *10*(6), 739.
3. Lykov, A.V. (1952). *Teoriya teploprovodnosti.* Moskva: Gostekhteoretizdat.
4. Pivnyak, G., Dychkovskyi, R., Bobyliov, O., Cabana, E., & Smoliński, A. (2018). Mathematical and Geomechanical Model in Physical and Chemical Processes of Underground Coal Gasification. Solid State Phenomena, (277), 1-16. https://doi.org/10.4028/www.scientific.net/SSP.277.1
5. Russo, Yu.V. (1957). Teplovye poteri v bokovykh porodakh pri podzemnoy gazifikatsii tonkikh pologikh i naklonnykh kamennougol'nykh plastov. *Podzemnaya gazifikatsiya ugley*, (5).
6. Kreynin, Ye.V. (2004). *Netraditsionnye termicheskie tekhnologii dobychi trudnoizvlekayemykh topliv: ugol', uglevodorodnoe syr'ye.* Moskva: IRTS Gazprom.
7. Tishkov, V.V. (2014). Otsenka vodopritoka v kanal podzemnogo gazogeneratora pri izmenenii parametrov pronitsayemosti massiva v usloviyakh Dneprobassa. *Mining of Mineral Deposits*. *8*(4), 409-413. https://doi.org/10.15407/mining08.04.409
8. Sadovenko, I.O., Timoshchuk, V.I., & Tishkov, V.V. (2010). Doslidzhennia vplyvu napruzheno-deformovanoho stanu vmishchuiuchykh hirskykh porid na ikh filtratsiini vlastyvosti pry pidzemnii hazyfikatsii vugilnykh plastiv v umovakh Dniprobasu. *Naukovyi Visnyk Natsionalnoho Hirnychoho Universytetu*, (3), 32-33.

9. Sadovenko, I.A., Polyashov, A.S., & Inkin, A.V. (2004). Eksperimental'nye issledovaniya mekhanizma fil'tratsii produktov gazifikatsii. V *Materialakh Ukrainsko-Polskoho Forumu Hirnykiv* (pp. 598-603).
10. Mironenko, V.A. (1983). *Dinamika podzemnykh vod.* Moskva: Nedra.
11. Semenenko, D.K., Russa, Yu.V., & Ovchinikov, V.M. (1959). Gazopronitsaemost' vygazovannogo prostranstva, zapolnennogo oshlakovannymi porodami. *Podzemnaya gazifikatsiya ugley*, (4), 19-21.
12. Lomtadze, V.D. (1952). *Metody laboratornykh issledovaniy fiziko-mekhanicheskikh svoystv peschanykh i glinistykh gruntov*. Moskva: Gosgeolizdat.
13. Skafa, P.V. (1960). *Podzemnaya gazifikatsiya ugley*. Moskva: Gosgeolizdat.
14. Sotskov, V.O., Demchenko, Yu. I., Salli, S.V., & Dereviahina, N.I. (2017). Optimization of parameters of overwoked mining gallery support while carrying out long-wall face workings. *Naukovyi Visnyk Natsionalnoho Hirnychoho Universytetu*, (6), 34-40.
15. Fomichov, V., Sotskov, V., Pochepov, V., & Mamaikin, O. (2018). Formation of a calculation model determining optimal rate of stoping face movement with a large deformation of a rock massif. *ARPN Journal of Engineering and Applied Sciences*, *13*(7), 23 81-2389.
16. Sotskov, V., & Saleev, I. (2013). Investigation of the rock massif stress strain state in conditions of the drainage drift overworking. *Annual Scientific-Technical Collection – Mining of Mineral Deposits*, 197-202. https://doi.org/10.1201/b16354-36
17. Malashkevych, D., Sotskov V., Medyanyk V. & Prykhodchenko D. (2018). Integrated Evaluation of the Worked-Out Area Partial Backfill Effect of Stress-Strain State of Coal-Bearing Rock Mass. *Solid State Phenomena*, (277), 213-220. https://doi.org/10.4028/www.scientific.net/ssp.277.213

E3S Web of Conferences **60**, 00036 (2018) https://doi.org/10.1051/e3sconf/20186000036
Ukrainian School of Mining Engineering

Substantiating operational parameters of load-haul-dumpers in the context of irregular-shaped underground mine workings

Leonid Shyrin[1], and *Ivan* Iniutkin[1*]

[1]National Mining University, Department of Transport Systems and Technologies, 19 Yavornytskoho Ave., 49005 Dnipro, Ukraine

Abstract. Objective of the paper is to substantiate scientifically parameters and design of energy-efficient transportation and operational schedules for ore masses transfer by means of new-generation load-haul-dumpers (LHDs) adapted to real mining environment at different development stages. Features concerning formation of cargo traffics and components of mining as well as tunneling systems in the context of uranium deposits mining have been studied. Mathematical model to identify operational indices of load-haul-dumpers has been developed. Layout features of the spaced pin-connected frame construction have been used to develop mathematical models of disturbed motion. Potential sources to improve the efficiency of load-haul-dumpers in the context of mining intensification have been analyzed.

1 Introduction

In the current context of uranium deposits mining in Ukraine, previously implemented projects of their development are corrected towards phased increase in production capacities of operating ore mines at the expense of implementation of the advanced technical and process design details and systems of innovative self-propelled facilities. High economic operational performance of the world mining enterprises engaged in transportation and technological schemes mastering with the use of multi-purpose load-haul-dumpers has been the objective factor to make such design solutions [1]. However, introduction of high-tech equipment into traditional schemes of underground transport without scientific and technological arguments for its productive use in terms of specific environment of uranium deposits makes it impossible to achieve planning variables.

First of all, that depends on the following: disproportion between performances of interdependent subsystems of underground transport arises mostly in the process of operation of diverse self-propelled facilities differing in their functional areas, designs, productivity, and energy consumption. Significant difference in performance indices of the available high-tech self-propelled facilities and those being implemented needs the improvement of traditional transportation and technological schemes as well as their

* Corresponding author: injutkin@gmail.com

adaptation to the variable mine environment in the process of primary mining and stoping intensification [2].

Comparative evaluation of operation parameters of foreign-made self-propelled facilities, applied in the context of domestic enterprises, with the performance of similar equipment abroad has helped identify a number of problems being typical for enterprises applying innovative high-tech equipment. It has been determined that basic factors, restricting large-scale implementation of innovative load-haul-dumpers, are as follows: low maintenance level; lack of production diagnosing system; and imperfection of obsolete techniques to determine performance parameters of energy-intensive self-propelled facilities in terms of complicated operational environment.

Taking into account the latter, the necessity to make expert evaluation of available transportation and technological schemes for ore transfer within irregular-shaped underground mine workings has arisen as well as the necessity to carry out specific research to determine adaptation capacities of innovative LHDs in terms of gradual increase in mine output and further improvement of design solutions.

Earlier research has determined that, in terms of mining intensification, a number of mining and geological, technical, technological, and organizational factors effect the performance of self-propelled face equipment as well as formation of traffic flows from stopes and development faces. Hence, current systems of underground transport should rely upon the use of high-tech vehicles, involve maximally changes in technologies aimed at uranium extraction, and be geared to energy-and resource-saving. Fundamental difference of transportation and technological systems with the help of high-tech LHDs is that traditional schemes of ore delivery from stopes using self-closed self-propelled facilities with uncompleted bonds are replaced by multi-purpose equipment with innovative characteristics [3]. Intensification of the first mining and second mining and the use of energy-intensive vehicles under complicated mining and geological conditions need adequate scientific and research support.

In this connection, scientific and research substantiation of performance parameters of self-propelled facilities as well as the development of highly adaptive operation schedules of underground transport on the basis of high-tech LHDs under real mine conditions are topical problem which solution will improve sufficiently both energy saving and efficiency of underground uranium ore mining.

2 Performance features of load-haul-dumpers in terms of irregular-shaped underground mine workings

Specific conditions of uranium deposit development predetermine different alternatives concerning integration of load-haul-dumping and transportation systems. Traditional steplike schemes of underground transport used in ore mines of the industry include different types of bottom hole drilling and loading assemblies as well as site systems of loading equipment with self-propelled cars, load-haul-dumpers with load-carrying shovels and load-haul-dump machines with load-carrying bodies. In the context of main haulage road, locomotive haulage of ore mass and cargo delivery are mainly used. Application area of one or another variant of transportation-technological scheme depends upon available extraction systems and sections of load-haul and conveyor entries.

While ore mining at 480 m levels and deeper, design solutions of the staged plant developments involve gradual transition to transportation and technological schemes with the use of high-tech LHDs and dumpers to hoist rock mass by means of runaways. Intermediate stage of the activities involves solution of a system of interdependent multicomponent problems determining operational parameters of innovative bucket LHDs and principles of resource- and energy-saving while uranium ore mining.

Following scientific and research problems are basic ones:

– analysis of features concerning formation of cargo flows as well as mining and tunneling systems at different stages of uranium ore deposit development;

– determination of operation parameters and stability of LHDs within irregular-shaped mine workings;

– identification and evaluation of potential sources to improve the efficiency of innovative LHDs in the context of mining intensification;

– substantiation of design parameters of transportation entries providing the efficient use of high-tech self-propelled facilities.

Idea of the paper is to substantiate scientifically operating principles, parameters and design of non-multipurpose transportation and technological schemes to develop uranium deposits relying upon the use of innovative self-propelled LHDs providing intensification of mining operations and energy-saving in the context of gradual ore development.

The objective achieving involves the problem to be solved: it is necessary to analyze parameters of *a mine working-LHD* transportation and technological system in the process of ore transfer within a stope and mine sites of haulage level.

Integrated research technique has been applied to solve the problems. The technique involves processing and generalizing of scientific sources concerning the object problem, carrying out of theoretical studies as well as in-mine testing of the technical solutions being developed.

Lateral stability of LHD was analyzed in the process of its motion within underground mine workings with variable hypsometry of a path. The analysis was based upon the use of mathematical modeling methods [4]. Operational parameters of *a mine working-LHD* technological system were determined in accordance with the results of in-mine tests of LHD indices using methods of mathematical analysis and statistics. Simulation with the use of standard software was applied to substantiate adaptation criterion of LHD and to evaluate its effect on energy-and resource-saving. The program of the experiments generalized the simulation results and calculations of actual data concerning the performance of mine transportation and technological systems in the context of different operation stages.

It should be noted that development of uranium ore deposits with the help of self-propelled facilities unites such linked transportation and technological processes as: well drilling for ore breaking; ore mass drawing and its removing from chambers (vehicle loading and removal from a stope); its transfer to a haulage level and transportation to a shaft station with following surface haulage and delivery to a processing plant.

In the context of mining intensification, technological schemes for ore transportation rely basically on the use of LHDs manufactured by well-known world producers.

If operational practices and resource tests concerning such machines are not available, the necessity for scientific substantiation with the consideration of limiting factors and specific character of mining arises.

Specific factors of vein deposits effecting the efficiency of LHDs are characterized considerably by ore body occurrence features and a drill and blast tunneling method applied to construct both temporary and development mine workings [5]. According to a pattern for the temporary and development workings construction, drilling operations are performed through the ore body with wall rock cutting. As a result, loading-hauling mine workings are of complicated configuration, i.e. they are non-linear in the plan. Processing of statistic data on the changes in the direction of the mine workings under construction has shown that probable variations of their dynamic axis within 15° in terms of sites with 50 lengths are 90-95%. Fig. 1 demonstrates actual profile and a plan of a development mine working constructed by means of drill and blast tunneling method under mining and geological conditions of "Smolinska" mine.

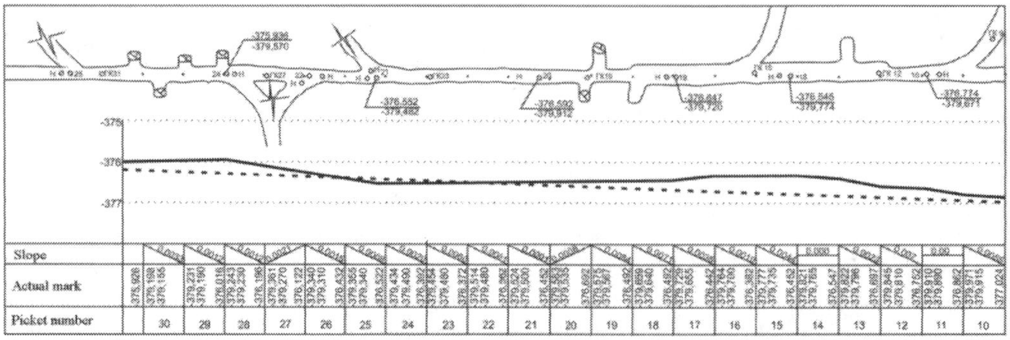

Fig. 1. Actual profile and a plan of haulage mine working constructed by means of drill and blast tunneling method.

Underground investigations have helped determine that changes in direction and floor profile of loading-haulage and development mine workings as well as ground roughness and road covering quality effect significantly the technical state and operation parameters of LHDs. To evaluate operational parameters of LHD performance, track profile of its advance from loading point to ore passageway was divided into typical sections within ore-passageway level, on the decline and level haulage mine workings. Longitudinal vibration (galloping) and limiting values of LHD body lateral deviation as for the design axis, resulting in deformation of metal structures, redistribution of loads applied for semi-frame, and deceleration, have been registered in the process of the self-propelled facilities motion within haulage mine workings [6]. The research data have demonstrated that efficiency of high cost of energy-consuming innovative LHDs is supported by high velocities, minimum unproductive time during a shift, and corresponding service level.

Thus, substantiation of such rational energy parameters of a bucket LHD as the increased pressure while ore mass loading, movement velocity and time (both with load and without it) involved coordination quantitative and qualitative indices determining interaction of a *mine working-LHD* transportation and technological scheme. Analysis of parameters of the system components interaction involved consideration of following factors: soil status and slopes of mine workings within a zone of ore mass loading; ratio of oversize yield (i.e. more than 1000 mm); distance form a stope to ore drawing; section of haulage mine workings; length of slopes and rises within the path; irregularities of floor and walls of haulage mine working; availability of side tracks etc. Research [7] has determined that effect of the parameters on innovative LHD performance is demonstrated specifically when it performs operations aimed at ore mass loading, haulage, and dumping to ore passageway. Such mining factors as thickness and slope angle of ore bodies, stability of wall rocks as well as parameters of extraction technique being applied and its components determine specific cost to carry out the processes. It should also be noted that while forming technological schemes for ore transportation, drilling and blasting parameters, a technique for the broken ore drawing, geometry of sub-standard ore fragments as well as methods of their secondary crushing are significant too. According to recommendations, oversize yield and optimum value of the fragment average dimension should be determined according to the criterion of minimum expenditures connected with drilling and blasting operations and delivery. Taking into consideration the fact that drilling and blasting operations parameters also identify section, shape, and operating conditions of haulage mine workings, indices of explosion effect upon interaction conditions of *mine working-LHD* transportation and technological system should also be taken into consideration. Up till now, such research has not been carried out.

Earlier studies have proved that increase in the delivery distance decreases the efficiency of self-propelled facilities; however, the fact is controversial since a type of a path and its state predetermine the machine velocity as well as its efficiency [8]. It depends on the following: on short paths, a vehicle cannot produce high velocity since it should accelerate and break; average speed is higher in the context of longer paths. Average velocity decreases on the irregular-shaped path due to the necessity of frequent maneuvering. Moreover, the facilities operate more efficiently within horizontal mine workings to compare with up-hill motion or down-hill one. The world practices concerning performance of such LHDs confirm: their velocity is much higher if a floor is leveled or surfacing (i.e. chippings, bitumen concrete, or concrete) is available. At the same time, such activities involve extra charges; thus, their expediency should be substantiated by means of feasibility analysis.

In this context, while performing a design process of high-adaptive systems of self-propelled facilities to extract and deliver ore under real conditions of mining, each enterprise should solve problems of optimum ratio between drilling and blasting operations, operational parameters of LHDs, and parameters of haulage mine workings.

3 Results of mathematical modeling of operational parameters of LHDs

In the context of world practices, current technological schemes of ore delivery by means of self-propelled facilities rely upon the use of either bucket LHDs or dumper-LHD systems [9]. In this case, the self-propelled machines as well as their systems are used to load ore mass from open face floor and from bottom ore drawing and sublevel ore drawing from special mine workings.

Generally, frame of pneumatic-tired LHD has two-sectional design variant and two pin-connected joints. One pin has vertical axis providing the frame horizontal articulation; another pin involves vertical axis making it possible to perform mutual transverse turn of the section.

Possible stability loss within transverse plane of each section independently of one another is characteristic feature of the pin-connected machines. It results from the availability of feathering hinge and mutual arrangement of two pin-connected joints.

As it is known [10], a design with the spaced vertical and horizontal hinges is rational for pneumatic-tired LHDs. In this case, ladling mechanism is in an assembly section in turn being mounted on the front axle either with rigid suspension or with flexible one. A body is mounted on a balancing bridge.

Below you can find calculations of allowable velocity of pneumatic-tire LHDs with the achievement of their operational parameters as well as typical dimensions of a mine working.

Characteristic feature of LHD is as follows: having spaced pin-connected joints, it can vibrate progressively within transverse plane relatively to longitudinal axis passing through feathering hinge. Fig. 2 demonstrates calculation model of a load-haul-dumper (front view).

Consider that kinematic perturbations act only from the right forehand in the direction of a load-haul-dumper travel. Perturbation equation is obtained in the form of Lagrange's equation of second kind [11]:

$$E_K = \frac{d}{dt}\left(\frac{dE_K}{d\dot{q}_j}\right) - \frac{dE_K}{d\dot{q}_j} = \frac{dE_P}{d\dot{q}_j} + Q_j, \tag{1}$$

where E_K is kinetic energy of the system, J; E_P is potential energy of the system, J; $\dot{q}_j$ is generalized coordinate, m (rad); Q_j is generalized force of nonconservative origin, N (H·m).

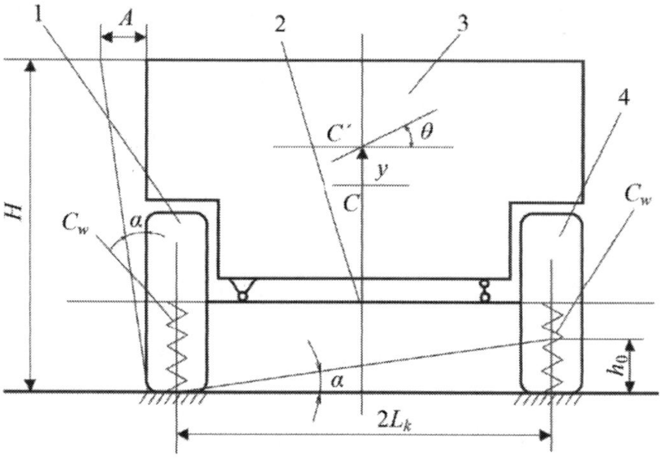

Fig. 2. Calculation model of a load-haul-dumper (front view): 1 – right wheel; 2 – continuous bridge; 3 – body; and 4 – left wheel.

In this case, $\dot{q}_1 = y$, $\dot{q}_2 = \theta$ where y is vertical displacement of the system mass centre, and θ is turning angle of a machine within transverse plane.

Kinetic energy of the system is:

$$E_K = \frac{m_c \dot{y}^2}{2} + \frac{I_c \dot{\theta}^2}{2} + \frac{m_c \upsilon^2}{2}, \tag{2}$$

where m_c is total weight of the machine, kg; I_c is integral inertia of the machine relative to longitudinal axis passing through the machine mass centre, kg·m^2; and υ is the machine velocity, m/s.

Potential energy of the system is implemented through air power within pneumatic wheels. It is:

$$E_P = \frac{C_w}{2}(y - L_k\theta)^2 + \frac{C_w}{2}(y - L_k\theta - h_L)^2, \tag{3}$$

where C_w is radial stiffness of the machine pneumatic wheel, N/m; L_k is one-half of the machine wheel track, m; h_L is a pitch of a left wheel of the continuous bridge on the sinusoidal profile irregularities, m.

In general, height of the irregularity can be expresses as follows:

$$h_L = h_0 \sin\frac{2\pi X}{L_0}, \tag{4}$$

where h_0 is amplitude value of a mine path profile irregularity, m; x is an absciss approximating a function of floor irregularities, m; and L_0 is length of a sinusoid approximating a function of floor irregularities, m.

$x = \upsilon t$ if the machine moves; then a pitch of a right wheel is:

$$h_L = h_0 \sin \frac{2\pi X \upsilon}{L_0} t. \tag{5}$$

While neglecting nonconservative forces, we substitute (in the y coordinate) the expression of kinetic and potential energy into Lagrange's equation of second kind (1). Thus, combining the obtained expressions, we form differential equation one:

$$m_c \ddot{y} + 2C_w y = C_w h_L . \tag{6}$$

To obtain differential equation two, we repeat the operation in the θ coordinate:

$$m_c \ddot{\theta} + 2C_w L_k^2 \theta = C_w L_k h_L . \tag{7}$$

Rearrange equations (6) and (7) taking into consideration expression (5):

$$m_c \ddot{y} + 2C_w y = C_w h_0 \sin \frac{2\pi \upsilon}{L_0} t ; \tag{8}$$

$$I_c \ddot{\theta} + 2C_w L_k^2 \theta = C_w L_k h_0 \sin \frac{2\pi \upsilon}{L_0} t . \tag{9}$$

The addends, describing free vibrations of the load-haul-dumper, are determined relying upon the initial conditions; they almost nullify after a while. Forced vibrations remain. They are described as follows:

$$y = \frac{C_w h_0}{m_c \left| P_y^2 - \dfrac{4\pi^2 \upsilon^2}{L_0^2} \right|} \sin\left(\frac{2\pi \upsilon}{L_0} t - \psi_1 \right); \tag{10}$$

$$\theta = \frac{C_w L_k h_0}{I_c \left| P_\theta^2 - \dfrac{4\pi^2 \upsilon^2}{L_0^2} \right|} \sin\left(\frac{2\pi \upsilon}{L_0} t - \psi_2 \right). \tag{11}$$

After proper cyclic frequency replacing, we obtain a condition of LHD instability within a transverse plane during asymmetric disturbance:

$$\frac{2C_w L_k^2}{I_c} - \frac{4\pi^2 \upsilon^2}{L_0^2} = 0 . \tag{12}$$

Using equation (12) we obtain critical velocity υ_D when pneumatic-tire LHD with stiff suspension becomes unstable in the context of the disturbed motion within transverse plane:

$$\upsilon_D = \frac{L_k L_0}{\pi} \sqrt{\frac{C_w}{2I_c}} . \tag{13}$$

Fig. 3 shows dependences of allowable LHD velocity in accordance with expression (13) after processing with the help of standard mathematical methods and approximation.

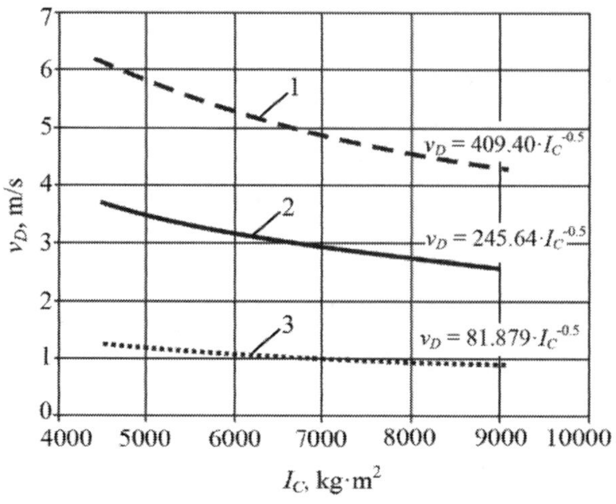

Fig. 3. Dependence of allowable LHD velocity in terms of various lengths sinusoidal wave approximating a function of road irregularities: 1 – L_0 = 0.1 m; 2 – L_0 = 0.3 m; 3 – L_0 = 0.5 m.

Fig. 3 demonstrates clearly the fact that in the context of the disturbed motion within transverse plane, allowable velocity of pneumatic-tired LHD with stiff suspension (v_D) varies according to exponential law depending upon a value of inertia moment ($I_c - 0.5$).

4 Conclusions

The outcomes of the research have made it possible to analyze operation conditions of LHD, and their layout diagrams; moreover, the expediency to use the spaced scheme of pin-connected joints when rear and front sections are interconnected only by means of a lagging hinge. In this context, flapping hinge is applied for center-pin-supported rear axle. Layout features of the spaced pin-connected frame construction have been used to develop mathematical models of disturbed motion, if only antisymmetrical kinetic disturbances are available since they are the basis to select geometry of mine workings.

The developed mathematical models involve both static and dynamic criteria in the process of a mine working geometry selection; they are rated for the development of innovative self-propelled facilities while applying flexible suspension of continuous bridges and stabilizer bars.

Thus, operation parameters of LHD with flexible suspension, helping determine critical velocity within irregular-shaped underground mine workings, have been obtained.

The authors highly appreciate the assistance and consultations rendered by Vadym Kosarenko, Director of "Novokonstiantynivska" and "Smolinska" mines of "Skhidnyi hirnycho-zbahachuvalnyi kombinat" SE.

References

1. Khomenko, O., & Maltsev, D. (2013). Laboratory research of influence of face area dimensions on the state of uranium ore layers being broken. *Naukovyi Visnyk Natsionalnoho Hirnychoho Universytetu*, 31-37.
2. Shirin, A., Rastsvetaev V., & Morozova, T. (2012). Estimation of reliability and capacity of auxiliary vehicles while preparing coal reserves for stoping. *Geomechanical Processes during Underground Mining: School of Underground Mining 2012*, 105-108. https://doi.org/10.1201/b13157-18

3. Shyrin, A., Rastsvetaiev, V., & Inyutkin, I. (2017). Operativno-proizvodstvennoe upravlenie tekhnicheskim sostoyaniem transportno-tekhnologicheskikh skhem gornorudnykh predpriyatiy. *Visnyk Kryvorizkoho Natsionalnoho Universytetu*, 38-43.
4. Kovalevska, I., Symanovych, G., & Fomychov, V. (2013). Research of stress-strain state of cracked coal-containing massif near-the-working area using finite elements technique. *Annual Scientific-Technical Collection – Mining of Mineral Deposits*, 159-163. https://doi.org/10.1201/b16354-28
5. Shirin, L., Korovyaka, Y., & Tokar, L. (2011). Justification of design parameters of compact load-haul dumper to mine narrow vein heavy pitching deposits. *Technical and Geoinformational Systems in Mining*, 85-92. https://doi.org/10.1201/b11586-15
6. Bartashevski, S.E., Strashko, V.A., Shirin, L.N., & Shumrikov, V.V. (2001). Mathematical Models of Work-cycle Time of Ladleman Loading Machine. *Vibration in Methods and Technology*, *3*(19), 46-49.
7. Shirin, L., & Iniutkin, I. (2009). Osobennosti formirovaniya transportno-tekhnologicheskikh sistem podzemnyh rudnikov s uchetom adaptatsionnykh vozmozhnostey samokhodnogo oborudovaniya. *Naukovyi Visnyk Natsionalnoho Hirnychoho Universytetu*, 66-68.
8. Shirin, L.N., Korovyaka, E.A., & Shirin, A.L. (1998). Model Analysis of Adaptive Capacity of Load-haul Dumpers to Mine Narrow Vein Deposits. *Inter-agency Collection of Scientific Papers of IGTM of the NAS of Ukraine*, (6), 67-73.
9. Kalnitskiy, Ya. & Filimonov, A. (1974). Samokhodnoe pogruzochnoe i dostavochnoe oborudovanie na podzemnykh rudnikakh. Moskva: Nedra.
10. Malinovskiy, E. & Gaytsgori, M. (1974). Dinamika samokhodnykh mashin s sharnirnoy ramoy. Moskva: Mashinostroenie.
11. Beihul, O., & Kolesnyk, I. (2011). Dynamika ta mitsnist metalurhiinykh mashyn. Dniprodzerzhynsk: Dniprodzerzhynskyi derzhavnyi tekhnichnyi universytet.

Technological aspects of technogenic disturbance liquidation in the areas of coal-gas deposits development

Volodymyr Medianyk[1*], and *Oleksii* Cherniaiev[2]

[1]National Mining University, Department of Underground Mining, 19 Yavornytskoho Ave., 49005 Dnipro, Ukraine
[2]Institute for the Design of Mining Enterprises, National Mining University, 19 Yavornytskoho Ave., 49005 Dnipro, Ukraine

Abstract. The objective of this work is to develop the technological measures for liquidation of technogenic disturbances that have arisen within the residential areas. Based on the results of geophysical studies, the causes and factors of Earth's surface caving occurrence that arose in the result of natural-technogenic processes activation in underworked massif and are caused by the soil subsidence under buildings and constructions, have been determined. The solution of the problem was carried out with the use of methods for observing the natural pulsed electromagnetic field of the Earth and electric tomography in the area adjacent to the place of caving. The conclusions about the possible causes of the caving trough formation and its further development have been formulated, as well as the technological scheme for its liquidation have been developed. The developed new technological scheme allows conducting the effective liquidation works in conditions of proximity to the residential buildings and maximally reduces these scopes of works at ecological and protective measures.

1 Introduction

Annually, from the Earth's interior in Ukraine, more than a billion cubic meters of the mined rock is extracted in the form of minerals and waste rock after preparatory mine workings development, which leads to the formation of cavity with appropriate sizes. In the area of old mines and underground mine workings location, as a result of natural-technogenic processes, the cavity is formed in disturbed rock massif after some extended period of time.

The specificity of the underground mining is in the fact that together with shut-down of ventilation and water pumping, the shifts activation takes place in the underworked rock massif due to its moistening. It causes additional deformations of buildings and constructions located in the area of liquidated mining industry facilities. In addition, water flooding can lead to hazardous methane gas release into the atmosphere without

* Corresponding author: vmedyanyk1975@gmail.com

reclamation operation of underground mine workings.

The process of coal mines closure has intensified since the middle of the last century. The process of withdrawing mine workings from production has become more rapid, and in comparison with previous years in significant quantities. This is explained both by growing role of liquid and gaseous energy sources, large-scale implementation of nuclear energy, and by the fact that mining operations in traditional coal mining countries have reached such depths, where even with the application of new imported mining equipment and technologies, the coal mining costs often are not covered by revenues from its implementation [1].

The presence of many influencing factors, such as inability to control the processes occurring in the disturbed rock massif after mine working closure, the lack of models for formation and forecasting of the Earth's surface displacements leads to the cavity formation in regions significantly overloaded by coal mining enterprises [2 – 5].

It is a well-known fact that as a result of underground mining, there is a shift, deformations, and subsidence of rock massif. As a result of minerals mining under the overlying rocks the cavities are formed, and at the Earth's surface the subsidence troughs are formed. This is the initial rock massif disturbance and its regularities are well studied.

Thus, the necessity of liquidation of vertical mine workings is obvious. It is necessary to carry out the works of filling the mine shaft void space by noncombustible material, to construct the floors (floor slab panels) or stoppings, to use other means to shut-down the aerodynamic connection with operating mine workings, as well as to ensure the long-term stability of underground constructions and structures (fastening) for preventing the Earth's surface dangerous deformations.

The caving occurs suddenly and is unpredictable due to the fact that no one controls the rock massif state in the locations of old mines and mine workings. To predict the secondary disturbances of rock massif over the old mine workings, it is necessary to know the causes of the displacement processes manifestations (activation).

The objective of the work is to develop the technological measures for liquidation of technogenic disturbances that have arisen within the residential area in Dobropillia City (Ukraine). On the example of this object, the conditions of its formation have been considered and the technical solutions for the disturbed rock massif restoration, close to the natural state, have been proposed.

It is important to note that the technogenic disturbances liquidation in difficult conditions of soil slip processes should be carry out with account of the technologies that meet up-to-date requarements: use of special means and methods of performance of works, which will ensure a minimum impact on neighboring facilities, including mine workings that are in a stressed state [6]; use of rational schemes for waste rocks transportation, which minimize the specific energy intensity of the transportation processes of mined rock [7 – 9]; conducting recultivation and revitalization of disturbed areas until the most environmentally acceptable landscape territories, as well as their regeneration [11, 12]. It should also be noted the necessity to comply with ecological aspects of work performance with account of clean technologies development (Clean High Technologies) [13, 14].

2 Methodology of project development

To determine the causes and factors of technogenic disturbance occurrence that arose as a result of natural-technogenic processes activation in the underworked rock massif and have caused the soil subsidence under buildings and constructions, the methods for observing natural pulsed electromagnetic field of the Earth (NPEFE), electric tomography (ET), dismantling and cleaning water wells by subdivision of Ministry of Emergency Situations (MES), have been used.

 https://doi.org/10.1051/e3sconf/20186000037

The application of the NPEFE observation method is substantiated by SNiP 1.02.07-87, and protected by authors' certificates and the State Patent of Ukraine [15].

The observations were carried out with the use of microprocessor-based indicator of spontaneous pulsed electromagnetic radiation of the Earth, a receiving antenna with a ferrite core and a system of amplification and registration blocks.

The NPEFE observation was carried out in accessible and safe areas on the territory of Dobropillia (Ukraine). The experimental observations were made in accordance with DBN A.2.1-1-2014. The observation profiles were located along the entire area in parallel to the streets, at a distance of 2.5 m from each other, bending round the permanent structures (apartment houses), auxiliary structures (cellars, sheds). The profiles position was related to the contours of buildings, fences, and the power transmission line supports. In total, 411 physical points have been worked out on this territory.

The processing of the measured values consists in interpolating the data for a regular network. The performance of two-dimensional filtration and drawing a scheme of NPEFE flow density are shown on Fig. 1.

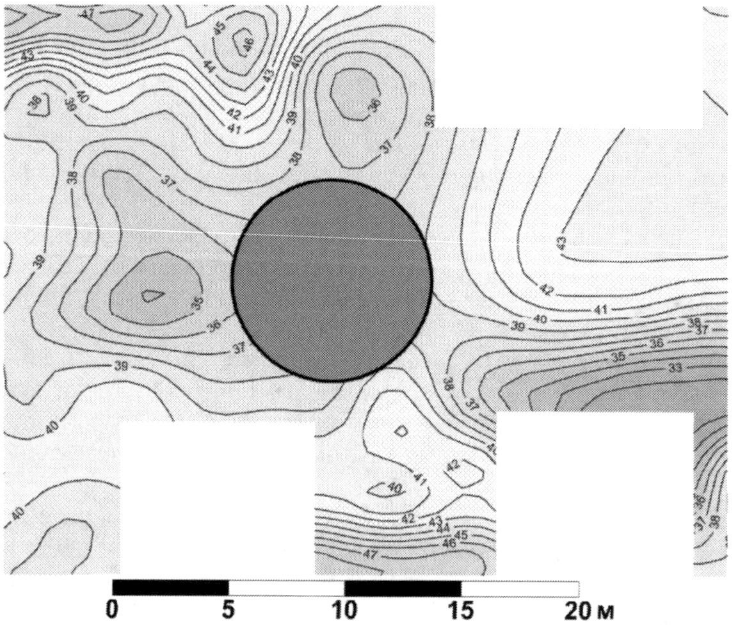

Fig. 1. Scheme of NPEFE flow density in a caving zone.

The application of electric tomography method was carried out in accordance with [16]. Electric tomography is a complex which includes both field observation techniques and the technology of processing and interpreting field data. Its feature is the repeated use of the same positions of electrodes, fixed on the observation profile, as a power supply and measuring instruments.

To study geoelectric sections, the use of electric tomography is a necessary criterion for reliable interpretation. Such a complex structure is usually characteristic of ore zones and zones with tectonic disturbances, landslides, bulked and artificial soils in urban areas, with a steep layers jump and with the occurrence of karst.

The fieldwork was carried out in accordance with the requirements of the electric inspection instruction by the method of resistance with the use of equally dipole axial installation. The measuring device EIN-209m was used as a receiver of an electromagnetic field at a frequency of 1.22 Hz in the receiving dipole. The dimensions of the receiving and feeding dipoles were 5 m. The generator GER-1/300 was used as a power source.

According to the results of field measurements, a geoelectric section has been received (Fig. 2).

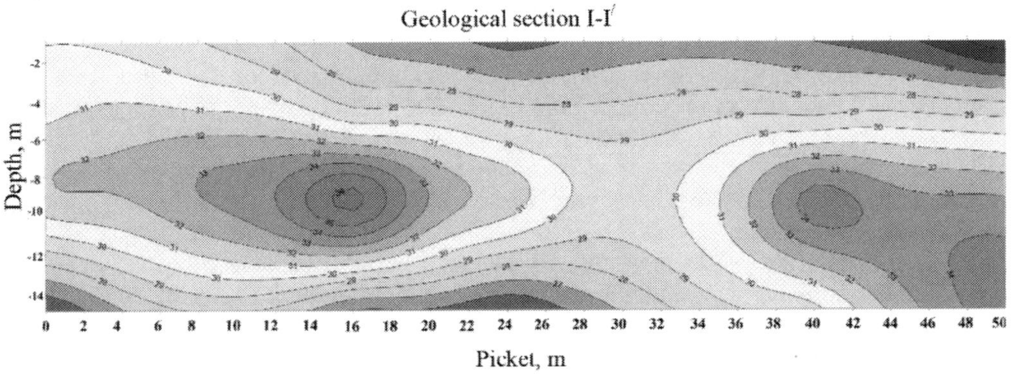

Fig. 2. The results of work by electric tomography method.

3 Results and discussion

3.1 The results of the NPEFE observation

The Fig. 1 shows the scheme of the NPEFE flow density at the territory adjacent to the caving zone. As can be seen from the scheme, the NPEFE flow density ranges within the limits of 37 – 51 conventional units (c.u) and has a mosaic form in most of the surveyed area. In the area of the north-eastern corner of the building, an insignificant linearly elongated anomalous zone of reduced values can be observed, which can be explained by the existing leaks from the sewer thread. In the neighboring territory, a cavity formation has arisen, which has not a significant difference by a value of flow density. This may indicate that there was not a significant moistening of soils in the zone of cavity formation.

3.2 The electric tomography results

The Fig. 2 shows the geoelectrical section according to the observations. There is a steady increase in the value of specific resistance from 26 to 36 Ohm·m. In such a case, the upper part of the section – topsoil layer is characterized by the lowest values. At a depth of 3 – 4 m in the zone of sandy loam occurrence the values of more than 30 Ohm·m are observed that indicates the water-free topsoil layer. There are no moistening zones throughout the length of the profile.

3.3 The results of visual examination

The residential area was built after 1963. All the residential buildings of this area are located on the underworked territory of the former RKKA Mine. When carrying out a visual examination at the moment of work beginning, there was an increase in the subsidence cavity (the diameter increased to 10 m). The cavity zone is identified by topsoil layer and sandy loam. The walls of the prospecting shaft are fastened by wooden prop stays. According to local residents the air pits, in some cases, are used as septic tanks.

With participation of detachment 19 of the Ministry of Emergency Situation of Dopropillia Town, a diagnosis of a collar of shaft fastening and the state of surrounding rocks has been carried out. With their participation, the actual dimensions of the air pit have been determined, where the technogenic disturbance has arisen and caving trough has been formed. It was established that the prospecting shaft has a wooden fastening with dimensions 2.8×2.8 m, the surface area along the caving contour is 52.8 m^2 and the volume of caving sinkhole is 330 m^3.

It was established that the process of caving sinkholes formation is multifaceted and depends on many factors. With that, the cavity formation process is possible in four different types. An advance prevention of cavities is possible only by observing the state of massif after the mine workings closure for a long period of time. The example of the cavity formation in Dobropillia City shows processes of displacements and karstification, leading to fall of large rock masses and the Earth's surface caving. This means the shafts (pits) liquidation without full backfilling, but fastened by concrete, armored concrete, and other types of mine fastening, with capping thickness of up to 45 m on the basis of assessment of the actual shaft fastening and the long-term stability of shaft bottom massif.

Analyzing the obtained results and considered standard technology for vertical mine workings liquidation [3, 4, 17, 18], the following conclusions have been made. The normative technological scheme for vertical mine working liquidation has to be carried out with the use of floor slab panel placed at parent rock level, with subsequent backfilling of the upper part of the shaft and construction of the second-floor slab.

The vertical mine workings liquidation technology consists of five main stages: preparation for filling; determination and outlining the boundaries of hazardous backfilling operations zone; surface preparation for filling; operation of pit shaft filling; completion of the final works. A waste rock of extinct mine dumps, sometimes more valuable materials extracted in quarries (breakstone, sand, gravel), as well as quick-cooled metallurgical slags are used for filling. The filling operations of the flooded shafts have to be carried out with the use of water-proofed material. It is forbidden to use a slag block (after disassembling of buildings) for cavity filling due to its toxicity.

The geologic data of prospecting the shaft location shows the presence of parent rocks at a depth more than 45 m. It means that floor slab panel position at the level of parent rocks can cause an increase of caving trough up to 100 m. But, in light of the fact, that technogenic disturbance has arisen in close proximity to residential buildings, the vertical mine working liquidation according to the normative technological scheme, will be cost-intensive.

Based on conclusions, the decision has been made to improve the liquidation technology of vertical mine working with a simultaneous liquidation of technogenic disturbance located in restricted sanitary protected zone of residential buildings. To achieve the solution of this problem, it is proposed to made a wedged choke with a pipe in the axis center of the liquidated mine working, with the subsequent void space filling (Fig. 3).

The quality control of backfilling material (grain size, density, etc.) and its volume have important implication while filling operations.

It is provided to mount the steel bearing frame (1) which will rest on the collar of prospecting shaft and cavity wall surfaces (Fig. 3) with further formation of a wedged choke. After concrete hardening (2), the caving trough is filled with rock-sand-sandy loam material in a volume of 300 m^3 (4). In two months, the tamping and additional rock packing is carried out after slump of backfilling material under the action of environmental conditions, and the permanent floor slab panel installation is carried out (5).

In order to prevent the accumulation and explosion of methane in the prospecting shaft, it is necessary to drill a well and carry out the backfilling of mine working. In this view, the most appropriate means for backfilling is a conveyor, which ensures a slow and uniform flow of material, its packing and interaction with prospecting shaft fastening. When used a backfilling with hardening materials the pulp is supplied into the vertical mine working through pipes with diameter of 150 – 200 mm, in volume of 300 – 320 m^3. The technology of vertical mine working liquidation is shown in Fig 4.

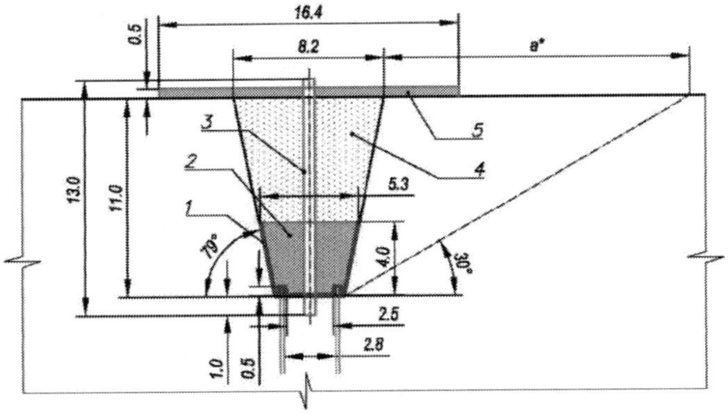

Fig. 3. The scheme of caving trough liquidation: 1 – steel bearing frame for collar of prospecting shaft, 2 – wedged choke 5.2×5.2 m (concrete M 400), 3 – steel pipe 377×9 (State Standards of Ukraine 8732), 4 – rock-sand-sandy loam material, 5 – armored concrete floor 16.4×16.4 m (*h* = 0.5 m), *a** – bankette of possible displacement (*a* = 17 m).

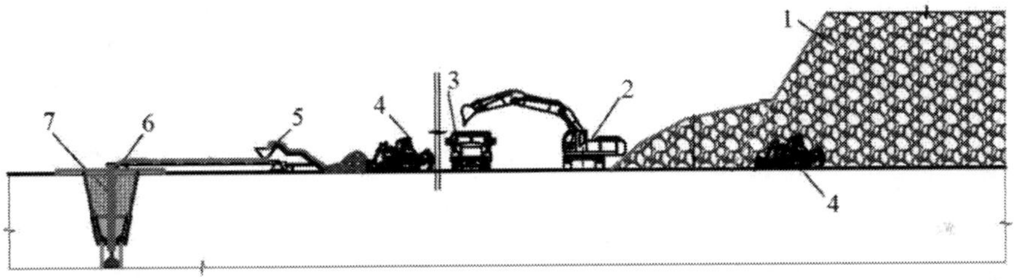

Fig. 4. The technological scheme of vertical mine working liquidation: 1 – waste-rock pile of "Almazna" Mine, 2 – mechanical shovel, 3 – dumping truck, 4 –bulldozer, 5 – stage loader, 6 – scraper conveyor, 7 – liquidated caving trough.

At the final stage, after slump of backfilling material, the covering of cavity collar and prospecting shaft is carried out. In world practice, the shaft collar is bridged over by a reinforced concrete slab with a focus on fasteners or poles hammered to the parent rocks. The construction of the slab, its thickness, cross sections, and bearing capacity are determined by practice and calculations. It is necessary to strengthen the rock massif preliminary in case of unsatisfactory prediction of collar stability. To exclude the possibility of rainwater and floodwater intake into liquidated prospecting shaft through collar, it should be protected by a drainage ditch of 2.5 m and with cross section 0.5 m^2.

If it is difficult to implement the drainage of storm water and floodwater through the construction of a ditch, then it is advisable an application of an insulating shield of clay with a thickness of 1 m and an area by 1.5 times larger than the cross section of a shaft or a pit. In the future, according to the project, it is necessary to control constantly the backfilling material shrinkage and the release of methane on the surface of the yard.

4 Conclusions

In accordance with the research results, observations and their comparison with visual examination results, the following conclusions can be made:

– the NPEFE observations in the cavity zone have shown that at the moment of examination there is no formed zone of soils moistening;

– based on the electric tomography method results, it was established the absence of loose, water-flooded zones located structurally below the cavity zone;

– in case of nonfulfillment of the measures for liquidation of the vertical shaft, the cavity zone will develop further, caused by soaking of overlaying rock massif and atmospheric precipitation, and this may result in the destruction of mine working support.

Based on the results of performed geophysical studies for identifying the cause and factors of the occurrence of technogenic disturbances that arose in the result of natural-technogenic processes activation in underworked massif and are caused by the soil subsidence under buildings and constructions, a technological scheme has been developed for caving trough liquidation in the conditions of restricted sanitary protected zone.

The developed new technological scheme allows conducting the effective liquidation works in conditions of proximity to the residential buildings and maximally reduces the scopes of works for caving trough liquidation due to the presented above technological system.

This work was carried out within the framework of a project of the technological scheme development for liquidation of caving trough in Dobropillia City (Ukraine), and was developed by the contract No. 010110-17. The authors are grateful to the management and employees of the detachment 19 of the Ministry of Emergency Situations in Dobropillia City for assistance in the investigation and detection of anthropogenic disturbance, as well as to Dr. Sc., prof. Pinkovskyi H.S. and Dr. Sc., prof. Bondarenko V.I, the management of DTEK "Dobropilliavuhillia" for providing scientific advice and recommendations.

References

1. Pivnyak, G.G., & Shashenko, O.M. (2015). Innovations and safety for coal mines in Ukraine. *Naukovyi Visnyk Natsionalnoho Hirnychoho Universytetu*, (6), 118-121.
2. Bondarenko, V.I., Rudenko, H.K., & Medyanik, V.Yu. (2017). *Ugol'naya shahta*. Dnipropetrovsk: Natsionalnyi hirnychyi universytet.
3. Bondarenko, V., Kovalevs'ka, I., & Fomychov, V. (2012). Features of carrying out experiment using finite-element methodat multivariate calculation of mine massif – combined support system. *Geomechanical Processes During Underground Mining*, 7-13. https://doi.org/10.1201/b13157-3
4. Petlovanyi, M.V., Lozynskyi, V.H., Saik, P.B., & Sai, K.S. (2018). Modern experience of low-coal seams underground mining in Ukraine. *International Journal of Mining Science and Technology*. Article in press https://doi.org/10.1016/j.ijmst.2018.05.014
5. Chetverik, M.S., Bubnova, E.A., & Soroka, Yu.N. (2013). Prichiny obrazovaniya i napravleniya likvidatsii provalov ot gornykh vyrabotok. *Heotehnichna mekhanika*, (111), 193-205.
6. Strilets, O., Pcholkin, G., & Oliferuk, V. (2015). Monitoring of mass blasting seismic impact on residencial buildings and constructions. *Theoretical and Practical Solutions of Mineral Resources Mining*, 441-443.
7. Shirin, L.N., Denishchenko, A.V., Yurchenko, O.O., & Mikhalev, D.V. (2012). Methodology for determination of rope vehicles energy consumption. *Naukovyi Visnyk Natsionalnoho Hirnychoho Universytetu*, (4), 464-469.
8. Cherniaiev, O.V. (2017). Systematization of the hard rock non-metallic mineral deposits for improvement of their mining technologies. *Naukovyi Visnyk Natsionalnoho Hirnychoho Universytetu*, (5), 11-17.
9. Shyrin, A., Rastsvetaev, V., & Morozova, T. (2012). Estimation of reliability and capacity of auxiliary vehicles while preparing coal reserves for stoping. *Geomechanical Processes during*

Underground Mining: School of Underground Mining 2012, 105-108. https://doi.org/10.1201/b13157-18

10. Huang, F., Luo, X., & Liu, W. (2017). Stability Analysis of Hydrodynamic Pressure Landslides with Different Permeability Coefficients Affected by Reservoir Water Level Fluctuations and Rainstorms. *Water*, *9*(7), 450. https://doi.org/10.3390/w9070450
11. Gorova, A., Pavlychenko, A., Kulyna, S., & Shkremetko, O. (2012). Ecological problems of post-industrial mining areas. *Geomechanical Processes During Underground Mining: School of Underground Mining 2012*, 35-40. https://doi.org/10.1201/b13157-7
12. Gorova, A., Pavlychenko, A., Borysovs'ka, O., & Krups'ka, L. (2013). The development of methodology for assessment of environmental risk degree in mining regions. *Annual Scientific-Technical Collection – Mining of Mineral Deposits*, 207-209. https://doi.org/10.1201/b16354-38
13. Falshtynskyi, V.S., Dychkovskyi, R.O., Saik, P.B., Lozynskyi, V.H., & Cabana, E. (2017). Formation of thermal fields by the energy-chemical complex of coal gasification. *Naukovyi Visnyk Natsionalnoho Hirnychoho Universytetu*, (5), 36-42.
14. Lozynskyi, V., Dychkovskyi, R., Saik, P., & Falshtynskyi, V. (2018). Coal Seam Gasification in Faulting Zones (Heat and Mass Balance Study). *Solid State Phenomena*, (277), 66-79. https://doi.org/10.4028/www.scientific.net/SSP.277.66
15. Bilyi Y.S., Dovbnich, M.M., Kuzina, G.P., & Stovas, G.M. (2004). *Sposib vydilennia na ploshchi dilianok neodnoridnoi budovy hruntovoho masyvu*. Patent No. 64221, Ukraine.
16. DBN A.2.1-1-2014. (2014). *Inzhenerno-geologicheskie izyskaniya dlya stroitelstva*. Kyiv: Ukrayinskyi derzhavnyi holovnyi naukovo-doslidnyi i vyrobnychyi instytut inzhenerno-tekhnichnykh i ekolohichnykh vyshukuvan.
17. SOU 10.1.05400632.1:2004. (2004). *Proekt likvidatsii vuhilnykh shakht Ukrainy*.
18. Malashkevych, D., Sotskov V., Medyanyk V. & Prykhodchenko D. (2018). Integrated Evaluation of the Worked-Out Area Partial Backfill Effect of Stress-Strain State of Coal-Bearing Rock Mass. *Solid State Phenomena*, (277), 213-220. https://doi.org/10.4028/www.scientific.net/ssp.277.213

Physical simulation of integrated protective structures

Serhii Nehrii[1*], *Tetiana* Nehrii[1], and *Hanna* Piskurska[2]

[1]Donetsk National Technical University, Department of Mineral Deposits, 2 Shybankova Sq., 85300 Pokrovsk, Ukraine
[2]Donetsk National Technical University, Department of Language Training, 2 Shybankova Sq., 85300 Pokrovsk, Ukraine

Abstract. An effective way of protecting development workings by integrated rock wall structures is considered. The necessity of improving this method and its experimental verification is substantiated. Integrated rock wall structures are experimentally tested under laboratory conditions. On the basis of the results of physical modeling, parameters of integrated structures have been specified in terms of impact factors; their yield strength is also specified. The condition for ensuring the stability of the integrated protective structure, which allows calculating appropriate rock wall width, is determined.

1 Introduction

Under conditions of coal seam mining by underground method, the question of maintaining the stability of mining operations is becoming rather relevant, since it effects mineral extraction schedule and mine safety [1 – 4]. A lot of studies abroad and in Ukraine are devoted to this issue. In particular, theoretical and practical results of research on the stability of rock contour of workings are given in terms of impact factors. In most cases, they are based on mine field observations [5 – 10, 17], physical modeling [10 – 14], numerical methods [5, 10, 12, 15, 18, 19], etc.

New ways of protecting workings are being designed and existing ones are being improved by means of well-known research methods. For example, methods of physical modeling and numerical methods are widely used to study stress-strain state of an array around mine workings and to substantiate technological parameters [5, 10 – 12, 16, 19, 20]. Mine observations are mostly used to obtain raw data for designing and testing the technologies under study [6 – 8, 17, 20].

Currently, more and more studies are being devoted to the development of effective resource-saving means of protecting development workings, which are built behind the longwall face on the line between the working and the mind-out space. Most widespread protective methods do not produce the desired results because they have a limited scope of use, and their yielding is high. Besides, they are rather costly to implement. For example, workings protection by means of ordinary rock is not always effective because it has a high degree of shrinkage. On the other hand, these technologies have the following advantages:

* Corresponding author: serhii.nehrii@donntu.edu.ua

firstly, rock is left in the mine; secondly, natural materials are used for the construction. This allows reducing the cost of protective means and transportation costs. Along with this, miners' wages increases because of considerable labour intensity. Therefore, in order to ensure the effectiveness of the protective means by using ordinary rocks, it is necessary to eliminate these shortcomings. This is possible when the necessary amount of rock is reduced in structures due to the use of restraining elements. Therefore, the issue of improving the traditional means of protecting workings and developing new ones is relevant and can be solved by means of proven methods and techniques.

2 Reference to scientific and practical results

The traditional way of protection by means of ordinary rock is constructing a rock wall along the working on the border with the mined-out space. In this case mining rock or special deading rock is used. But this is the least efficient technology, because its implementation requires a large amount of rock and significant labor costs. In addition, it has the yield strength of 40 to 65%, which negatively affects the state of the working.

Currently, the Department of Mineral Deposits in Donetsk National Technical University of Ministry of Education and Science of Ukraine is working on developing and improving the means of protection. In most cases, these are effective and low-cost technologies based on the use of ordinary rocks and restraining elements. These include the integrated protective structures, which means that two bag walls filled with fine-friction rock are laid out behind the longwall face on the line between the working and the mind-out space, and ordinary rock of any fraction is placed in the cavity between them (Fig. 1).

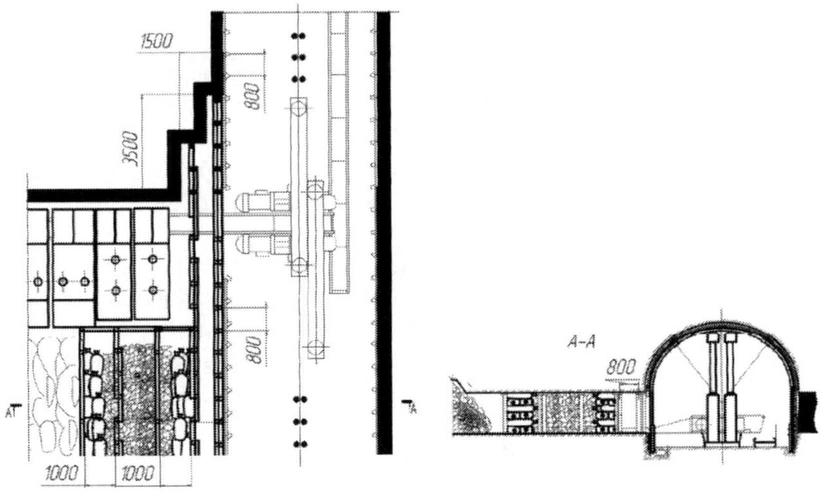

Fig. 1. Technological scheme of lining the final part of the longwall face and erecting the integrated protective rock structure [22].

According to the results of previous laboratory and analytical studies the parameters of this technology have been determined [21, 22]. But a small number of parameters which varied in a limited range have been considered in laboratory models. Besides, theoretical calculations also need practical confirmation. That is, it would be advisable to carry out a research phase to improve this technology and prove the adequacy of theoretical results in practice.

Thus, the purpose of the research is to justify the parameters of an integrated protective rock structure on the basis of the results of physical modeling.

3 Determining the parameters of integrated protective structures

In order to achieve this aim, the research phase was based on laboratory testing of rock constructions, which were models of the suggested protective means and had different linear parameters. That is, if we take as a ground the fact that the minimum required width of the rock wall should be equal to 8 seam height, but not less than 5 m [23], the minimum required conditions for maintaining the operational state of workings will be provided. Therefore, additional measures to improve the protective means will not be required. However, if it is necessary to reduce the structure yielding strength or the rock volume, provided that the state of the working is maintained or improved, the use of such restraining surfaces as bag walls is recommended.

Previously, it was found out that walls should contain two bags in a row laid perpendicular to each other along the height of the protective structure [21], but by changing the rock width between these walls, it is possible to change their bearing capacity and yielding strength, as well as affect the stability of the working. The rock wall width can be calculated by the equation [22]:

$$x = \frac{tg^2\left(45^\circ - \frac{\varphi}{2}\right)h}{2f} \ln \frac{\sigma_y}{P_b}, \tag{1}$$

where h is height of the structure, m; σ_y is stress over the protective structure caused by the weight of the rocks held by the structure, MPa; f is coefficient of rock friction within the structure against the material of its shell; φ is angle of internal friction of rocks, degrees; P_b is stress occurring when ordinary rocks contact the bag wall, MPa.

The integrated structure involves erection of walls made of bags and rock packs, but from the equation (1) it follows that at $\sigma_y = P_b$ the rock pack is not required, and this construction will be similar to the method of protecting by rock walls [21, 24]. In the case when $\sigma_y > P_b$, it will be insufficient to make use of rock pack only, so it is necessary to fill space between them with ordinary rocks, i.e. to construct a rock wall. The width of the rock pack may be from several tens of centimeters to several meters, depending on the impact factors.

To determine the reaction of the bag wall P_b, we will present a calculation scheme for the redistribution of forces along the width of the protective structure (Fig. 2). It is evident that under the action of forces, there occur extrusion forces caused by rock weight forces σ_y in the rock pack toward the walls, while P_b forces withstand them. In turn, P_b forces are caused by the friction force of the bags between themselves against the rocks of the hanging layer and the ground because of the force action σ_y and the weight of the bags. Thus, σ_y affects both rock structure destroying and its holding.

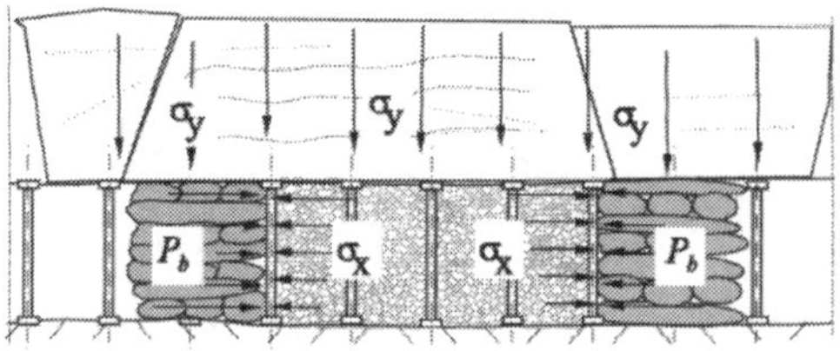

Fig. 2. The scheme of force distribution along the width of the integrated protective structure.

The P_b can be determined experimentally by studying the separated rock bag walls using structural models (Fig. 3). For this purpose, cotton bags filled with fine fraction rock were made in the scale of modeling M1:25, according to the similarity conditions [25]. The length of the bags was 45 mm and the diameter was 12.5 mm. In the rows of the walls the bags were perpendicular to each other along the required height. The width of the rock walls was 45 mm, which equaled 1.13 m in full size. Their length was 120 mm (3.0 m in full size), and the height depended on the number of rows of bags in the wall. It was 40 (1.0 m), 65 (1.63 m) and 90 mm (2.25 m) for 3, 5 and 7 rows respectively.

Several models of different height were made. They were tested on a hydraulic press in turns. For this purpose, they were squeezed with a pressing force of 0.5 kN, then the bags were pushed horizontally in different points of the wall with the help of a metal rod and a dynamometer (Fig. 3).

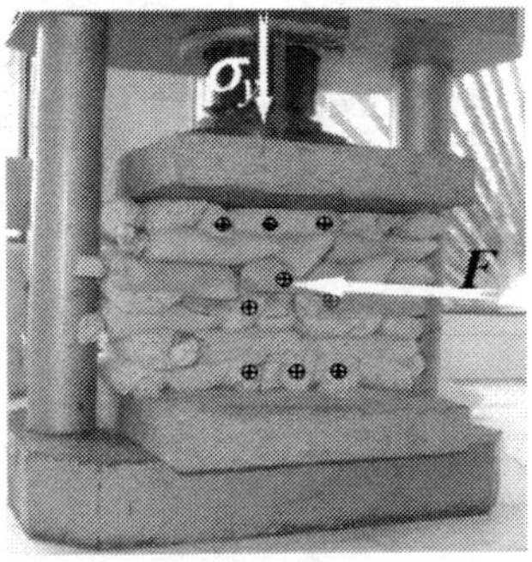

Fig. 3. Representation of the rock wall made of bags.

In each wall, when pushing the bag, the dynamometer readings were recorded at the moment of shift in a specified location. Then the construction returned to its starting position and measured again. Since the pushing force F is equal to the maximum reaction of the wall P_b, it is possible to determine the relative index ψ if we know σ_y and F. The index shows the portion of wall reaction in the total loading value σ_y, that is

$$\psi = \frac{P_b}{\sigma_y} . \tag{2}$$

Subsequently, with this indicator the equation (1) will be as follows:

$$x = -\frac{tg^2\left(45^\circ - \frac{\phi}{2}\right)h}{2f}\ln\psi , \tag{3}$$

which implies the condition for the stability of the construction:

$$\frac{x}{h} \geq -\frac{tg^2\left(45^\circ - \frac{\phi}{2}\right)}{2f}\ln\psi . \tag{4}$$

Having obtained the results of numerical measurements, we were able to establish the dependence of the indicator ψ on the height of the structure (Fig. 4), which implies that the middle row of the construction has the minimum value, and the lowest one has the maximum value. But if we compare graphs showing different amounts of bags along the height, the values of the indicator ψ in the middle rows of different models are almost identical and differ by no more than 5%. For the upper and lower rows the pattern is similar and the deviations do not exceed 9 and 6%, respectively.

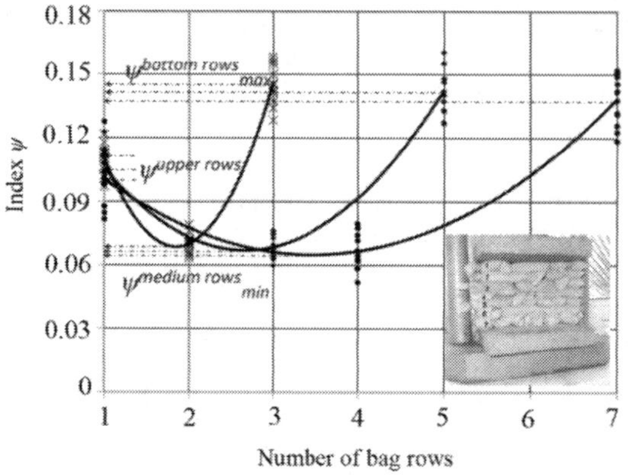

Fig. 4. Graphs showing dependence of the relative index ψ on the rock wall height expressed by number of bag rows.

To calculate x by the equation (3) in order to ensure the stability of the structure, it is necessary to assume the minimum value of the index $\psi_{min} = 0.065$, which is obtained for the middle row, regardless of the number of bags in the row. Following from the equation (3) and taking into account this index, the condition (4) will take the following form:

$$\frac{x}{h} \geq \frac{tg^2\left(45^\circ - \frac{\phi}{2}\right)}{0.73f}. \tag{5}$$

To confirm experimentally the obtained results, the following stage of structural modeling was carried out on the scale M 1:25. 19 models were made; they differed from each other in terms of width of the rock pack between the bag walls (Fig. 5). The bags were made of cotton fabric and filled with quartz sand. For the narrowest model, the rock pack width was calculated according to the equation (3) and was 1.22h, so the width of this model was 3.5h, including the rock pack 1.3h wide. Additionally, several models containing rock walls without bags were made (Fig. 6), as well as models made of rock walls 1.1h and 2.2h wide (where h is the initial height of the model) (Fig. 7).

Rock with the same fraction composition was placed in the rock packs. The constructions presupposed installing wooden racks, laying bags and burying the rock (Fig. 8).

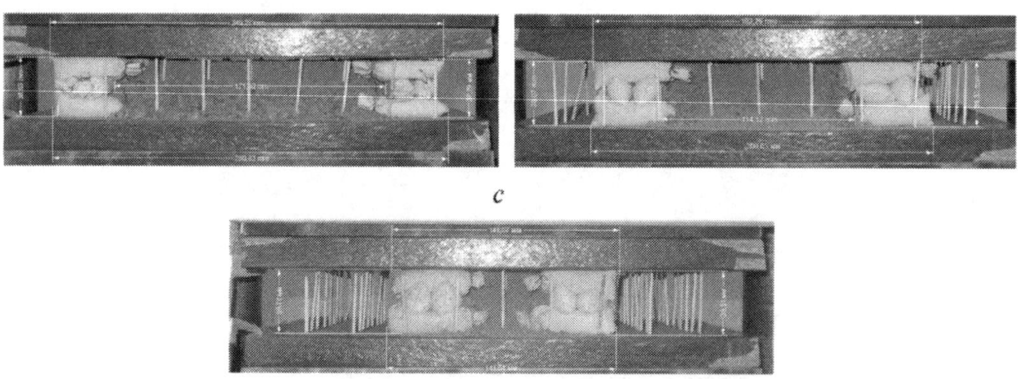

Fig. 5. Representation of the models of integrated structures made of bag walls and rock pack (a, b, c – rock pack width, respectively, 4.3h, 2.8h and 1.3h).

https://doi.org/10.1051/e3sconf/20186000038

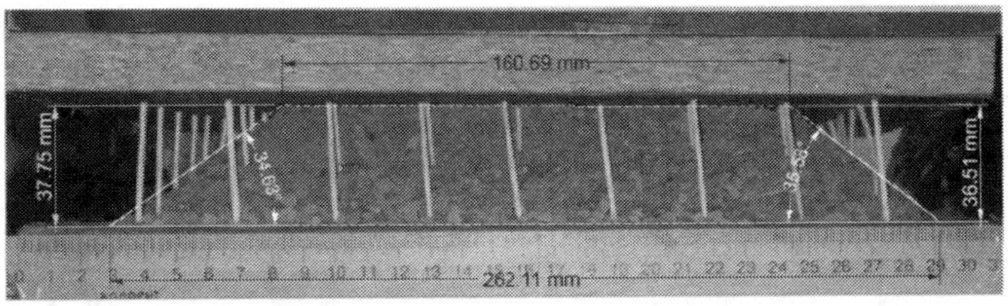

Fig. 6. Representation of a rock pack model.

Fig. 7. Representation of rock wall models 1.1*h* (a) and 2.2*h* (b) wide.

Fig. 8. Representation of the models before and after testing.

After the models were prepared, their testing was carried out. For this purpose, by means of the hydraulic press *П-50*, the models were squeezed in turns and their linear parameters were fixed with the electronic caliper gauge *INTERTOOL MT-3006*. Photographic evidence was also taken with *Canon PowerShot A710* digital camera. Press load was measured by means of the integrated universal measuring system *SI-9v2 "Universal"* displaying values of the current load and its application speed. Several photographs of the modeling stages are shown in Figure 9.

While the models were tested, there were some changes in the linear parameters of the structures. In particular, all of them, depending on the load, decreased in height; in models without walls, the rock pack partially extended and its width increased; in wall models the bags were compacted and there were some holed bags.

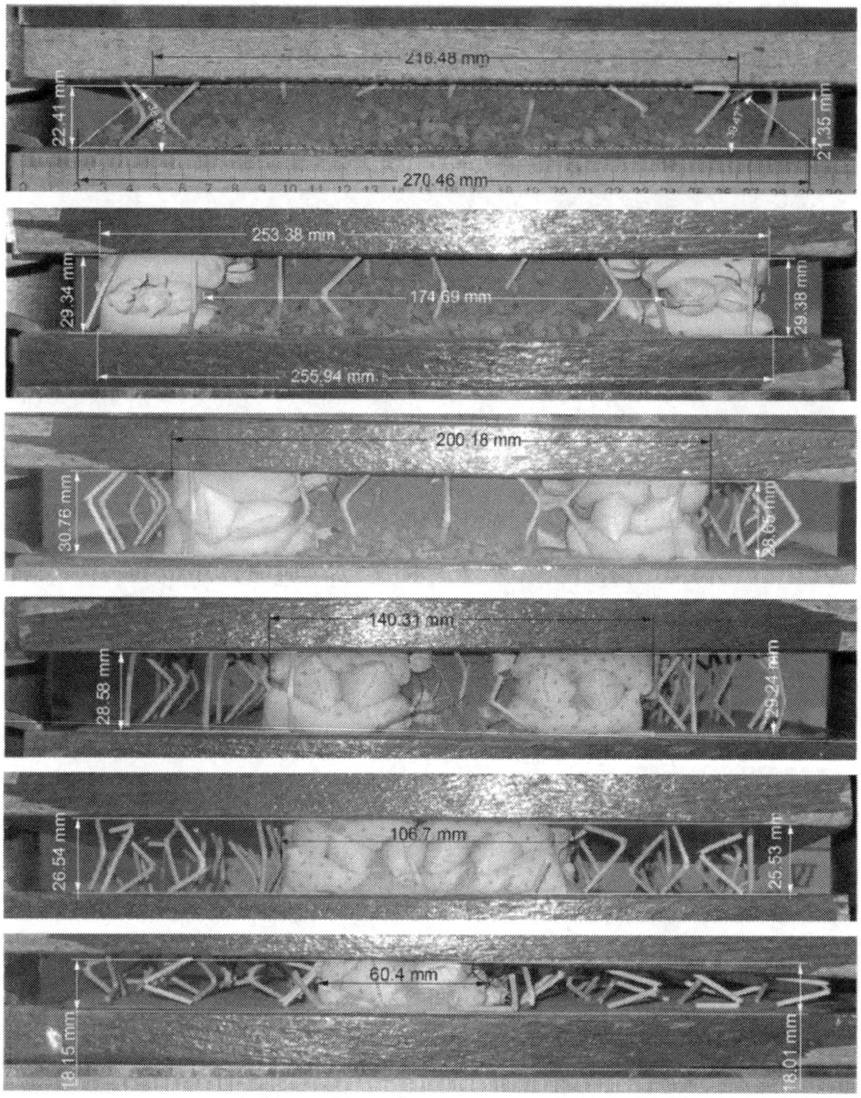

Fig. 9. Representation of the final modeling stage.

The limit load value was calculated by the formula [26]:

$$g_m = g_f \frac{\gamma_m}{\gamma_f} \frac{l_m}{l_f}, \tag{6}$$

where g_f is rock pressure value in field; (γ_m/γ_f) is the ratio of bulk weight of field materials and the model; (l_m/l_f) is the geometric simulation scale, i.e. at $\gamma_m/\gamma_f = 1$ and $l_m/l_f = 0.04$ – $g_m = 0.04g_f$. Thus, at depth of 1000 m, assuming that the working contour stress can exceed 2 times the stress in virgin ground [22], the g_m should be 2 MPa, and taking into account the bearing surface area, the load intensity should be 180 kN. Therefore, the boundary load was 200 kN.

Having measured shrinkage of structures in models, we determined their relative yielding strength depending on the load intensity (Fig. 10). Generally, the models yielding at maximum load reached 52%, depending on the particular design, and this value corresponded to the rock wall 1.1h wide. Widening the rock wall twofold to 2.2h ensured the yielding of 36%, which was almost equal to the rock pack yielding (38%). It was concluded that the use of rock legs only is not effective enough.

Testing of integrated structures made of rock pack and bag walls showed that in models of different width the value of yielding was almost the same and at maximum load it was 26%, which is almost 40% less than when protected by rock pack and a rock wall 2.2h wide. That is, it wasn't reasonable to increase the rock pack width by more than 1.3h. Thus, for the simulated conditions, the width of the integrated construction was 3.5h, including the rock pack 1.3h wide, which corresponds to the analytical results of the study and confirms the correctness of the previously obtained dependence of the rock pack width in the integrated structure on impact factors. But the equation (3) presupposes the condition of preserving the equilibrium state of running ground in the rock pack, that is, at the estimated width of 1.3h there should not have been any roof subsidence, and, in our case, it is equivalent to 0.26h. This is explained by the fact that at the first stage of the loading in the rock pack there is no bearing core and the structure intensively changes its shape. At the second stage, when the construction yielding is exhausted, it works when there is the core in the rock pack. The justification of the mechanism for forming the core in the construction requires further research. In general, the obtained results allowed us to clarify the parameters of the integrated protective structure and confirmed the adequacy of the analytical dependence of the rock pack width in this structure on the impact factors.

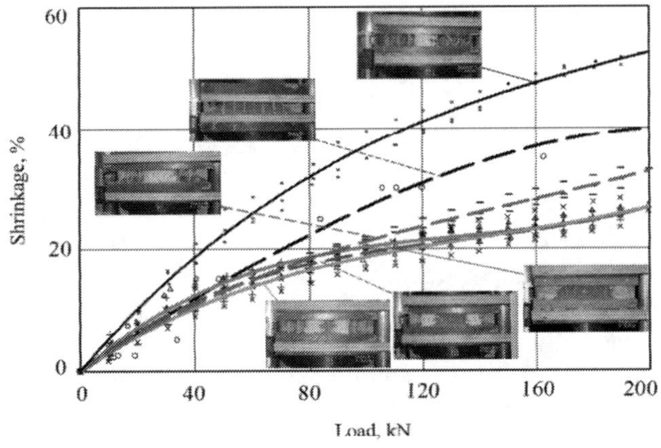

Fig. 10. Graphs showing dependence of protective structures shrinkage amount on load intensity under various construction parameters.

4 Conclusions

The conducted research made it possible to draw the following conclusions:

– the correlation of the rock wall reaction to its total load value is established;

– the condition for ensuring the stability of the integrated protective structure is substantiated;

– the dependence of the amount of protective structures shrinkage on the load intensity under various parameters of structures (rock pack, rock walls and integrated structures) is determined;

– it was established that there is yielding value of 26% in the integrated protective structure made of rock walls and rock pack between them;

– the analytical dependence of the rock pack width in the integrated protective structure on impact factors is specified, by means of which it is possible to determine the minimum required rock pack width for effective maintenance of top rock;

– it is established that exceeding the minimum required rock pack width in the integrated structure is not appropriate.

The authors are grateful to the staff of the Department of Geotechnologies and Labor Protection of the Industrial Institute of Donetsk National Technical University for the assistance in conducting laboratory research.

References

1. Bulat, A.F., Popovych, I.M., Vivcharenko, O.V., & Krukovskyi, O.P. (2014). Tekhnolohiia ankernoho kriplennia hirnychykh vyrobok na shakhtakh Ukrainy: stan i perspektyvy. *Uhol Ukrainy*, (2), 3-7.
2. Stanisław Prusek, Sylwester Rajwa, Aleksander Wrana, & Alicja Krzemień. (2017). Assessment of roof fall risk in longwall coal mines. *International Journal of Mining, Reclamation and Environment*, *31*(8), 558-574. http://doi.org/10.1080/17480930.2016.1200897
3. Gregory M. Molinda, Christopher Mark, & D. Debasis. (2001). Using the coal mine roof rating (CMRR) to assess roof stability in U.S. coal mines. *J Mines Met Fuels* (India), 314-321. https://www.cdc.gov/niosh/mining/works/coversheet894.html
4. Pivniak, H.H., Tabachenko, M.M., Dychkovskyi, R.O., & Falshtynskyi, V.S. (2015). *Keruvannia ryzykamy v hirnychodobuvnii diialnosti*. Dnipropetrovsk: Natsionalnyi hirnychyi universytet.
5. Małkowski Piotr, Ostrowski Łukasz, & Bachanek Piotr. (2017). Modelling the Small Throw Fault Effect on the Stability of a Mining Roadway and Its Verification by In Situ Investigation. *Energies*, *10*(12), 2082. http://doi.org/10.3390/en10122082
6. Piotr Małkowski, Łukasz Ostrowski, & Piotr Bachanek. (2017). The impact of the low throw fault on the stability of roadways in a hard coal mine. *Studia Geotechnica et Mechanica*, *39*(1), 63-72. http://doi.org/10.1515/sgem-2017-0006
7. Qiangling Yao, Xuehua Li, Fan Pan, Teng Wang, & Guang Wang. (2016). Deformation and Failure Mechanism of Roadway Sensitive to Stress Disturbance and Its Zonal Support Technology. *Shock and Vibration*, (6), 1-14. http://doi.org/10.1155/2016/1812768
8. Weijian Yu, Weijun Wang, Xinyuan Chen, & Shaohua Du. (2015). Field investigations of high stress soft surrounding rocks and deformation control. *Journal of Rock Mechanics and Geotechnical Engineering*, (7), 421-433.
9. Frith, R., Reed, G., & McKinnon, M. (2017). Fundamental principles of an effective reinforcing roof bolting strategy in horizontally layered roof strata and three areas of potential improvement. In *17th Coal Operators' Conference* (pp. 149-170). Australia: University of Wollongong.
10. Hapieiev, S.M., Khoziaikina, N.V., Tereshchuk, R.M., & Kovalenko, V.V. (2016). *Resursozberihaiuchi tekhnolohii upravlinnia stiikistiu protiazhnykh vyrobok vuhilnykh shakht*. Dnipro: Natsionalnyi hirnychyi universytet.
11. Minghe Ju, , Xuehua Li, Qiangling Yao, Shengyou Liu, Shun Liang, & Xiaolin Wang. (2017). Effect of sand grain size on simulated mining-induced overburden failure in physical model tests. *Engineering Geology*, 1-14. http://doi.org/10.1016/j.enggeo.2017.05.015
12. Nehrii, S., & Nehrii, T. (2017). Conditions of use trapezoidal security installations. *Journal of Donetsk Mining Institute*, *2*(41), 7-18 http://doi.org/10.31474/1999-981x-2017-2-7-18
13. Rui Gao, Bin Yu, Hongchun Xia, & Hongfei Duan. (2017). Reduction of Stress Acting on a Thick, Deep Coal Seam by Protective-Seam Mining. *Energies*, *10*(8), 1209. http://doi.org/10.3390/en10081209
14. Nan Zhou, Jixiong Zhang, Hao Yan & Meng Li. (2017). Deformation Behavior of Hard Roofs in Solid Backfill Coal Mining Using Physical Models. *Energies*, *10*(4), 557. http://doi.org/10.3390/en10040557

15. Demin, V.F., Fofanov, O.B., Demina, T.V., & Yavorskiy, V.V. (2017). Anchoring of development workings in a zone of influence of mining in case of the level anchoring system. IOP Conference Series: *Materials Science and Engineering*, (177), 12041. http://doi.org/10.1088/1757-899X/177/1/012041
16. Solodiankin, A.V., Mashurka, S.V., Dudka, I.V., & Kuziaeva, O.A. (2015). Provision stability local workings for reusing under the state enterprise "Coal company "Yuzhnodonbasskaya №1". *Up-to-date resource- and energy- saving technologies in mining industry*, *1*(15), 96-105.
17. Khalimendik, Yu.M., Bruy, A.V., & Baryshnikov, A.S. (2013). Issledovanie deformirovaniya porod krovli vyemochnykh vyrabotok. *Transactions of UkrNDMI NAN Ukraine*, *13*(1), 21-30.
18. Bondarenko, V., Gryadushchiy Yu., Kovalevskaya, I., & Simanovich, G. (2009). Metodika issledovaniya vliyaniya poperechnogo secheniya plastovoy vyrabotki vida krepi na peremescheniya eye kontura. *School of Underground Mining 2009*, 23-29.
19. Kozhushok, O., & Ilyashov, M. (2009). Matematicheskoe modelirovanie raboty kombinirovannoy okhrannoy sistemy pri podderzhanii vyemochnogo shtreka za lavoy. *School of Underground Mining 2009*, 185-193.
20. Solodiankin, O.V., Dudka, I., Tereshchuk, R., & Hryhoriev, O. (2017). *Okhorona pidhotovchykh vyrobok, shcho vykorystovuiut povtorno, v umovakh antratsytovykh shakht*. Dnipro: Natsionalnyi hirnychyi universytet.
21. Nehrii, S. (2016). Tests of rock bands with limited suppleness. *Journal of Donetsk Mining Institute*, *2*(39), 54-60.
22. Nehrii, S., & Nehrii, T. (2017). Determination of parameters of rock bands with limited suppleness. *Transactions of Kremenchuk Mykhailo Ostrohradskyi National University*, 1/2017 (102), 50-57.
23. SOU 10.1.00185790.011:2007. (2007). *Pidhotovchi vyrobky na polohykh plastakh. Vybir kriplennia, sposobiv i zasobiv okhorony*. Kyiv: Donetskyi vuhilnyi instytut.
24. Khazipov, I.V. (2009). *Razrabotka sposobov sozdaniya iskusstvennykh porodnykh sooruzheniy dlya okhrany povtorno ispol'zuemykh vyrabotok*. Rh.D. DonNTU.
25. Yanko, S.V., Kasian M., Petrenko, Yu, & etc. (2003). *Suchasni problemy provedennia ta pidtrymannia hirnychykh vyrobok hlybokykh shakht*. Donetsk: DUNVHO.
26. Borisov, A.A. (1964). *Raschety gornogo davleniya v lavakh pologikh plastov*. Moskva: Nedra.

Development of a three-parameter model of the shoe brake contact interaction with the drum in mine hoisting machine

Kostiantyn Zabolotnyi[1], *Oleksand* Zhupiiev[1], and *Artur* Molodchenko[1*]

[1]National Mining University, Mining Machines & Engineering Department, 19 Yavornytskoho Ave., 49005 Dnipro, Ukraine

Abstract. For safe operation of mine hoisting machines (MHM) in the mining industry, it is necessary to provide high constructive reliability of brake systems, in particular, brake systems based on the block brake. The contact interaction of the brake with the translational movement of the shoes applied in MHM, remains insufficiently studied. In particular, it is necessary to develop a technique for accurate determination of the braking moment and forces in the elements of the brake linkage, as well as to study the nature of the pressure distribution along the brake beam. The purpose of this article is to simulate the contact interaction of the MHM brake taking into account the coefficient of friction, the ratio of the flexural stiffness to the longitudinal stiffness of a beam, as well as the ratio of the transverse stiffness of a lining to the flexural stiffness of a beam. The application of the developed model will help to reduce the maximum contact pressure in the shoe brakes of the mine hoisting machines and will allow more accurate calculation of the braking moment value compared to existing methods.

1 Introduction

The hoists are considered to be the most important installation among all the machinery used in the mining industry. An emergency situation in the process of the hoisting machine operation actually stops all the mining works in the mine, which leads to huge financial losses. This determines the high requirements to the technical level of mine hoisting machines and the quality of their manufacturing.

Novokramatorskyi Mashynobudivnyi Zavod (NKMZ), one of the largest production and research complexes in Europe, up to now produces drum-type hoisting machines, which are often equipped with shoe brakes. A lot of well-known scientists have been involved into the designing of the braking device for MHM, but because of the imperfection of the previously used computing facilities, the very important parameters for the interaction

* Corresponding author: artamaranth0.0@gmail.com

between the brake beam, lining and the brake rim, have not been studied. For example, the problem of calculating the brakes of mine hoisting machines has recently been considered by Barecki Z. and Sciezska S.F. [1]. However, in their mathematical model of the contact interaction between the brake shoe and the drum, they did not take into account the influence of tangential forces on the distribution of the contact pressure, and the calculation given as an example concerns a particular machine without specifying brake parameters, in addition, the article published on this topic does not contain any generalized dependences.

To calculate the braking parameters for large hoisting machines produced by the NKMZ with the translational movement of the brake shoes, the method described in the works of Davydov B.L., Karpyshev N.S., and Fedorova Z.M. has been used so far. Here, a hypothesis is made about the sinusoidal law of contact pressures distribution, which is consistent with the assumption of the absolute stiffness of the brake beam [3, 7, 10].

The traditional mathematical model of Davydov B.L. for calculating the distributed tangential and normal forces acting on the brake beam of the MHM shoe brake is based on the assumption that these forces are distributed independently of the lining friction against the drum and the flexural stiffness of the beam. This assumption cannot be considered valid, and therefore the results of determining the specified loads in the construction of a shoe brake can lead to significant errors and the creation of further emergency situations.

The results of the computational experiment carried out by means of the SolidWorks Simulation Program on the study of a shoe brake for the CR-4x3/0.7 type mine hoisting machine with translationally moving shoes, make it possible to construct a U-shaped pressure distribution curve.

Many researchers of contact interaction of shoe brakes with a drum, in particular, Day A.J., Newcomb T.P., Harding P.R., Huang YU.M., Shyr J.S., Tirovic M., Barecki Z. and Sciezska S.F., using in their works the method of finite elements, finite-difference method and the boundary element method, have proved the invalidity of the hypothesis about absolute stiffness of a beam.

The calculations, given in works of Day A.J., with the use of the finite elements method relate to the specific case where the brake beam is modeled as a sequence of beams with different cross sections [4-6].

In the calculations of Huang Y.M. and Shir J.S. the boundary element method was used without analysis of the basic parameters influencing the contact interaction [8].

In the works of Zabolotnyi K.S., Kolosov D.L., Moskalova T.V. the techniques were described for the construction optimization of drums and brakes in the mine hoisting machines by determining the stress-strain state [11, 15, 17 – 20].

In further numerous works on improving the calculation technique for brakes, the dynamic and thermal processes were taken into account which proceed in the braking system elements. For example, in the works of Kyrychenko Y.O., Samusia V. I., Ilyn S.R. and Bondarenko V.I., the problems of hoisting plant dynamics during working and safety braking was considered. Most importantly, that the hypothesis of stiffness of the brake rim and beam, was not substantiated [2, 9, 12 – 14, 16].

2 Development of a three-parameter model of contact interaction

Fig. 1 shows the scheme of a part of the braking system for large-size MHM produced by NKMZ. By means of the brake linkage, the braking force is transferred to the upper 4 and lower 5 rods. The center of the brake beam rotates on a vertical pillar 3. The brake shoe consists of a brake beam 1 and a set of linings 2. These linings interact with the drum 6 during braking. Thus, it can be assumed that the brake shoe is affected by two couplings: a vertical pillar and a drum.

The exact calculation of the forces arising in the brake shoe is possible only with the use of numerical methods (the methods of finite or boundary elements), but in order to reveal the basic laws of contact interaction, it is required to create a simplified analytical model of a structure. It has been decided to represent it in the form of a circular bar with the constant cross section, fixed from vertical displacements in the central point. According to the accepted assumption, the same external forces act on the bar in the horizontal rods.

The physical model of the brake lining has been defined as an array of elastic bodies, like Winkler foundation, which do not interact with each other, work in compression and transfer the distributed frictional forces arising between the brake drum and the brake shoe, calculated for the limited equilibrium state in accordance with Coulomb's law (Fig. 2).

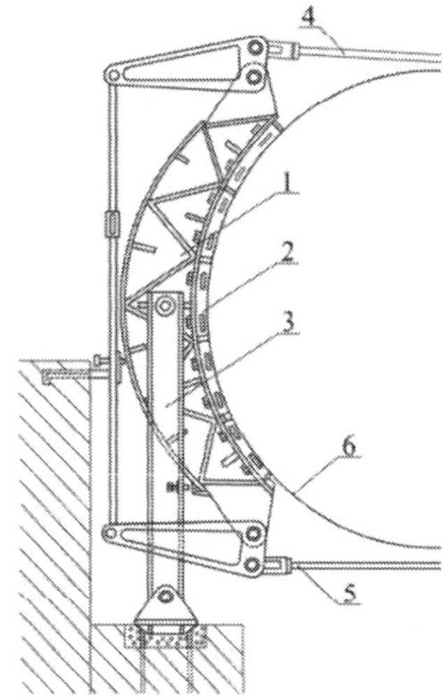

Fig. 1. Scheme of a part of the MHM brake system.

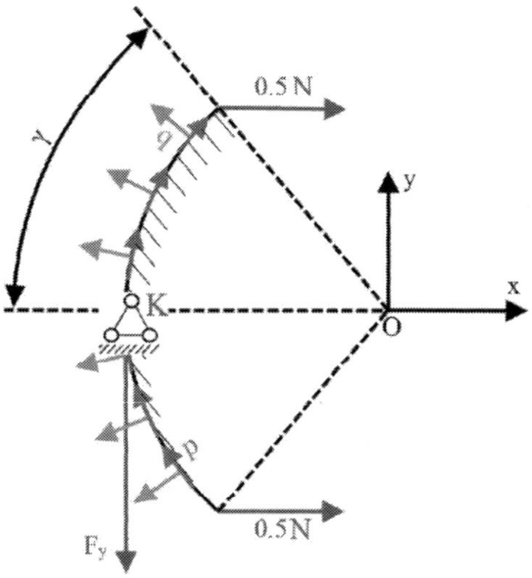

Fig. 2. Scheme of the forces arising in the brake beam.

The condition has also been accepted that the drum rotates clockwise, and the angle of wrap of the brake shoe is equal to 2γ. Having replaced the couplings on reactions, we determine the concentrated vertical force F_y in the central point of the bar, as well as the distributed forces q and p of the drum action on the lining. In view of small thickness of the beam, we consider that these forces are applied to the neutral axis of the bar.

The action of the horizontal movable bearing in the brake system causes the occurrence of a force jump (vertical reaction of the bearing), which ensures the discontinuity of the longitudinal force function, excluding the possibility to consider the bar as an element of structural mechanics. Therefore, in the model for calculation, it is provided that the bar consists of two parts: the upper and lower ones.

The mathematical model for calculation of forces in the brake shoe includes the number of the equations presented below.

Equilibrium equations for the lower ($i = 0$) and upper ($i = 1$) parts of the bar have the following form:

$$\frac{dY_i}{d\varphi} = A_i Y_i, \tag{1}$$

where $Y_{i,1} \equiv T_i$, $Y_{i,2} \equiv Q_i$, $Y_{i,3} \equiv M_i$, $A_i \equiv \begin{bmatrix} Y_{i,2} + p_i R \\ -Y_{i,1} + q_i R \\ Y_{i,2} R \end{bmatrix}$, $T_i(\varphi)$ is the longitudinal force; $Q_i(\varphi)$ is the shear force; $M_i(\varphi)$ is the bending moment; φ is the current angular coordinate; R is the radius of the neutral line of the brake beam.

The equation for calculating the distributed normal forces that correspond to the model of a structure with an elastic base, can be represented as follows:

$$q_i(\varphi) = \kappa w_i(\varphi); \quad (i = 0,\ 1), \tag{2}$$

where w_i (φ is the depth of beam deflection; E_l is the elasticity modulus of the lining material; H_l is the thickness of the lining; B_b is the width of the brake field; κ is the transverse stiffness of the lining, defined by the following expression:

$$\kappa = E_l \frac{B_b}{H_l}. \tag{3}$$

The equation describing the distributed friction force according to Coulomb's law has the following form:

$$p_i(\varphi) = f\kappa w_i(\varphi); \quad (i = 0,\ 1). \tag{4}$$

We formulate Hooke's law for determining the bending moment, that is

$$M_i(\varphi) = -\frac{EI}{R}\frac{d\theta_i(\varphi)}{d\varphi}; \quad (i = 0,\ 1). \tag{5}$$

where E is the elasticity modulus of the beam material; I is the inertia moment of the beam section; θ_i is the angle of the beam rotation.

Here we take into account the following kinematic dependence:

$$\theta_i(\varphi) = \frac{1}{R}\left(\frac{dw_i(\varphi)}{d\varphi} + v_i(\varphi)\right); \quad (i = 0,\ 1), \tag{6}$$

and Hooke's law for determining the longitudinal tension, that is

$$T_i(\varphi) = \frac{EF}{R}\left(\frac{dv_i(\varphi)}{d\varphi} - w_i(\varphi)\right). \tag{7}$$

where F is the cross-sectional area of the beam.

Both bars are connected in a place designated in the scheme by the point K, in which the value $\varphi = 0$, they are fixed there from vertical displacements. Thus, the state of the construction corresponds to the following dependences:

$$v_0(0) = 0; \qquad v_1(0) = 0; \qquad w_0(0) = w_1(0);$$
$$\theta_0(0) = \theta_1(0); \qquad M_0(0) = M_1(0); \qquad Q_0(0) = Q_1(0).$$

At the upper and lower ends of each beam, zero values of the bending moments are set, as well as the tangential and radial components of the horizontal forces in the rods, that is

$$T_0(-\gamma)=0.5N\sin(\gamma); \qquad T_1(\gamma)=0.5N\sin(\gamma);$$
$$M_0(-\gamma)=0; \qquad M_1(\gamma)=0;$$
$$Q_0(-\gamma)=0.5N\cos(\gamma); \qquad Q_1(\gamma)=-0.5N\cos(\gamma).$$

Now we turn to the determination of dimensionless quantities, namely:

$$v_i*=\frac{v_i}{\mu R};$$
$$w_i*=\frac{w_i}{\mu R};$$
$$\theta_i*=\frac{\theta_i}{\mu};$$
$$T_i*=\frac{T_i}{\mu EF};$$
$$M_i*=\frac{RM_i}{\mu EI};$$
$$Q_i*=\frac{R^2 Q_i}{\mu EI},$$

where $\mu=\dfrac{0.5N}{EF}$ is a multiplier factor proportional to the applied force.

To reduce the volume of the record, we transform the listed expressions into a vector-matrix form.

The vector of undetermined values is a transposed line, that is:

$$y_i=\left(v_i* \quad w_i* \quad \theta_i* \quad T_i* \quad M_i* \quad Q_i*\right)^T.$$

Then the equation corresponding to the mathematical model will be as follows:

$$\frac{dy_i}{d\varphi}=Ay_i,$$

where

$$A=\begin{pmatrix} 0 & 1 & 0 & 1 & 0 & 0 \\ -1 & 0 & 1 & 0 & 0 & 0 \\ 0 & 0 & 0 & 0 & -1 & 0 \\ 0 & -f\lambda\zeta & 0 & 0 & 0 & -\zeta \\ 0 & 0 & 0 & 0 & 0 & -1 \\ 0 & -\lambda & 0 & \dfrac{1}{\zeta} & 0 & 0 \end{pmatrix}$$

where λ is the relative transverse stiffness, defined as follows:

$$\lambda=\frac{\kappa R^4}{EI},$$

ζ is the relative flexural stiffness, which is calculated by the following formula:

$$\zeta = \frac{EI}{EFR^2}.$$

To solve the formulated system of linear differential equations with constant values of the coefficients, Euler's method is used. When used for the MHM braking system, the following dependences are valid:

$$n_0 = \alpha_0 + i\beta_0; \qquad n_1 = \alpha_0 - i\beta_0;$$
$$n_2 = \alpha_1 + i\beta_1; \qquad n_3 = \alpha_1 - i\beta_1;$$
$$n_4 = 0; \qquad n_5 = p_i(\varphi);$$

$$z_0(\varphi) = \exp(\alpha_0\varphi)\left[\mathrm{Re}\left(\Phi^{\langle 0\rangle}\right)\cos(\beta_0 t)\varphi - \mathrm{Im}\left(\Phi^{\langle 0\rangle}\right)\sin(\beta_0\varphi)\right];$$

$$z_1(\varphi) = \exp(\alpha_0\varphi)\left[\mathrm{Re}\left(\Phi^{\langle 0\rangle}\right)\sin(\beta_0\varphi) + \mathrm{Im}\left(\Phi^{\langle 0\rangle}\right)\cos(\beta_0\varphi)\right];$$

$$z_2(\varphi) = \exp(\alpha_1\varphi)\left[\mathrm{Re}\left(\Phi^{\langle 2\rangle}\right)\cos(\beta_1\varphi) - \mathrm{Im}\left(\Phi^{\langle 2\rangle}\right)\sin(\beta_1\varphi)\right];$$

$$z_3(\varphi) = \exp(\alpha_1\varphi)\left[\mathrm{Re}\left(\Phi^{\langle 2\rangle}\right)\sin(\beta_1\varphi) + \mathrm{Im}\left(\Phi^{\langle 2\rangle}\right)\cos(\beta_1\varphi)\right];$$

$$z_4(\varphi) = \frac{\Phi^{\langle 4\rangle}}{\left(\Phi^{\langle 4\rangle}\right)_0};$$

$$z_5(\varphi) = \exp(\tau\varphi)\frac{\Phi^{\langle 5\rangle}}{\left(\Phi^{\langle 5\rangle}\right)_0},$$

where n is the eigenvalues of the matrix A; Φ is the eigenvectors of the matrix A.

If the coefficient of friction is zero, then the zero root becomes reiterated, namely:

$$n_5 = 0.$$

This makes it necessary to find out the eigenfunction as a linear one, that is

$$y(\varphi) = C + D\varphi; \quad D = A(C + D\varphi); \; D = AC; \qquad AD = 0;$$

$$D = (1, \quad 0, \quad 1, \quad 0, \quad 0, \quad 0)^T;$$

$$C = \left(0, \quad \frac{1}{1+\lambda\zeta}, \quad 0, \quad \frac{\lambda\zeta}{1+\lambda\zeta}, \quad -1, \quad 0\right)^T.$$

In this case

$$z_4(\varphi) = D;$$

$$z_5(\varphi) = C + D\varphi.$$

The solution of the differential equations system corresponds to a linear combination of fundamental functions with undetermined values of the coefficients, that is

$$y_i(\varphi) = \sum_{k=0}^{5}\left[S_{6i+k}\left(z(\varphi)_k\right)\right]; \quad (i = 0,\ 1).$$

To determine these coefficients, the boundary conditions given below (Table 1) are used. These conditions are assigned in the form of the following system of algebraic equations:

$$GS = B.$$

To calculate the coefficient values, the function $S = \text{lsolve}\ (G,\ B)$ built into the MathCAD program is used.

We designate the ratio of the contact pressure to the pressure caused by the same force N acting on an absolutely rigid section of the same area ($F = BR2\gamma$) by the adjusted pressure value Ξ. We denote by χ its maximum value.

Table 1. Boundary conditions of the mathematical model for determining the forces acting on the brake beam.

Boundary conditions	Physical meaning
$G_{0,k} = \left[z(0)_k\right]_0$	$v_0(0) = 0$
$G_{1,k+6} = \left[z(0)_k\right]_0$	$v_1(0) = 0$
$G_{2,k} = \left[z(0)_k\right]_1$; $G_{2,k+6} = -\left[z(0)_k\right]_1$	$w_0(0) = w_1(0)$
$G_{3,k} = \left[z(0)_k\right]_2$; $G_{3,k+6} = -\left[z(0)_k\right]_2$	$\theta_0(0) = \theta_1(0)$
$G_{4,k} = \left[z(0)_k\right]_4$; $G_{4,k+6} = -\left[z(0)_k\right]_4$	$M_0(0) = M_1(0)$
$G_{5,k} = \left[z(0)_k\right]_5$; $G_{5,k+6} = -\left[z(0)_k\right]_5$	$Q_0(0) = Q_1(0)$
$G_{6,k} = \left[z(-\gamma)_k\right]_3$; $B_6 = \sin(\gamma)$	$T_0(-\gamma) = 0.5N\sin(\gamma)$
$G_{7,k} = \left[z(\gamma)_k\right]_3$; $B_7 = \sin(\gamma)$	$T_1(\gamma) = 0.5N\sin(\gamma)$
$G_{8,k} = \left[z(-\gamma)_k\right]_4$	$M_0(-\gamma) = 0$
$G_{9,k+6} = \left[z(\gamma)_k\right]_4$	$M_1(\gamma) = 0$
$G_{10,k} = \left[z(-\gamma)_k\right]_5$; $B_{10} = \dfrac{\cos(\gamma)}{\zeta}$	$Q_0(-\gamma) = 0.5N\cos(\gamma)$
$G_{11,k+6} = \left[z(\gamma)_k\right]_5$; $B_{11} = -\dfrac{\cos(\gamma)}{\zeta}$	$Q_1(\gamma) = -0.5N\cos(\gamma)$

Fig. 3 shows the distribution curves of the adjusted values of the contact pressure according to different values of the parameters λ and ζ. Here we see that the parameter ζ influences the pressure distribution only in case when $\lambda > 100$. The nature of the distribution depends essentially on the relative transverse stiffness of the structure. Thus, if $\lambda > 10$, then the distribution curve has a U-shaped form with a clear-cut edge effect, but in case of λ value increasing it becomes asymmetric. When $\lambda \leq 10$, the character of the curve is sinusoidal, according to which the maximum values of the contact pressure are observed in the center of the brake shoe. Close distribution to the uniform distribution of contact pressures, taking into account the considered values of the parameter ζ occurs if $\lambda \approx 10$. Neglect of the frictional forces when calculating the distribution of the contact pressure leads in the sequel to an error of the obtained maximum pressure values of 1.3%, if $\zeta = 0.002$, and 11% when $\zeta = 0.07$.

In order to consider both types of distribution, we use the above described adjusted maximum value of the contact pressure Ξ. Since in Fig. 3 the adjusted values of the contact pressure in specific points are indicated, and the value of the parameter λ exceeds the value of the parameter ζ, we plot the graph of the maximum pressure dependence as a continuous function λ (Fig. 4).

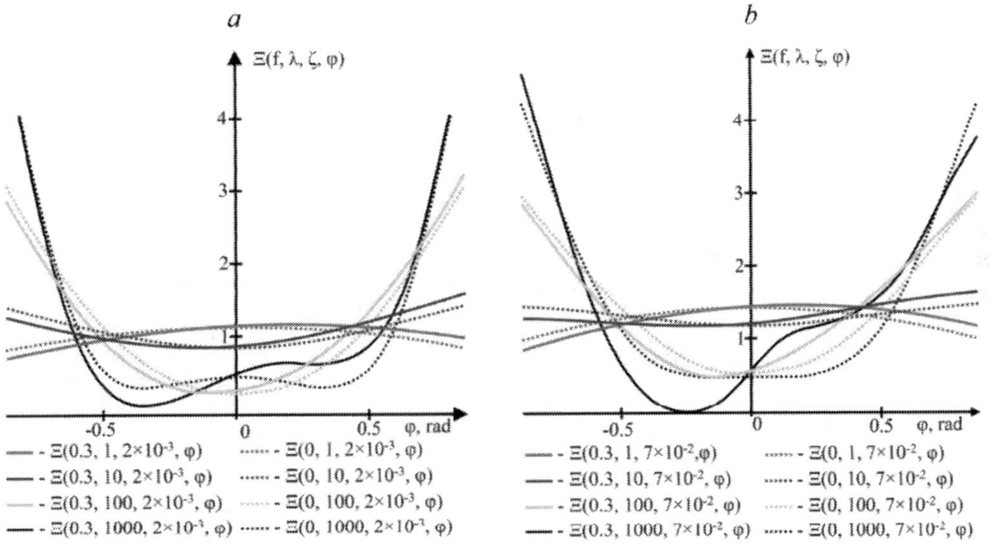

Fig. 3. The distribution curves of contact pressure values along the arc of the brake shoe: $a - \zeta = 0.002$, $b - \zeta = 0.07$.

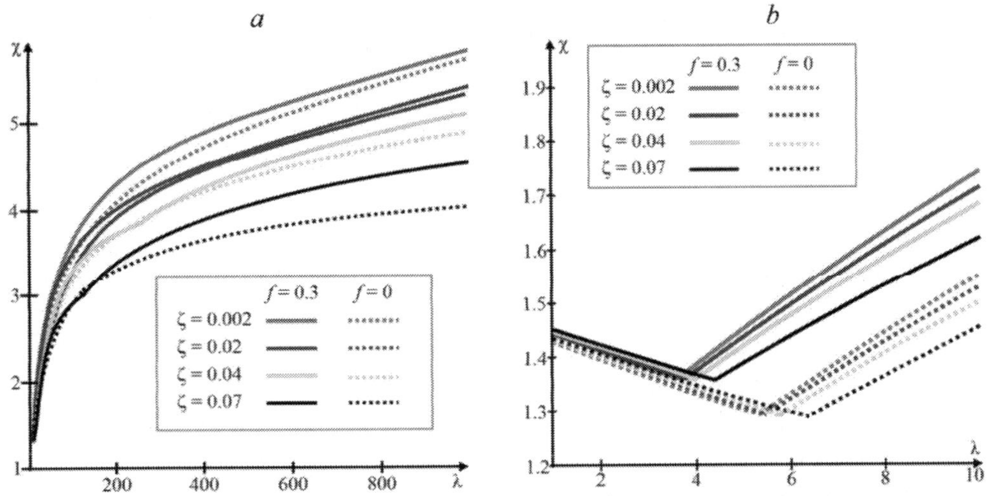

Fig. 4. Graph of the maximum contact pressure dependence on the relative transverse stiffness, at various values of the parameter ζ: a – the general graph view, b – the enlarged fragment of the graph.

As follows from the data in Fig. 4, in the range of the parameter $\lambda > 4$ values, the dependence of the maximum contact pressure on the relative transverse stiffness is a monotone function with a decreasing angle of its curve slope, which corresponds to the edge effect of the U-shaped distribution curve. If the values of the parameter $\lambda < 4$ (which corresponds to the central part of the sinusoidal curve), then this dependence is a monotonically decreasing function, close to the linear one.

In Fig. 5, for greater clearness, the curves of the level of contact pressure maxima are shown, depending on different values of the relative transverse stiffness and the relative longitudinal stiffness of the structure (dashed lines indicate the level curves reflecting the values, which were calculated without considering the influence of friction forces on the contact pressure distribution).

As the data analysis in Fig. 5 has shown, if the value of the parameter $\lambda < 150$, then the curves are close to vertical straight lines and reflect insignificantly the dependence on the ζ values. With an increase in the value of the parameter λ, they acquire a parabolic shape, and at $\lambda > 800$, they show the dependence close to a straight line at an angle of 45°.

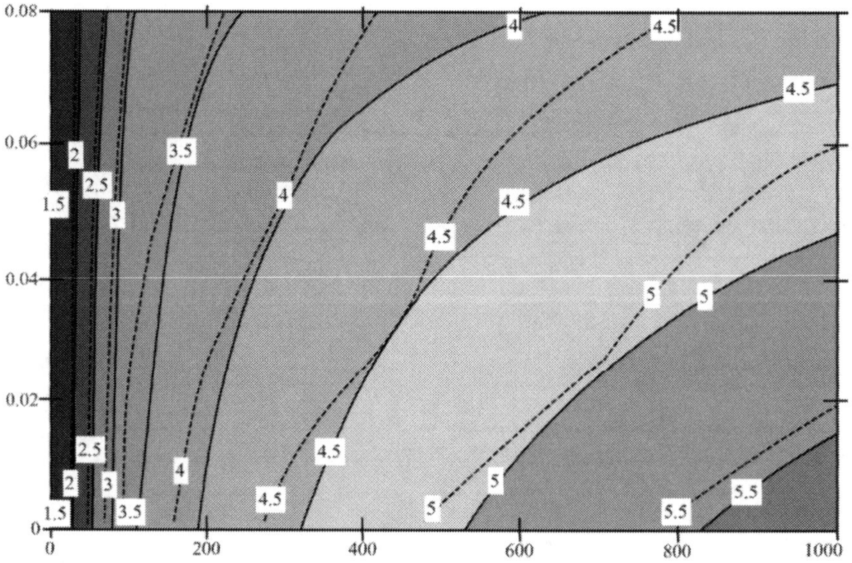

Fig. 5. The level curves of the adjusted contact pressure maxima depending on different λ and ζ parameters values.

To carry out the numerical experiment, the characteristics of the brake beam of the CR-4x3/0.7 machine with the following parameters have been chosen:

– radius of the neutral line of the brake beam $R = 2.26$ mm;
– width of the brake field $B_l = 400$ mm;
– half of the grip hold arc of the brake drum $\gamma = 50°$;
– thickness of the brake beam $H = 400$ mm;
– thickness of the lining $H_l = 80$ mm;
– elasticity modulus of the beam material $E = 2.1 \cdot 10^{11}$ Pa;
– elasticity modulus of the lining material $E_l = 3 \cdot 10^{8}$ Pa;
– coefficient of friction $f = 0, 0.3$;
– relative stiffness of the brake beam $\lambda = 109.4$;
– relative flexural stiffness of the brake beam $\xi = 2.61 \cdot 10^{-3}$.

The graph of the contact pressure calculation results is shown in Fig. 6, where the red color indicates the curve reflecting the calculation by the model without consideration of the effect of friction on the contact pressure distribution (hereinafter 'without friction'), the blue color – the curve with account of frictional forces and the green color indicates the curve constructed by calculations using the model of Davydov B.L.

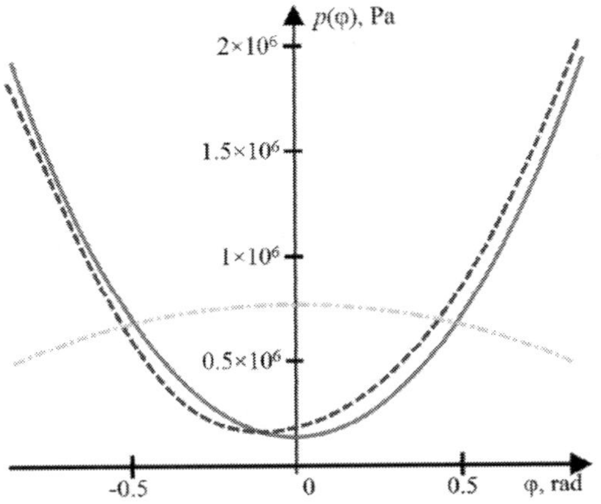

Fig. 6. Graph of the contact pressure values distribution along the arc of the brake beam.

Having transformed the equations describing the distributed normal forces in the beam, we obtain the following dependences:
– braking moment:

$$M_b = fB_bR^2 \int_{-\gamma}^{\gamma} q(\varphi)d\varphi;$$

– total force in the horizontal rods:

$$N_x = B_bR \int_{-\gamma}^{\gamma} q(\varphi)\cos(\varphi)d\varphi;$$

– force in the vertical pillar:

$$N_y = B_bR \int_{-\gamma}^{\gamma} q(\varphi)\sin(\varphi)d\varphi.$$

The force value in the horizontal rods applied to this machine is 699 kN, used in the calculations based on the model of Davydov B.L. [3].

Then the braking moment calculated with the use of 'no friction' model is equal to $7.72 \cdot 10^5$ N·m, with account of friction – $7.87 \cdot 10^5$ N·m, but the braking moment calculated by the model of Davydov B.L. – $6.98 \cdot 10^5$ N·m.

The deviation of this parameter value towards the decrease when calculating with the use of simplified model is 1.94%, but in the calculations by the model of Davydov B.L. – 11.3%.

As in the previous case of the mentioned models use, the force in the vertical pillar has the following values: $2.75 \cdot 10^5$ N; $3.48 \cdot 10^5$ N; $2.75 \cdot 10^5$ N respectively. The error in determining this indicator with the use of both models, simplified and Davydov B.L., is 26.5%.

3 Conclusions

For the first time, a physical model of the MHM brake shoe has been developed, in which the linings are modeled as an array of elastic bodies like Winkler foundation. These bodies work in compression and transfer the distributed tangential load (frictional force) arising between the brake drum and the brake shoe; the load is calculated for the ultimate equilibrium state in accordance with Coulomb's law; and the brake beam is represented as a circular bar with a constant cross-section, which has the finite tensile stiffness, the bar is mounted on a vertical pillar and interacts with the brake drum through the brake lining loaded with a distributed normal and tangential load, which simulates the contact interaction of the brake shoe and the drum; at the same time, the vertical pillar has been modeled as a movable hinge pivot located in the middle of the circular bar.

For the first time, a three-parameter mathematical model for the determination of the tangential and normal forces acting on the brake beam has been developed, in which the flexural stiffness and tensile stiffness of the beam, as well as the transverse stiffness of the lining under the influence of different friction forces values on the contact pressure distribution, have been taken into account. The model contains six equilibrium equations for the elementary sections of a circular bar on an elastic foundation, which are under the action of distributed friction forces, calculated according to Coulomb's law; equations describing Hooke's law as applied to the bending moment and longitudinal force; twelve boundary conditions for determining the values of the cutting and longitudinal force, the bending moment along the bar edges, the continuity of all displacements and forces, with exception of the longitudinal one; while the longitudinal displacement in the place of setting the movable hinge pivot is equal to zero.

By the method of solving the problem of finding the eigenvalues, the dimensionless quantities of all variables and forces in the form of linear combinations of eigenfunctions have been determined. Using a linear system of equations for coefficients determination of these combinations obtained with account of the boundary conditions, we have formulated the system of linear equations and calculated the contact pressure, the braking moment, the forces in the rods and in the pillar.

For the first time, the dependences of the adjusted maximum pressure value on the relative transverse stiffness of the brake lining and the relative longitudinal stiffness of the brake beam have been established.

The most uniform distribution of the contact pressure is achieved in the range of variation of the relative transverse stiffness of the beam from 1 to 10. At lower values of this parameter, the nature of the distribution curve has a sinusoidal form when the maximum values of the contact pressure are observed in the center of the brake shoe, and the higher values are distributed along the U-shaped curve with a clear-cut edge effect. Neglect of the frictional forces when calculating the distribution of the contact pressure causes an error in calculation of the maximum pressure value equal to 1.3%, with a relative flexural stiffness of 0.002 and 11% when the relative flexural stiffness is 0.07.

If the values of the relative transverse stiffness are below 150, the curves of the maximum pressure level are close to vertical straight lines and reflect insignificantly the dependence on the values of the relative flexural stiffness. With an increase in the relative transverse stiffness parameter, the curves acquire a parabolic shape, while when the relative transverse stiffness exceeds 800, the curves of this dependence are close to straight lines at an angle of 45°.

By the example of the shoe brake operation of the CR-4x3/0.7 mine hoisting machine, it has been proved that the distribution of the contact pressure acting upon the brake beam, which is calculated by the method of Davydov B.L. and obtained with the use of proprietary technology, has a fundamental difference: in the first case of the given distribution – a sinusoidal, and in the second case, stated in this work, it is a U-shaped parabola with a clear-cut edge effect. In this case, the value of the maximum contact pressure, calculated by the model of Davydov B.L., has turned out to be by 2.5 times lower.

The error in the braking moment determination with the application of the model of Davydov B.L. is 11.3%, and the efforts in the vertical pillar is 26.5%.

Authors express gratitude to Yurii Ovchynnikov, the Lead Designer of the Novokramatorskyi Mashynobudivnyi Zavod for providing help and consultations during the work performance.

References

1. Barecki, Z., & Scieszka, S.F. (1987). A Mathematical Model of the Brake Shoe and the Brake Path System. *N&O JOERNAAL*, 13-17. http://dx.doi.org/10.1115/1.3258977
2. Bondarenko, V.I., Samusya, V.I., & Smolanov, S.N. (2005). Mobile lifting units for wrecking works in pit shafts. *Gornyi Zhurnal*, 99-100.
3. Davydov, B.L. (1949). *Raschet i konstruirovanie shakhtnykh pod'yemnykh mashin*. Moskva: Ugletekhizdat.
4. Day, A.J. (1991). Drum brake interface pressure distributions. *Proc Instn Mech Engrs*, 127-135. http://dx.doi.org/10.1243/PIME_PROC_1991_205_161_02
5. Day, A.J., Harding, P.J., & Newcomb, T.P. (1979). A finite element approach to drum brake analysis. *Proceedings of the Institution of Mechanical Engineers*, (193), 400-406. http://dx.doi.org/10.1243/PIME_PROC_1979_193_043_02
6. Day, A.J., Tirovic, M., & Newcomb, T.P. (1991). Thermal effects and pressure distributions in brakes. *Proceedings of the Institution of Mechanical Engineers*, (205), 199-205. http://dx.doi.org/10.1243/PIME_PROC_1991_205_171_02
7. Fedorova, Z.M. (1961). *Sbornik primerov i zadach po rudnichnym pod'yemnym ustanovkam*. Moskva: Gosudarstvennoe nauchno-tekhnicheskoe izdatel'stvo literatury po gornomu delu.
8. Huang, Y.M., & Shyr, J.S. (2002). On Pressure Distributions of Drum Brakes. *Journal of Mechanical Design*, (124), 115-120. http://dx.doi.org/10.1115/1.1427694
9. Ilin, S.R., Samusia, V.I., Ilina, I.S., & Ilina, S.S. (2016). Influence of dynamic processes in mine hoists on safety exploitation of shafts with broken geometry. *Naukovyi Visnyk Natsionalnoho Hirnychoho Universitetu*, (3), 42-47.
10. Karpyshev, N.S. (1968). *Tormoznye ustroystva shakhtnykh pod'yemnykh mashin*. Moskva: Nedra.
11. Kolosov, D., Dolgov, O., & Kolosov, A. (2013). The stress-strain state of the belt on a drum under compression by flat plates. *Annual Scientific-Technical Collection – Mining of Mineral Deposits*, 351-357. http://dx.doi.org/10.1201/b16354-65
12. Kyrychenko, Y.O., Samusia, V.I., & Kyrychenko, V.Y. (2012). Software development for the automatic control system of deep water hydrohoist. *Geomechanical Processes During Underground Mining: School of Underground Mining 2012*, 81-86. http://dx.doi.org/10.1201/b13157-14
13. Kyrychenko, Y.O., Samusia, V.I., Kyrychenko, V.Y., & Goman, O. (2012). Experimental investigation of aeroelastic and hydroelastic instability parameters of a marine pipeline. *Geomechanical Processes During Underground Mining: School of Underground Mining 2012*, 163-167. https://doi.org/10.1201/b13157-28
14. Kyrychenko, Y.O., Samusya, V.I., Kyrychenko, V.Y., & Romanyukov, A.V. (2013). Experimental investigation of aero-hydroelastic instability parameters of the deepwater hydrohoist pipeline. *Middle East Journal of Scientific Research*, 530-534. http://dx.doi.org/10.5829/idosi.mejsr.2013.18.4.12426
15. Moskalyova, T.V., & Polushyna, M.V. (2015). Cross way of fastening steel ropes to a single-drum mine hoisting plant with the location of pulleys on the same axis. *Naukovyi Visnyk Natsionalnoho Hirnychoho Universitetu*, 56-62.
16. Samusia, V.I., Oksen, Y.I., & Radiuk, M.V. (2013). Heat pumps for mine water waste heat recovery. *Annual Scientific-Technical Collection – Mining of Mineral Deposits*, 153-157.

http://dx.doi.org/10.1201/b16354-27

17. Zabolotny, K., & Panchenko, E. (2010). Definition of rating loading in spires of multilayer winding of rubberrope cable. *New Techniques and Technologies in Mining*, 223-229. https://doi.org/10.1201/b11329-38
18. Zabolotny, K., Zhupiev, O., & Molodchenko, A. (2015). Analysis of current trends development of mining hoist design engineering. *Theoretical and Practical Solutions of Mineral Resources Mining*, 175-178. https://doi.org/10.1201/b19901-32
19. Zabolotnyi, K., Zhupiiev, O., & Molodchenko, A. (2017). Development of a model of contact shoe brake-drum interaction in the context of a mine hoisting machine. *Mining of Mineral Deposits*, *11*(4), 38-45. https://doi.org/10.15407/mining11.04.038
20. Zabolotnyi, K., Zhupiiev, O., & Molodchenko, A. (2018). The effect of stiffness of shoe brake elements on the distribution of contact pressures. *Naukovyi Visnyk Natsionalnoho Hirnychoho Universitetu*, *2*(164), 39-46.

E3S Web of Conferences **60**, 00040 (2018) https://doi.org/10.1051/e3sconf/20186000040
Ukrainian School of Mining Engineering

Recommended and experimental values of the earth surface shift troughs parameters when conducting the stope works

Mykhailo Filatiev[1*], and *Elvira* Filatieva[2]

[1]Donbas State Technical University, Mining Department, 84 Peremohy Ave., 93100 Lysychansk, Ukraine
[2]Donbas State Technical University, Department of Ecology, Life Safety and Labor Protection, 84 Peremohy Ave., 93100 Lysychansk, Ukraine

Abstract. The correspondence of angular and linear parameters of the undermined rocks displacement and the crustal movement to their values determined according to the normative document has been established. Based on the surveying observations over the reference points movement on the earth surface and the stope face advance, the empirical curve of the trajectory of the earth surface maximum subsidence is determined from the points of maximum subsidence. The curve of the maximum subsidence of the earth surface points when removing the stope face from the face entry has been experimentally established, which allowed to determine the location of the point of the beginning of the crustal movement and the dimension of the stope mine working corresponding to this situation. The coordinates of the characteristic points of the earth surface subsidence above the stope face have been determined. A unified method has been developed for determining the angular and linear parameters of the undermined rocks displacement for the two stages of shift troughs formation on the earth surface. The nonconformities have been established between the recommended parameters and the experimental data normative documents during the anthracite seams mining.

1 Introduction

At present time there is no general computational scheme of the link between the parameters of the crustal movement and the undermined rocks in the normative document [1]. The determination of the degree of the recommended parameters compliance with their experimental values is an important task. When considering the impact of coal seams mining on the earth surface, the two stages of shift troughs formation, conditionally independent among themselves, are taken into account. The first is connected with the beginning of the stope works and their development till the formation of the shift trough flat bottom. The second stage is after the flat bottom formation (complete undermining). The general computational scheme for determining the angular and linear parameters of rocks displacement and the crustal movement when changing from the first stage to the second is

* Corresponding author: mfilatev@gmail.com

not taken into account [1]. For this reason, there may be nonconformities between the calculated parameters determined for different stages of stope works performance. To remove the errors of this type, it is recommended [2] to predict the displacement processes using an additional element of the computational scheme – the trajectory of the maximum subsidence of the earth surface points, linking it with the other current parameters of the shift troughs.

In the case of incomplete undermining of the earth surface, according to [1], the maximum subsidence angle (θ) is determined regardless of the values of complete displacement angles (ψ_1, ψ_2 and ψ_3). By determination according to [1], the complete displacement angles are internal angles relatively to the mined-out space, formed on the vertical sections along the main cross section of the trough by the seam plane and by lines connecting the boundaries of the mine working with the boundaries of the flat bottom of the trough. Until the complete undermining of the earth surface, the flat bottom of the trough is absent, for this reason, for this case, there are no recommendations [1] to determine the angles ψ_1, ψ_2 and ψ_3. In the computational scheme [3, 4] until the complete undermining, the angles ψ_1, ψ_2 and ψ_3 in terms of physics are formed by connecting the lines of the mine working boundaries with the points of maximum subsidence of the earth surface (η_m). The set of such points corresponds to the trajectory curve (8) of the maximum subsidence of the earth surface (Fig. 1). At each point of maximum subsidence of the earth surface, theoretically, the lines segments characterizing the maximum subsidence angle and complete displacement angles should intersect. As an example (Fig. 1), the position of the point A_0 corresponding to the beginning of the crustal movement is determined by the angles ψ_{01}, ψ_{02} and θ_0 (the angle ψ_{01} is not indicated on the chart).

The point K is determined by the angles ψ_1, ψ_2, θ_K and its position corresponds to the beginning of the trough flat bottom formation ($\eta_m = \eta_0$). After the trough flat bottom formation, the crustal movement depends only on the stope face advance.

The characteristic points of the curve of subsidence dynamics above the stope face are: A – corresponds to the beginning of the crustal movement; O – is located in the alignment with the stope face and serves as the starting point on the abscissa axis; B – the beginning of the active stage; C – the maximum subsidence rate and the bend point of the curve; D – the end of the active stage and the beginning of the decay stage; F – the beginning of the residual effect of undermining.

The purpose is to establish the correspondence between the angular and linear parameters of the undermined rocks displacement and the crustal movement with their values determined in accordance with the normative document.

On the basis of the surveying observations over the reference points movement on the earth surface and the stope face advance, the empirical curve (8) of the trajectory of the earth surface maximum subsidence is determined from the maximum subsidence points. Having accepted the value $\eta_m = 0$ in the equation $\eta_m = f_1(L_p)$, the distance L_p is calculated relative to the projection of the fixed wall of the first cut working. It determines the position of the point A_0 on the earth surface (Fig. 1). According to the experimental data, the removal of the stope face from the face entry L_H is also determined, which corresponds to the position of the point A_0. To do this, $\eta_m = 0$ is accepted in the equation $\eta_m = f_2(L)$ and the desired value $L = L_H$ is evaluated. Knowing the position of the point A_0 on the earth surface and the distance L_H, the angles of complete displacements ψ_{01}, ψ_{02} corresponding to them and the maximum subsidence angle θ_0, are graphically determined (Fig. 1).

Similarly, by the current positions of the maximum subsidence points and the corresponding distances from the stope face to the first cut working the subsequent values of the angles of complete displacement and maximum subsidence are determined.

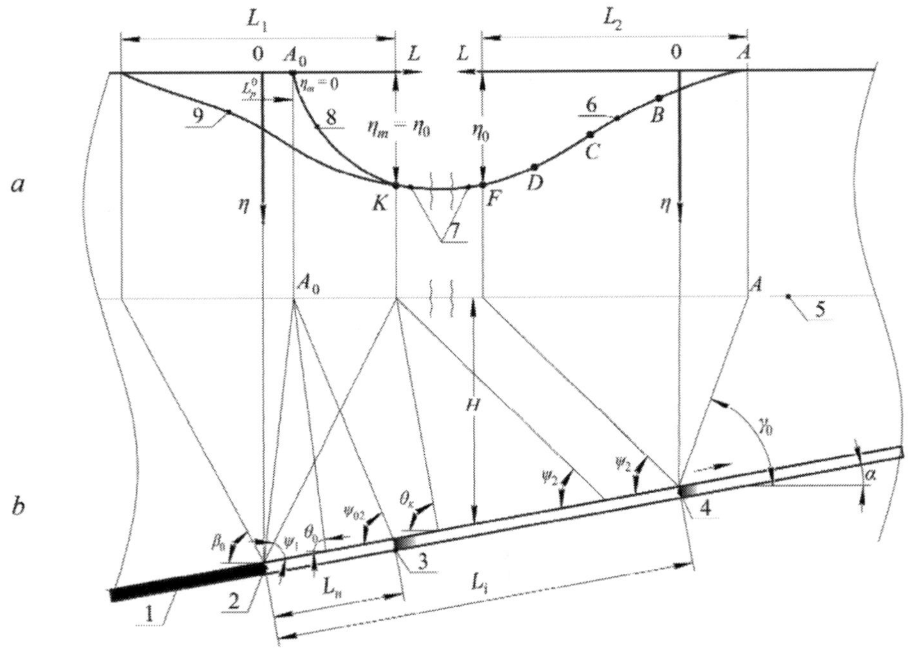

Fig. 1. The scheme of shift troughs formation of the earth surface (a) and its compliance with the stope mine workings parameters (b); 1 – the mined seam; 2 – the first cut working; 3 – the position of the stope face relative to the face entry, at which the crustal movement begins ($\eta_m = 0$); 4 – the position *i* of the stope face after a complete undermining of the earth surface (formation of the trough flat bottom); 5 – the earth surface; 6 – the dynamic half-trough above the stope face; 7 – the flat bottom of the shift trough; 8 – the trajectory of the maximum subsidence points of the earth surface; 9 – the stationary half-trough from the side of the face entry; β_0, γ_0 – the boundary angles, from the side of the seam gradient and the seam rise, respectively; ψ_1, ψ_2 – the angles of complete displacements, from the side of the seam gradient and the seam rise, respectively; ψ_{02} – the angle of complete displacements, corresponding to the beginning of the crustal movement when removing the stope face from the first cut working at a distance of L_H; θ_0, θ_K – the angles of maximum subsidence, when reaching the earth surface rocks ($\eta = 0$) by the displacement processes and the flat bottom formation ($\eta_m = \eta_0$), respectively; H – the average depth of conducting the stope works; α – the incidence angle of the seam; η, η_m – the subsidence and the maximum subsidence of the earth surface, respectively; η_0 – the depth of the trough flat bottom; A_0, A – the points of crustal movement beginning when removing the stope face from the face entry and ahead of it, respectively; B, C, D, F – the characteristic displacement points under the influence of the stope face advance; ⟶ – the direction of stope face advance.

The chart (Fig. 1), shows an example of determining the angular parameters for the end point *K*. It characterizes the end of the first stage of the shift trough formation on the earth surface and indicates the change to the second stage. At the second stage, the crustal movement is characterized by a half-trough (6), which is determined on the basis of surveying observations. The characteristic points (*A*, *B*, *C*, *D*, *F*) are calculated for this curve. The position of points *A* and *F* on the earth surface is determined by the boundary angle (γ_0) and the angle ψ_2 of complete displacements, respectively (Fig. 1).

When having available experimentally determined curves (6, 8, 9), which characterize the earth surface displacement shift trough and parameters of stope works development, it is possible to compare them with the recommended values according to the normative document [1] for the entire period of the extraction site operation at the two stages of shift troughs formation.

2 The research results

To achieve this purpose, the results of surveying observations conducted in different mining-and-geological and mining conditions have been involved into consideration. Most of the known experiments were carried out in the Western Donbas mines (Ukraine). In particular, the changes in the parameters of the stope mine workings and the crustal movements in the conditions of the "Stepova" Mine have been studied quite thoroughly [5]. For this object, almost all the parameter values necessary for the implementation of the computational scheme have been determined experimentally (Fig. 1). Statistical analysis of the initial experimental data [5] has shown that the trajectory curve of the maximum subsidence of the earth surface points (η_m) can be most accurately described by the exponential dependence:

$$\eta_m = a - b \cdot \exp(c \cdot L_p) \text{, mm,} \qquad (1)$$

where a, b, c are the empirical coefficients for the studied object.

Similarly, the value η_m changes when the stope face (L) is removed from the face entry:

$$\eta_m = a_1 - b_1 \cdot \exp(c_1 \cdot L) \text{, mm.} \qquad (2)$$

Empirical coefficients of equations (1, 2) $a = 853$, $b = 1273$, $c = -0.1$ and $a_1 = 853$, $b_1 = 1550$, $c_1 = 0.026$ have been obtained on the basis of experimental data processing (Table 1).

Table 1. Experimental values of the parameters L_p, L, η_m in the conditions of the "Stepova" Mine according to [4] and their corresponding angles θ, ψ_1 and ψ_2.

No.	The position of the stope face and the corresponding number of the shift trough	The distance from the point of maximum subsidence to the projection of the face entry, L_p, m	The distance between the stope face and the first cut working t, L, m	The maximum subsidence of the earth surface, η_m, mm	The angle of maximum subsidence of the earth surface, θ, degree	Angles of complete displacements	
						ψ_1, degree	ψ_2, degree
1	–*	4	23	0	84	97	93
2	3	0	27	64	83	90	81
3	4	6	41	255	83	83	77
4	5	14	58	478	83	79	73
5	7	24	83	693	82	74	66
6	8	31	97	740	82	71	63
7	9	41	116	780	82	66	59
8	11	50	130	812	82	62	58
9	16	46	145	820	76	64	52
10	23	53	156	828	78	62	55
11	34	54	181	836	72	62	49
12	37	55	188	836	70	61	42
13	38	55	208	844	66	61	38
14	40	59	219	852	65	59	37
15	41	64	231	852	64	57	35
16	42	66	259	852	58	56	30
17	43	69	286	868	54	55	28

*Note: parameters $L_p = 4$ m and $L = 23$ are determined from the condition $\eta_m = 0$ according to equations 1 and 2.

The position of the point A_0 on the earth surface (Fig. 1) has been determined from the condition $\eta_m = 0$:

$$L_p^0 = \frac{\ln(a/b)}{c} = \frac{\ln(853/1273)}{0.10} = 4 \text{ m.} \qquad (3)$$

The parameter L_H is determined at $\eta_m = 0$ according to the equation 2:

$$L_H = \frac{\ln(a_1/b_1)}{c_1} = \frac{\ln(853/1550)}{0.026} = 23 \text{ m}. \quad (4)$$

The experimentally determined parameters of the shift trough are shown in the graph (Fig. 2a), and the corresponding dimensions of the stope mine workings are shown in Fig. 2b.

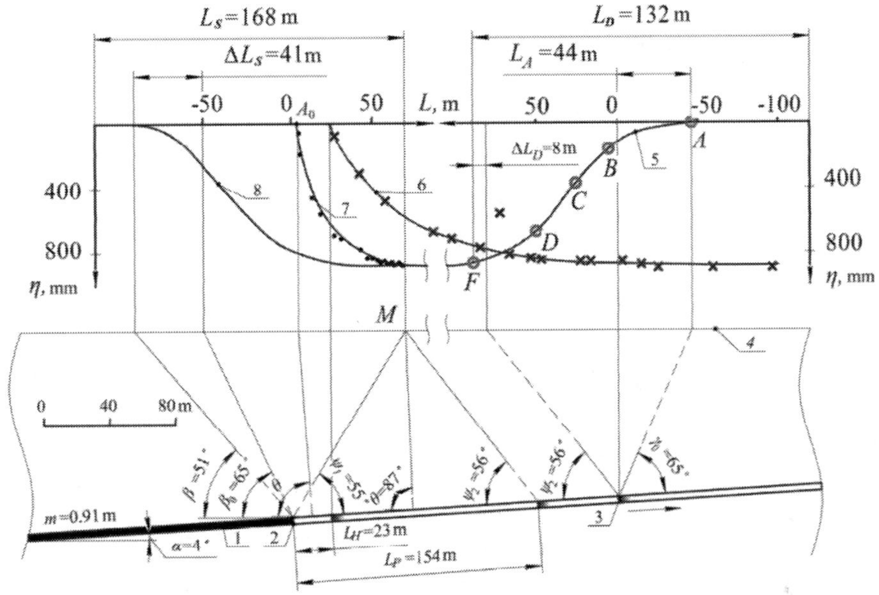

Fig. 2. The results of experimental observations of the crustal movement during its undermining in the conditions of the "Stepova" Mine [5] (a) and the corresponding characteristic dimensions of the stope mine working and angular parameters of rocks displacement (b); 1 – the mined seam; 2 – the stationary face of the face entry; 3 – the position of the stope face after the flat bottom formation; 4 – the earth surface; 5 – the curve of the dynamic half-trough above the stope face according to the results of the averaged experimental data of reference points 25, 30, 35, 40 and 45 subsidence [5]; 6 – the curve of maximum subsidence of the earth surface points depending on the distance between the first cut working and the stope face ($\eta_m = f_2(L)$); 7 – the curve of maximum subsidence of the earth surface points depending on the distance to the face entry projection ($\eta_m = f_1(L_p)$); 8 – the stationary half-trough according to the experimental data [5]; m – the thickness of the mined seam; α – the incidence angle of the seam; β – the boundary angle determined from the dimension of stationary half-trough; β_0, γ_0 – the boundary angles in accordance with the recommendations [1]; ψ_1, ψ_1 – the angles of complete displacements [1]; θ – the angle of maximum subsidence of the earth surface [1]; L_S, L_D – the lengths of the stationary and dynamic half-troughs, respectively; $-\Delta L_S$, ΔL_D, – the divergence of results in determining the dimensions of half-troughs according to the surveying observations on the earth surface and in the angles, as recommended [1]; A, B, C, D, F – the characteristic points of the dynamic half–trough above the stope face; •, × – the experimental data; ⟶ – the direction of the stope face advance.

Their comparison shows the correspondence in most cases of the recommended displacement angles [5] to the parameters of the troughs. The angle of the maximum subsidence of the earth surface ($\theta = 87°$) has determined quite accurately the position of the point of the crustal movement beginning. At the time of the trough flat bottom formation, when removing the stope face from the face entry by 154 m, the straight line segments which characterize the complete displacement angles ($\psi_1 = 55°$ and $\psi_2 = 56°$) and the angle of maximum subsidence of the earth surface ($\theta = 87°$), have intersected approximately, in one point (M). Point M (Fig. 2) corresponds to the beginning the flat bottom formation of the earth surface shift trough and the change to the second stage of its formation. For this reason, further analysis involved data on the reference points 25, 30, 35, 40 and 45 subsidence (Fig. 3). The results of these observations were systematized, taking as the origin of coordinates the moment of intersection by the stope face of the vertical line drawn through the point (reference point) of the observation (Fig. 4).

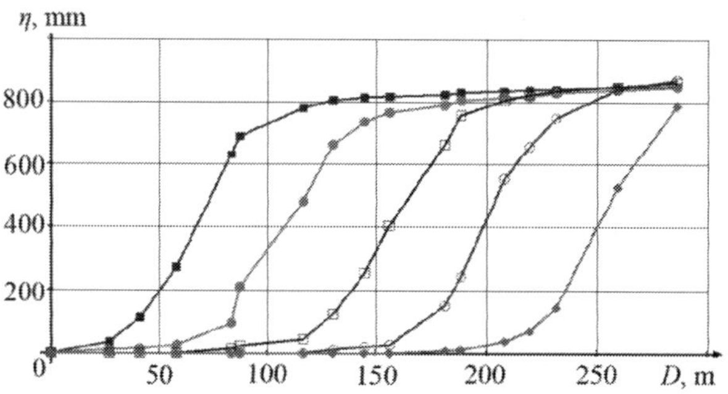

Fig. 3. The subsidence of reference points (η) on the earth surface with removal (L) of the stope face from the face entry in "Stepova" Mine [5]: ■, ●, □, ○, ♦ – the experimental data of reference points 25, 30, 35, 40 and 45 subsidence, respectively.

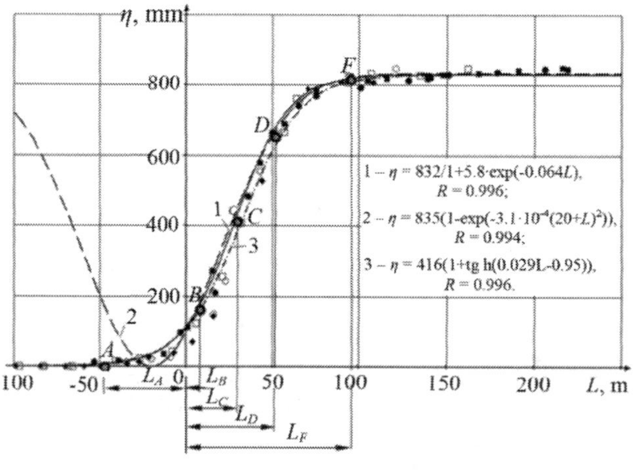

Fig. 4. The subsidence (η) of the earth surface points during the stope works performance in "Stepova" Mine with respect to the projection position of the stope face (L) according to [5]: 1, 2, 3 – the approximating curves corresponding to the logistic, the exponential dependences and the function of hyperbolic tangent; A, B, C, D, F – the characteristic points of the curve of the subsidence dynamics of the earth surface according to the logistic equation; L – the distance from the point (reference point) of observation to the line projection of the stope face; R – the correlation ratio; ■, ●, □, ○, ♦ – the experimental data of reference points subsidence, respectively, 25, 30, 35, 40 and 45.

It is known [3, 4, 6, 7], that the curve of the subsidence (η) of the earth surface points under the influence of the stope face advance can be described by logistic, exponential or hyperbolic tangent dependences. The processing of experimental data for determining the empirical coefficients of these equations, has been performed by the method of least squares (Table 2). Using their numerical values, we have found the boundaries of the stages of the earth surface subsidence (positions of points A, B, C, D, F). The points A and F

characterize the dimensions of the dynamic half-trough ($L_D = 32$ m) above the stope face (Fig. 2). The boundary angle $\gamma_0 = 65°$ [1] completely corresponds to the position of point A, and the position of point F determines accurately the complete displacements angle ($\psi_2 = 56°$). The difference in the determination of the dynamic half-trough boundary (ΔL_D) does not exceed 8 m, which accounted for approximately 6% deviation.

Table 2. The results of determining the coordinates of the characteristic points of the earth surface subsidence [5] in its undermining by the "Stepova" Mine.

The results of processing and the characteristic points (Fig. 4)	Mathematical functions					
	Logistic dependence		Exponential dependence		Hyperbolic tangent	
	$\eta = \frac{a}{1 + b \cdot \exp(-c \cdot L)}$		$\eta = \eta_\kappa\left(1 - \exp[-\beta_2 (L + L_0)^2]\right)$		$\eta = n_1[1 + \tanh(n_2 L + n_3)]$	
Correlation ratio	0.996		0.994		0.986	
Empirical coefficients	$a = \eta_K = 832$; $b = 5.8$; $c = 0.064$		$\beta_2 = 3.1 \cdot 10^{-4}$; $L_0 = L_H = -20$; $\eta_K = 835$		$n_1 = 416 = 0.5\eta_K$; $n_2 = 0.029$; $n_3 = -0.95$	
A	$L_A = L_H = -44$ m	$\eta_A = 0$ mm	$L_A = L_H = -20$ m	$\eta_A = 0$ mm	$L_A = L_H = -46$ m	$\eta_A = 0$ mm
O	$L_o = 0$ m	$\eta_o = 122$ mm	$L_o = 0$ m	$\eta_o = 97$ mm	$L_o = 0$ m	$\eta_o = 108$ mm
B	$L_B = 7$ m	$\eta_B = 175$ mm	$L_B = 10$ m	$\eta_B = 201$ mm	$L_B = 10$ m	$\eta_B = 175$ mm
C	$L_c = 27$ m	$\eta_C = 416$ mm	$L_c = 20$ m	$\eta_C = 328$ mm	$L_c = 33$ m	$\eta_C = 416$ mm
D	$L_D = 48$ m	$\eta_D = 657$ mm	$L_D = 50$ m	$\eta_D = 649$ mm	$L_D = 55$ m	$\eta_D = 665$ mm
F	$L_F = L_K = 88$ m	$\eta_F = 815$ mm	$L_F = L_K = 112$ m	$\eta_F = 818$ mm	$L_F = L_K = 100$ m	$\eta_F = 815$ mm

The maximum deviations (up to 30%) have been obtained between the values of the boundary angles from the side of the seam gradient. According to recommendations [1], $\beta_0 = 65°$ and according to experimental data [5], $\beta_0 = 51°$. Such a divergence led to a difference in the determination of the stationary half-trough $\Delta L_C = 41$m.

The experimental data [5] made it possible to estimate changes in the angle of complete displacements (ψ_1) from the side of the first cut working when removing (L) the stope face. After the square of mined-out space has formed with the side equal to the length of the longwall $L_l = 150$ m, the angle ψ_1 changed slightly in the range of 55 – 62° (Table 1), which corresponds to the recommendations [1]. The angle of complete displacements (ψ_2) from the side of the stope face, after the formation of the square of mined-out space, continued to decrease to 28° (see Table 1).

This caused doubts about the reality of such a change, since the values of L_p and angle ψ_1 remained almost constant. This is explained by the fact that the value of complete displacements angle $\psi_2 \approx 55°$ was formed after the formation of the square of mined-out space, and the stope face advance did not influence its further formation. This is confirmed by insignificant earth surface subsidence during this period (from 828 to 868 mm), which were possible with the rocks compaction. The total subsidence during this period was less than 5% of the maximum for the entire period of the extraction site operation. Such subsidence could not cause a change in the angles of complete displacements, and the results indirectly confirm the scientific position [8] on the constancy of the rocks displacement zone with discontinuity after the formation of the square of the mined-out space.

As the second object for consideration, the facilities have been chosen for the anthracite layer K_5' mining by P.L. Voikova Mine (SE "Sverdlovantratsyt", Ukraine) [9, 10]. The choice of these facilities is conditioned by the significant differences in their mining and geological conditions if compared to the conditions of the stope works performance by "Stepova" Mine. These differences were in the conducting the stope works at a depth of about 700 m and in the more coherent rocks enclosing anthracite seams. This approach makes it possible to establish differences in the formation of the earth surface shift troughs parameters.

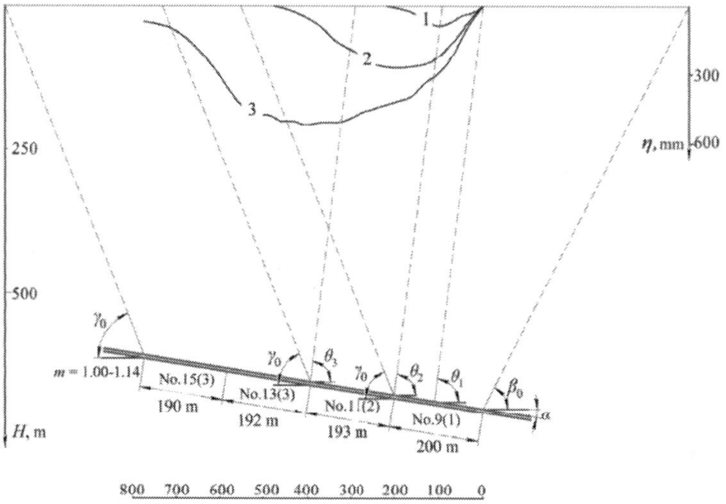

Fig. 5. The graphs of the earth surface subsidence after mining the extraction pillars by P.L. Voikova Mine according to [9]: No.9(1), No.11(2), No.13(3), No.15(3) – the numbers of the longwalls, the order of their mining and the troughs of the earth surface displacement corresponding to them; m – the seam thickness, m; H – the depth of the stope works performance, m; $\alpha = 9°$ – the angle of the seam gradient; η – the earth surface subsidence, mm; $\beta_0 = 63°$ – the boundary angle on the seam gradient; $\gamma_0^1 = \gamma_0^2 = \gamma_0^3 = 68°$ – the boundary angles on the seam rise and their location after mining the longwalls No.9, No.11 and double longwalls No.13 and No.15, respectively; $\theta_1 = \theta_2 = \theta_3 = 83°$ – the angles of maximum rocks subsidence and their location after mining the longwalls No.9, No.11 and double longwalls No.13, No.15, respectively.

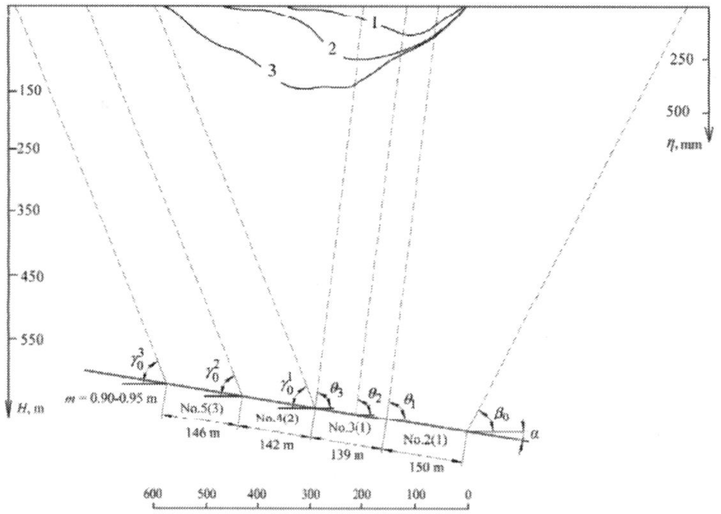

Fig. 6. The graphs of the earth surface subsidence after mining the extraction pillars by P.L. Voikova Mine according to [9]: No.2(1), No.3(1), No.4(2), No.5(3) – numbers of the longwalls, the order of their mining and the troughs of the earth surface displacement corresponding to them; m – the seam thickness, m; H – the depth of the stope works performance, m; $\alpha = 9°$ – the angle of the seam gradient; η – the earth surface subsidence, mm; $\beta_0 = 63°$ – the boundary angle on the seam gradient; $\gamma_0^1 = \gamma_0^2 = \gamma_0^3 = 68°$ – the boundary angles on the seam rise and their location after mining the double longwalls No.2, No.3 and single longwalls No.4 and No.55; $\theta_1 = \theta_2 = \theta_3 = 83°$ – the angles of maximum rocks subsidence and their location after mining the double longwalls No.2, No.3 and longwalls No.4, No.5, respectively.

On the anthracite seam K_5', two groups of extraction sites have been mined, with four longwalls in each group [9, 10]. In the first group, two single longwalls of 200 and 190 m in length have been mined consistently, and then two more double longwalls with a total length of 382 m. Based on the results of such type of mining, the parameters of the shift troughs for the mined out space of 200, 393 and 775 m width, have been established (Fig. 5). In the second group, the seam has been mined, at first, by two double longwalls with a total length of 289 m. Then, two more longwalls with length of 142 and 146 m have been mined consistently. For the longwalls of the second group the parameters of three shift troughs of the earth surface for the mined-out out space with width 289, 431 and 577 m, respectively, have been experimentally determined (Fig. 6).

An example of the anthracite seam mining has shown that the method [1] of geometrical determination of the angular and linear parameters of shift troughs of the earth surface does not correspond to the experimental data. This indicates the need for additional surveying observations when mining the anthracite seams.

3 Conclusions

The developed scheme for the formation of shift troughs of the earth surface made it possible to combine the displacement parameters of two stages.

In the conditions of the Western Donbas (Ukraine), angular and linear parameters of the undermined rocks displacement and the crustal movement in most cases correspond to the recommendations of the normative document. The significant differences were obtained only in the determination of the boundary angles from the side of the face entry.

When mining the anthracite seams, the angular and linear parameters of the undermined rocks displacement and the crustal movement do not comply with the recommendations of the normative document. This situation indicates the need for additional studies using the parameters of the earth surface shift troughs, including the use of the curve of maximum subsidence of the points.

This work was conducted within the project “Development of the forecasting and control system for the dynamics of methane content in mine workings” (State registration No. 0114U004419).

References

1. GSTU 101.00159226.001-2003. (2004). *Pravyla pidrobky budivel i pryrodnykh obiektiv pry vydobuvanni vuhillia pidzemnym sposobom*. Kyiv: Ministerstvo palyva ta enerhetyky Ukrainy.
2. Filat’ev, M.V., Antoshchenko, N.I., & Pyzhov, S.V. (2016). Formirovanie dinamicheskoy i statsionarnoy polumul’d sdvizheniya zemnoy poverkhnosti pro otrabotke ugol’nykh plastov. *Zbirnyk naukovykh prats Donbaskoho derzhavnoho tekhnichnyoho universitetu*, *1*(45), 15-24.
3. Dubovik, A.I., Filat’ev M.V., & Filat’eva, E.N. (2017). *Inzhenernaya geomekhanika pri otrabotke ugol’nykh plastov*. Lysychansk: Donbaskyi derzhavnyi tekhnichnyi universitet.
4. Filatiev, M., & Laguta, A. (2017). Experimental determination of angle values of the rocks full displacement when undermining them by breakage headings. *Mining of Mineral Deposits*, *11*(4), 111-116. https://doi.org/10.15407/mining11.04.111
5. Larchenko, V.G. (1998). Vliyanie podzemnoy razrabotki ugol’nykh plastov na sostoyanie zemnoy poverkhnosti. *Vestnik Mezhdunarodnoy akademii nauk ekologii i bezopasnosti*, *4*(12), 39-41.
6. Filatiev, M. (2017). Determination of cross-correlation dependences between the parameters of swallies of the earth surface and the movement of underworked rocks. *Naukovyi Visnyk Natsionalnoho Hirnychoho Universytetu*, (3), 43-48.

7. Filatiev, M. (2017). Analytical determination of co-ordinates of characteristic points of settling of earthly surface above a cleansing coalface. *Naukovyi Visnyk Natsionalnoho Hirnychoho Universytetu*, (1), 33-37.
8. Lobkov, M.I. (2012). *Rozvytok naukovykh osnov prohnozu obvalennia porid pokrivli pry vyimanni lavoiu polohoho plasta*. Ph.D. Instytut fizyky hirnychykh protsesiv.
9. Borzykh, A.F., Gorovoy, Ye.P. (1999). Vliyanie shiriny vyrabotannogo prostranstva na aktivizatsiyu sdvizheniya uglenosnogo massiva. *Ugol' Ukrainy*, (9), 26-30.
10. Filatiev, M. (2017). Effect of rocks displacement activation on the formation of the surface trough during anthracite seams extraction. *Mining of Mineral Deposits*, *11*(2), 91-95. https://doi.org/10.15407/mining11.02.091

E3S Web of Conferences **60**, 00041 (2018) https://doi.org/10.1051/e3sconf/20186000041
Ukrainian School of Mining Engineering

Selective underground mining of complex structured ore bodies of Kryvyi Rih Iron Ore Basin

Vsevolod Kalinichenko[1], *Serhii* Pysmennyi[1*], *Nataliia* Shvaher[2], and *Olena* Kalinichenko[1]

[1]Kryvyi Rih National University, Department of Underground Mining of Mineral Deposit, 11 Matusevycha Ave., 50027 Kryvyi Rih, Ukraine
[2]Kryvyi Rih National University, Department of Labor and Law, 11 Matusevycha Ave., 50027 Kryvyi Rih, Ukraine

Abstract. Selective mining of complex structured ore bodies will enable increased iron content in the mined ore mass in underground mining at Kryvyi Rih iron ore basin applying the existing mining systems. The article determines the way the grade changes in a mine block of a complex structured ore body when applying bulk and selective mining as well as factors influencing indices of extracting ore mass from the block. The investigation results prove that it is reasonable to mine complex structured ore deposits with barren inclusions applying selective mining if the iron content in the block varies from 64 to 67% and horizontal thickness of the barren inclusion makes 4 – 12 m. At that, the thickness ratio should not exceed 0.3. Stoping methods should be applied to receive high indices of ore mass extraction. There is determined an ore/barren inclusion-to-ore body thickness ratio considering the dilution factor that confirms practicability of selective mining of complex structured ore bodies. Selective mining of complex structured ore deposits of Kryvyi Rih iron ore basin applying open stoping enables increasing the mined ore mass quality without considerable capital and operating costs and, consequently, staying on the market and enhancing the environmental situation in the region.

1 Introduction

Ore deposits of Kryvyi Rih iron ore basin are represented by single and parallel-contiguous bodies consisted of complex structured ore bodies. The number of the areas within the deposits makes about 8% on average, i.e. over 100 mln t of the balance rich iron ore reserves of Kryvyi Rih iron ore basin. The horizontal thickness of barren/ore inclusions (BOI) varies from 5 to 12 m with the Protodiakonov hardness ratio of 8 – 16 points which is average 4 – 6 points higher than that of the main ore massif.

When mining deposits represented by complex structured ore bodies (CSOB) by traditional underground methods, the iron content in the mined ore mass decreases by 3 – 6% in relation to the main grade with ore losses being 1.5 – 2.0 times larger as compared with the standard. This results in decreased iron content in the mined ore mass and loss of the global

* Corresponding author: psvknu@gmail.com

market. Selective mining of the mineral from blocks enables mining it with the iron content in the ore mass of over 62% and thus holding market positions.

Mining of complex structured ore bodies is treated in a number of works dealing with determining regularities of mining indices, rock pressure manifestations, mining sequence and determining parameters of main structural elements of mining systems [1 – 5]. To settle resource saving issues, a complex approach to building the concept of managing environmental and economic systems should be applied considering functional elements and using principles of organizational and technical management [6 – 9].

Efficiency of mining deposit reserves are influenced by mining and geological conditions. The main factors of successful mining CSOB include: stoping sequence, rock pressure, work intensity, number and stability of pillars, level height, location of stopes and pillars in ore bodies of the main strike.

When dealing with the efficiency of underground mining at the Kiruna mine (Sweden), it was proposed to grade the ore mass according to iron content under the ground. For this purpose, the required number of opening main openings were built. Later on the company had to deal with the problem of mining objects' stability [10, 11]. Currently, there are developed techniques of controlling the state of the ore massif disturbed by underground mining operations [12 – 14]. However, due to technological peculiarities of underground mining of Kryvyi Rih iron ore deposits implementation of the techniques will cause escalation in mining costs and decrease of annual productivity.

Practice of Kryvbas underground mines proves that efficiency of mining complex structured deposits is influenced by stoping sequence, thickness and strength of the interbedded layer (barren inclusion), mining systems [2, 15, 16]. When mining CSOB by room and pillar methods, the number of pillars should be minimal as they are stress concentrators and complicate conditions of further mining. Determining shear and de-stressed zones in mining parallel bodies has proved that intensity of rock pressure in enclosing rocks of the hanging wall is lower than that in underlying rocks of the footwall [17 – 19].

The opposite conclusion is made in [20 – 21], i.e. advance mining from the side of the hanging wall reduces rock pressure in main strike layers. Such controversial conclusions are caused by the fact that the above investigations were conducted under different conditions and at different depths. In [12] various zones of rock pressure changes were studied that is conditioned by advance mining of one of the layers, temporary delays in works and their spatial relationship. It is determined that inclined ore bodies cross impact decreases when their dips increase.

The results of investigating optimization of ore raw materials mining and treatment enable the conclusion that efficiency indices of controlling concentration of ore raw materials significantly depend on accuracy of process parameters [22]. In most cases when developing methods of non-destructive control of ore material characteristics, electromagnetic, ultrasonic and radiometric methods are applied.

The authors of [23 – 26] consider it reasonable to use stoping for mining ore bodies of the main strike and bulk caving for the parallel strike. The current flowsheets and standards are developed for mining CSOB with the horizontal thickness of the barren inclusion of 35 m and more, ores and rocks should be strong and stable. However, what minimal thickness should be mined by the selective method still remains understudied.

2 Methods

To solve the problem of escalating the iron content and reducing ore losses in mining deposits with complex structured ore bodies, it is expedient to apply the approach proposed in [27, 28]. The authors suggest modernization of traditional concentrating processes through introducing hydrometallurgical and chemical treatment which enhances concentration efficiency due to use of other kinds of energy. Such modernization of concentrating is based on force impacts on a substance during disintegration in the activator and does not consider processes of underground mining.

The critical analysis of issues of development and processing of useful minerals enables the following conclusions.

1. Most researchers propose to increase the iron content in the mined ore mass through application of an underground or surface concentrating complex that will inevitably increase mining costs and result in loss of the world market.

2. To increase the iron content in the mined ore mass at the first stage through application of the resource saving selective mining in order to exclude the concentrating process. At that, traditional methods involving concentration combines are suggested for mining ore bodies with the horizontal thickness of barren/ore inclusions of less than 12 m.

Thus, when mining CSOB, there occurs a necessity of enhancing the resource saving technology to provide escalation of the iron content at selective mining of deposits depending on mining systems applied. So, it is essential to determine how barren inclusion thickness influences indices of ore mass extraction.

3 Results and discussion

According to the suggested classification, efficiency of this or that applied mining system in underground mining of iron ore deposits consists of the following: indices of ore extraction from a block containing BOI; the grade of the mined rock mass; mining costs [29].

The world market requires saleable products with the grade of the mined ore mass of not less than 62%. The European market offers considerably lower prices for saleable products with the iron content of less than 62%, otherwise they are not competitive at all.

The permissible weight dilution of ore when mining a complex structured ore body by the bulk mining method is determined according to the formula:

$$D = \frac{\Sigma Fe_{p.i} - Fe_{\min}}{\Sigma Fe_{p.i} - \Sigma Fe_{BOI.i}}, \qquad (1)$$

where D is the permissible dilution , unit fractions; $Fe_{p.i}$ is the grade in a block mined, %; $Fe_{\min}$ is the minimum permissible grade in the ultimate dose of the drawn ore mass, %; $Fe_{BOI.i}$ is the grade in BOI, %.

Dependencies of the permissible dilution variations in the caved ore according to (1) when mining CSOB of rich ores by open stopes; mass induced caving of BOI and ore and backfilling are presented in Fig. 1.

Fig.1 demonstrates that decrease of the grade in the mined ore mass with the standard dilution from 63 to 62% (by 1%) is not possible for the considered mining methods applied in underground mines as the permissible dilution is significantly lower.

When applying mining systems with backfilling dead areas, dilution of ore makes average 5%; the permissible weight dilution when mining CSOB varies from 1.5 to 3.7% and the iron content in BOI grows from 0 to 40%. In case of the iron content in the ore massif of 64 or 65%, the iron content in BOI should make 27 and 7% respectively at the standard dilution of the mined ore mass.

When mining CSOB by open stoping systems, the mined ore mass with the content of 62% at the ore content in the ore massif of lower than 64% cannot be obtained as the standard dilution makes 9% on average. The required grade (62%) in the mined ore mass with the iron content in the ore massif of 65, 66 and 67% can be obtained when the iron quality in BOI makes more than 34, 23 and 12% respectively.

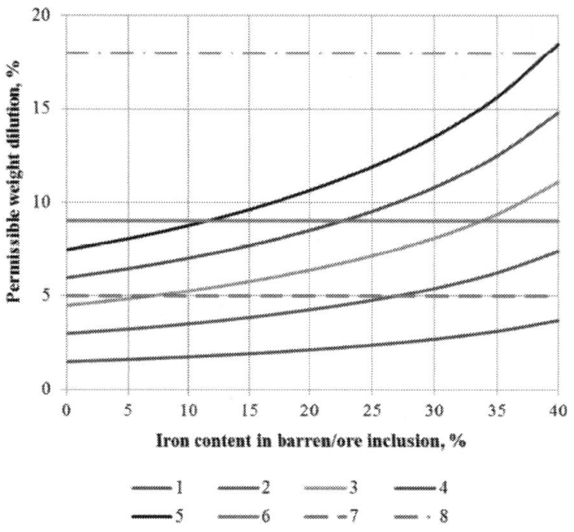

Fig. 1. Dependencies of the permissible dilution in the ultimate dose of the drawn ore mass on the iron content in BOI and the rock massif when mining rich ores with the grade of 67%: 1, 2, 3, 4, 5 – the iron content in the rock massif of 67, 66, 65, 64 and 63% respectively; 6, 7, 8 – standard dilution of the mined mineral when applying open stoping, mass induced caving of BOI and ore and backfilling respectively.

Thus, application of mass caving in mining complex structured ore bodies is limited significantly. To secure the grade in the mined ore mass at the level of 62%, the iron content in BOI and ore massif must be greater than 40 and 67% respectively. The system with backfilling the dead area is one of the most efficient when mining CSOB as it allows the ore mass with iron content of 62% regardless the iron content in BOI at bulk mining.

It should be noted that the obtained indices only reflect quantitative characteristics as the ore-to-BOI volume ratio is not considered.

In mining complex structured ore bodies the basic condition of mining a block is the average iron content in the mined ore mass. In bulk mining of CSOB the average iron content in a mine block is determined according to the formula:

$$Fe_{av} = \frac{\Sigma\left(Fe_{p.i} \cdot Q_{p.i}\right) + \Sigma\left(Fe_{BOI.i} \cdot Q_{BOI}\right)}{Q_b}, \tag{1}$$

where Fe_{av} is the average useful component content in a mine block, %; $Q_{p.i}$ is the ore reserve in a block with the iron content of over 62%, kt; $Q_{BOI.i}$ is the ore reserve in BOI located within the mine block, kt; Q_b is the reserve of the block, kt.

In general, the ore reserve in a block or BOI is determined by the formula:

$$Q_b = L \cdot h \cdot M \cdot \gamma, \tag{2}$$

where Q_b is the ore reserve, t; L is the block length along the strike, m; h is the height of a block level (sub-level), m; M is horizontal thickness of the ore body, m; γ is the specific weight, t/m^3.

When analyzing (3) which is part of (2), it should be noted that the block length along the strike and the height of its level (sub-level) are constant values. So, after conversion the average iron ore content in the block at bulk mining looks like:

$$Fe_{av} = \frac{\sum Fe_{p.i} \cdot m_{p.i} \cdot \gamma_p + \sum Fe_{BOI.i} \cdot m_{BOI} \cdot \gamma_{BOI}}{\sum m_{p.i} \cdot \gamma_p + \sum m_{BOI} \cdot \gamma_{BOI}}, \quad (3)$$

where $m_{p.i}$ is the horizontal thickness of ore bodies in a mine block, m; m_{BOI} is the horizontal thickness of BOI, m; γ_o, γ_{BOI} is the specific weight in the ore body and the barren/ore inclusion respectively, t/m^3.

Let us analyze the impact of the block or BOI thickness on the average iron content in a mine block when the iron content in ore and in BOI makes 67 and 30% respectively. The results of calculating the average ore content in a mine block of CSOB according to (4) are given in Fig. 2.

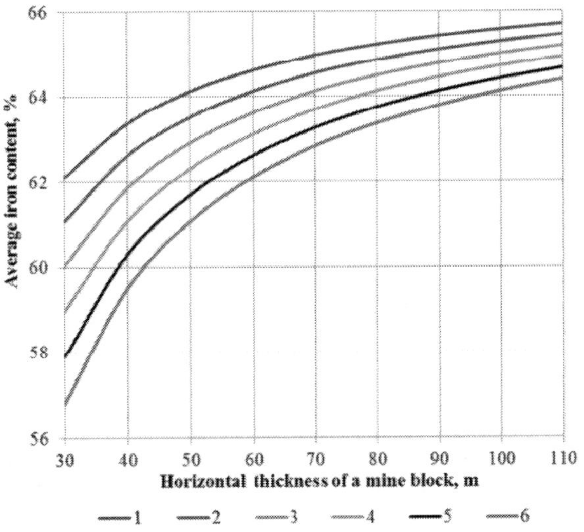

Fig. 2. Dependencies of the average iron content in the ore mass of CSOB on the horizontal thickness of a mine block and thickness of BOI at bulk mining of a block: 1, 2, 3, 4, 5, 6 – horizontal thickness of BOI of 5, 6, 7, 8, 9 and 10 m respectively.

Fig. 3 shows that the average iron content in a block is significant impacted by the horizontal thickness of BOI. Increase of BOI thickness from 5 to 10 m and the iron content in ore of 67% and that in BOI of 30% decreases the average content from 62 to 57% at the block thickness of 30 m and from 65.5 to 64.5% at the block thickness of 110 m.

It should be noted that (4) is true when the block is mined by traditional methods with bulk mining without considering the mining system. The iron content in the ore mass considering indices of ore extraction when mining a block by the corresponding mining system is determined by the formula:

$$Fe_{mom} = Fe_{av} - \frac{(Fe_{av} - Fe_{rock}) \cdot D}{100}, \quad (4)$$

where Fe_{mom} is the useful component content in the mined ore mass, %; Fe_{rock} is the grade in enclosing rocks of the footwall and the hanging wall, %; D is dilution of the mined ore mass at mining the block by the corresponding system, %.

Fig. 3 presents dependencies of the varying iron content in the mined ore mass at the iron content in: ore – of 65%, BOI – of 30%, enclosing rocks – of 32% and BOI horizontal thickness of 8 m.

Fig. 3 demonstrates that increase of the mine block horizontal thickness from 30 to

110 m and constant BOI thickness of 8 m raises the average quality of the ore mass in the block from 57 to 63% (curve 7). It should be noted that to provide the iron quality of 62% in the massif, CSOB thickness must be 75 m and more. Iron ores are mined by various systems with certain ore dilution indices at ore drawing. Fig. 4 (curves 1, 2, 3) shows that to provide the iron quality at the level of 62% is not possible due to high dilution indices.

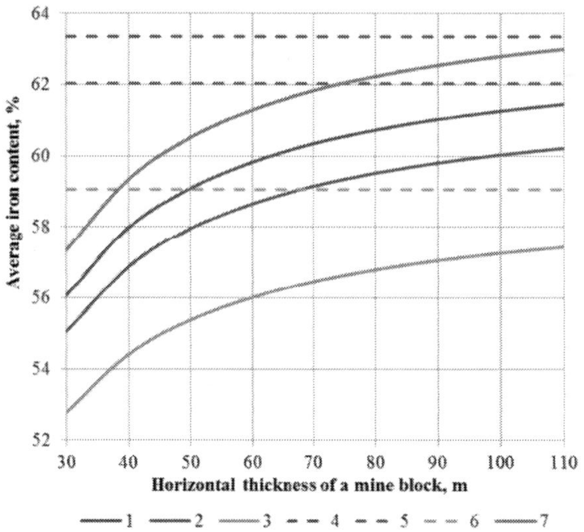

Fig. 3. Dependencies of the average iron content on the horizontal thickness of a mine block at BOI thickness of 8 m: 1, 2, 3 – dilution of the ore mass of 5, 9 and 18% respectively at bulk mining of a block; 4, 5, 6 – dilution of the ore mass of 5, 9 and 18% respectively at selective mining of a block; 7 – average iron content in a mine block at bulk mining.

At bulk mining by mass caving, the thickness increase from 30 to 110 m reduces the quality of iron in the rock mass from 59 – 63% to 53 – 57%. Thus, the required quality of iron in the mined mass can be obtained through selective mining of a mine block. Fig. 3 (curves 4, 5) shows that application of stoping systems raises iron grade in the mined ore mass up to 62%.

The mass caving system cannot provide the level of the required quality of iron in the mined rock mass at 62% due to high dilution of the ore of 18% (Fig. 3, curve 6).

The maximal permissive thickness of BOI considering the horizontal thickness of a mine block, the mineral grade in the ore massif and the average quality in a block is determined by the expression:

$$\left[m_{BOI}\right]=\frac{\Sigma\left(Fe_{p.i}\cdot m_{p.i}\right)\cdot\gamma_p-\Sigma\left(Fe_{\min}\cdot m_{p.i}\right)\cdot\gamma_p}{\left(Fe_{\min}-Fe_{BOI.i}\right)\cdot\gamma_{BOI}}\rightarrow\max, \qquad (5)$$

where $[m_{BOI}]$ is the critical thickness of BOI, %.

Fig. 4 presents dependencies of the critical thickness of BOI at bulk mining of rich or lean ores on the horizontal thickness of a mine, the content of iron in ore, BOI and the mined ore for rich ores of 62% and for lean ores of 26%.

The dependencies in Fig. 4 show that with the increase of the content of iron in BOI to 40%, its the critical thickness rises from 1.8 to 12.0 m at mining rich ores. When the horizontal thickness of a mine block rises from 30 to 80 m, it is reasonable to apply selective mining if the BOI thickness does not exceed 2 – 8 m with the iron content in the barren/ore inclusion of 20%.

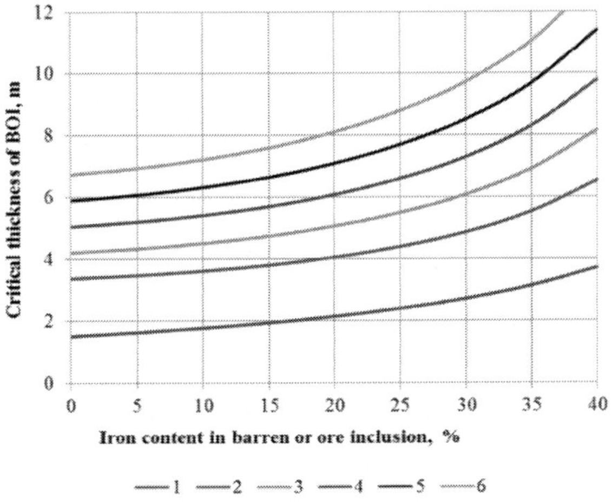

Fig. 4. Dependency of the critical thickness of BOI on the iron content at bulk mining of a mine block (with the content of ore in the ore massif of 65%, in the mined ore mass of 62%) and on horizontal thickness of the ore body: 1, 2, 3, 4, 5, 6 – horizontal thickness of a mine block represented by CSOB of 30, 40, 50, 60, 70, 80 m respectively.

However, (6) does not consider BOI horizontal thickness and dependency of indices of working out a block by a mining system. So, the critical thickness of BOI at bulk mining of a block considering its horizontal thickness, the grade and the dilution ratio is determined as follows:

$$\begin{cases} \left[m_{BOI}\right] = \dfrac{\Sigma\left(Fe_{p.i} \cdot m_{p.i}\right) \cdot \gamma_p - \Sigma\left(Fe_{\min} \cdot m_{p.i}\right) \cdot \gamma_p}{N_{BOI} \cdot \left(Fe_{\min} - Fe_{BOI.i}\right) \cdot \gamma_{BOI}} \cdot (1-D), \\ m_{BOI} \le \left[m_{BOI}\right], \end{cases} \qquad (6)$$

where $[m_{BOI}]$ is the critical thickness of BOI which provides the corresponding extraction indices, m.

Thus, it has been determined that the width of BOI and a mine block directly influences the content of iron in the mined ore mass.

Analysis of (4) demonstrates that besides a block thickness and minable width and the miningextracting ore from a block at mining CSOB is determined according to the formula:

$$K_{ext} = \frac{\Sigma M_{p.i}}{M_b} \cdot \frac{1-B}{1-3} = \left(1 - K_{m.BOI}\right) \cdot \frac{1-B}{1-D}, \qquad (7)$$

where K_{ext} is the ore extraction ratio, unit fractions; $K_{m.BOI}$ is the barren/ore inclusion-to-block thickness ratio, unit fractions.

Fig. 5 presents the dependencies of the thickness ratio on BOI width, Fig. 6 shows the same on the thickness of a block at bulk mining.

Fig. 5 demonstrates that increase of the block thickness from 40 to 80 m decreases the thickness ratio from 0.30 to 0.06; increase of the barren/ore inclusion thickness from 5 to 12 m escalates the thickness ratio from 0.06 to 0.30.

Equating the BOI-block thickness ratio to the average mined ore-to-ore body iron content ratio enables obtaining the basic condition (9) in which selective mining of CSOB is practicable:

$$K_{m.BOI} \geq \frac{Fe_{om}}{Fe_{ob}} \cdot (1 - D), \quad (8)$$

where Fe_{om} is the average content of iron in the mined ore at bulk stoping and BOI from the block, %; Fe_{ob} is the average content of iron in ore bodies around BOI in the block, %.

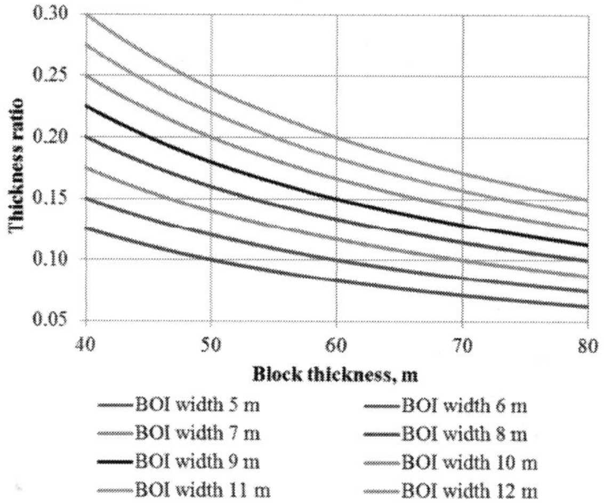

Fig. 5. Dependency of the thickness ratio on the width of the barren/ore inclusion.

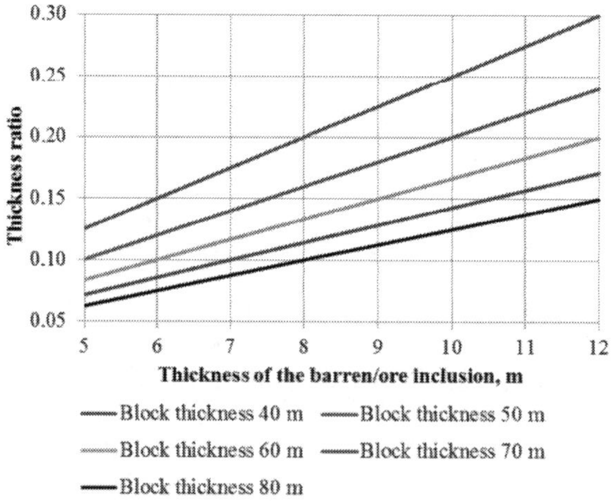

Fig. 6. Dependency of the thickness ratio on the block thickness.

4 Conclusions

The investigations conducted results in the conclusion that selective mining of complex structures ore bodies enable increase of the content of iron in the mined ore mass. However, high extraction indices can be reached by applying certain mining systems in current mining and geological conditions. It is practicable to mine ore bodies containing barren inclusions of 4 to 12 m by systems that are applied at Kryvbas underground mines.

Thus, selective mining applied at blocks represented by CSOB will enable enhancing mined ore mass indices, keeping the position on the world market and decreasing mining and processing costs.

This work was conducted within the project "Determination of regularities of the stress-strain state of rocks disturbed by workings with the purpose of developing resource-saving ore mining technologies" (State registration No. 0115U003179).

References

1. Fedko, M.B., Kolosov, V.A., Pismennyi, S.V., & Kalinichenko, Ye.V. (2014). Economic aspects of change-over to TNT-free explosives for the purposes of ore underground mining in Kryvyi Rih basin. *Naukovyi Visnyk Natsionalnoho Hirnychoho Universytetu*, (4), 79-84.
2. Shchelkanov, V.A., Khivrenko, O.Ya., & Khivrenko, V.O. (2001). Analiz slozhnostrukturnykh zalezhey Krivbassa. *Razrabotka rudnykh mestorozhdeniy*, (75), 30-35.
3. Pysmennyi, S.V. (2017). Metodyka vyznachennia aktyvnoi zony sklepoutvorennia na konturi pidzemno-transportnoi vyrobky pry kombinovanii rozrobtsi zalizorudnykh rodovyshch. *Visnyk Natsionalnoho Tekhnichnoho Universytetu "Kharkivskyi Politekhnichnyi Instytut"*, *16*(1238), 99-106.
4. Stupnik, N., Kalinichenko,V., Kolosov, V., Pismennyi, S., & Shepel, A. (2014). Modeling of stopes in soft ores during ore mining. *Metallurgical and mining industry*, (3), 32-36.
5. Lavrinenko, V.F., & Lysak, V.I. (1991). Uroven' udaroopasnosti porod na glubokikh gorizontakh shakht Krivbassa. *Razrabotka rudnykh mestorozhdeniy*, (52), 30-37.
6. Morkun, V., & Tron, V. (2014). Ore preparation energy-efficient automated control multi-criteria formation with considering of ecological and economic factors. *Metallurgical and Mining Industry*, (5), 8-10.
7. Morkun, V., Morkun, N., & Pikilnyak, A. (2015). Adaptive control system of ore beneficiation process based on Kaczmarz projection algorithm. *Metallurgical and Mining Industry*, (2), 35-38.
8. Andreev, B.M., Brovko, D.V., & Khvorost, V.V. (2015). Determination of reliability and justification of object parameters on the surface of mines taking into account change-over to the lighter enclosing structures. *Metallurgical and mining industry*, (12), 378-382.
9. Morkun, V., Morkun, N., & Tron, V. (2015). Formalization and frequency analysis of robust control of ore beneficiation technological processes under parametric uncertainty. *Metallurgical and Mining Industry*, (5), 7-11.
10. Dineva, S., & Boskovic, M. (2017). Evolution of seismicity at Kiruna Mine. *Proceedings of the Eighth International Conference on Deep and High Stress Mining. Australian Centre for Geomechanics*, 125-139. https://papers.acg.uwa.edu.au/p/1704_07_Dineva/
11. Biruk, Y., & Mwagalanyi, Hannington. (2010). Investigation of Rock-fall and Support Damage Induced by Seismic Motion at Kiirunavaara Mine. *Department of Civil, Environmental and Natural Resources Engineering*, 74. urn:nbn:se:ltu:diva-58466
12. Lutsenko, I., Fomovskaya, E., Koval, S., & Serdiuk, O. (2017). Development of the method of quasioptimal robust control for periodic operational processes. *Eastern-European Journal of Enterprise Technologies*, *4*, (2(88)), 52-60.
13. Lutsenko, I., Fomovskaya, O., Konokh, I., & Oksanych, I. (2017). Development of a method for the accelerated two-stage search for an optimal control trajectory in periodical processes. *Eastern-European Journal of Enterprise Technologies*, *3*, (1(87)), 47-55.
14. Lutsenko, I., Tytiuk, V., Oksanych, I., & Rozhnenko, Zh. (2017). Development of the method for determining optimal parameters of the process of displacement of technological objects. *Eastern-European Journal of Enterprise Technologies*, *6*, (3(90)), 41-48.
15. Khomenko, O., Sudakov, A., Malanchuk, Z., & Malanchuk, Ye. (2017). Principles of rock pressure energy usage during underground mining of deposits. *Naukovyi Visnyk Natsionalnoho Hirnychoho Universytetu*, (2), 35-43.
16. Stupnik, N.I., Kalinichenko, V.A., Kolosov, V.A., Pismenniy, S.V., & Fedko, M.B. (2014). Testing complex-structural magnetite quartzite deposits chamber system design theme. *Metallurgical and mining industry*, (2), 89-93.

17. Vladyko, O., Kononenko, M., & Khomenko, O. (2012). Imitating modeling stability of mine workings. *Geomechanical Processes during Underground Mining: School of Underground Mining 2012*, 147-150. https://doi.org/10.1201/b13157-26
18. Khomenko, O., & Maltsev, D. (2013). Laboratory research of influence of face area dimensions on the state of uranium ore layers being broken. *Naukovyi Visnyk Natsionalnoho Hirnychoho Universytetu*, (2), 31-37.
19. Stupnik, N.I., Fedko, M.B. Pismenniy, S.V., & Kolosov, V.A. (2014). Development of recommendations for choosing excavation support types and junctions for uranium mines of state-owned enterprise SkhidHZK. *Naukovyi Visnyk Natsionalnoho Hirnychoho Universytetu*, (5), 21-25.
20. Khomenko, O., Sudakov, A, Malanchuk, Z, & Malanchuk, Ye. (2017). Principles of rock pressure energy usage during underground mining of deposits. *Naukovyi Visnyk Natsionalnoho Hirnychoho Universytetu*, (2), 35-43.
21. Carusone, O., & Hudyma, M. (2017). Variations in apparent stress and energy index as indicators of stress and yielding around excavations. In *Underground Mining Technology 2017* (pp. 205-218). Canada: Australian Centre for Geomechanics.
22. Golik, V., Komashchenko, V., & Morkun, V. (2015). Innovative technologies of metal extraction from the ore processing mill tailings and their integrated use. *Metallurgical and Mining Industry*, (3), 49-52.
23. Kononenko, M., & Khomenko, O. (2010). Technology of support of workings near to extraction chambers. *New Techniques and Technologies in Mining*, 193-197. https://doi.org/10.1201/b11329-32
24. Tarasyutin, V.M. (2015). Geotechnology features of high quality martite ore from deep mines of Kryvyi Rih basin. *Naukovyi Visnyk Natsionalnoho Hirnychoho Universytetu*, (1), 54-60.
25. Khomenko, O. (2012). Implementation of energy method in study of zonal disintegration of rocks. *Naukovyi Visnyk Natsionalnoho Hirnychoho Universytetu*, (4), 44-54.
26. Khomenko, O., Kononenko, M., & Myronova, I. (2013). Blasting works technology to decrease an emission of harmful matters into the mine atmosphere. *Annual Scientific-Technical Collection – Mining of Mineral Deposits*, 231-235. https://doi.org/10.1201/b16354-43
27. Morkun, V., Gubin, G., Oliinyk, T., Lotous, V., Ravinskaia, V., Tron, V., Morkun, N., & Oliinyk, M. (2017). High-energy ultrasound using to improve the quality of iron ore particles purification in the process of its enrichment. *Eastern-European Journal of Enterprise Technologies*, *6*, (12(90)), 41-51.
28. Plevako, V., Potapov, V., Kycenko, V., Lebedynecj, I., & Pedorych, I. (2016). Analytical study of the bending of isotropic plates, inhomogeneous in thickness. *Eastern-European Journal of Enterprise Technologies*, *4*, (7(82)), 10-16.
29. Stupnik, N., Kalinichenko, V., & Pismennyi, S. (2013). Pillars sizing at magnetite quartzites room-work. *Annual Scientific-Technical Collection – Mining of Mineral Deposits*, 11-15. http://dx.doi.org/10.1201/b16354-4

AUTHOR INDEX

AUTHOR INDEX